International Conference for Free and Open Source Software for Geospatial (FOSS4G 2015)

OSGeo Journal Volume 15, Issue 1

Seoul, South Korea
14-19 September 2015

Editors:

Venkatesh Raghavan

ISBN: 978-1-7138-1901-1

Printed from e-media with permission by:

Curran Associates, Inc.
57 Morehouse Lane
Red Hook, NY 12571

Some format issues inherent in the e-media version may also appear in this print version.

OSGeo
9450 SW Gemini Dr. #42523
Beaverton, Oregon
United States
97008

www.osgeo.org

Additional copies of this publication are available from:

Curran Associates, Inc.
57 Morehouse Lane
Red Hook, NY 12571 USA
Phone: 845-758-0400
Fax: 845-758-2633
Email: curran@proceedings.com
Web: www.proceedings.com

TABLE OF CONTENTS

Free and Open Source Software for Geospatial (FOSS4G) Conference Proceedings

Volume 15 *Seoul, South Korea* Article 1

2015

Volume Introduction Letter

Venkatesh Raghavan
Osaka City University, Japan

Recommended Citation

Raghavan, Venkatesh (2015) "Volume Introduction Letter," *Free and Open Source Software for Geospatial (FOSS4G) Conference Proceedings*: Vol. 15 , Article 1.
DOI: https://doi.org/10.7275/R5CR5RJQ
Available at: https://scholarworks.umass.edu/foss4g/vol15/iss1/1

From the Editor...

Greetings! This Conference Proceedings is a collection of outstanding papers submitted to the Academic Program of the International Conference for Free and Open Source Software for Geospatial (FOSS4G), 14th to 19th September 2015 in Seoul, South Korea.[1]

This was the first time that the international FOSS4G conference took place in Asia. The Academic Track Committee, chaired by Venkatesh Raghavan, Osaka City University, Japan, and Kwang Woo Nam, Kunsan National University, South Korea, achieved - together with the authors - a remarkable contribution to the conference with more than 30 academic papers. Their efforts and results are highly appreciated. The acknowledgment and gratitude is also due to the Committee Members:

Maria Antonia Brovelli, Politecnico di Milano, Italy
Massimiliano Cannata, SUPSI, Switzerland
Eun Mi Chang, Ziinconsulting Inc., South Korea
Serena Coetzee, University of Pretoria, South Africa
Kimberly Durante, Stanford University, USA
Nimalika Fernando, Sri Lanka Institute of Information Technology, Sri Lanka
Heui Chae Jin, Baekseok University, South Korea
Youngok Kang, Ewha Womans University, South Korea
Hwahwan Kim, Chonnam National University, South Korea
Javier Lopez, Javeriana University, Colombia
Peter Lwe, German National Library of Science and Technology, Germany
Helena Mitasova, North Carolina State University, USA
Markus Neteler, Fondazione Edmund Mach, Italy
Byoung-Woo Oh, Kumoh National Institute of Technology, South Korea
K. S. Rajan, International Institute of Information Technology, Hyderabad, India
Xianfeng Song, University of Chinese Academy of Sciences, China
Jung-Sup Um, Kyungpook National University, South Korea
Tuong-Thuy Vu, University of Nottingham, Malaysia campus, Malaysia
Samsung Lim, University of New South Wales, Australia
Ale epek, Czech Technical University in Prague, Czech

We hope readers enjoy these proceedings, and are inspired to continue to build and grow the academic program generally, and work to develop standard academic sub-tracks - such as Geospatial Modeling, Spatial Database research, Geovisualization, and Environmental Science Applications in FOSS4G conferences in the future. We also hope that future editors of regional and global FOSS4G conferences will take advantage of the review and publishing system we have used[2]. Future editors should contact editor-in-chief Charlie Schweik[3] for instructions and access.

For the publication work of academic papers on the Scholarworks Journal site the support of Erin Jerome, PhD, University of Massachusetts, Amherst, and Muhammad Yahya, Stuttgart Universtiy of Applied Sciences, is highly appreciated.

Sincerely,

Venkatesh Raghavan, FOSS4G Academic Committee Chair

[1]http://2015.foss4g.org
[2]http://scholarworks.umass.edu/foss4g
[3]cschweik AT pubpol DOT umass DOT edu

Free and Open Source Software for Geospatial (FOSS4G) Conference Proceedings

Volume 15 *Seoul, South Korea* Article 3

2015

Solar Energy Resource Assessment Using R.SUN In GRASS GIS And Site Suitability Analysis Using AHP For Groundmounted Solar Photovoltaic (PV) Farm In The Central Luzon Region (Region 3), Philippines

Ben Hur Pintor
University of the Philippines Diliman

Eula Fae Sola
University of the Philippines Diliman

Justine Teves
University of the Philippines Diliman

Loureal Camille Inocencio
University of the Philippines Diliman

Ma. Rosario Concepcion Ang
University of the Philippines Diliman

PO-13

SOLAR ENERGY RESOURCE ASSESSMENT USING R.SUN IN GRASS GIS AND SITE SUITABILITY ANALYSIS USING AHP FOR GROUND-MOUNTED SOLAR PHOTOVOLTAIC (PV) FARMS IN THE CENTRAL LUZON REGION (REGION 3), PHILIPPINES

Ben Hur Pintor[1], Eula Fae Sola[1], Justine Teves[1], Loureal Camille Inocencio[1], and Ma. Rosario Concepcion Ang[1]

[1]Phil-LiDAR 2: Project 5 – REMap, University of the Philippines Diliman
Quezon City 1101, Philippines
Email: bhs.pintor@gmail.com

ABSTRACT

In the study, the solar energy resource in the Central Luzon Region (Region 3), Philippines was determined using r.sun – a topography-based solar radiation model implemented in GRASS GIS – and suitable sites for the installation of ground-mounted solar photovoltaic (PV) farms were identified using the Analytic Hierarchy Process (AHP) to determine the weights of different physical, environmental, socio-economic, risk, and constraint criteria.

For the resource assessment, the inputs to r.sun used in the study consisted of freely available data that include: an SRTM (90m resolution) Digital Elevation Model (DEM) and monthly average Linke turbidity coefficients available from the SoDA (Solar Radiation Database) webservice (www.soda-is.com). Daily solar radiation data from eight (8) measuring stations throughout the region were gathered. Readings from six (6) stations were used to interpolate monthly clear-sky index rasters while the readings from the remaining two (2) stations were used to validate the modelled monthly average Global Horizontal Irradiation (GHI) computed by r.sun.

For the site suitability analysis, different criteria rasters were created and combined using weighted overlay to generate a suitability map for ground-mounted solar PV farms in the region.

From the results, the monthly average GHI in the region computed by r.sun ranged from 3706.8 Wh/m^2-day in December to 6021.0 Wh/m^2-day in May with an annual average GHI of 4727.12 Wh/m^2-day indicating a good amount of resource potential. High GHI values were observed for the summer months of March to May (Mean: 5640.26 Wh/m^2-day) while the cold and rainy season ranging from July to December showed relatively lower values (Mean: 4298.98 Wh/m^2-day). The Mean Absolute Error (MAE) and Mean Absolute Percentage Error (MAPE) between the measured and modelled GHI were 352.88 Wh/m^2-day and 8.53%, respectively, with the lowest error in March (73.94 Wh/m^2-day, 1.44%) and the highest in August (844.01 Wh/m^2-day, 21.65%). In fact, the model performed well for the months of January to June (MAE: 192.18 Wh/m^2-day, MAPE: 3.83%) and slightly poorer for July to December (MAE: 512.824 Wh/m^2-day, MAPE: 13.22%).

For further study, other data sources and inputs can be looked into to improve the accuracy of the resource assessment and site suitability analysis. Aside from this, the use of more solar radiation recording stations for validation is preferred in order to better validate the results of r.sun and its applicability for solar energy resource assessment in the Philippines.

1. INTRODUCTION

1.1 Background and Study Area

Solar energy is quickly gaining popularity as a choice for small-scale and large-scale power generation in the Philippines through the use of solar photovoltaic (PV) panels and solar PV farms which have been found to be robust, scalable, and largely sustainable (Nguyen and Pearce, 2010). As of the end of April 2015, the country's Department of Energy

(DOE) has awarded a total of 82 Grid-Use Solar Energy Projects under the Renewable Energy (RE) Law with a total potential capacity of 1,749.53 MW with 42 more projects pending approval for an additional potential capacity of 1,520.14 MW. Among the different sources of renewable energy in the country, the values for solar in terms of the number of projects and total potential capacity rank 2nd only to hydro power (DOE, Awarded Solar Projects as of 30 April 2015). However, solar energy is still very much under-appreciated and under-utilized, accounting for only 0.02% share of the total gross power generation of the country in 2014 (DOE, 2014 Philippine Power Statistics). In order to effectively utilize solar energy as a source of power, especially for large-scale applications such as solar PV farms, the reliable estimation of the solar radiation received in an area is necessary because even though the solar radiation hitting the top of the earth's atmosphere is relatively constant, the radiation that reaches the Earth's surface varies due to factors such as the location, the time, the effects of terrain, and the attenuation caused by the atmosphere. Furthermore, in order to identify locations suitable for setting-up large-scale solar PV farms, several other factors aside from the available solar radiation come into play and need to be considered. The accurate identification of these factors and how they affect the suitability of a location for solar PV farms is important because solar energy projects involve different stakeholders such as the government, the environment, the developers, and the consumers. Finding a site that meets the criteria set forth by the different stakeholders is as important as finding one with high resource potential.

In this study, the solar energy resource in the Central Luzon Region (Region 3), Philippines is assessed using the r.sun model implemented in GRASS (Geographic Resources Analysis and Support System) GIS. Specifically, the monthly average and annual average Global Horizontal Irradiation (GHI) received by the region are computed. In addition to calculating the solar energy resource in the region, site-suitability analysis is performed to find suitable locations for setting-up ground-mounted solar PV farms. Different physical, environmental, socio-economic, risk, and constraint criteria are first identified from related literature as well as interviews with stakeholders and experts. The relative weights of these criteria in terms of how much they affect the suitability of a site for building solar PV farms are determined using the Analytic Hierarchy Process (AHP). Criteria values are standardized and aggregated using Weighted Linear Combination or Weighted Overlay.

The study area is the Central Luzon Region (Figure 1) located between 14° 21' 50" and 16° 31' 48" latitude and 119° 47' 06" and 122° 16' 23" longitude, situated near the heart of the Luzon Island of the Philippine Archipaelago. It has seven (7) provinces – Aurora, Bataan, Bulacan, Nueva Ecija, Pampanga, Tarlac, and Zambales – covering a total land area of 18,230.80 km^2 composed of mountains, extinct and active volcanoes, as well as vast flat farmlands (EMB and ICETT, Green Framework for Innovative Strategy (GFIS) for Sustainable Consumption and Production, 2008). As of April 2015, there are twelve (12) awarded solar projects in the region under the RE Law. One of these projects have already began commercial operation with an installed capacity of 10MW. The remaining eleven have an additional potential capacity of 407 MW. The total potential capacity in the region once all the solar projects are in operation (417 MW) accounts for more than 40% of the expected 985.91 MW capacity of solar projects in the entire Luzon Island (DOE, Awarded Solar Projects as of April 30, 2015).

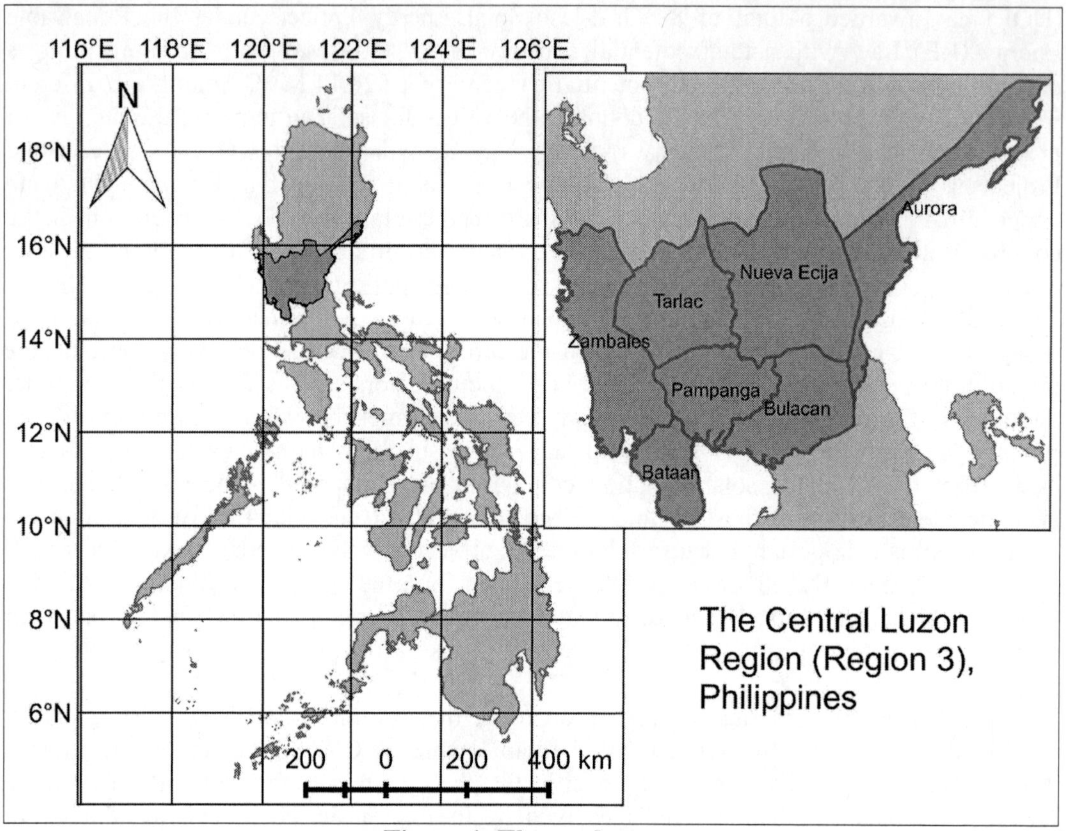

Figure 1. The study area

1.1 Solar Radiation Modelling and r.sun

The solar radiation hitting the top of the earth's atmosphere is relatively constant with an accepted value of 1367 W/m^2 but the radiation that reaches the earth's surface varies depending on spatial and temporal factors. There are three groups of factors that determine the interaction of solar radiation with the earth's atmosphere surface (Hofierka and Suri, 2002).

1. the earth's geometry, revolution, and rotation (declination, latitude, solar position)
2. terrain (elevation, surface inclination and orientation, shadowing)
3. atmospheric effects (scattering, absorption) by:
   1. gases (air molecules, ozone, etc.),
   2. solid and liquid particles (aerosols including non-condensed water), and
   3. clouds (condensed water)

The first group determines the available extraterrestrial radiation and can be precisely calculated. The second group considers the effects of topography and can also be modelled with high precision. For the third group, the elevation above sea level determines the attenuation due to the atmosphere's thickness. The attenuation caused by gas particles is given by the relative optical mass and optical thickness – both of which can be calculated at a good level of precision. The Linke turbidity coefficient can be used to describe the effects of solid and liquid particles but due to the dynamic nature of this coefficient, it cannot be modelled at a high level of accuracy. Lastly, the effect of clouds, which are the greatest attenuants, is very difficult to model. As such, simple empirical techniques are often used to estimate the attenuation caused by clouds.

Most solar radiation models compute for solar radiation under clear (cloudless) skies. This is done by disregarding the effects of clouds altogether. However, in reality, actual days with clear skies are hard to come by. In fact, a good assessment of solar energy resource should always account for the effects of clouds. To do this, the clear-sky index (Kc) is often used. The clear-sky index is a value that relates the modelled clear-sky radiation with the actual real-sky radiation measured on the ground (Hofierka and Suri, 2002). There are three ways to estimate Kc (Nguyen and Pearce, 2010):

1. The ratio between the measured and modelled solar radiation.

$$Kc = \frac{radiation_{measured}}{radiation_{modeled}} \tag{1}$$

2. An emperical formula relating cloudiness (C measured in Oktas) to Kc.

$$Kc = 1 - 0.75\left(\frac{C}{8}\right) \tag{2}$$

3. Derivation from cloud index values of satellite images.

The first method is the simplest and most commonly used. Using this method, it is necessary to have points where the values for the measured and modelled solar radiation are known. It is important to note that the Kc for the beam and diffused solar radiation is not equal to the Kc for global solar radiation beacause the ratio of diffuse to global radiation changes relative to cloudiness and are thus computed separately.

$$Kc_{global} = \frac{global_{measured}}{global_{modeled}} \tag{3}$$

$$Kc_{beam} = \frac{beam_{measured}}{beam_{modeled}} \tag{4}$$

$$Kc_{diffuse} = \frac{diffuse_{measured}}{diffuse_{modeled}} \tag{5}$$

The product of the clear-sky radiation and the clear-sky index is the real-sky radiation.

$$global_{real} = Kc_{global} \times global_{modeled} \tag{6}$$

$$beam_{real} = Kc_{beam} \times beam_{modeled} \tag{7}$$

$$diffuse_{real} = Kc_{diffuse} \times diffuse_{modeled} \tag{8}$$

The r.sun module implemented in GRASS GIS is a topography-based solar radiation model based on the European Solar Radiation Atlas (ESRA) model. It runs in two modes. Mode 1 computes for the solar incidence angle and solar irradiance (W/m^2) at a specific time of day while Mode 2 computes for the insolation time and solar irradiation (Wh/m^2-day) for a specific day of the year. Both modes compute for the global solar radiation and its three components – beam, diffused, and ground-reflected solar radiation – based on time, location, as well as surface and atmospheric conditions. Its inputs are an elevation raster and the day of the year. Location values are provided by latitude and longitude rasters or internal computations. Shadowing is computed internally or through the use of horizon raster maps outputted by the r.horizon module. Other parameters such as the slope, aspect, Linke turbidity coefficients, and albedo can be provided as constants, rasters, or have default values. The outputs are raster maps for beam, diffused, ground-reflected, and global solar irradiance or irradiation. Raster maps for the solar incidence angle for Mode 1 and insolation time for Mode 2 are also outputted (GRASS Development Team, 2015).

As a solar radiation model, the strengths of r.sun are (Hofierka and Cebecauer, 2008):

1. it accounts for the effects of topography on incoming solar radiation;
2. it can use rasters as inputs making it highly scalable;
3. the source-code is editable (open-source); and
4. its implementation in GRASS GIS enables it to integrate spatial and non-spatial data and provides for a wide variety of pre-processing, processing, and analysis tools.

1.3 The Analytic Hierarchy Process (AHP)

The Analytic Hierarchy Process (AHP) is a multi-criteria decision making approach developed by T.L. Saaty where factors or criteria are arranged in a hiearchic structure (Saaty, 2000). In AHP, priorities or weights are generated by decomposing the decision making steps into the several steps: (Saaty, 2008)

1. Definition of the problem.
2. Structuring of the decision hierarchy with the goal at the top and different objectives at the lower hierarchy.
3. Constructing a set of pairwise comparison matrices. Elements in the same hierarchical level are compared to one another using a scale of numbers (Table 1).
4. Generating the weights of each element of the decision-making hierarchy based on the comparisons.

Table 1. The fundamental scale of absolute numbers (Saaty, 2008)

Intensity of Importance	Definition	Explanation
1	Equal Importance	Two activities contribute equally to the objective
2	Weak or slight	
3	Moderate Importance	Experience and judgement slightly favour one activity over another
4	Moderate plus	
5	Strong importance	Experience and judgement strongly favour one activity over another
6	Strong plus	
7	Very strong or demon-strated importance	An activity is favoured very strongly over another; its dominance demonstrated in practice
8	Very, very, strong	
9	Extreme importance	The evidence favouring one activity over another is of the highest possible order of affirmation
Reciprocals of above	If *i* has value *x* when compared to *j*, then *j* has value *1/x* when compared to *i*.	A reasonable assumption
1.1 – 1.9	If the activities are very close	May be difficult to assign the best value but when compared with other contrasting activities the size of the small numbers would not be too noticeable, yet they can still indicate the relative importance of the activities.

2. OBJECTIVES, SCOPE, AND LIMITATIONS

2.1 Objectives

The objectives of the study are:

1. To assess the solar energy resource in the Central Luzon region using r.sun.
2. To validate the results of the model and determine if the appropriateness of its use in a tropical setting like the Philippines.
3. To identify possible sites for installing ground-mounted solar PV farms in the region based on different criteria.

2.2 Scope and Limitations

The study will only compute for the global component of solar radiation or the Global Horizontal Irradiation (GHI) in the Central Luzon region. The resolution of the output solar radiation maps are limited by the DEM used which is the Shuttle Radar Topography Mission (SRTM) DEM whose horizontal resolution is 90m. The computation of the real-sky radiation and its subsequent validation is also hindered by the number of solar radiation measuring stations in the region.

3. REVIEW OF RELATED LITERATURE

As a solar radiation model, r.sun is typically used in European countries or those at higher latitudes since it is based on the ESRA model (Nguyen and Pearce, 2010; Hofierka and Cebecauer, 2008; Kryza et al., 2013). It has been applied in large-scale estimation and assessment of solar radiation in Canada (Nguyen and Pearce, 2010), Slovakia (Hofierka and Cebecauer, 2008), and Poland (Kryza et al., 2013) with good and promising results in terms of accuracy.

4. METHODOLOGY

4.1 Solar Energy Resource Assessment

The datasets used in the solar energy resource assessment include a Shuttle Radar Topography Mission (SRTM) DEM (90 meter horizontal resolution) available at PhilGIS (www.philgis.com) and monthly average Linke turbidity coefficient values downloaded from the Solar Radiation Database(SoDA) webservice (www.soda-is.com). The DEM is used to compute for the slope and aspect in the region using GRASS' r.slope.aspect module. It also used to create horizon rasters using r.horizon. The Linke turbidity coefficient values are used to interpolate monthly average Linke turbidity rasters using the v.surf.rst (Regularized Spline with Tension) module.

Daily solar radiation readings from Bureau of Soils and Water Management (BSWM) sensors throughout the region were downloaded for the computation of the clear-sky index and validation of the modelled solar radiation values. To ensure the quality of the solar radiation data used, only those that met different quality control criteria in terms of the completeness of the readings were used for solving the measured monthly average GHI in the region. For each month, a minimum of eight (8) sensors with solar radiation data were available. Two (2) were chosen for validation of the modelled GHI while the remaining sensors were used for the interpolation of the monthly clear-sky index. From the solar downloaded solar radiation data, twenty-four (24) point vector files were created (one each month for Kc interpolation and another for validation).

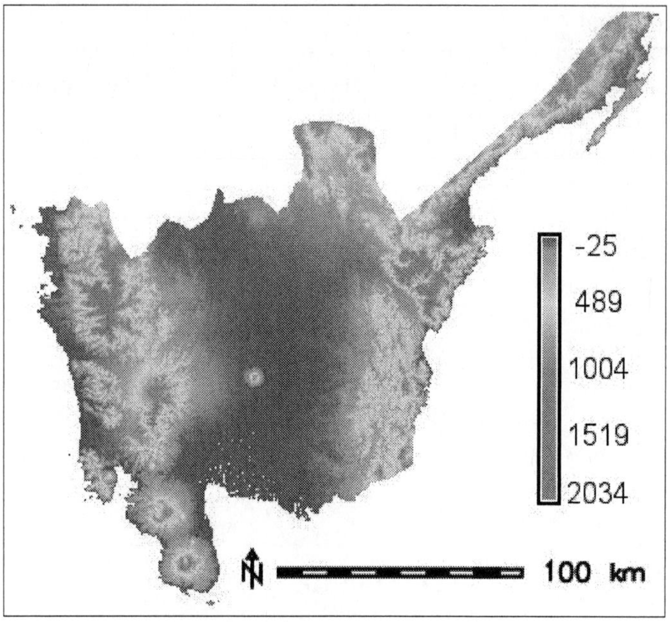

Figure 2. SRTM DEM of Region 3

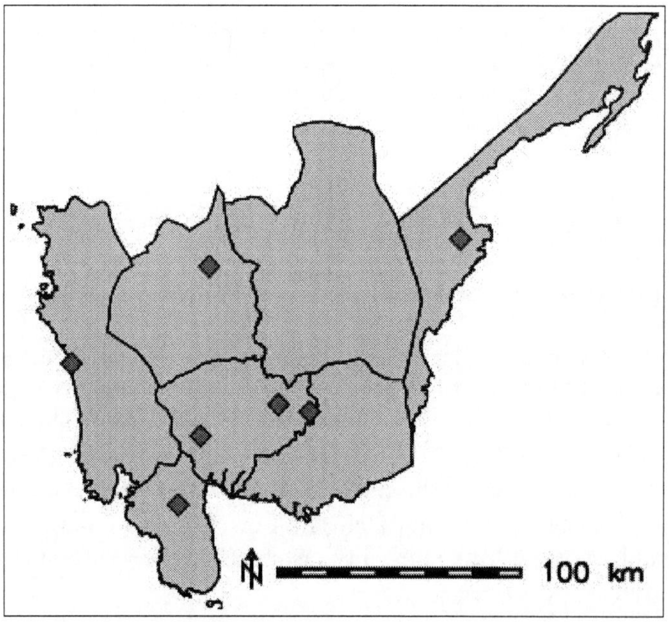

Figure 3. BSWM Solar Sensors for Kc Interpolation

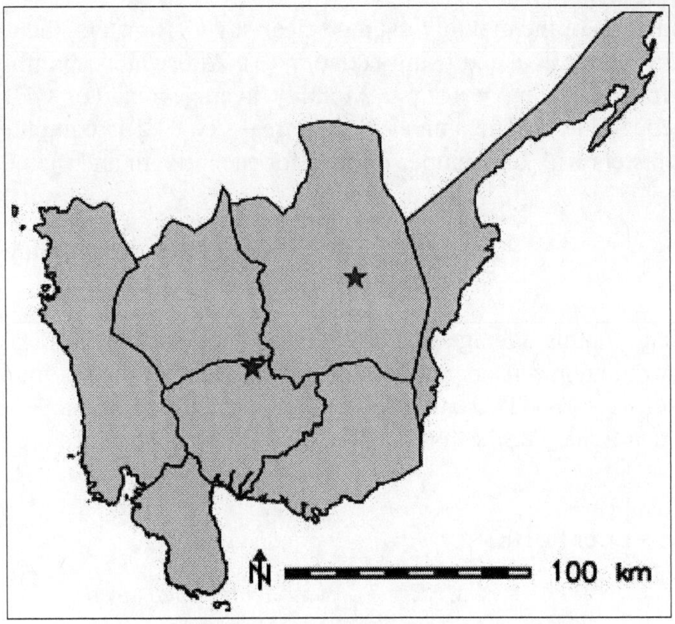

Figure 4. BSWM Solar Sensors for Validation

In order to reduce the number of model runs from 365 to 12, the Julian day and declination of the recommended average day for each month (Duffie and Beckman, 1991) was used to calculate the monthly average clear-sky GHI in the region. The values for these average days are provided below (Table 2).

Table 2. The average days for each month (Duffie and Beckman, 1991)

Month	Average Day of the Month			
	Day	Julian day	Declination (degrees)	Declination (radians)
January	17	17	-20.92	-0.3651228795
February	16	47	-12.95	-0.2260201381
March	16	75	-2.42	-0.0422369679
April	15	105	9.41	0.1642354826
May	15	135	18.79	0.3279473664
June	11	162	23.09	0.4029965243
July	17	198	21.18	0.3696607356
August	16	228	13.45	0.2347467844
September	15	258	2.22	0.0387463094
October	15	288	-9.60	-0.1675516082
November	14	318	-18.91	-0.3300417616
December	10	344	-23.05	-0.4022983926

After computation of the monthly average clear-sky GHI rasters, the clear-sky index for the BSWM sensors were calculated using equation (3). Afterwhich, monthly clear-sky index rasters were interpolated using v.surf.rst. Monthly average real-sky GHI rasters are then calculated using equation (6). The annual average real-sky GHI is computed by multiplying the monthly GHI rasters with the number of days for each month and then dividing the result by 365.

$$GHI_{real,annual} = \sum_{i=jan}^{dec} GHI_{real,i} \times number\ of\ days_i \quad (9)$$

The modelled monthly average real-sky GHI are compared with measured GHI values from the two solar radiation sensors not included in the clear-sky index interpolation in order to validate the results of r.sun. These two sensors are located in Statiion 916 in Pampanga and Station 929 in Nueva Ecija. For the comparison, the following statistics are computed (Zhang et al, 2013):

1. Mean Bias Error (MBE)

$$MBE = \frac{1}{n}\sum_{i=1}^{n}(modeled - measured) \quad (10)$$

2. Mean Absolute Error (MAE)

$$MBE = \frac{1}{n}\sum_{i=1}^{n}|modeled - measured| \quad (11)$$

3. Mean Absolute Percent Error (MAPE)

$$MBE = \frac{1}{n}\sum_{i=1}^{n}\frac{100 \times |modeled - measured|}{measured} \quad (12)$$

4. Root-Mean Square Error (RMSE)

$$RMSE = \sqrt{\frac{\sum_{i=1}^{n}(modeled - measured)^2}{n}} \quad (13)$$

The workflow for the solar energy resource assessment is shown below (Figure 2).

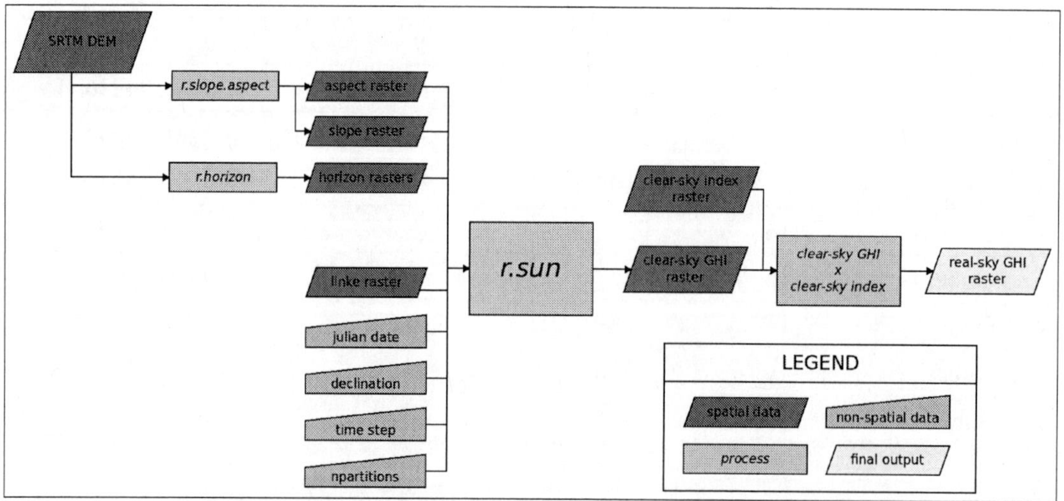

Figure 5. Solar Energy Resource Assessment Workflow

4.2 Site Suitability Analysis

Interviews with stakeholders and experts from the government, industry, environment sector, and the academe were held in order to determine the different criteria, their hierarchy, and their respective weights. From these interviews, the criteria were divided into factors and constraints. Factors are given weights and, when standardized, they have values ranging from 0 to 1. Meanwhile, constraints are used to filter the areas and have values of 0 or 1.

Table 3. Hierarchy of Criteria (Factors)

Level 0	**Suitability for Ground-Mounted Solar PV Farms**				
Level 1	Resource	Non-Resource			
Level 1	Annual Average GHI	Physical	Environmental	Socio-Economic	Risk
Level 2		*Slope* *Aspect* *Land Cover* *Proximity to water bodies*	*Protected areas* *Key biod-iversity areas*	*Proximity to airports* *Proximity to grid* *Proximity to transportation networks* *Proximity to built-up areas*	*Landslide*

Table 4. Constraint Criteria

Constraint
IP locations *Cultural heritage sites* *Infrastructures* *Areas with high flood susceptibility*

The first level of the hierarchy divides the factors into resource and non-resource criteria. The non-resource criteria are then divided into four groups: Physical, Environmental, Socio-economic, and Risk.

The Physical criteria deal with the effects of the physical space on the suitability of the site including the slope and aspect of the area, its land use, and how close it is to a water source. Relatively flat areas that are south facing are preferred since these areas require less earthworks for setting up ground-mounted solar PV's. In the same vein, bare and barren areas are preferred over heavily forested ones. Lastly, the proximity to a water source is important for a steady supply of water to the solar PV plant. The Environmental criteria deal with the possible effects of the PV plant on the environment. Sites that farther away from protected and key biodiversity areas are preferred. The socio-economic criteria take into account the possible costs in putting up a ground-mounted solar PV farm in an area. Thus, sites that are closer to the grid and transportation networks are considered as better options. Lastly, the effect of landslides on the suitability of a site are considered by the Risk criteria. The standardization of the values of each criteria was done using thresholds provided by the stakeholders and literature. For the resource criteria, the stakeholders agreed on a threshold of 4000 Wh/m^2-day as being suitable.

Upon determination of the criteria hierarchy, the assignment of weights for each of the criteria was performed using AHP. Eleven (11) respondents from different sectors including the policy-makers, environmentalists, academe, and industry were interviewed for the determination of criteria weights using AHP with the following results (Table 5).

Table 5. Criteria Weights using AHP

	Criteria	Weight
Level 1	Available resource	0.826
	Non-resource factors	0.174

	Criteria	Weight
Level 2 *Non-resource factors*	Physical	0.378
	Environmental	0.166
	Socio-economic	0.258
	Risk	0.198

	Criteria	Weight
Level 3 *Physical*	Slope	0.383
	Aspect	0.271
	Land Use	0.248
	Proximity to water bodies	0.098

	Criteria	Weight
Level 3 *Environmental*	Proximity to protected areas	0.500
	Proximity to key biodiversity areas	0.500

	Criteria	Weight
Level 3 *Socio-economic*	Proximity to built-up areas	0.167
	Proximity to transportation networks	0.168
	Proximity to the electricity grid	0.570
	Proximity to airports	0.095

	Criteria	Weight
Level 3 *Risk*	Landslide susceptibility	1.000

Weighted overlay analysis is used in order to aggregate the standardized criteria maps and produce the suitability map for ground-mounted solar PV farms in the region. The resulting site suitability map has values between 1 and 0 with 1 being the most suitable and 0 being the least suitable.

5. RESULTS AND DISCUSSION

5.1 Monthly Average and Annual Real-sky Global Horizontal Irradiation (GHI)

The measured and modelled monthly average Global Horizontal Irradiation values for Stations 916 and 929 and their comparison are shown in the figures (Figure 4, 5) and table (Table 6) below.

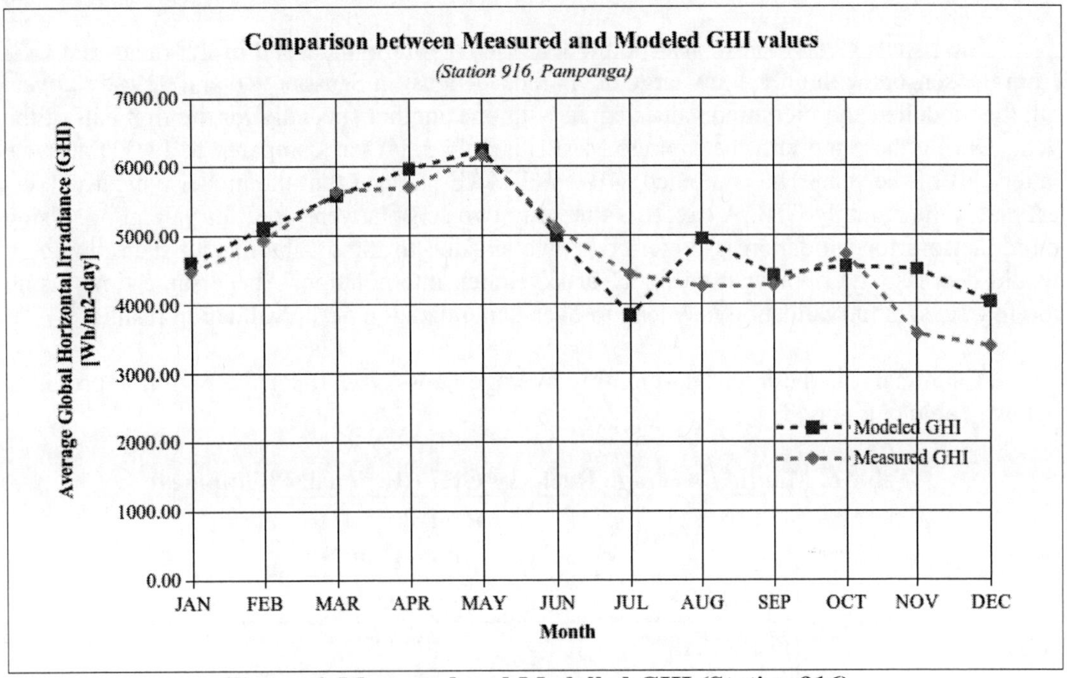

Figure 6. Measured and Modelled GHI (Station 916)

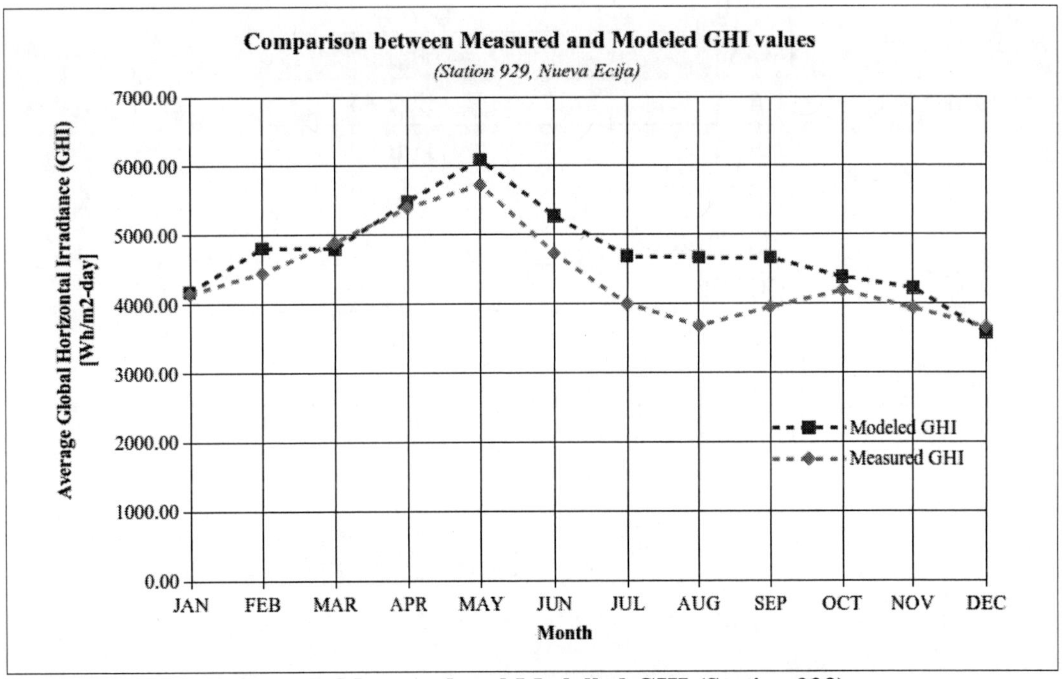

Figure 7. Measured and Modelled GHI (Station 929)

Table 6. Validation statistics of modelled GHI values

Mean Bias Error (MBE)	260.39 Wh/m^2-day
Mean Absolute Error (MAE)	352.50 Wh/m^2-day
Mean Absolute Percent Error (MAPE)	8.53%
Root-Mean Square Error (RMSE)	456.65 Wh/m^2-day

The results clearly show that the modelled GHI follow the trend of the measured GHI from the sensors with only a few large deviations (August in Sensors 916 and 929), but over-all, the modelled and measured values agree with one another specially for the first half of the year. During these months, the average MAPE is only 3.84% as compared to 13.32% for the latter half of the year. The computed MBE and MAE indicate that the model typically over-estimates the monthly GHI. A caveat is that only two sensors were used for validation. Using more sensors for validation is preferred, however, due to the availability of data, doing so would decrease the number used for clear-sky index interpolation. At the same time, using too few sensors for validation may lead to over-generalization of the validation results.

A summary of the modelled monthly average real-sky GHI for the region is provided below (Table 6, Figure 8).

Table 6. Monthly Average Real-sky GHI (Region 3, Philippines)

Month	**Real-sky GHI (Wh/m^2-day)**
JANUARY	4376.65
FEBRUARY	4929.56
MARCH	5226.91
APRIL	5672.79
MAY	6021.07
JUNE	4721.45
JULY	4301.99
AUGUST	4513.60
SEPTEMBER	4474.51
OCTOBER	4443.39
NOVEMBER	4353.56
DECEMBER	3706.80
ANNUAL	4727.12

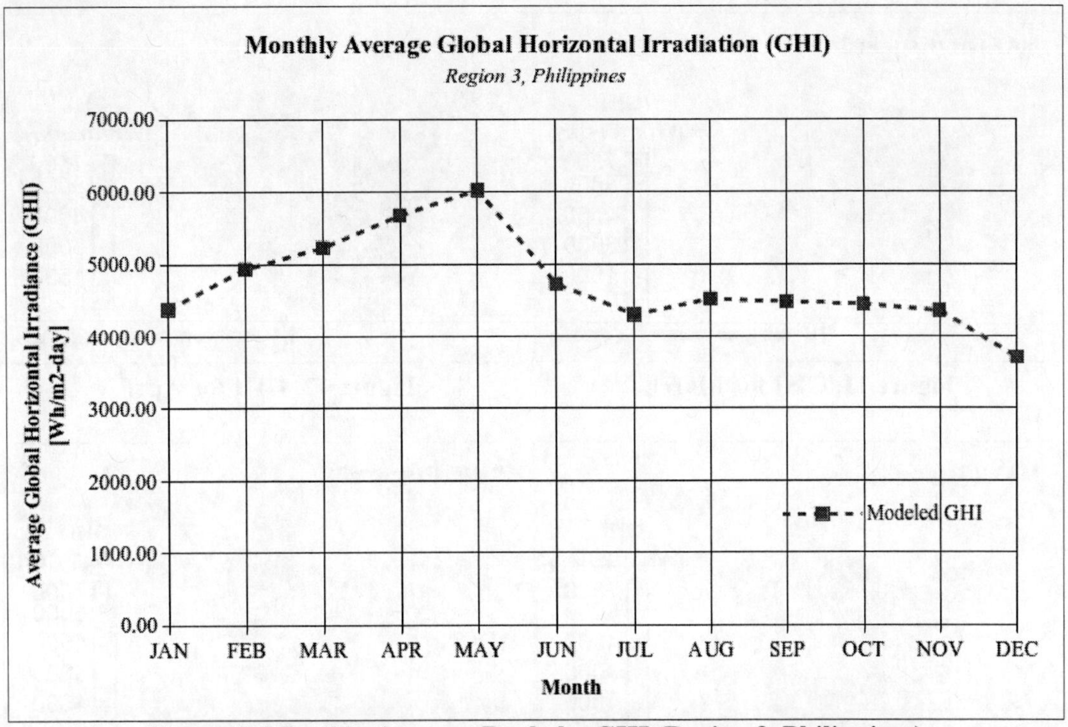

Figure 8. Monthly Average Real-sky GHI (Region 3, Philippines)

The trend in the global solar radiation received by the region is as follows: increasing rom January to May, decreasing from June onwards with values from July to November being similar, and sudden drops in June and December. This trend is somewhat expected since the months from March to May are considered as summer months in the coutry with the hottest and clearest days coming in May. The months of July to August are considered as wet and rainy months so a decrease the received solar radiation druing these months is not surprising.

The computed annual average real-sky GHI of 4727.12 Wh/m^2-day indicates a good amount of solar energy resource in the area. This is especially true from February to May whose average is 5475.16 Wh/m^2-day. For the remaining months of the year, the average is 4359.60 Wh/m^2-day.

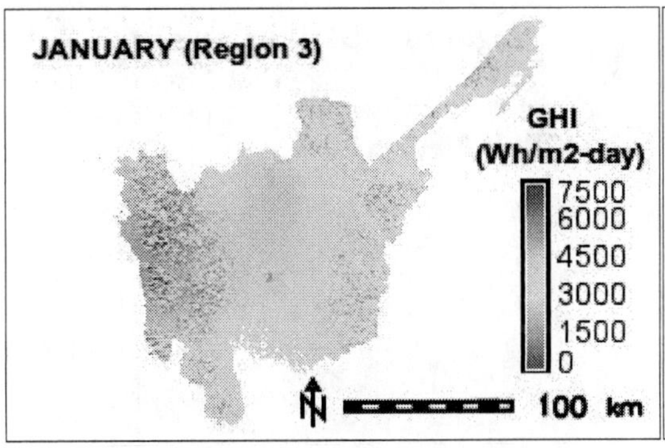

Figure 9. GHI for January

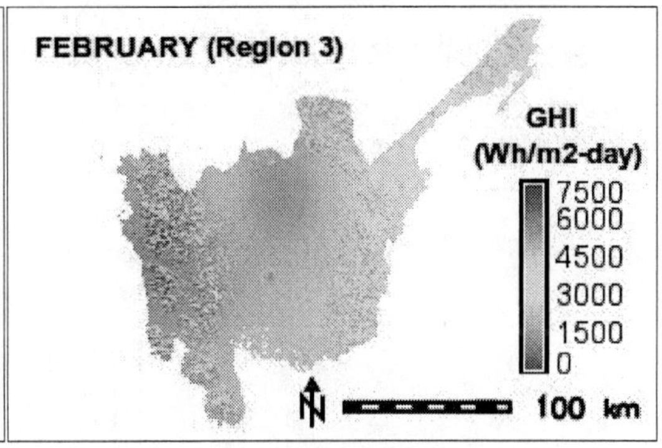

Figure 10. GHI for February

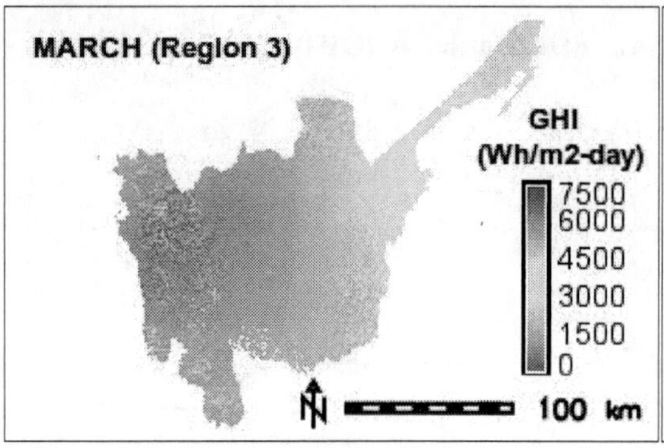

Figure 11. GHI for March

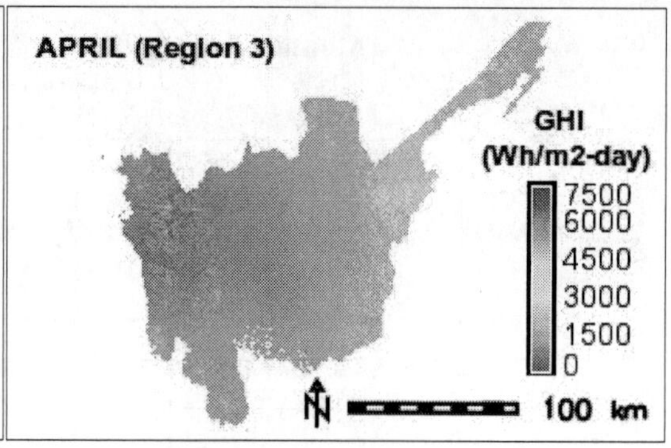

Figure 12. GHI for April

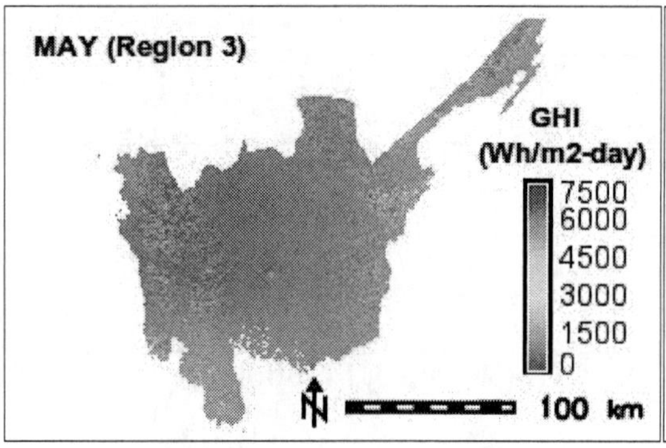

Figure 13. GHI for May

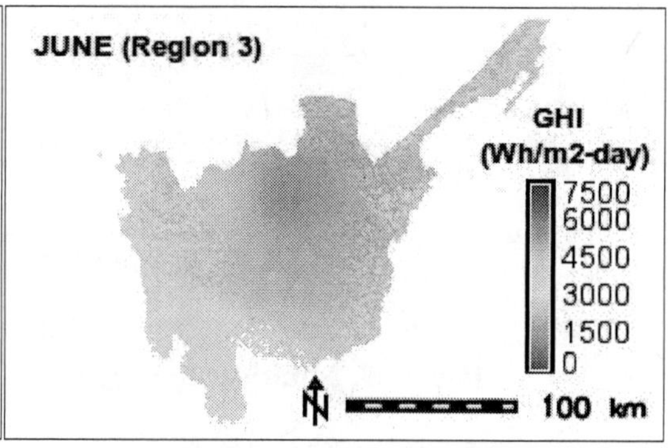

Figure 14. GHI for June

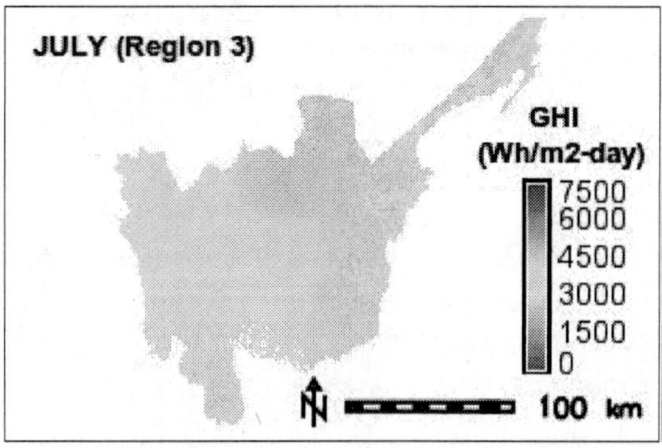

Figure 15. GHI for July

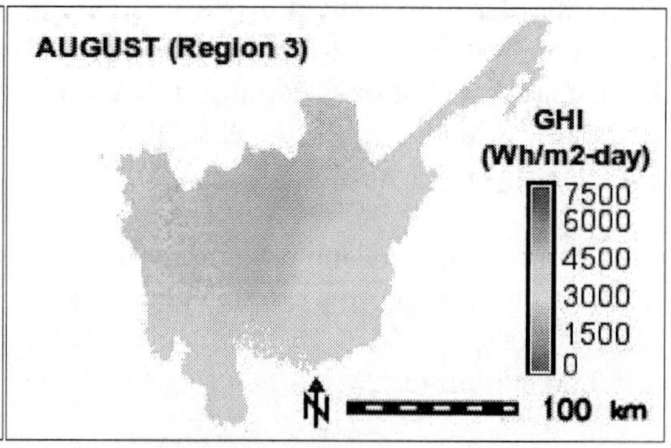

Figure 16. GHI for August

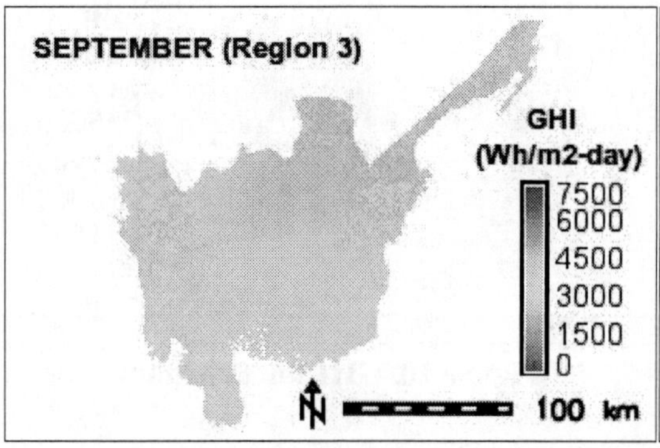

Figure 17. GHI for September

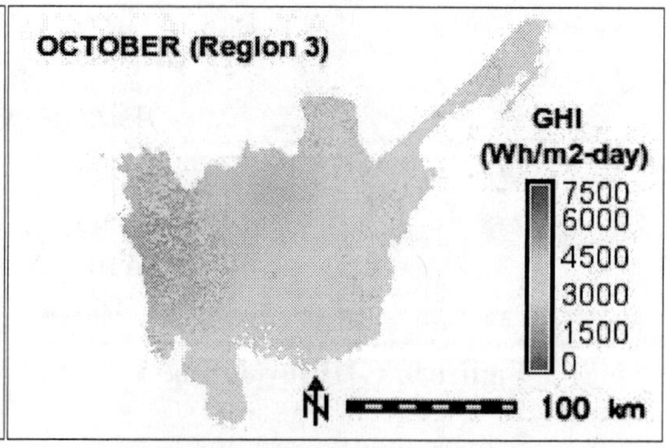

Figure 18. GHI for October

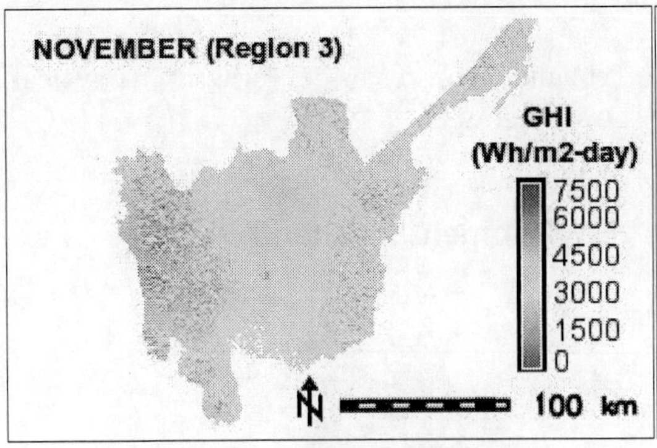

Figure 19. GHI for November

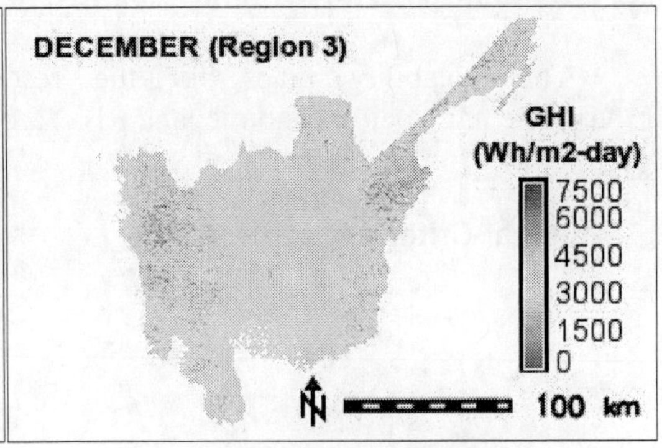

Figure 20. GHI for December

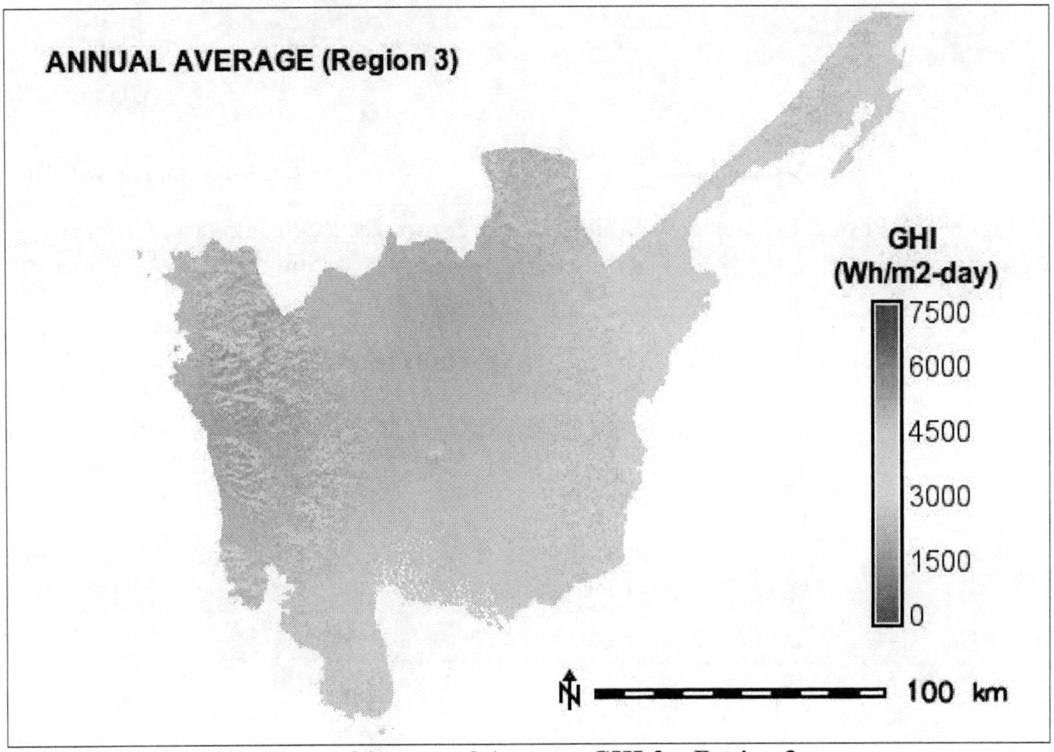

Figure 21. Annual Average GHI for Region 3

In terms of spatial variation, the western and northern parts of the region typically receive more solar radiation.

From the results, the monthly average GHI in the region computed by r.sun ranged from 3706.8 Wh/m^2-day in December to 6021.0 Wh/m^2-day in May with an annual average GHI of 4727.12 Wh/m^2-day indicating a good amount of resource potential. High GHI values were observed for the summer months of March to May (Mean: 5640.26 Wh/m^2-day) while the cold and rainy season ranging from July to December showed relatively lower values (Mean: 4298.98 Wh/m^2-day). The Mean Absolute Error (MAE) and Mean Absolute Percentage Error (MAPE) between the measured and modelled GHI were 352.88 Wh/m^2-day and 8.53%, respectively, with the lowest error in March (73.94 Wh/m^2-day, 1.44%) and the highest in August (844.01 Wh/m^2-day, 21.65%).

5.2 Suitable Sites for Ground-Mounted Solar PV Farms

The suitability maps for the resource and non-resource criteria (Physical, Environmental, Socio-economic, and Risk) are shown below.

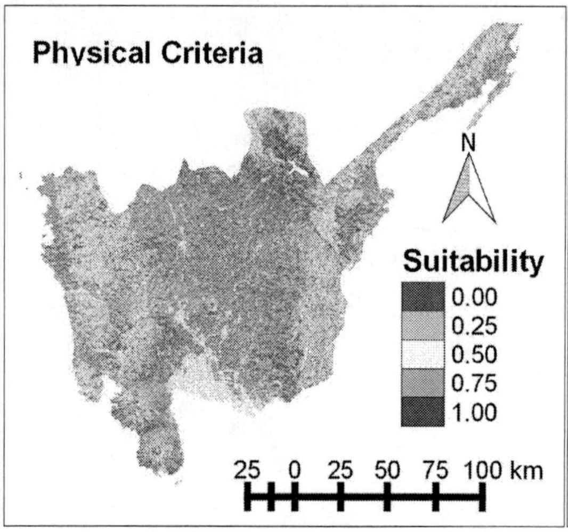

Figure 22. Physical Criteria Suitability

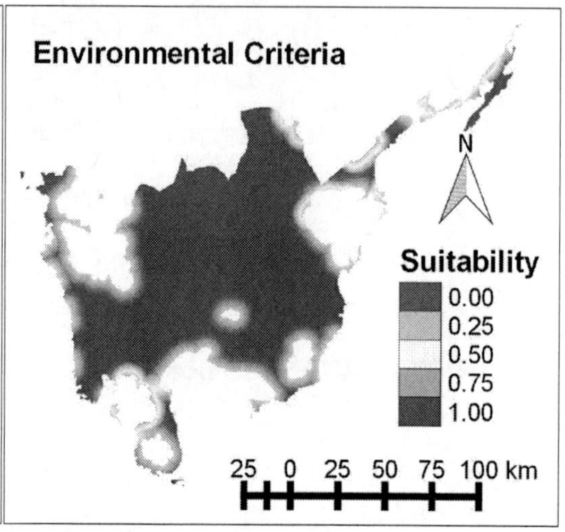

Figure 23. Environmental Criteria Suitability

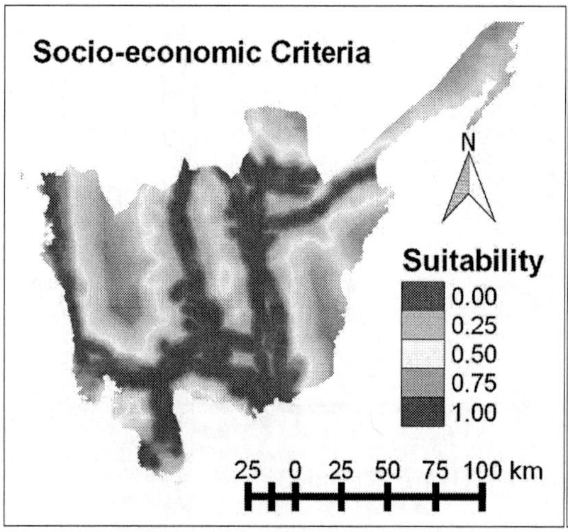

Figure 24. Socio-economic Criteria Suitability

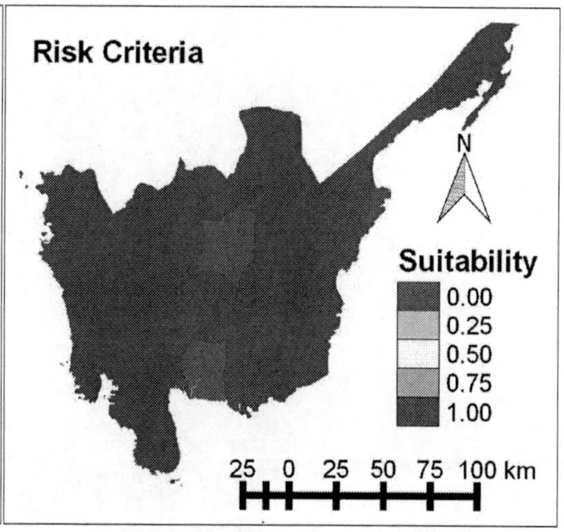

Figure 25. Risk Criteria Suitability

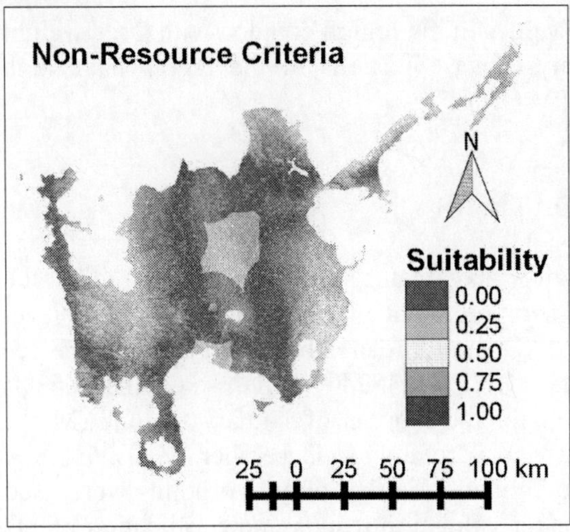

Figure 26. Non-Resource Criteria Suitability

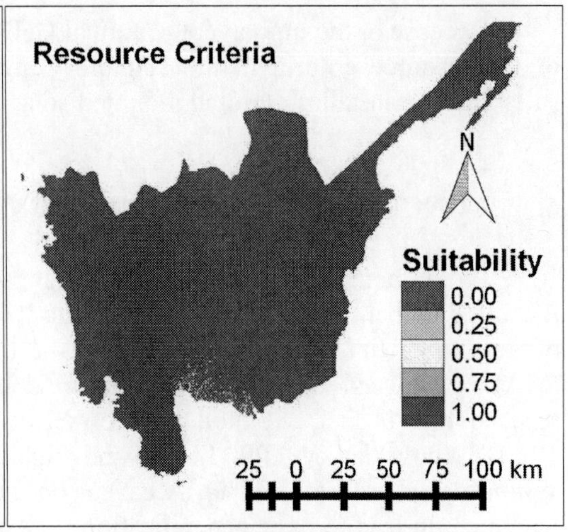

Figure 23. Resource Criteria Suitability

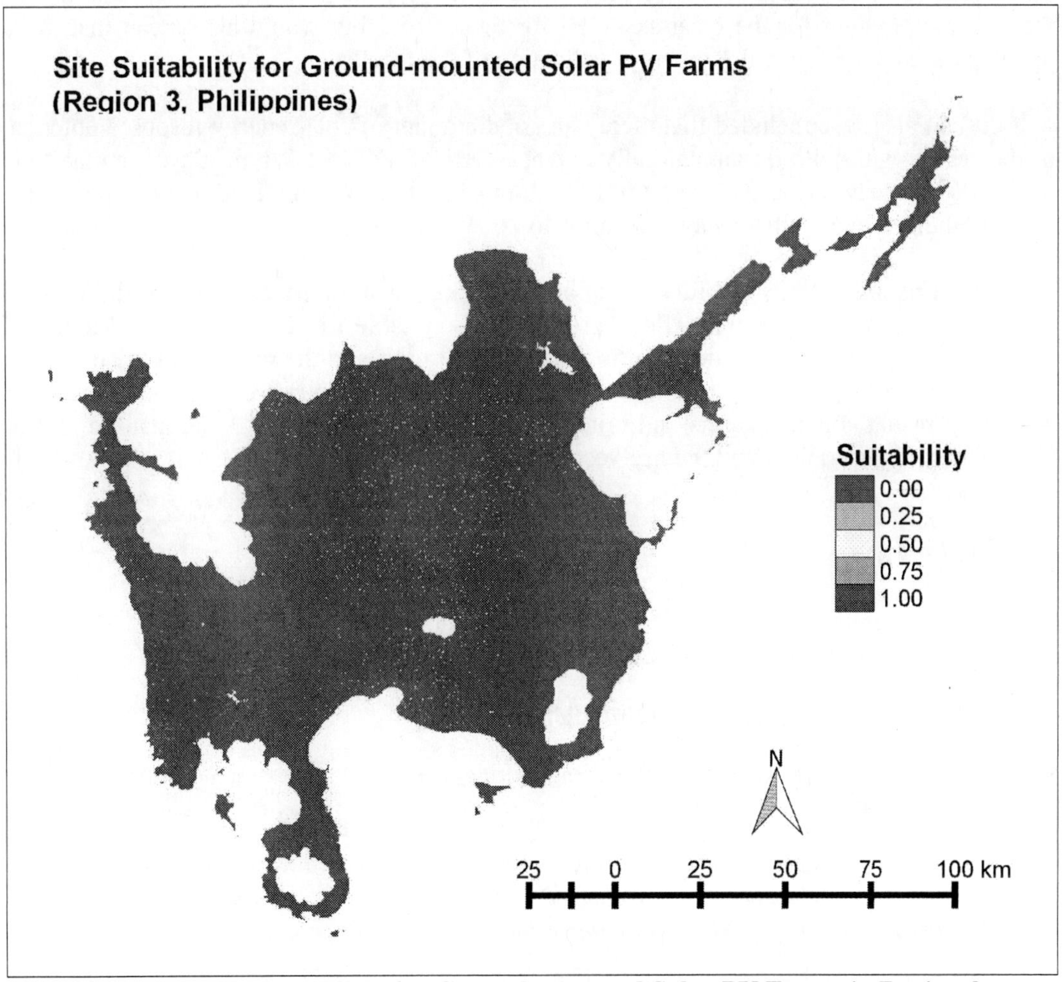

Figure 28. Site Suitability for Ground-mounted Solar PV Farms in Region 3, Philippines

Because of the high average annual GHI values in the region coupled with large weight of the resource criteria in the suitability analysis, vast portions of the region have high suitability for installing ground mounted solar PV farms.

6. CONCLUSIONS AND RECOMMENDATIONS

From the validation results, we can safely say that r.sun was able to model the incoming global solar radiation in the Central Luzon region at an acceptable, if not high, level of accuracy with a Mean Absolute Error (MAE) and Mean Absolute Percentage Error (MAPE) between the measured and modelled GHI of 352.88 Wh/m^2-day and 8.53%, respectively. In fact, the model performed well for the months of January to June (MAE: 192.18 Wh/m^2-day, MAPE: 3.83%) and slightly poorer for July to December (MAE: 512.824 Wh/m^2-day, MAPE: 13.22%). A caveat on the validation is that only two points were used which could lead to over-generalization. For better validation results, more validation points are needed without compromising the number of points for interpolation of the clear-sky index values. One such way to do this is to compute for the Kc rasters of nearby regions and areas and use these values to interpolate the Kc rasters for the Eastern Visayas region. This would entail solving for the clear-sky GHI for those areas but would also mean that more points within the region will remain for validation of the results.

It can also be concluded that there is a good amount of solar energy resource potential in the region with with an annual daily average GHI of 4727.12 Wh/m^2-day, ranging from 3706.8 Wh/m^2-day in December to 6021.0 Wh/m^2-day in May. This is specially true for the summer months of March to May (Mean: 5640.26 Wh/m^2-day).

In terms of the site suitability, it can be seen that almost the entire region aside from the constraint areas (protected areas, flood prone, etc) is suitable for installing ground mounted solar PV farms. This can be attributed to the high GHI values in the region and the relatively large weight of the resource criteria as compared to the non-resource criteria. For better site suitability results, other datasets and criteria can be added. Fuzzy AHP can also be looked into or the weights assignment themselves can be changed.

7. REFERENCES

GRASS Development Team, 2015. GRASS 7.0 Users Manual. Open Source Geospatial Foundation, USA. Electronic document: http://grass.osgeo.org/grass70/manuals/

GRASS-Wiki. http://grasswiki.osgeo.org/wiki/Main_Page

Duffie, J.A. and Beckman, W.A. 1991. Solar Engineering of Thermal processes, Second ed. John Wiley & Sons.

Hofierka, J. and Cebecauer, T. 2008. "Spatially distributed assessment of solar resources for energy applications in Slovakia." *Acta Facultatis Studiorum Humanitatis et Naturae Universitatis Prešoviensis. Prírodné vedy, Folia Geographica 12*, pp. 97-114.

Hofierka, J. and Suri, M. 2002. "The solar radiation model for Open Source GIS: implementation and application". *Proceedings of the Open source GIS - GRASS users conference 2002 - Trento, Italy, 11-13 September 2002.*

Kryza, M. et al. 2010. "Spatial information on total solar radiation: Application and evaluation of the r.sun model for the Wedel Jarlsberg Land, Svalbard." *Polish Polar Research, vol. 31 no. 1*, pp. 17-32.

Nguyen, H.T. and Pearce, J.M. 2010. "Estimating Potential Photovoltaic Yield with r.sun and the Open Source Geographical Resources Analysis Support System." *Solar Energy 84*, pp. 831-843.

Suri, M. and Horierka, J. 2004. "A new GIS−based solar radiation model and its application to photo-voltaic assessments." *Transactions in GIS 8* (2): pp. 175–190.

Zhang, J. et al. 2013. "Metrics for Evaluating the Accuracy of Solar Power Forecasting." *Conference Paper: NREL/CP-5500-60142.*

Free and Open Source Software for Geospatial (FOSS4G) Conference Proceedings

Volume 15 *Seoul, South Korea* Article 4

2015

Leightweight Urban Computation Interchange (LUCI) System

Lukas Treyer
Department of Architecture, ETH Zurich

Bernhard Klein
Future Cities Laboratory, Singapore-ETH Centre

Reinhard König
Department of Architecture, ETH Zurich

Christine Meixner
Department of Architecture, ETH Zurich

Recommended Citation

Treyer, Lukas; Klein, Bernhard; König, Reinhard; and Meixner, Christine (2015) "Leightweight Urban Computation Interchange (LUCI) System," *Free and Open Source Software for Geospatial (FOSS4G) Conference Proceedings*: Vol. 15 , Article 4.
DOI: https://doi.org/10.7275/R5RX998D
Available at: https://scholarworks.umass.edu/foss4g/vol15/iss1/4

PO-22

LEIGHTWEIGHT URBAN COMPUTATION INTERCHANGE (LUCI) SYSTEM

Lukas Treyer[1], Bernhard Klein[2], Reinhard König[1] and Christine Meixner[1]

[1]Department of Architecture, ETH Zurich
Wolfgang-Pauli-Str. 27, 8093 Zurich, Switzerland
Email: {treyer, reinhard.koenig, meixner}@arch.ethz.ch
[2]Future Cities Laboratory, Singapore-ETH Centre
1 CREATE Way, #06-01 CREATE Tower, Singapore 138602, Singapore
Email: klein@arch.ethz.ch

ABSTRACT

In this paper we introduce LUCI, a Leightweight Urban Calculation Interchange system, designed to bring the advantages of calculation and content co-ordination system to small planning and design groups by the means of an open source middle-ware.The middle-ware focuses on problems typical to urban planning and therefore features a geo-data repository as well as a job runtime administration, to coordinate simulation models and its multiple views.The described system architecture is accompanied by two exemplary use cases, that have been used to test and further develop our concepts and implementations.

Figure 1. Transition Workshop at FCL in Singapore in April 2015 showing several views of the CPlan tool backed by LUCI imitating a planning workshop.

1. INTRODUCTION

Among many new developments in urban planning, from a technical perspective it is foremost the increasing amount of data from simulation of urban phenomena that render the process of urban planning increasingly complex. This process often encompasses a wide variety of tools ranging from data acquisition, urban modelling, interactive simulations and 3d visualization. Integrating these tools to close the design-evaluation cycle is still in reality doomed by an integration through data files and manual format conversions.

Design-evaluation cycles require nearly real-time, complementary views on a planning site (Baldonado et al., 2000). To not only create such views but also coordinate and synchronize their content as well as the corresponding calculation processes, we envision a design-simulation pipeline where highly specialized tools interact with each other to perform complex tasks. The benefit of such a pipeline for urban planning raises with the number of potential tools that can be coordinated with each other. This pipe-line is usually achieved with a middleware that assembles and coordinates software components through a messaging system. As an additional layer of abstraction a middleware has several advantages like easing the software development, establishing data standards and easing software maintenance. Individual components can be exchanged on the fly without a requirement for managing new soft-ware releases. This fits the volatile urban planning requirements that vary a lot from one project to another. Moreover, the middleware approach covers well the increasing demand of scalable computing solutions to verify or calibrate complex urban models with big data. As standard middleware solutions impose a rather strict design process that contrast the diverging requirements in urban planning, we decided to implement our own middleware solution that we call Lightweight Urban Computation Interchange (LUCI).

The aim of LUCI is to simplify the middleware approach to a stage where planners are able to install and run it at the ease of any other desk-top application. We think this simplicity is necessary to enable architects, planners and stakeholders to plan smart cities or a smart neighborhoods, which in turn potentially fosters a bottom up and participation culture – an ever-growing topic in (western) modern urban planning. In the language of GIS this means we are looking for a way to bring the advantages of a network oriented GIS server to the desktop. This should enable users to a) run simulations nearly in real-time for a planning workshop on many machines distributed in a local network as shown in Figure 1 or b) to communicate with citizens over the web or to c) collect citizens feedback or knowledge for design tasks.

After a brief introduction about urban planning challenges, we outline recent trends in middleware development focusing on urban planning aspects. In Section 3 we will present the layered architecture of the LUCI middleware, its key functional features and provide additional information about its Desktop, Web and data interface. After this technical part in section 4 we describe two cases in which LUCI has been used and tested already. The use cases range from a multi-screen environment in which we tried to equip an interactive planning workshop scenario with not only complementary but coordinated views, to web applications that communicate to LUCI with web sockets.

2. RELATED WORK

There is no best middleware solution as they are typically designed for specific purposes, targeting problems they were supposed to solve. Urban planning tasks usually involve several disciplines and require the processing of relatively complex models in the background. Off-the-shelf middleware solutions (Gorton et al., 2003) therefore are of limit use, as they do not provide support for adequate data formats or planning functions. On the other side industrial planning suites like ESRI's ArcGIS product family may be too restrictive to support future planning concepts that require more flexible middleware solutions.

The two dominant integration styles in urban planning software are a) sequentially forming a digital workflow chain or b) following publish-subscribe model in a parallel way, where events are instantly propagated to all interested parties.

First research on digital workflow chains for urban planning have been conducted already for several years. They follow a similar principle in which collaborating tools are connected through Internet networks finally forming a seamless integrated urban planning pipeline. For instance, data acquisition, urban modelling, simulation and visualization tools can be linked so that a completely interactive exploration of design alternative becomes a reality. As Janssen mentions, this integration has to happen in a user friendly and flexible way to make this useful technology available to wider range of urban planners. He proposes a workflow configuration engine to manage tool collaborations within urban planning pipeline. Achieving a close design-evaluation cycle is especially important in design workshops as value raises with the capability to test proposed design solutions within a given workshop session. Since such proprietary tools rarely support the same data standard this middleware category often comes with data transformers. Programmable mapping functions can be used to manage the data translation process. The reality however shows that this task can become very complex and time consuming as data structures often may miss specific entries, have inconsistent columns or simply divergent conventions across sources.

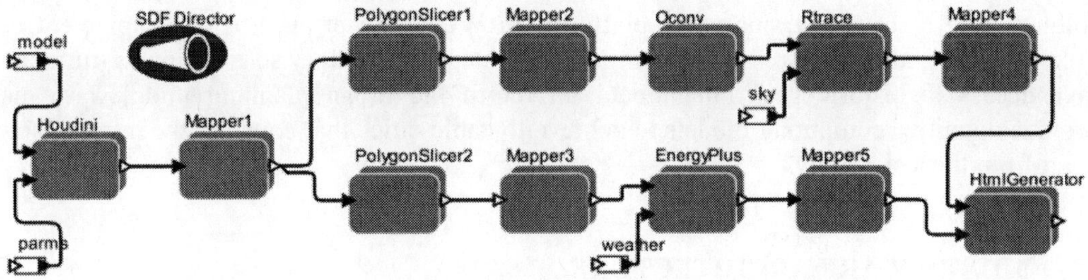

Figure 2. Kepler workflow (Janssen et al., 2014).

Publish subscribe systems enable the asynchronous linking of urban models and tools leading to richer view on the entire urban planning scenario. A good example is the Urban Strategy framework from TNO in the Netherlands. Urban Strategy accelerates and improves urban spatial planning by making information from linked, high-tech computer models available interactively. The models simulate traffic, air quality, noise, external safety, sustainability, groundwater, costs and other aspects of our physical surroundings. Urban Strategy links a central database containing data on the built environment (derived from local authority datasets) to independent computer models. The exchange between these models occurs via a separate communication layer. This enables changes in the results of the models to be immediately calculated into the other models. Changes can be made to the database containing data on the built environment: roads can be closed, houses built, transport routes altered, etc. The effect of these changes is calculated nearly real-time and made visible in Urban Strategy.

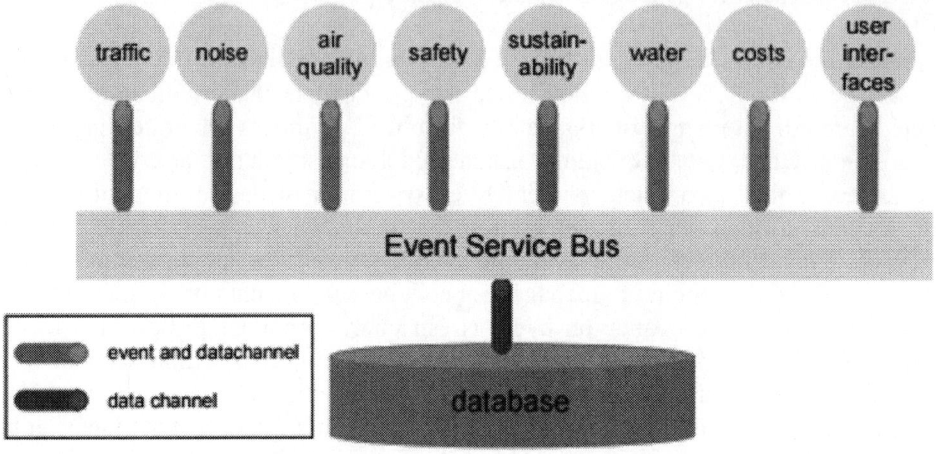

Figure 3. Software architecture of Urban Strategy (Hofman et al., 2011).

It is important to note that smart city platforms often offered as a cloud computing solution (see a survey of smart city platforms in Welington et al., 2013) complement our urban planning middle-ware by providing sensor data (participatory sensing in real-time) and give access to historic government data. In return the urban planning middleware may provide cognitive computing models to achieve liveable cities that consider the human factor apart from the technology.

3. MIDDLEWARE ARCHITECTURE

As shown in the related work section, there are quite a few middleware systems focusing on urban planning, see also Al-Jaroodi et al, 2003, Borst et al., 2009, Romualdo-Suzuki, 2013 and Janssen et al., 2014. We introduce an open geospatial middleware for urban planning that includes a geo-data repository and an asynchronous job management engine that allows a supervised execution of modeling, simulation and visualization tasks in a network. What our solution focuses on is simple configurability and usability, parallel computation of urban simulations and flexible web integration.

It is implemented combining the Message Broker model with a simple Server-Client architecture. We use MQTT for notification and a separate TCP socket for the content ex-change. Similar to MQTT the content exchange sockets remain open as long as the client is connected. Content messages follow a fairly simple protocol, in which messages consists of a JSON header followed by binary data if needed. The JSON header must contain one of the three keywords "action", "result", "error", where action and error are strings and result a JSON object.

3.1. Terminology

The name "action" corresponds to a terminology we define in LUCI as follows. "Actions" are small pieces of code used for the administration of data, simulation and data conversion tasks. They are supposed to run very fast and when called from a client they run synchronously. The client only gets an answer once the action is finished. "Services" on the other hand is the name for asynchronous tasks supposed to run for a longer time that would not be conceived as responsive by a user. "Convertors" are a class in-between running synchronously at the moment but being de-signed to run asynchronously with only a few modifications. "Actions", "Services" and "Converters" are part of a plugin system, i.e. dynamically (re-) loaded at runtime. Thus a restart of LUCI that would reset all socket connections is not necessary. As shown in Figure 2 there is also a category for "Database" plugins. Even though not reloadable at runtime this allows LUCI to have different database adaptors, which in turn enables users to work with their preferred database. Apart from the plugin structure, Figure 2 also shows the basic idea of the data structure. We map service in- and outputs to separate tables. "Scenarios" denote the main unit in which the geo-data repository is divided into. They are shared among the services. For more information on the data structure refer to Section 3.4.

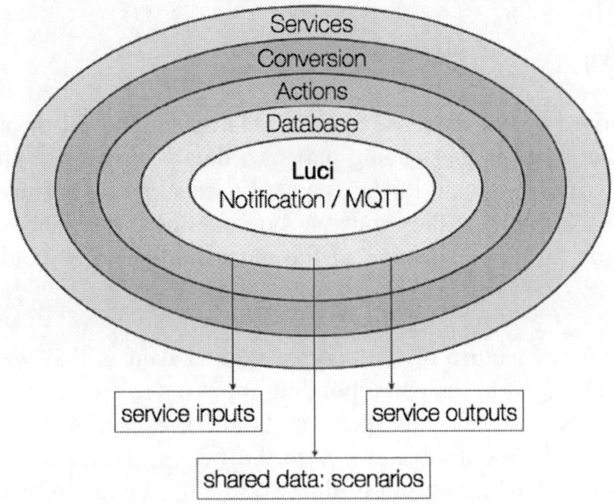

Figure 4. LUCI's pluggable architecture and data structure.

3.2. Graphical User Interface(s)

Administrative controls including an interactive console are available from a Desktop Sys-tem Tray Menu as shown in Figure 3. Nevertheless LUCI can also be configured to run headless without any GUI. By now, the most important control available from this menu is the interactive console. It is intended for developers to send actions to LUCI as raw JSON strings most likely for testing reasons.

Other commands available from that menu are Starting and Stopping LUCI, open a PDF documentation (Open Specification) and a few more, as listed in Figure 3.

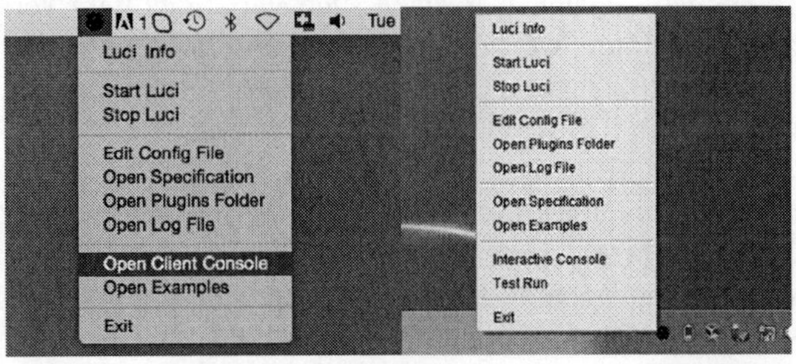

Figure 5. Desktop System Tray Menu (OSX/Win7).

3.3. Web Interface

A major highlight of the LUCI middleware is the fact that it can be embedded in any web system through web sockets. This opens the door for a wide variety of HTML5 web applications. We use ActiveMQ as a MQTT broker (see section 3.9 Job Management / MQTT). Upon starting up the broker, it also starts a jetty webserver to support MQTT over websockets. This jetty instance also serves LUCI's webcontent. In the future the web interface should develop into the main instance from which a user can administrate and monitor LUCI, its service instances and scenarios. Among others we envision a flow diagram to interactively visualize service instances and perhaps other parts of a LUCI scenario. As a showcase of the capabilities of LUCI's capabilities offered by websockets, please refer to section 4.2 "Teaching the Unkown".

3.4. Data Structure

LUCI is supposed to run database-agnostic. This is achieved through mostly standard SQL code and all the agnostic parts being part of a database specific plugin in the database layer (see Figure 2). At the moment Postgres and H2 are supported. The data structure can be subdivided into two main topics: the inputs and outputs of the ser-vices as well as the geo-data repository, which in the perspective of the individual services could also be termed as "shared data".

The most important feature of the service related data is that we need to be able to relate every generated output to the corresponding input data. We solve this with timestamps: The services can operate on either only their input data or as well their input data plus shared data. Therefore, in an additional table we store the newest available timestamp from input table, shared geo-data and the newer out of the two in a third column, which at the same time also is being used as the identifier of a call to a service. This call-ID then is being used to identify the service outputs; the outputs table holds a column call-ID, which is the newest timestamp of its inputs.

The data structure of the shared data is driven by the goal to make query code as simple as possible and put the complexity into the insert and update code. As introduced in the terminology section (3.1), the shared data is organized in scenarios. Among seven predefined columns a scenario is defined by its attributes and an optional coordinate system identifier. In other words a scenario defines a project space in which all geometry shares the same attributes. Besides self-explanatory attributes like "geomID", "geom", a general purpose "flag", "userID", "timestamp" a scenario holds:

- a "batchID": LUCI among other use cases is being used for an evolutionary optimization and evaluation process in which many different variations of the same scenario are created. To not to mistake such variations with versions planned to be implemented in the future, we call those variations "batches".
- a "layer": Similar to layers in CAD applications a layer in LUCI is exclusive. So, geometry cannot be part of two layers.

For each scenario four tables are being created. Besides the main table holding the current state of a scenario, there is also a history table with a nearly identical structure as the main table. The only addition is a timestamp column to store the time at which a record from the main table has been deleted. The two remaining tables will be used for versioning in the future development of LUCI.

3.5. Remote Services

Among a few "local" services that reside in plugins/services folder a major benefit of using LUCI results out of its network orientation. It features distributed computing, load balancing. "Remote" services are a key element in the implementation of the parallelization and/or distributed computation capabilities of LUCI. They are characterized by two main attributes: Firstly, from a client's perspective remote services are indistinguishable from local services when being called. Secondly, any client can register as a service.

Upon registration a service describes its inputs and outputs. The inputs of any future call to that service will be verified by LUCI using this input description. The input/output description is very similar to the capabilities of web process services (WPS). In the future we could even think of converting the description to WPS when exposing the available services to the web.

Since the data is extensively being marked with timestamps, LUCI is able to send only the updates in the scenario data and input parameters since the last execution. Therefore the "get_scenario and" "get_service_inputs" actions both support the concept of time-range-selections, which is basically a parameter of the action call consisting of one of the keywords "from", "before", "until", "after" and a timestamp as the value. To make use of this partial data extraction, the remote service must implement some sort of data cache to which the updates can be added.

3.6. Communication Protocol

Messages in LUCI consist of a JSON header and optional binary attachments. The first 16 bytes of all messages encode the length of each header and attachments with an 8 bytes big endian number respectively. This is crucial since connections are not being closed, but remain open during a session (web sockets) or until the connection gets closed by either client or server (TCP/IP). The attachments part can contain multiple byte arrays. All of them must be described in the JSON header by a streaminfo object; a JSON object with predefined structure and keywords. If processing of the header fails, using the informaiton oft he first 16 bytes, all subsequent bytes can still be read, which clears the socket for the next message.

At the moment messages can be sent through TCP/IP and web sockets. Parallel messages are not allowed, so each message must be answered before the next message can be sent. This shifts the complexity of parallelization away from the client to LUCI and the term call ID remains free for services. As mentioned in the terminology section (3.1), we distinguish between actions and services. Actions are similar to remote procedure calls with the exception of not having a call ID. Messages always call an action by using the "action" keyword. Any message in LUCI must either contain one of the keywords "action", "error", "result", which "action" and "error" holding a string-value, and "result" a json object.

Actions themselves are plugins similar to local services, database adapters, or data converters. LUCI comes with a standard set of actions, which can be extended or adapted to the specific needs of a project, just as services can. In section 4.2 we show an example of how LUCI can be adapted to special needs by implementing dedicated services and / or adapting actions and converters.

3.7. Converters

Converters are plugins that call predefined functions of database adapters to store geometry in the scenario table. Supported formats so far are:

- GeoJSON
- Shapefile
- OSM-JSON (read-only)

DXF and other formats closer related to CAD are on the task list for future development. Converters must not only translate the information from one format to the database, but also implement a few features specific to LUCI:

- Attribute Mapping: a JSON object being part of the stream info object that tells the converter which attribute (e.g. ID) should be mapped to the seven predefined attributes (e.g. geomID) described in the data structure section.
- Delete_list: a property being stored in the format itself that tells the converter which elements should be deleted from the table, i.e. moved to the history table. E.g. in GeoJSON the delete_list is a property of a feature that holds no geometry.

3.8. Job Management / MQTT

Jobs, in LUCI being called service instance, can be run synchronously or asynchronously. In case it should be run asynchronously the service instance must be created first in order to retrieve a SObjID. As discussed in the data structure section (3.4) services can have inputs and outputs, which they define at runtime. Upon instance creation all input parameters of a service instance are being stored to the data-base. Whenever the service is being run, its inputs are loaded from the database. In theory the service can be re-run as many times as desired. Still, the service can store the outputs that belong to one single call ID (see section 3.4) only once. Since the call ID is always equal to the newest timestamp of services inputs re-running services only makes sense, if one of its input parameters has changed.

To listen for such changes we use the Message Queue Transport Transfer (MQTT) protocol, a publish-subscribe framework. It was developed by IBM, is open source and builds on top of TCP/IP and web sockets. It is referred to as the protocol for the Internet of things. A LUCI service instance can either subscribe one of its inputs to the output of another service instance or subscribe the instance as a whole to the termination of other services, which will cause the instance to run immediately after another service instance has finished. With this setup service instances can be represented in a flow diagram, which is the intention of the configuration interface mentioned in section 3.3. Using MQTT enables client applications to run previously created service instances simultaneously with one publication to MQTT. Furthermore, it enables them to monitor all service instance related activity.

Synchronous calls cannot be called through MQTT, but they must be called through the run action built-in to LUCI. All service inputs and outputs are not transferred to the data-base but directly the (remote) service and back to the client. The run action will wait until the service completes.

4. USE CASES

4.1. Transition Workshop, Singapore

Equipped with an early prototype of LUCI we conducted a workshop called "Design Space Exploration for Urban Compaction" at the SmartGeometry conference held at Chinese University Hong Kong in 2014 (see Koenig et al., 2014). The concept of multiple coordinated views (Baldonado et al., 2000) representing the coordinated parallel simulations was very well received. As shown in Figure 5 for transition workshop of the Future Cities Laboratory of the Singapore ETH Centre in April 2015 we refined those ideas and were able to present a system with synchronized views across several machines in the network using MQTT's publish-subscriber mechanism.

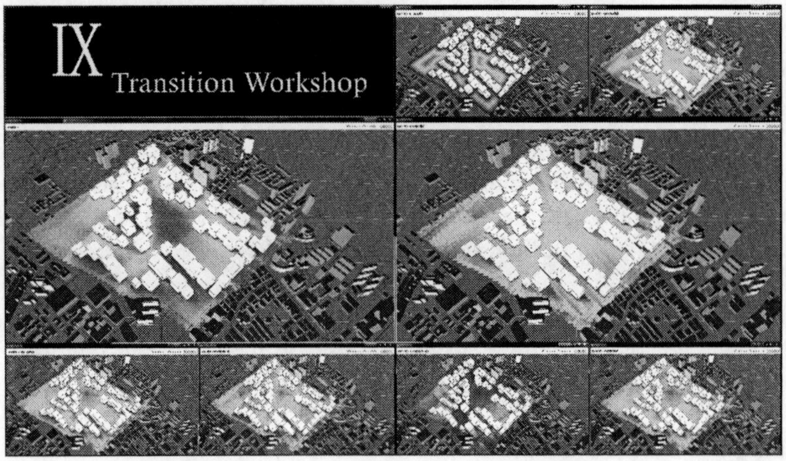

Figure 10. Coordinated & multiple views on multiple screens (machines).

In the context of the transition workshop we assessed the applicability of the LUCI system by using an example scenario in the district Rochor in Singapore. To prove that our approach works in an existing urban context we assume that this district needs to be completely re-planned. This exemplary area in Asia emphasizes the urgent need for fast and comprehensive planning systems. Necessary data on the street network was taken from Open Street Map, and information about neighboring built structures in 3D was available from the Future Cities Laboratory of the Singapore ETH Centre.

We used the LUCI system in combination with an evolutionary multi-criteria optimization method (EMO) for creating road networks with defined centrality characteristics, such as integration or choice for defined locations (Koenig et al., 2013). Inside the blocks of the road networks we generate building layouts with defined densities, taking into account specific properties of the open space qualities measured by Isovist fields (Koenig et al., 2014). The evaluation mechanisms were coupled as services with the EMO. Figure 11 illustrates how the resulting planning system can be used to help an urban planning proposal.

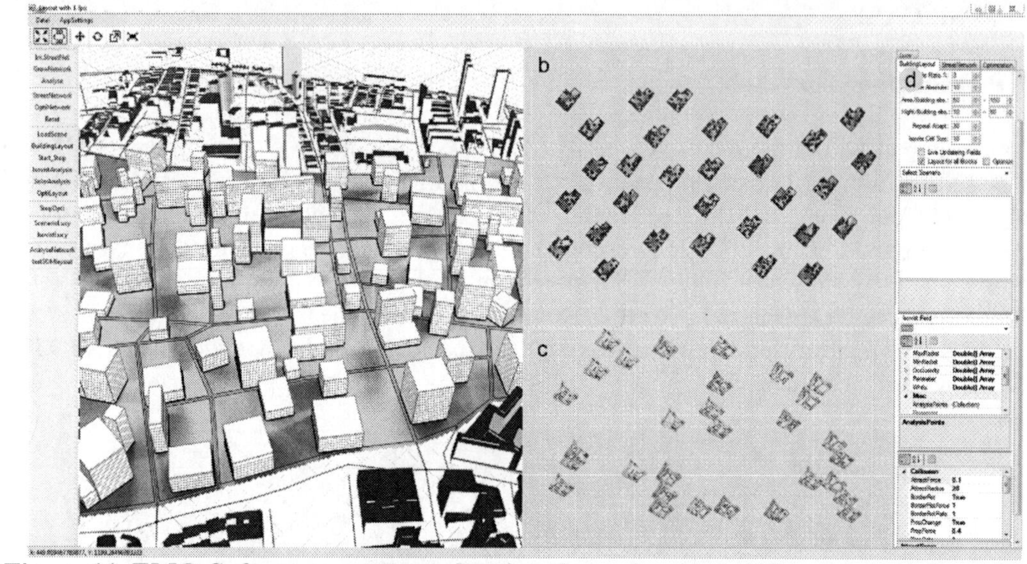

Figure 11. EMO-Software prototype showing the main areas of the user interface: (a) a 3D view com-bines one solution out of each archive, design solutions of the archives for (b) buildings layouts and (c) street networks, and (d) fields for the user input of size of population, number of generations, etc.

This kind of computational planning process we can call evidence-based planning. It helps the designer not only to eliminate potentially problematic configurations but also to meet the original design requirements.

4.2. Teaching the Unknown, Zurich

"Teaching the Unknown" is the title of a contribution we made to an exhibition by Moon Kyungwon & Jeon Joonho at the "Migros Museum für Gegenwartskunst" in Zurich from August 29th until November 8th 2015. The title of the exhibition was "News from Nowhere: Zurich Laboratory". The art piece consists of a touch screen table that shows a website running on LUCI. The website shows the images that were handed in by roughly 1000 students as a part of their exercises in a Massive Open Online Course (MOOC) on future cities held by Prof. Dr. Schmitt (Schmitt et al. 2014).

The website uses a map from Mapbox being semi-transparently laid over an image. The location where the image was taken is marked with a marker and an additional text box showing the description to the image. In an additional gaming mode visitors must select the city in which a randomly chosen image was taken. The last mode shows the rank of the player and the cities.

The students sent in 2083 images. For the game we selected 240 images from 120 cities. In LUCI each image is stored as a point with the city name, the image, the user name and the description being an attribute of that point. In order to speed up the client-side script (javascript), we wrote a small service that aggregates all selected images from the same city, so that in the Javascript we must not check for images with the same location. A second service ranks the cities based on the number of correct guesses. The player score is put together by the average distance to the correct city markers from the markers chosen by the player. A third service counts how many players achieved a better score, which is then used to display the rank of the player in the overall player history of the exhibition.

Figure 13. Screenshot of the game statistics mode showing city and player score.

While this project was mainly used to stabilize LUCI's data structure, one could think of LUCI's job management capabilities through which services could run automatically after a students upload more images in a future MOOC.

5. CONCLUSION AND OUTLOOK

We focus on small groups of planners or researchers that want to collaborate with each other and perhaps scale their collaboration to a larger group of participants at a later stage. As mentioned, with such a focus the simplification of the middleware approach becomes crucial. Simplification towards a desktop application for us means to avoid as many enterprise frameworks as possible that would hinder potential users to quickly develop a custom action, service or converter. We provide simple POJO-like access to LUCI that should allow users such as planners or researchers to easily adapt LUCI to their needs in a transparent way. The integration of an H2 database adaptor takes away the hassle of installing a PostGIS database for people unfamiliar with GIS databases such as architects or planners.

In the future we plan to integrate LUCI with the "Tool Library", a solution that makes view and tool configurations in a multi-screen situation room accessible and controllable from a web-based application. In this regard we would also like to create an editor that lets users put together service instances in a flow diagram. Furthermore, we could think of exposing LUCI's service descriptions to the web following the WPS standard as well as adding support for DXF and other NURB-based geometry formats. We also would like to add support for transformation matrices, so that the communication with regular 3D editors could be streamlined, since not the whole geometry would need to be sent through the network, but only its transformations. And as mentioned in the data structure section, the data structure has been designed with versioning in mind. A future implementation of this will extend LUCI's capabilities even more.

6. ACKNOWLEDGMENTS

We would like to thank Daniel Zünd, Alessandro Forino, Artem Chirkin, and Patrick Janssen for the many inspiring discussions and helpful ideas on LUCI.

7. REFERENCES

Al-Jaroodi, J., Mohamed, N., Jiang, H., and Swanson, D., 2003. *Middleware infrastructure for parallel and distributed programming models in heterogeneous systems,* Parallel and Distributed Systems, IEEE Transactions on, vol. 14, pp. 1100-1111.

Baldonado, M. Q. W., Woodruff, A., and Kuchinsky, A., 2000. *Guidelines for Using Multiple Views in Information Visualization, in Advanced Visual Interfaces*, pp. 110-119.

Borst, H., Salomons, E., Lohman, W., Zhou, H., and Miedema, H., 2009. *Urban strategy: noise mapping in instrument for interactive spatial planning, in 8th European Conference on Noise Control*, EURONOISE 2009, Edinburgh, PART 3, 31, 1-9.

Gorton, I., Liu, A., and Brebner, P., 2003. *Rigorous Evaluation of COTS Middleware Technology*, IEEE Computer Society, Vol 36, Issue 3, pages: 50 - 55.

Hofman, W., Lohman, W. and Schelling, A., 2011. *A Flexible IT Infrastructure for Integrated Urban Planning*, Journal of Theoretical and Applied Electronic Commerce Research, Vol 6 ISSUE 1, pp. 16-25.

Janssen, P., Stouffs, R., Chaszar, A., Boeykens, S. and Toth, B., 2014. *Custom Digital Workflows with User-defined Data Transformations via Property Graphs*, presented at the Design Computing and Cognition DCC.

Koenig, R., Standfest, M., & Schmitt, G., 2014. *Evolutionary multi-criteria opti-mization for building layout planning: Exemplary application based on the PSSA framework.* In E. M. Thompson (Ed.), Proceedings of the 32nd eCAADe Conference - Volume 2 (pp. 567–574). Newcastle upon Tyne; UK.

Koenig, R., Standfest, M., Treyer, L., Wei, Z. and Klein, B., 2014. *Design Space Exploration for Urban Compaction.* Available: http://smartgeometry.org/index.php?option=com_content&view=article&id=259&Itemid=201

Koenig, R., Treyer, L., & Schmitt, G., 2013. *Graphical smalltalk with my optimization system for urban planning tasks.* In R. Stouffs & S. Sariyildiz (Eds.), Proceedings of the 31st eCAADe Con-ference – Volume 2 (pp. 195–203). Delft, Netherlands.

L. Romualdo-Suzuki, A. Finkelstein, and D. Gann, 2013. *A middleware framework for urban data management*, in Proceedings of the 2013 ACM conference on Pervasive and ubiquitous computing adjunct publication, pp. 1359-1362.

Schmitt, G., Hebel D., Koenig R., Klein B., Brülisauer M., Tapias E., Zhong Ch., 2014. *Future Cities*, Retrieved from: https://www.edx.org/course/future-cities-ethx-fc-01x-0

Welington M. da Silva, Gustavo H. R. P. Tomas, Kelvin L. Dias, Alexandre Alvaro, Ricardo A. Afonso and Vinicius C. Garcia, 2013. *Smart Cities Software Architectures: A Survey*, in Proc. Of 28th Symposium On Applied Computing.

Free and Open Source Software for Geospatial (FOSS4G) Conference Proceedings

Volume 15 *Seoul, South Korea* Article 5

2015

CROWD-2-CLOUD – Remote Sensing Land Cover Verification With Crowd-Sourcing Data

Moataz Ahmed
School of Geography, University of Nottingham

Dai Huynh
School of Geography, University of Nottingham

Darshana Wickramasinghe
School of Geography, University of Nottingham

Tuong-Thuy Vu
School of Geography, University of Nottingham, Malaysia campus (UNMC)

Recommended Citation

Ahmed, Moataz; Huynh, Dai; Wickramasinghe, Darshana; and Vu, Tuong-Thuy (2015) "CROWD-2-CLOUD – Remote Sensing Land Cover Verification With Crowd-Sourcing Data," *Free and Open Source Software for Geospatial (FOSS4G) Conference Proceedings*: Vol. 15 , Article 5.
DOI: https://doi.org/10.7275/R5WQ020P
Available at: https://scholarworks.umass.edu/foss4g/vol15/iss1/5

PO-11

CROWD-2-CLOUD – REMOTE SENSING LAND COVER VERIFICATION WITH CROWD-SOURCING DATA

Moataz Ahmed[1],Dai Huynh[2],Darshana Wickramasinghe[3],Tuong Thuy Vu[4]

[1234]School of Geography, University of Nottingham
43500 Semenyih, Selangor Darul Ehsan, Malaysia
Email[1]: khgx4mad@nottingham.edu.my,[2]isxcdh@nottingham.ac.uk,
[3]khgx4dcs@nottingham.edu.my,[4]tuongthuy.Vu@nottingham.edu.my

ABSTRACT

Nowadays, advanced remote sensing technologies provide huge amount of Earth Observation (EO) data timely. Growing quickly in terms of size and structure, EO data require a new way of handling and processing as it is considered big data. Cloud-computing platform proved to be a reliable and scalable platform that suits various user demands in remote sensing data processing. To verify the ambiguity of information derived solely from remote sensing, ground data is vital. The only way to keep pace with big remote sensing data is to exploit the crowdsourced data, which has been recently proposed elsewhere. In this study, we developed a prototype of an integrated location based service on top of cloud computing platform to detect land cover features and engage the crowd of volunteers during training and verification process. Relying on open-source tools, the proposed system provides location-based data collection and satellite image classification. The prototype was tested over the rapid on-going landscape surrounding the University of Nottingham, Malaysia campus. More advanced functions will be developed and a full system will be deployed and tested in further study.

1. INTRODUCTION

Remote Sensing technologies have a wide scope of applications. Urban planning, agriculture, environmental management, or climate change, remote sensing products are the indispensable sources of information in our daily life activities (Robila 2006). In the past, lesser remote sensing data were available and limited in spatial, spectral and temporal resolutions. This limitation makes it difficult to analyze data and derive the proper information about the Earth surface (Ramapriyan 2013).

Until recently, the picture has changed. As the number of operating satellites increase periodically and each of which has a specific mission to achieve with different types of attached sensors. Furthermore, advanced sensor technologies generate higher spectral and spatial resolution images than before, which can contribute effectively in better understanding of the Earth as an integrated system. Different remote sensing datasets become more complicated in terms of structure and size. Therefore, bigger volume of data received from satellites, more storage and computing power is required to analyze these datasets(Christophe 2011).

High-performance computing system is obviously a solution and cloud-computing platform is the today system for big data. Cloud computing provides unlimited storage capacity and elastic computing services in form of Software As A Service, (SAAS) Platform

As A Service (PAAS) and Infrastructure As A Service (IAAS) models. Also, Cloud Computing can reduce the costs, as it's possible to pay as use and on demand (Xiaoqiang 2010), instead of purchasing advanced hardware and software licenses to process these big data. Moreover, the scalability options could make it easier to scale up or down the number of processors required to achieve EO data processing and analyzing.

Earth observation agencies are moving toward cloud computing platforms to host and deliver their remote sensing data like ESA Helix Nebula (Helix Nebula Online), NASA NEX (NASA NEX Project Online) and MODIS Azure (MODISAzure Online). The established system so far is merely a storage platform whereas few data processing services have been deployed to utilize the efficiency of cloud-computing platform. Various applications and user demands require different ways to derive the information from satellite EO data and it seems complicated to provide a general processing service. Let's work on the best resolution available 30-m Landsat data set; the land-cover product seems a generally required product. Land-cover classification processing service can be developed, but a critical issues remains, validation. The challenge here is how to validate such big data? We cannot rely on a small team of experts conventionally, but a bigger group of people.

Recent advances in positioning technologies, location-based services, broadband Internet connections, and Web 2.0 allow every human to easily become an intelligent sensor and collect geospatial data accurately (Goodchild, 2007b). Goodchild also states that surprisingly among those six billion sensors exist thousands of volunteers who are willing to contribute, share, and update their collected geographic data. This new innovation of generating geographic information is termed *volunteered geographic information – VGI* (Goodchild, 2007a), or *crowdsourcing*. Thereafter, we have observed the rise of crowdsourcing and its applications in various areas including land cover (Comber *et al.*, 2013, Foody and Boyd, 2012, Fritz *et al.*, 2012).

This paper reports our first step of application development taking advantage of current advancements in remote sensing data evolution and the rise of using crowdsourced data and its applications. Here, we utilized cloud computing elastic architecture to build supervised classification service with crowd-based accuracy assessment model. The second section describes the development steps and the system architecture starting from data acquisition till the classification results, while the third section explains the initial results and and fourth section presents the further development and future work to be done.

2. Methodology

This paper proposes a cloud-based framework utilizing crowdsourcing data to validate ambiguity land cover information derived from remote sensing supervised Support Vector Machine (SVM) classification algorithm. For this purpose, we present an open-source mobile application, *Crowd-2-Cloud,* which is used to collect crowd-sourced land cover data. The crowd-sourced database later has two important roles in assisting the validation of remote sensing land cover data: (i) as a training sample set and (ii) as a validation data set of our proposed classification method.

2.1 Data Acquisition

Crowd-sourced volunteers can use the mobile app to capture landscape photos embedded with Geo-location data. They also can decide which land cover features dominate in the photo. Finally, submitted data from the crowd will be stored on a cloud based Postgresql server using Nottingham University private cloud. Figure (1) shows a screenshot of the mobile app main page.

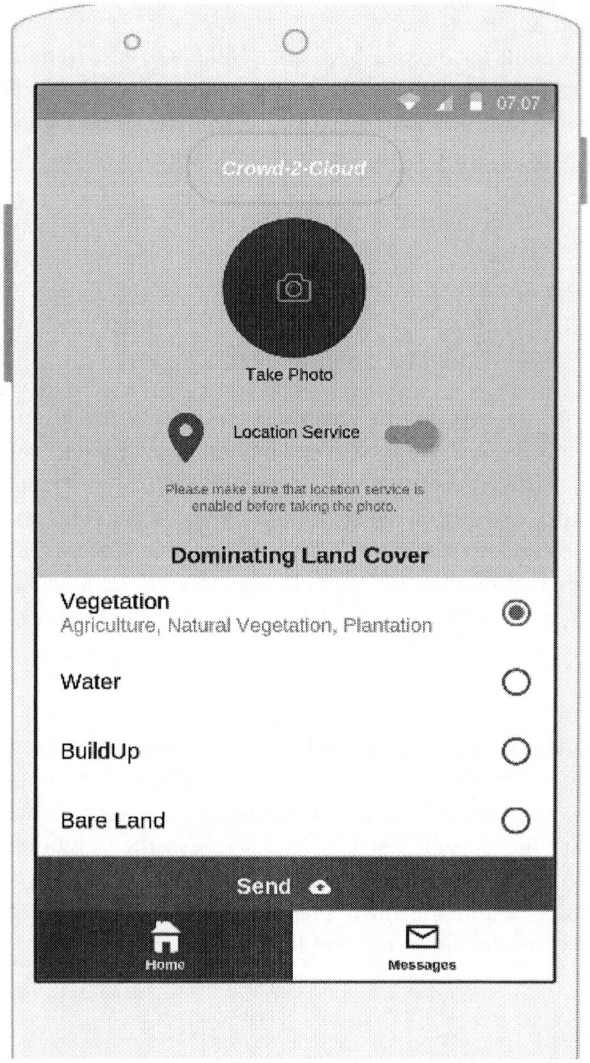

Figure 1. Screenshot of the mobile application

2.2 Processing

The overall system architecture consisting of all components is shown in Figure (2). The main components of the system are: 1) A mobile app to collect crowdsourced data from users (as shown in figure (1)). 2) A Postgresql Database Server on top of the cloud to receive the sent data from users. 3) A large computing node to run the supervised classification algorithm and publish results to be assessed using 10-fold cross-validation method.

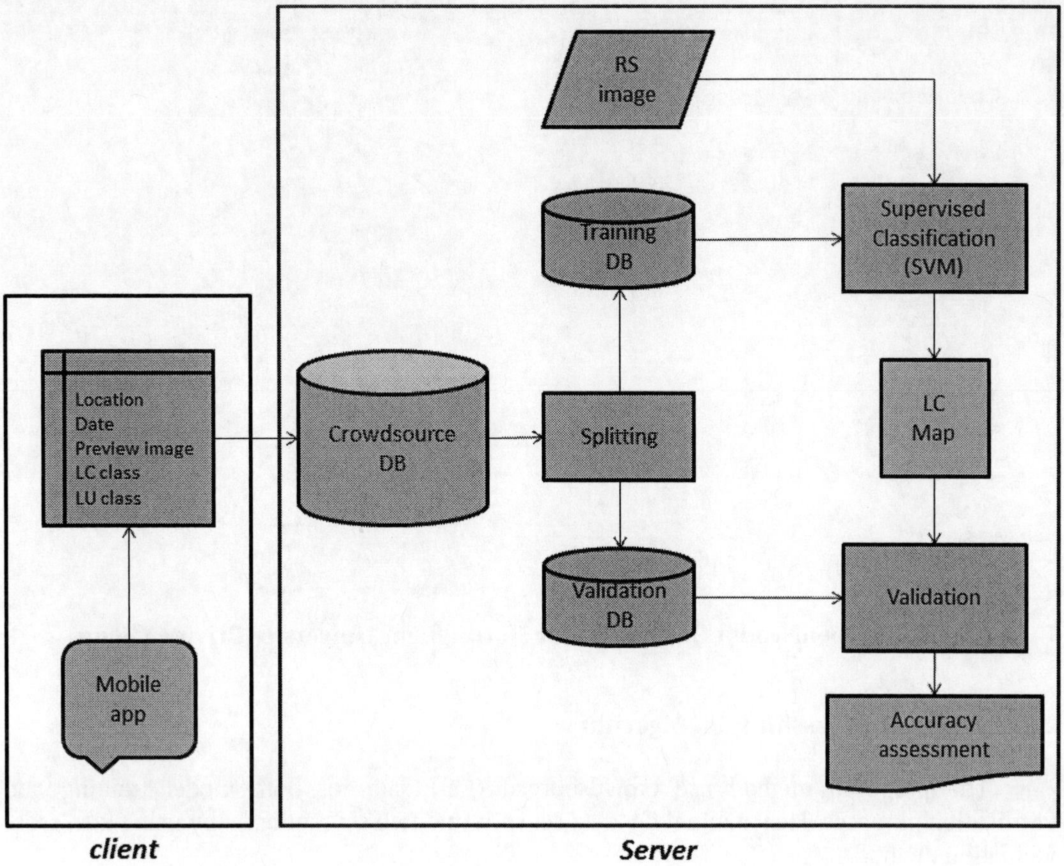

Figure 2: System Architecture and Data Flow Between Client and Server

2.2.1 Mobile Application

A simple Android-based mobile application is used for crowdsourced data collection from the users. The application's main page asks the user to take a photo and enables the location service to retrieve the user's current location. The app also asks the user to pick one choice as a dominating land cover class from list contains "Vegetation, Water, Built-up, Bare land" then the user can press "send" to upload the data to the server side which is a cloud-based Postgresql server.

2.2.2 Database Server

We set up a fully Open Source enabled database server configured with Postgresql 9.3 and Ubuntu 14.10 Server on top of Nottingham University private cloud infrastructure as shown in figure (3) from the dashboard of OpenNebula (OpenNebula Online). The aim from using Postgresql server is to store the crowdsourced data coming from the mobile app and then split it into two different databases. Using 10-fold cross-validation method (Ljumovi 2015), crowd-sourced data could be divided into training sample database and verification database.

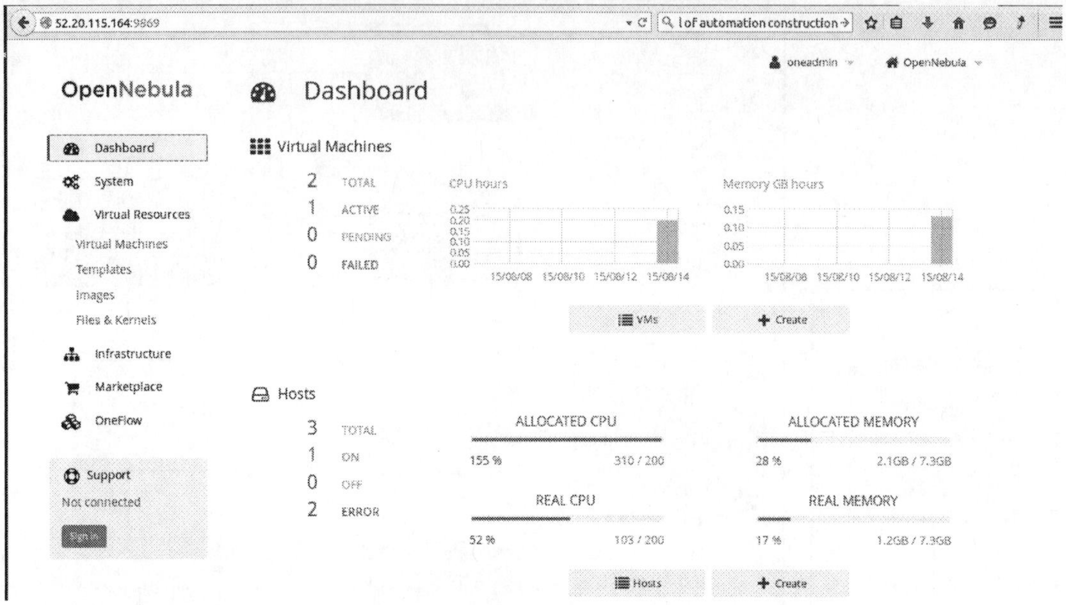

Figure 3: OpenNebula Dashboard for Nottingham University Private Cloud

2.2.3 Supervised Classification Algorithm

The proposed method use crowdsourced (CS) data for both model training and verification purposes. Hence initial CS data set split into two set randomly by using k-th cross validation method.

First set is considered as training samples. Pixel-based image classification methods are still applicable and well working on moderate resolution image like Landsat. The study adopted a proven SVM image classification algorithm for the image classification but it is extendable to any type of supervised learning methods. As a dynamic system, the collected training data are regularly updated and refined. The set of spectral signature library including temporal factor for a specific study area like Malaysia will be made available on our system. Though it is uneasy to have a big and properly distributed training data from begining but we assume that it would be sufficient in the long-run of the system operation thanks to the crowd contribution, In our initial development here for Malaysia study site, we proposed to classify the image into four dominant land cover classes: built-up, water, vegetation and bare-land.

The second set is validation data, used to validate the resulted classification maps acquired by SVM supervised learning (Liu 2014). The system extracts the classified land cover class of each validation points and calculates the confusion matrix by comparing classified and CS validation class. The accuracy assessment will be delivered in a form of a text document, containing overall accuracy, user and producer accuracy represented as percentage. The temporal factor needs to be taken into account in both training and validation via checking the time stamp of the crowdsourced photos and the satellite image acquisition date. Though it is not critical in tropical Malaysia but elsewhere.

3- Results

In this initial pilot version, we collected crowdsourced data by using the mobile application. The mobile application distributed over internally and gets the support of student and staff to collect the crowd sourcing data around Kuala Lumpur and surrounding city areas.

The resulted crowd source data contains around 60 locations with their dominating land cover description. Lesser number of samples selected as training points and used to train the SVM classification model and gets the classifier file. Figure 4 showing the training sample distribution over the testing area on top of Landsat8 image which going to be classified. Once the user selected the required satellite image, system automatically will run the SVM classifier and outputs the classified colored land cover classification image.

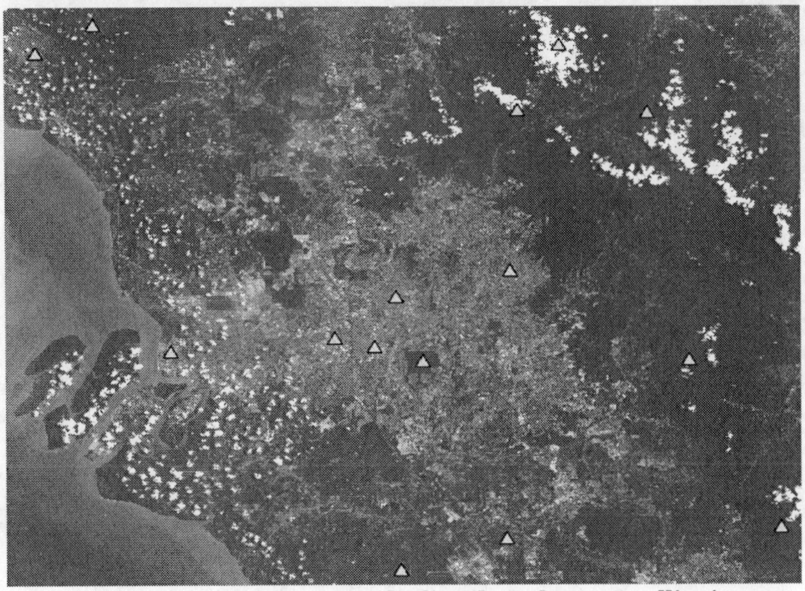

Figure 4. Training sample distributed over satellite image

Landsat 8 images used for classification with bands 2, 3, 4 and 5 only . Figure 5 shows the resulted SVM classification map with four dominant classes. There is few cloud classes in the satellite image and they are masked out using quality assessment band. Hence, cloud areas which are mixed with other land cover types have been minimized.

The accuracy assessment carried out using the rest of the CS data which contains 54 sample points for classification validation purpose. The system automatically extracts the corresponding land cover type in classified map and calculates the confusion matrix and Table 1 shows the resulted confusion matrix.

According to this result, the classification got 81.48% as an overall accuracy. Water classes accurately identified while Vegetation class also got the highest accuracy level however in some places it is mixed with bare-land and built-up areas. The built-up and bare-land areas also mixed with each other classes and got the lesser accuracy.

4- Conclusion

The development of land cover database met new challenge when Earth Observation (EO) data accelerate both in size and structure thanks to advanced remote sensing technologies. Although a deeper understanding of the Earth surface is made possible with a growing number of satellites in operation and higher spectral and spatial resolution images, such large volume of data require sufficient storage and computing power to handle. Cloud computing arises as a solution due to its advantages and reliability. This paper addressed the usage of cloud computing in detecting land cover features and data verification through this system architecture.

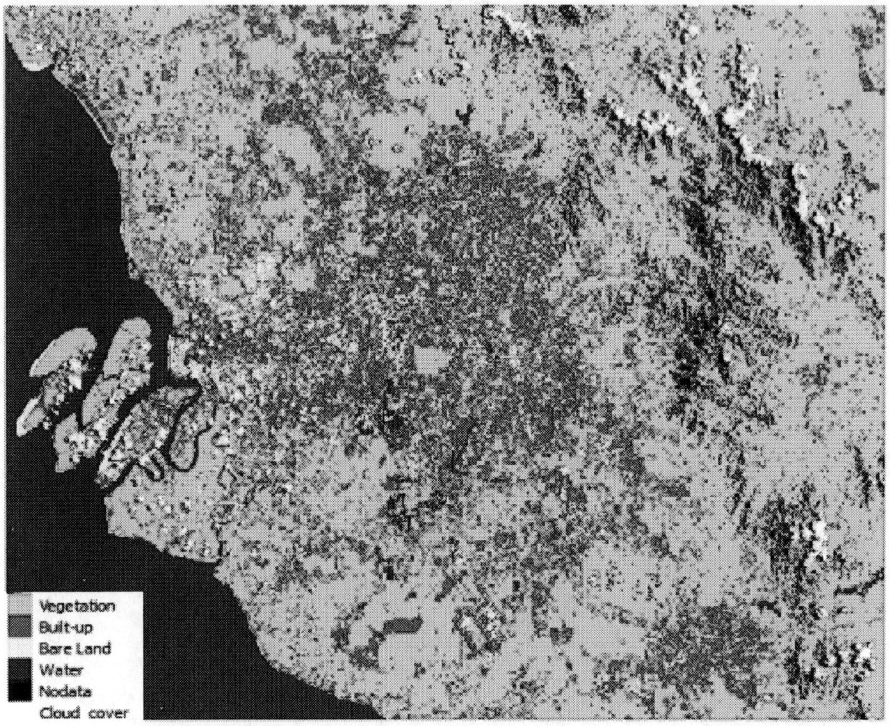

Figure 5. classified satellite image with four land cover classes

A mobile application is designed to collect data, and send it to the cloud based server for analysis which is supported by collected crowdsourcing data, machine-learning algorithms to identify land cover features from Landsat8 image. At this initial stage, implementation and trial test applied to the areas around Kuala Lumpur, Malaysia and future expansion will be reported in later research.

5- Future Work

This paper presented the initial steps of development which have been taken to provide land cover detection using Landsat 8 data sets and verification using crowd source data on top of cloud computing resources. Future work will include modifications to let the system work in 2 ways communication to let the users get notifications on their mobile application once a nearby spot detected to be changed from type of land cover to another. furthermore, we will provide a comparison study between using cloud computing and shared hosting solutions for hosting the service.

Table 1: Confusion Matrix

LC Class		Classified Image				Total	**Producer's Accuracy**
		Vegetation	Water	Built-up	Bare-land		
Crowd source data	Vegetation	13	0	1	1	15	86.66667
	Water	0	7	0	0	7	100
	Built-up	2	0	16	3	21	76.19048
	Barren	1	0	2	8	11	72.72727
Total		16	7	19	12	54	
User's Accuracy		81.25	100	84.21053	66.66667		**81.48148**

6- References

Comber, A., SEE, L., Fritz, S., Velde, M. V. D., Perger, C. & Foody, G. 2013 Using control data to determine the reliability of volunteered geographic information about land cover. *International Journal of Applied Earth Observation and Geoinformation,* 23**,** 37-48.

Foody, G. M. & Boyd, D. S. (2012) Exploring the potential role of volunteer citizen sensors in land cover map accuracy assessment. *Proceeding of the 10th International Symposium on Spatial Accuracy Assessment in Natural Resources and Environmental Sciences*.

Fritz, S., Mccallum, I., Schill, C., Perger, C., See, L., Schepaschenko, D., Velde, M. V. D., Kraxner, F. & Obersteiner, M. (2012) Geo-Wiki: An online platform for improving global land cover. *Environmental Modelling & Software,* 31**,** 110-123.

Goodchild, M. F. 2007a Citizens as Sensors: The World of Volunteered Geography. *GeoJournal,* 69**,** 211-221.

Goodchild, M. F. 2007b Citizens as Voluntary Sensors: Spatial Data Infrastructure in the World of Web 2.0. *International Journal of Spatial Data Infrastructures Research,* 2**,** 24-32.

Jackson, M. J., Rahemtulla, H. A. & Morley, J. 2010 The Synergistic Use of Authenticated and Crowd-sourced Data for Emergency Response. *The 2nd international workshop on validation of geoinformation products for crisis management (Valgeo)* 91-99.

S. A. Robila, "Use of Remote Sensing Applications and its Implications to the Society," in *2006 IEEE International Symposium on Technology and Society,* 2006, pp. 1–6.

H. Ramapriya*n* "Managing Big Data « Earth Imaging Journal: Remote Sensing, Satellite Images, Satellite Imagery." [Online]. Available: http://eijournal.com/print/articles/managing-big-data. [Accessed: 13-July-2015].

Christophe E, Michel J, Inglada J. Remote sensing processing: From multicore to GPU. *IEEE J. Sel. Top. Appl. Earth Obs. Remote Sens.* 2011;4(3):643-652. doi:10.1109/JSTARS.2010.2102340.

Y. Xiaoqiang and D. Yuejin, “Exploration of cloud computing technologies for geographic information services,” *2010 18th Int. Conf. Geoinformatics*, pp. 1–5, Jun. 2010.

“ESA SSEP Use Case | Helix Nebula.” [Online]. Available: http://www.helix-nebula.eu/usecases/esa-ssep-use-case. [Accessed: 17-July-2015].

“NASA NEX.” [Online]. Available: http://aws.amazon.com/nasa/nex/. [Accessed: 23-July-2015].

“MODISAzure - Microsoft Research.” [Online]. Available: http://research.microsoft.com/en-us/projects/modisazure/. [Accessed: 12-Aug-2015].

OpenNebula | Flexible Enterprise Cloud Made Simple. Available at: http://opennebula.org/. Accessed August 15, 2015.

Ljumovi M, Gmbh RB. Estimating Expected Error Rates of Random Forest Classifiers : A Comparison of Cross-Validation and Bootstrap. 2015:212-215.

Liu Q, Guo Y, Liu G, Zhao J. Classification of Landsat 8 OLI image using support vector machine with Tasseled Cap Transformation. In: *2014 10th International Conference on Natural Computation (ICNC)*. IEEE; 2014:665-669. doi:10.1109/ICNC.2014.6975915.

Free and Open Source Software for Geospatial (FOSS4G) Conference Proceedings

Volume 15 *Seoul, South Korea* Article 6

2015

Modeling Of Terminology Database In Field Of Geodesy And Geographic Information System

Zolzaya Lkhamsuren
Mongolian Geospatial Association

Oyuntsetseg Dash
School of Mining Engineering, Mongolian University of Science and Technology

Alimaa Jargalsaikhan
E-Open School, Mongolian University of Science and Technology

Ochirkhuyag Lkhamjav
Mongolian Geospatial Association

Recommended Citation

Lkhamsuren, Zolzaya; Dash, Oyuntsetseg; Jargalsaikhan, Alimaa; and Lkhamjav, Ochirkhuyag (2015) "Modeling Of Terminology Database In Field Of Geodesy And Geographic Information System," *Free and Open Source Software for Geospatial (FOSS4G) Conference Proceedings*: Vol. 15 , Article 6.
DOI: https://doi.org/10.7275/R51G0JGK
Available at: https://scholarworks.umass.edu/foss4g/vol15/iss1/6

PO-19

MODELING OF TERMINOLOGY DATABASE IN FIELD OF GEODESY AND GEOGRAPHIC INFORMATION SYSTEM

Zolzaya Lkhamsuren[1], Oyuntsetseg Dash[2], Alimaa Jargalsaikhan[3] and Ochirkhuyag Lkhamjav[1]

[1] Mongolian Geospatial Association
Room #208, Taiwan Cultural Center, Juulchiny Street-38, Ulaanbaatar, Mongolia
Email: zolzayamust@gmail.com, mongeoinfo@gmail.com
[2] School of Mining Engineering, Mongolian University of Science and Technology
P.O. Box 516/24, Ulaanbaatar 210524, Mongolia
Email: daoyunaa@must.edu.mn,
[3] E-Open School, Mongolian University of Science and Technology
P.O. Box 46/520, Ulaanbaatar 210524, Mongolia
Email: alimaaj@must.edu.mn

ABSTRACT

Recently in Mongolia, many techniques and methods are being applied to the surveying and mapping fields along with the development and innovation of geodetic surveying technology, more new terminology are being used in those area. Despite of the origin of the terminology that came from or which language it's being translated, those new words need to be interpreted by concerning the exact meaning of the word and what this word is actually referring to. However, in Mongolia, there's no legal framework including regulation and standards of terminology and professional word translation. Consequently, a single term is being used with various meanings in many organizations, companies and so on. And also an ambiguous meaning of the term has been aroused and leads to wrong definition and idea. Due to this, lack of use of the terminology influences negatively on not only surveying and mapping field, but social areas as well.

In order to solve this issue, study aims to create glossary database based on development of the terminology dictionary of geodesy and geographic information system through dynamic web design. In this study, many research methods have been carried out including, literature review based on the Dr. Damdinsuren (2013)'s "Glossary for Geodetic Terms", textbooks and other materials of surveying and geospatial in English, Mongolian and Russian, as well as in print and online dictionaries, surveying questionnaire on usability and quality of professional terminology and dictionary and their applications within academia, professors, students, civil servants and private companies' engineers, and SWOT analysis on printed and online dictionaries. Surveying questionnaire resulted that 20 per cent of 150 questionnaire use professional textbooks, which are printed in the Mongolian language, 45 per cent use Russian and 35 per cent use English sources. Glossary database on geodesy and geographic information system is established in three languages (Mongolian-English-Russian) using MySQL and PHP programming language. The users from geo-information industry will be able to develop and update the database based on the free and open source platform, particularly, online. This glossary database contains around 500 terminology including over 200 graphic illustrations. Frequent access of the terminology can be counted and web page includes links of relevant public organizations' and universities' web pages.

So, the study results that this glossary database on geodesy and geographic information system could provide specific meaning of the term and how this term can be used depending on the specific field via dynamic web-page. And it allows and facilitates accessibility of database anywhere and anytime free of charge meaning that removes ambiguity and saves users' precious time and money.

Keywords: Geodesy, Geo Information System (GIS), database, terminology, online glossary

1. INTRODUCTION

More techniques and methods are being applied to the surveying and mapping fields along with the development and innovation of geodetic surveying technology, more new terminology are being used in those area. Despite of the origin of the terminology that came from or which language it's being translated, those new words need to be interpreted by concerning the exact meaning of the word and what this word is actually referring to.

However, in our country, there's no legal framework including regulation and standards of terminology and professional word translation. Consequently, a single term is being used with various meanings in many organizations, companies and so on. And also an ambiguous meaning of the term has been aroused and leads to wrong definition and idea. Due to this lack of use of the terminology negatively influences on not only surveying and mapping field, but social areas as well.

2. PURPOSE OF STUDY AND OBJECTIVES

In order to solve this issue, research aims to create web-database of the terminology based on development of the terminology dictionary of geodesy and geographic information system through web design. In this research, many research methods have been carried out including the following,

- Literature review on terminology, translation rules and grammar and articles and books by linguistics;
- Questionnaire on usability and quality of professional terminology and dictionary and their applications within academia, professors, students, civil servants and private companies' engineers;
- SWOT analysis on printed and online dictionaries

Terminology database on geodesy and geographic information system has been established in three languages (Mongolian-English-Russian) based on one of the database management systems, MySQL and using PHP programing languages and dynamic web designing.

3. APPROACHES

Analysis of studies done at the end of "Geodesy, geographic information systems professional terminology database" system on the model for MySQL database (Figure 1) and PHP scripting language online access decided to be in a dynamic web site (Figure 2).

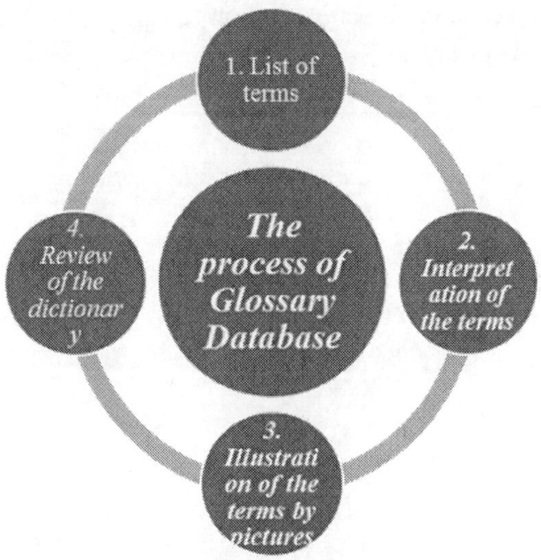

Figure 1. Model for the database of terms in field of surveyor and GIS

3.1 List of terms

The firstly, list of terms in Mongolian language by alphabetical order are created in spreadsheets, e.g Excel sheet, and it is converted into My SQL database.

3.2 Interpretation of the terms

4 In the many professional dictionary, the specific words highlight the main keywords of the process, function of geodetic and geospatial, but there are changed or updated and re-defined by the chronological periods and cutting edge technology. In this section, the following glossary books are used for the main translation of keywords, including:

- Demberel Baldan, 2010, "Glossary of Geodesy and Cartography References"
- Damdinsuren Avirmed, 2013, "Glossary of Geodetic Terms"

4.1 Illustration of the terms

The terms are illustrated by using pictures, graphs and icons, especially for the special names, equipment and tools. But terms that difficult to illustrated by the graphs and has ambiguous meanings are explained with their references.

4.2 Review of the dictionary

Finally, the glossary is reviewed and improved by the different level users and professionals of geodetics and GIS

Furthermore, the glossary is built online and free access to all levels of customers and specialist by user-friendly. PHP language for the online dictionary is used. The main reasons are chosen this web page language:

- PHP code use with HTML
- Abundant research of materials
- MySQL, ORACLE, ACCESS, MSSQL, connect to data flow management no problem.
- Open source code
- The class consists of funds that can be used independent of each other
- A lot of components to work with the database, fill in the table, testing and trial such as user data
- The completely based on object extraction technology
- Development always update

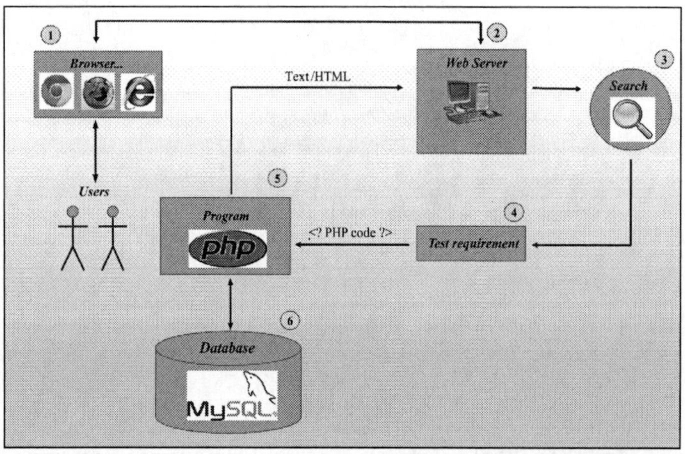

Figure 2. The Glossary System Data Flow Diagram

4. RESULTS

In this paper, survey was involved about 150 professionals, include students, who were studying geodetic survey, land management and cadastral, professors and lecturers from universities, government officials and private companies. (Figure 3).

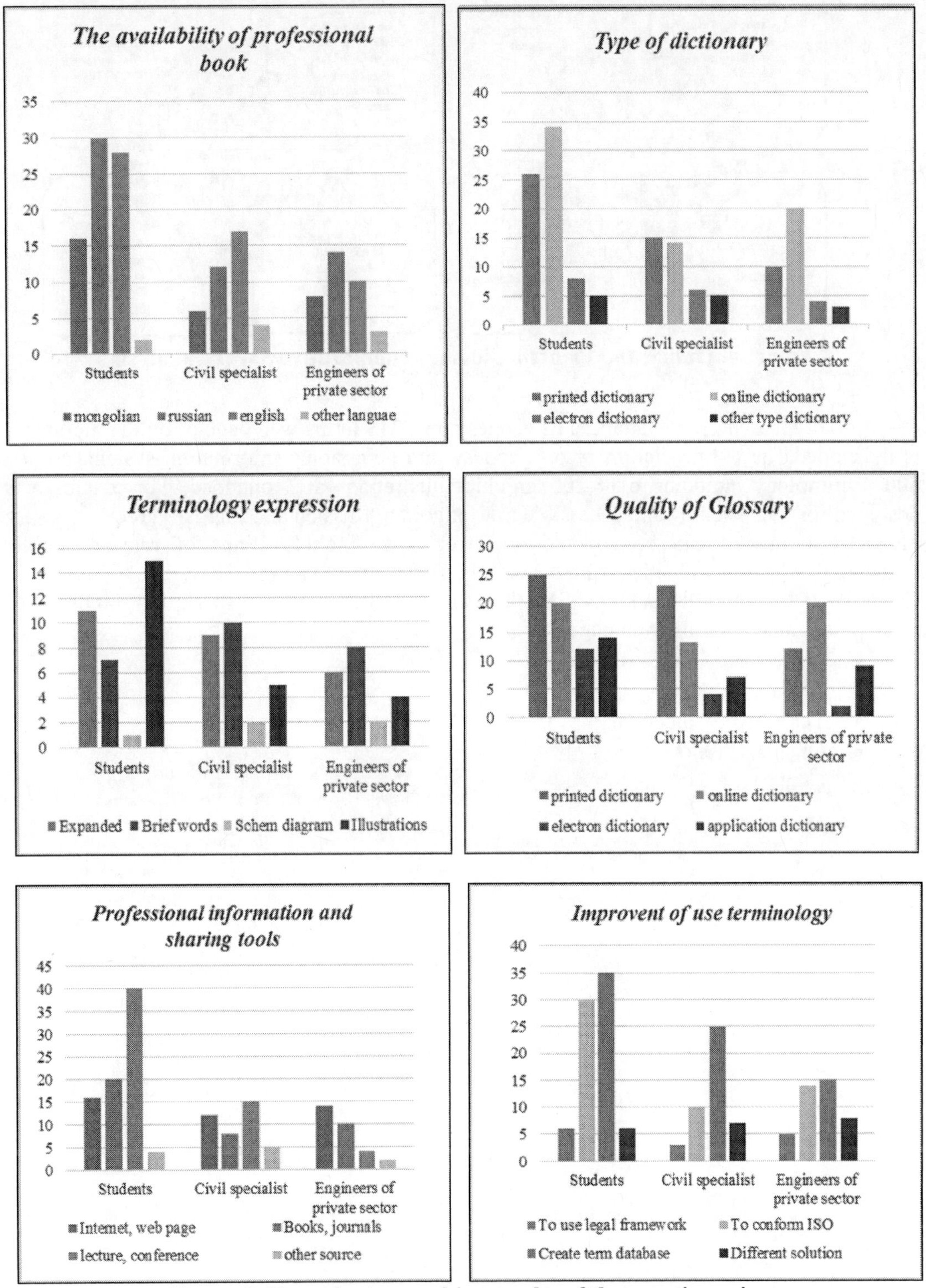

Figure 3. Transfer to graphics results of the questionnaire

Figure 4 shows the result of SWOT analysis of both professional dictionaries, printed and online, and also, it identifies that online glossary system is more useful, accessible to users.

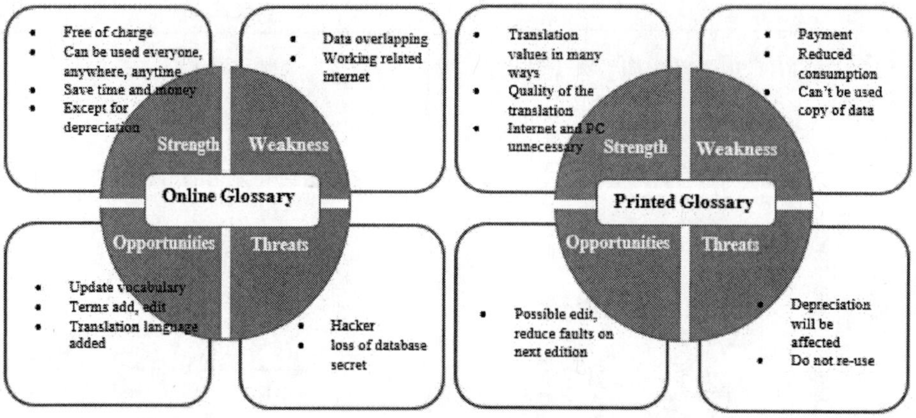

Figure 4. Online and Printed Glossary comparative SWOT analysis

Finally, based on the database of geodesy and GIS terms web page of online dictionary is developed.The online dictionary on geodesy and geographic information system, around 500 terminology including over 200 graphic illustrations are contained. This database is designed for user-friendly and accessible in Mongolian, Russian and English. Frequent access of the terminology can be counted and web page includes links of relevant public organizations' and universities' web pages (Figure 5). Here loading it is system "GEODESY, GEOGRAPHIC INFORMATION SYSTEM TERMINOLOGY ONLINE DICTIONARY" with 5 main menu per that web page.

- Home page
- About web page
- Terminology dictionary
- Open discussion
- Contact

Figure 5. Home page online dictionary show in browser

Generally, the online dictionary is designed for user-friendly, and list of terms is sorted in alphabetically orders, "А-Я", and each letter is linked to glossary.

Query in the database: It is possible to search terms by three languages, Mongolian, Russian and English, in the database, and it is shown close and related terms, together, and the link, end of terms, illustrate picture, graph, equation and others (Figure 6).

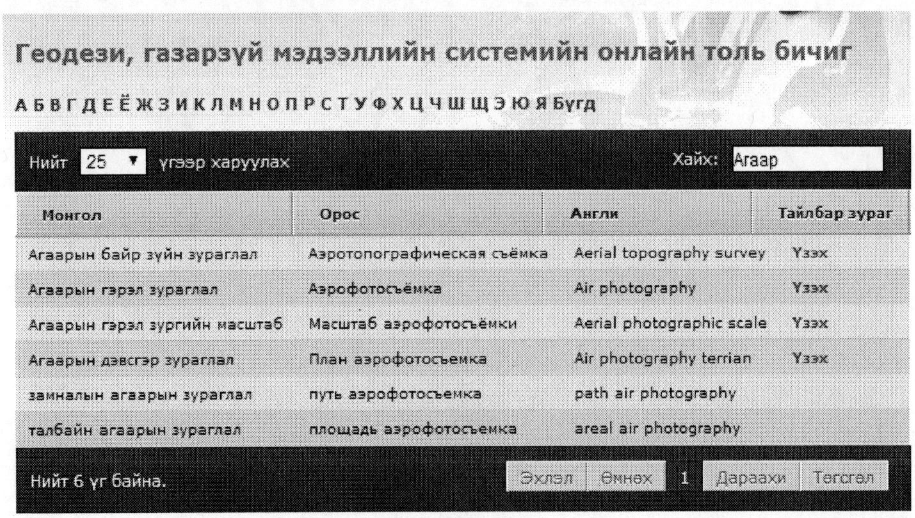

Figure 6.

Add terms in the database: Super user or admin of the database can do to add new terms or edits of glossary in Mongolian, Russian, and English language with explanation pictures or graphs (Figure 7).

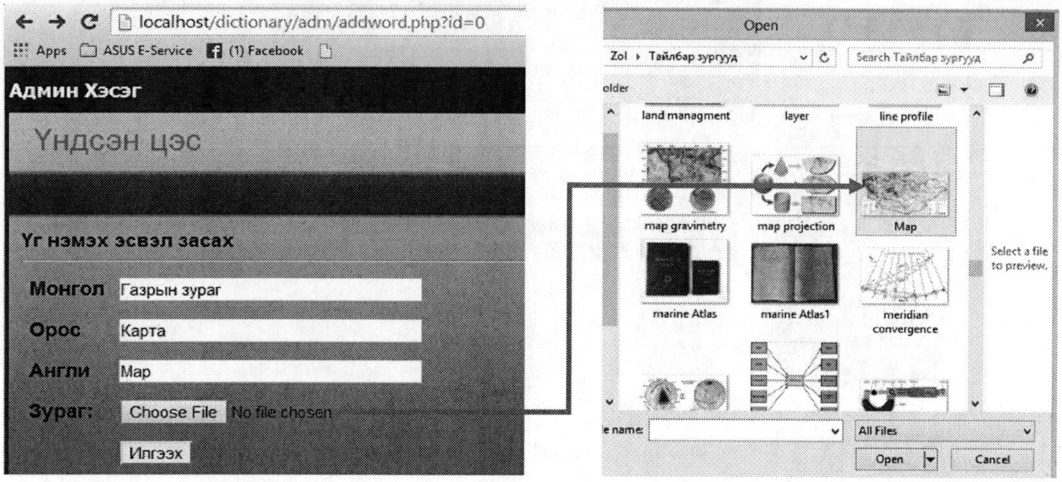

Figure 7.

5. DISCUSSIONS AND CONCLUSION

This research was conducted within the framework of the following tasks. Including:

1. Geodesy and Geographic Information System professional terminology know so as current level each sector delegation total 150 person taken survey questionnaire result

Access to the availability of professional books acts on the Mongolian language of 20% and 45% in Russian and English, while 35%

- 43% of the professionals making the translation dictionary online, printed dictionary (37%) and 20% use electronic dictionary.

Easy to use properly-end requirements, such as online dictionary of 45% and 35% of the vocabulary of print and electronic dictionary provides 20%

- 55% of business information on professional lectures and seminars conferences 25% of books, newspapers and magazines, and receive 20% of the Internet and other media that the figures released. Professional terms in future mother tongue key factors that influence the development and rational use up a database of information on known public record envisages that should be retained.

2. Research on the basis of both print and online glossaries of terms / SWOT / analysis of more than 600 terms made on the Mongolian-Russian-English My SQL database, create and explanation consistent with the requirements of the internet / search-filtering-choosing/ total Terms 30% or 200 words Publication additional sketches, diagrams, photographs included.

3. Terms word and picture directory of users in a clear and accessible, in accordance with their needs for geodesy, geographic information systems spread online dictionary of terms used in the form of PHP web programming language, you http://geomedeelel.mn/ uploaded on the Internet translation dictionary of terms of the system to access, regardless of time and space to be able to use free of charge to any user looking for work easier, precious time and money on energy and economic importance is considered to have concluded.

Therefore, terms should be translated according to the nature of the sentences. So the research result shows that terminology database on geodesy and geographic information system could provide specific meaning of the term and how this term can be used depending on the field to the users via web-page. Terminology database allows and facilitates users to access terminology database anywhere and anytime with no cost and integrate different sources of information into one.

6. REFERENCES

Damdinsuren A., 2013. *Geodetic Glossary of Terms*. Munkhiin Useg publish office. Ulaanbaatar,

Demberel B., 2009. *Geodesy, Cartography Vocabulary Reference*. "Sudar Munkh" the Publication office Ulaanbaatar

Luvsanjav Choi. Avarzed O., Ganbold Ayu., Dharmabal Lu., 2000 *Calculator English-Russian-Mongolian Glossary of Terms.* Academy of Science, Informatics Institute, Publication office Ulaanbaatar

Zunduisuren Ch., Dashpurev D., Engineering team MCS International LLC., 2013. *Encyclopaedic dictionary of science and engineering*. Ulaanbaatar

Avdai Ch., Enkhtuya D., 2000. *Methodology of conducting academic research*. "Chuluunbar" Publication office .UB

Tsevel Y., 1996. *Mongolian vocabulary*. Ulaanbaatar

ESRI. ArcGatalog-start. Admon Printing. UB .: 2007

Padmanabhan G., Mark R., Leipnik and Jeawan Yoon A., *Glossary of GIS Terminology*. Assistant Professor of Civil Engineering, North Dakota State University, Fargo ND 58105

Adilbat L., 2005. *Issues of applying linguistic tools for translation of electronic machines* /NMU, OUKhS, DOT-scientific conference publication brochure/ NMS Publication Unit.UB. 34-37page

Janlavchariv M., 2010. *To create terms of cadastral electronic dictionary* /MUST, MES, Academic conference in honor of Dr. B. Ishmend, Arvai Barhan printing industry. UB. (Page 37)

Sumiya Ts., 2004. *Theories of Dictionary. Mongolian National University, Research papers of FLCI* NMU Publication Unit. UB. 235 (06) 45-59 page

Gerelmaa G., *Assigned terms of information and communication experience and methodological issues. "Glossary,"* №12-13.UB .: 2011 page.19-23

Önörbayan Ts.,1999. *Issues of regulating scientific terms. " Glossary,"* No2. UB. page.31-39

Undraa D., Batzolboo B., 2013. *Basics of Database system roots.* "Munkhiin Useg" group. UB

Uyanga S., 2012*The basics of information systems.* "Bit Pryess" LLC. UB

Ganbat Ts., 2004. *Software quality management*, UB.

Dolgorsuren B., 2012 *Enhancing glossary of terms used for programming technology* / MUST, CSMI, programming technology professors team, master thesis /. UB

1. www.wikipedia.org accessed 2014.01.12
2. http://translate.google.com accessed 2014.01.15
3. www.yandex.ru accessed
4. http://www.geomongol.blogspot.com/ accessed 2014.02.01-10
5. http://translate.google.com accessed between March and June 2014
6. www.bolor-toli.com accessed between March and June 2014
7. http://mongoliantranslation.blogspot.com accessed 2014.03.21
8. http://en.globalglossary.org accessed 2014.03.22
9. http://ddelgersaikhan.blogspot.com/
10. http://www.tizag.com/mysqlTutorial/mysqldatabase.php accessed between March and June

11. http://support.esri.com/en/knowledgebase/GISDictionary/browse/ accessed between March and June 2014
12. http://toli.query.mn/ accessed between March and June 2014
13. http://geoearthmapping.com/ accessed between March and June 2014
14. http://www.php.net/ accessed 2014.04.18
15. http://www.mysql.com/ accessed 2014.04.20

Free and Open Source Software for Geospatial (FOSS4G) Conference Proceedings

Volume 15 *Seoul, South Korea* Article 7

2015

House Number Interpolation For Route Planning Applications

Stefan Funke
FMI, Universität Stuttgart 70569 Stuttgart, Germany

Sabine Storandt
University of Freiburg, Germany

House Number Interpolation For Route Planning Applications

Stefan Funke[1*], Sabine Storandt[2]

[1]*University of Stuttgart, Germany*
[2]*University of Freiburg, Germany*

Corresponding Author : Stefan Funke (funke@fmi.uni-stuttgart.de)

Abstract

Nowadays users expect door-to-door routing from a navigation system; for a given address including street name and house number this requires at least an approximate location on the routable part of the OSM data.

Of course, eventually the goal is to have every single building mapped and tagged in OSM with its housenumber. For most regions this is still a pipe dream, though. Often, only a small subset of the buildings have been mapped, and even if mapped, quite frequently their housenumber is missing. In particular buildings mapped automatically from aerial imagery typically lack more detailed tags. And even though there are explicit interpolation tags recommended to be used, they are not frequently used, hence many addresses with house numbers cannot be located even approximately.

In this paper we show how to automatically infer house numbers based on the data already present in OpenStreetMap. The result of our algorithm is a street graph where every road segment (part of a way between two consecutive nodes) is associated with a range/subset of house numbers. On one hand, this yields an interpolation of house numbers that are not explicitly present in the OSM data. On the other hand this allows route planning applications to perform door-to-door routing up to OSM road segment level (which seems sufficient given that a typical OSM road segment between consecutive nodes is rather short) without the need to look up house numbers in respective building data etc. This is an important advantage in particular in offline scenarios where the route planning data is stored locally e.g. on a mobile device.

For example, in a small (but rather well mapped) excerpt of the OpenStreetMap data of Germany, we have counted about 494k buildings, but only around 241k with a housenumber provided (either at the building outline/way or at an entrance node). For most of these buildings our methods allow for the association of the correct range of house numbers including these missing house numbers (and potentially housenumbers of still unmapped buildings). The addr:interpolation tag has only been used 81 times in this excerpt.

Our implementation is efficient enough such that a country-sized excerpt can be processed within few minutes (e.g. in preparation for offline storage on a mobile route planning device).

Academic Discipline and Sub-Disciplines : Open Source; geoinformatics; route planning; map

Keywords : Open Source; route planning; housenumbers; interpolation; data quality

Free and Open Source Software for Geospatial (FOSS4G) Conference Proceedings

Volume 15 *Seoul, South Korea* Article 8

2015

GIS Oriented Service Optimization Tool For Fecal Sludge Collection

Mohammad Dalower Hossain
Environmental Engineering and Management FoS, Asian Institute of Technology

Sarawut Ninsawat
Remote Sensing and Geographic Information Systems (RS&GIS) FoS, Asian Institute of Technology

Shulaxan Sharma
Remote Sensing and Geographic Information Systems (RS&GIS) FoS, Asian Institute of Technology

Thammarat Koottatep
Environmental Engineering and Management FoS, Asian Institute of Technology

Yuttachai Sarathai
Water Quality Management Bureau, Pollution Control Department

PO-16

GIS ORIENTED SERVICE OPTIMIZATION TOOL FOR FECAL SLUDGE COLLECTION

Mohammad Dalower Hossain[1], Sarawut Ninsawat[2], Shulaxan Sharma[3], Thammarat Koottatep[4], Yuttachai Sarathai[5]

[1]Environmental Engineering and Management FoS, Asian Institute of Technology (AIT),
P.O. Box 4, Klong Luang, Pathumthani 12120, Thailand
Email: hossain.dalower@gmail.com
[2] Remote Sensing and Geographic Information Systems (RS&GIS) FoS, Asian Institute of Technology (AIT),
P.O. Box 4, Klong Luang, Pathumthani 12120, Thailand
Email: sarawutn@ait.asia
[3] Remote Sensing and Geographic Information Systems (RS&GIS) FoS, Asian Institute of Technology (AIT),
P.O. Box 4, Klong Luang, Pathumthani 12120, Thailand
Email: shulaxan@gmail.com
[4]Environmental Engineering and Management FoS, Asian Institute of Technology (AIT),
P.O. Box 4, Klong Luang, Pathumthani 12120, Thailand
Email: thamarat@ait.asia
[5]Water Quality Management Bureau, Pollution Control Department,
Phahon Yothin Road, Phayathai, Bangkok 10400, Thailand
Email: s_yuttachai@hotmail.com

ABSTRACT

In developing countries most of the urban dwellers don't have access to sewer system. People are mostly using "on-site" systems such as septic tanks or pit latrines that need to be emptied periodically, as the densely built urban environment won't allow new pits to be dug every time they fill up. In the conventional fecal sludge collection systems, authorities are collecting the sludge from house to house and dump on the plant. Fecal sludge collection system is different from traditional vehicle routing and even from solid waste collection system in terms of dynamic collection points, urgency of getting the service and diversity of demand. Due to those vibrant factors authorities are facing proper networking and management problems. This research describes algorithms that can accommodate constraints and prioritized customers who need immediate service. The GPS log data of the fecal sludge collection truck that maintained by Nonthaburi Municipality, Thailand has been considered as the base data during the development of this application. Spatial analysis has been done using Geographic Information Systems (GIS).Tabu Search has been implemented in order to optimize. Basically two algorithms were produced for assisting fecal sludge collection systems. First algorithm was able to produce multiple trip for each vehicle if required considering all the customers having equal priority, time window. The second one was able to perform optimization that can accommodate priority along with the first one. Input for the algorithms were very simple; distance matrix having distance between each customers and plant, customer order with latitude, longitude, order unit, time window, priority and vehicles with capacity. Algorithms were able to produce better result than the actual operation or even from shortest path algorithm in term of distance. After optimization, efficiency of the algorithms were tested with the actual travelling distance. Travelling distance were reduced to half compare to actual cost and it ensured maximum utilization of vehicle capacity by allocating maximum number of customers in each route.

1. INTRODUCTION

Mostly in urban areas of developing countries, onsite sanitation systems (OSS) are used for the collection of the black water and (or) grey water where the separation of the solid and liquid fraction occurs. Examples of OSS include septic tanks, pit latrines, dry toilets, and aqua privies, (Strande *et al.*, 2014). The liquid contained in the septic tanks are drained into the sewer system and the solid parts are accumulated at the bottom. The solid parts that has been digested due to microbial activities and are settled at the bottom of the septic tanks is known as fecal sludge. In urban and rural areas of the developing countries OSS are widely used for collection and pretreatment of excreta. Usually, the untreated fecal sludge collected from these systems are disposed haphazardly into the urban and semi-urban environment which causes a serious threat to the environment and to the public health (Ingallinella *et al.*, 2002). Surface and groundwater pollution, transmission of excreta related infections, unpleasant smell are some of the effects in the environment and public health. OSS is cost effective but on the other hand there are several problems. One of the major problem is emptying (Furlong *et al.*, 2014). It cost handsome amount of money for emptying (ICEA, 2007), in addition to that it is hazardous. Different small scale emptying process have been developed but due to high cost couldn't sustain for long run. Another problem arises after emptying, where to dump and practice is simply dumped into the immediate environment (Furlong *et al.*, 2014). This makes high risk of contamination of pathogens to the environment and to the human. In larger cities, fecal sludge collection and haulage are faced with great challenges: emptying vehicles often have no access to pits; traffic congestion prevents efficient emptying and haulage; emptying services are poorly managed (Ingallinella *et al.*, 2002).

Fecal Sludge Management (FSM) service chain consists of household level users, private collection and transport (C&T) businesses that empty the fecal sludge with trucks from onsite septic tanks and transport it to the treatment facility. This process is complicated considering the facts regarding the distance between the customer and the service provider. Further, service provider also needs to offload the sludge to the plant (Dodane *et al.*, 2012). Serving as many customers as possible in the shortest possible time using available vehicles but optimizing travel cost is the major concern of C&T businesses. FSM collection and transportation system is different from traditional logistics and solid waste collection system. In FSM, collection points are dynamic e.g. everyday customers are different, demand is diverse and need the services immediately.

Geographic Information Systems (GIS) is one of the most advanced technology to capture, store, manipulate, analyze and display spatial data. GIS has been successfully used in a wide variety of applications, such as urban planning, transportation, natural resources management, health sciences, forestry, geology, natural disasters prevention and relief, and various aspects of environmental modelling and engineering (Brimicombe, 2003). GIS can be used in order to calculate route in real scenario. One way, turn restriction, tollway should be considered in this case as metaphor. GIS application in waste management is mostly focus on solid waste only. GIS has been applied on routing through selecting appropriate transfer station (Esmaili, 1972), suitable landfill sites (Despotakis *et al.*, 2007) and location of treatment sites (Adamides *et al.*, 2009; Zsigraiova *et al.*, 2009) for solid waste management. GIS may provide significant economic and environmental savings through the reduction of travel time, distance, fuel consumption and pollutants emissions (Johansson, 2006; Kim *et al.*, 2006; Sahoo *et al.*, 2005; Tavares *et al.*, 2008).

Vehicle Routing Problem (VRP) refers to a problem that can be defined as a set of spatially distributed customers with associated demands that needs to be served with number of vehicles having defined capacity. In that, the starting and the ending point of the routes are predefined and generally each vehicle starts from depot and ends at depot. The main objective of VRP is to minimize the travelling distance and number of vehicles used (Brandão, 2004). Similar to VRP, while managing fecal sludge collection there are a number of considerations to be made like; multiple customers, capacity of septic tank to be served, multiple vehicles, vehicle capacity limit, serving time window for customer etc. Also, the vehicles might have to go for multiple trips if necessary. Among the VRP algorithms Travelling Salesman Problem (TSP) is a classic combinatorial optimization problem. In this problem, a travelling salesman is expected to visit a set of cities. The objective of this problem is to find the shortest tour in which the travelling salesman visits each city once and returns to the starting city without considering vehicle capacity. To overcome shortcoming of TSP, Pham and Karaboga (2012) were applied Genetic Algorithms (GA), Tabu Search (TS), Simulated Annealing (SA), and Neural Networks (NN) to a TSP consisting of 50 cities to evaluate their performances. Among these, GA was able to generate best solution in terms of travelling distance (5.58) followed by TS (5.68), NN (6.61) and SA (6.80). On that particular test Tabu Search was close to best results but they hadn't evaluate the calculation time. Bajeh and Abolarinwa (2011) compared the calculation time between Genetic Algorithm and Tabu Search in solving scheduling problem. They found that Tabu Search can produce better timetables than Genetic Algorithm even at greater speed.

Hence, the requirement of effective application/tool that can generate viable routes that optimizes the cost but without compromising with the service time is always in demand. This research is focused on development of Tabu search based algorithm to generate optimal route for fecal sludge collection. The overall objective is geospatial route optimization for fecal sludge collection in complex urban environment considering multi-path, time windows, cost (distance) and vehicle capacity.

2. STUDY AREA AND DATA

Nonthaburi is over 400 year's old city, it was from the time when Ayutthaya was the capital. It is locates in northwest of Bangkok on the Chao Phraya river. It has six district (Figure 1) and it's a part of the greater Bangkok Metropolitan Area. The province has area of 622.30 square kilometers and total population of the province is 1,334,083 as of 2010 (National Statistics Office, Thailand). Per annum almost 9,000 cubic meter of the sludge is treated in the fecal sludge treatment plant. The liquid fertilizers produced from sludge was tested by Kasetsart University, Thailand and it was found safe to be used in the agriculture especially for the edible crops (USAID, 2010).

The study was based on fecal sludge collection vehicle GPS log data obtained from the company operated in Nonthaburi Municipality. Specifically, data between September 2013 and February 2014 of Nonthaburi Municipality was processed for the purpose. It has been found that four vehicles each with the capacity of six (6) cubic meters has been used for the fecal sludge collection. Nonthaburi province was suitable due to the availability of the data and testing of algorithm in real scenario. Cost were calculated in terms of distance only. Traffic conditions have not be considered due to unavailability of data. Beside the GPS log data, road network plays an important role for the vehicle routing. Actual road network data was purchased from NOSTRA (business unit of ESRI (Thailand) Company Limited). GSP log data were stored in PostgreSQL with PostGIS and pgRouting extension. PostGIS extension is required for spatial query whereas pgRouting is used for finding the distance that has been travelled by the vehicles during fecal sludge collection. Python scripting was used for calculating the actual cost, shortest path for the whole six months. Basic Tabu Search algorithm with time window was downloaded from github (https://github.com/pgRouting/pgrouting/tree/gsoc-vrppdtw/src/vrp_basic/src).

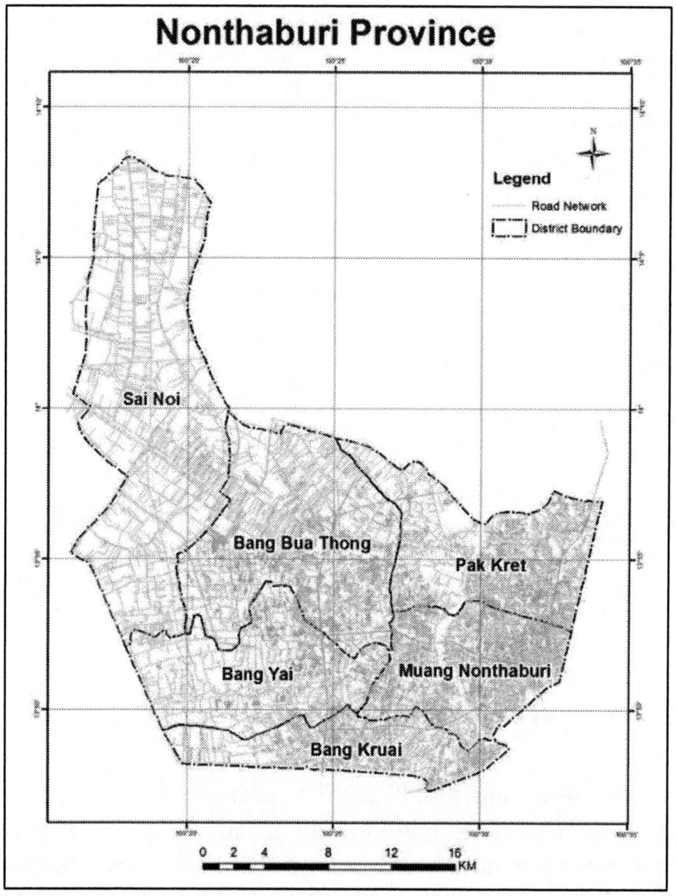

Figure 1. Study Area

3. METHODOLOGY

GPS log data was recorded in geographic coordinate system i.e. WGS 1984 datum. In order to calculate the distance in meter, it was necessary to project it into UTM system. QGIS was used to convert into shapefile where the projection system was changed to UTM and later stored in PostgreSQL with PostGIS extension. GPS log data was the combination of the way points and the customers' location. Hence, it was necessary to extract the customers' location. According to the expert interview, the distance between two septic tank (for different house) are more than 4 meters.

So, buffer of 2 meters was created around each GPS points. Buffer of nearby points (if within 4 meters) overlapped with each other were merged so that all the points within that buffer can be represented with the centroid for each overlapped buffers. Total time spent on the buffer area was calculated from all the points within it and if the time difference between two successive centroids was found to be ten minutes or more than that point was said to be the possible customer location. To remove the kinematic error of the GPS, time difference between the successive possible customers' was identified. If the time difference of five minutes or less was found then first points among the two was removed from the possible customers' list. Finally, all the points that lies within 50 meters from the plant location was excluded from the customer list. After identifying the customers, it was verified with the vehicle log book. Only in 10% cases, the number of customers' were not matched. Finally, python script was used for generating the customers' list from the GPS log data for each day for each vehicle.

Given a set of vehicles and the number of customers where the constraint being the vehicle capacity which cannot be exceeded during serving the customers. Initially from list of available vehicles, all were marked as "Untapped Vehicle" as they have never been used on that day. From the "Untapped Vehicles", a vehicle was assigned to Tabu Search and marked as Single Used Vehicle. From the list of customers, distance matrix was generated in PostgreSQL using pgRouting which indicate distance between each customers and plant. Customers order with priority if any, distance matrix, time window and assigned vehicle was inserted in Tabu Search. Based on distance, customer's priority and considering time window along with the capacity of the assigned vehicles, Tabu Search was generating optimized order of serving by that vehicle in single tour. After that those customer orders were deleted from the order list. It was checked whether there were some remaining customer or not. If there were, then it was checked whether there were any "Untapped Vehicle" or not. In case there was any Untapped Vehicle, it was assigned to the next tour and marked as Single Used Vehicle. Whenever there was not any, one Single Used Vehicle was assigned for the next tour and marked that as Used Vehicle. The same process was repeated until it serves all the customers or all vehicle became Used Vehicle. In this study it was considered that one vehicle can go for only two trip, one in the morning and another in the afternoon based on the actual timing. After getting the optimal order for each tour, Dijkstra shortest path in pgRouting was used to create routes considering turn restriction, one way etc.

The customer order table should have columns as shown in Table 1. The first column provide the customer ID where the first row is the details about the depot. 99999 was assigned as the plant ID. Xcoord (longitude), Ycoord (latitude) were the geographical coordinates of the customer location in UTM. The Demand column provides the demands that has been requested by the customers and the Ready_time and Due_time were used for the time windows. Ready_time was the starting time for the time window and Due_time was the end time. Service_time was used for the time required to provide the service to the customers and the Priority column described the priority and the equal priority customers. Service time was extracted from the actual data. It varies customer to customer based on septic tank size and time required for opening of septic tank, fixing the suction pipe etc. In Priority column, 1 denoted the customers that has been left on the previous day and need to be served on the following day whereas 0 denotes the customers that needs to be served on the same day. The starting time 0 was set to be 07:30 HRS in the morning and the 540 would be 16:30 HRS. While serving the last customer it was check whether the vehicle was able to reach the depot before closing time (16:30 HRS).

Table 1. Sample customer order table

Cust_no	Xcoord	Ycoord	Demand	Ready_time	Due_time	Service_time	Priority
99999	665126.082	1533589.78	0	0	540	0	0
124852	662428.1813	1530759.392	1500	0	360	29	0
124686	662382.3807	1535740.701	1500	0	360	25	0
124581	664704.7437	1533434.066	1500	0	360	12	1
124421	660665.2113	1534906.019	1500	0	360	27	0

Distance matrix was consisting of three columns as provided in Table 2. First column was the origin, second column was the destination and third column was the cost in terms of travelling distance. Origin, destination can be plant or customer. Vehicle table was much simple one. It has only two columns as presented in Table 3. First column as vehicle ID and second column was the capacity in liter.

Table 2. Sample distance matrix

To	From	Cost (Km)
124375	124686	4.25764331425731
124375	124421	0.337719708091727
124375	99999	6.91473205848734
124375	124852	7.41003837112014
124375	124581	6.43249171235125
124421	124581	6.28082480515041

Table 3. Sample vehicle table

V_ID	Capacity
126175	6000
12618	6000

After calculating optimized route it was compared with the actual cost and shortest path cost by keeping the serving order as it did practically. Actual cost was the calculated from the GPS log data which was the actual distance travelled by each vehicles. For simplicity, all points for each vehicle was filtered to see the route pattern. Due to the GPS error, points recorded were not in a straight line and not also aligned properly with the road network. All the GPS points for each day, for each vehicle was identified to calculated the actual distance that the vehicle has travelled in a particular day.

According to the experts, the average septic tank size for a residential household in Thailand is 1 to 2 cubic meter. Hence, with the capacity of 6 cubic meter a vehicle at a time can serve for at the max four customers (considering the capacity of the septic tank to be 1.5 cubic meters). After serving the customers, vehicle need to go back to the plant for emptying the sludge containing in the tank. And then it had to go again for the next trip for serving the other unserved customers. Shortest path problem is the problem of finding a path between two vertices (or nodes) in a graph so that the sum of the weights of its edges is minimized. The inbuilt Dijkstra algorithm from pgRouting was used for calculating the shortest path.

4. RESULT AND DISCUSSION

A total number of 1978 customers were served during the specified period and four vehicles were used for this purpose. The details of the customers served by each vehicle is shown in Table 4. The available implementation of vehicle routing assumes that the customers will be served within a single trip by each vehicle. However, simulating the available GPS log data with the current implementation, it has been observed that all the customers could not be served with a single trip. Hence, a need to modify the implementation such that it can incorporate multiple trip was generated and hence the existing code has been modified to facilitate the multiple trip.

Table 4. Total Number of Customers Extracted

Vehicle No.	Total Customer	Average Customer / day
12617	743	5
12618	406	6
12619	342	3
12829	487	5
Total Customers	1978	

Taking an example of customer distribution as of December 16, 2013, total five customers' needs to be served. The details of the attribute of the customers is shown in Figure 2 and the distribution of the customers is shown in Figure 3. From the figure, it was clear that five customers' were being served by one vehicle. After serving first three customers the vehicle had to come back to the plant to make the tank empty. The Actual cost and the shortest path cost for the particular date was 66.08 Km and 31.41 Km respectively whereas the cost was found to be 27.24 Km only for the same day using one vehicle and using the multi-trip.

start_time	end_time	id	eid	vlctl_gps_	v_id
09:09:27	09:29:57	124375	12441	20131216	12617
09:35:27	10:01:57	124421	12447	20131216	12617
10:56:17	11:07:17	124581	12460	20131216	12617
12:23:06	12:47:36	124686	12473	20131216	12617
13:49:36	14:19:05	124852	12491	20131216	12617

Figure 2. Attribute of customers as of December 16, 2013

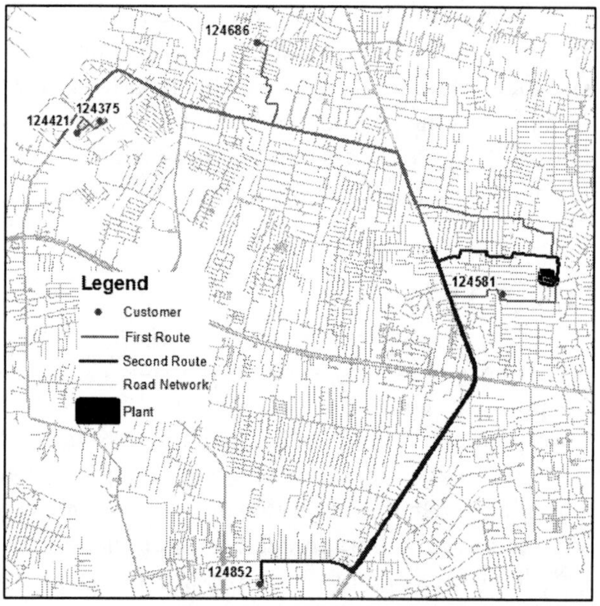

Figure 3. Customer distribution of customers as of December 16, 2013

The optimal route was obtained using the distance matrix calculated from the actual road network and was used in the algorithm. Two routes were generated, which were:

- Plant – 124375– 124421– 124686– 124581– Plant
- Plant – 124852 – Plant

The modified program was tested and verified again with some random sets of available customer locations for some day and the result shows that even with the multiple trip sometimes it was difficult to serve all the customers in a single day. Here, the term multiple trip would mean that a vehicle can go for at the maximum two trip in a day, one in the morning and one in the afternoon. Those customers who could not be served in the ith day have to be automatically prioritize for (i+1) th day. So, the algorithm needed further modification and enhancement such that the priority customers would be served with higher priority the following day. This has been implemented such that while choosing customers, the nearest priority customer from the depot was chosen as initial node and then subsequent nodes are chosen on the basis of shortest distance from the preceding node ultimately having at least one priority customer in one trip.

Taking an example of customer distribution as of December 12, 2013, total 11 customers' needs to be served. The details of the attribute of the customers is shown in Figure 4 and the spatial distribution of the customers is shown in Figure 5. From the figure, it is clear that 11 customers' was being served by three vehicles. The Actual cost and the shortest path cost for the particular date was 100.63 Km and 67.98 Km respectively whereas the cost was found to be 48.21 Km only for the same day using three vehicles. Here, in this example vehicle 12618, 12619 and 12617 have served seven, one and three customers respectively.

start_time	end_time	id	eid	vlctl_gps_	v_id
09:00:19	09:24:49	168	1729	20130912	12618
09:52:20	10:15:21	178	1830	20130912	12618
10:22:51	10:45:52	184	1891	20130912	12618
10:59:22	11:14:22	191	1948	20130912	12618
11:45:53	12:06:54	201	2053	20130912	12618
12:37:25	13:00:55	211	2161	20130912	12618
13:36:36	13:50:07	222	2253	20130912	12618
14:43:33	15:23:33	138	1468	20130912	12619
09:07:33	09:24:02	240	2437	20130912	12617
09:48:42	10:02:02	247	2504	20130912	12617
10:21:22	10:52:52	253	2582	20130912	12617

Figure 4. Attribute of customers as of September 12, 2013

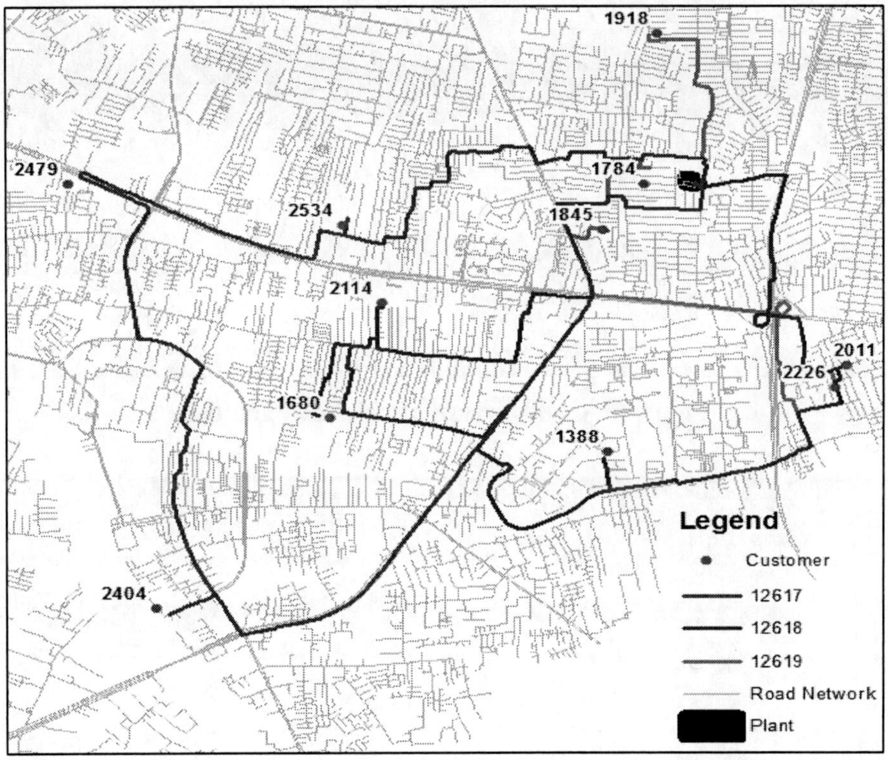

Figure 5. Customer distribution of customers as of September 12, 2013

Though, in real data there was no any priority customers, but to check whether the application provides the optimal route or not 2534, 1784, 2011, 2226 and 1388 customers were assumed as priority customers. All the customers have equal demands of 1,500 liters. Three vehicles with the capacity of 6,000 liters has been assumed.

The optimal route was obtained using the distance matrix calculated from the actual road network and was used in the algorithm. Three routes were generated:

- Plant – 1918 – 1784 – 1845 – 2226 – Plant for vehicle 12619
- Plant – 2114 – 1680 – 1388 – 2011 – Plant for vehicle 12618
- Plant – 2404 – 2479 – 2534 – Plant for vehicle 12617

Actual cost of travelling, the cost calculated using shortest path and the cost using route optimization was compared among each other to analyze the result (Figure 6). The result obtained by comparing the datasets for 176 days implies that the optimized cost was the lowest in majority of days (Figure 7(a)). The accumulated total cost in the long run also supports the same result. Moreover, the average distance travelled to serve each customer in case of route optimization was also seen lowest among them (Figure 7(b)).

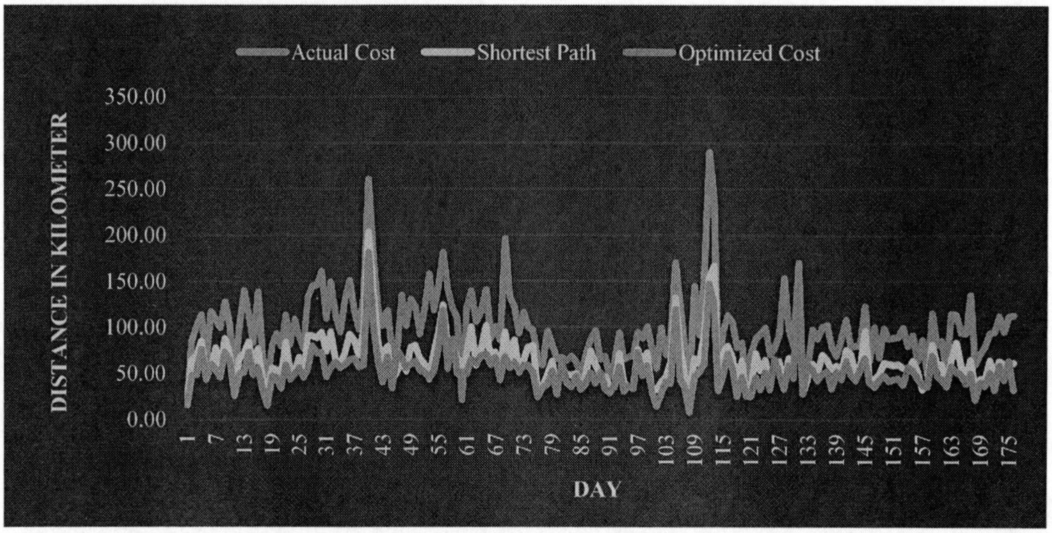

Figure 6. Comparison of cost for each day

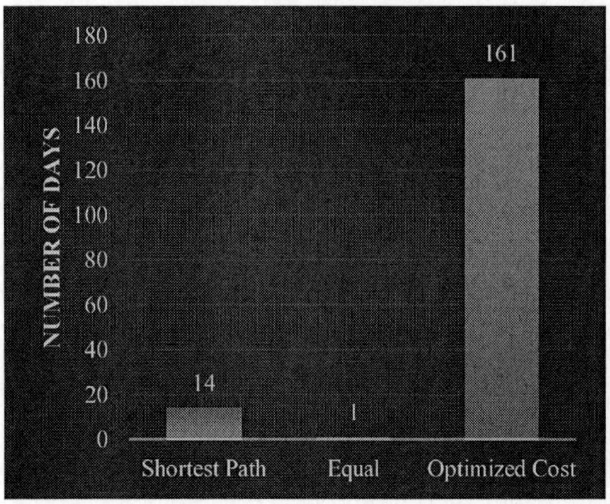

(a)

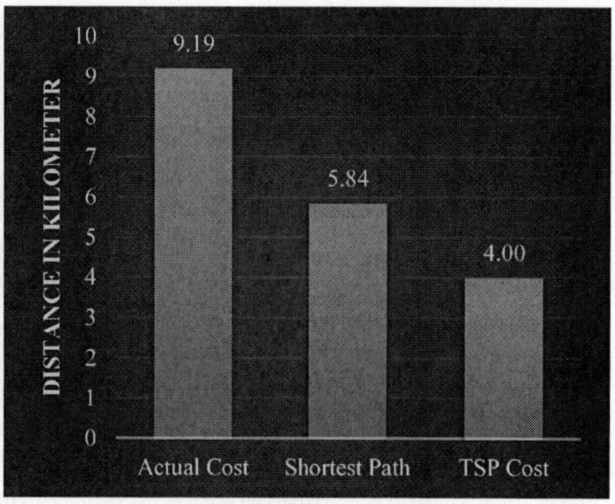

(b)

Figure 7. (a) Frequency of the best algorithm (b) Cost per route

Analyzing the number of customers, a vehicle has served per trip, it was identified that the vehicles were optimally used in case of tabu search based application as compared to the actual data (Figure 8). This was the important part of optimization to maximize resource utilization. In actual cases there were many cases where vehicles had returned to the plant with empty space. They could go to serve another customer before returned to the plant.

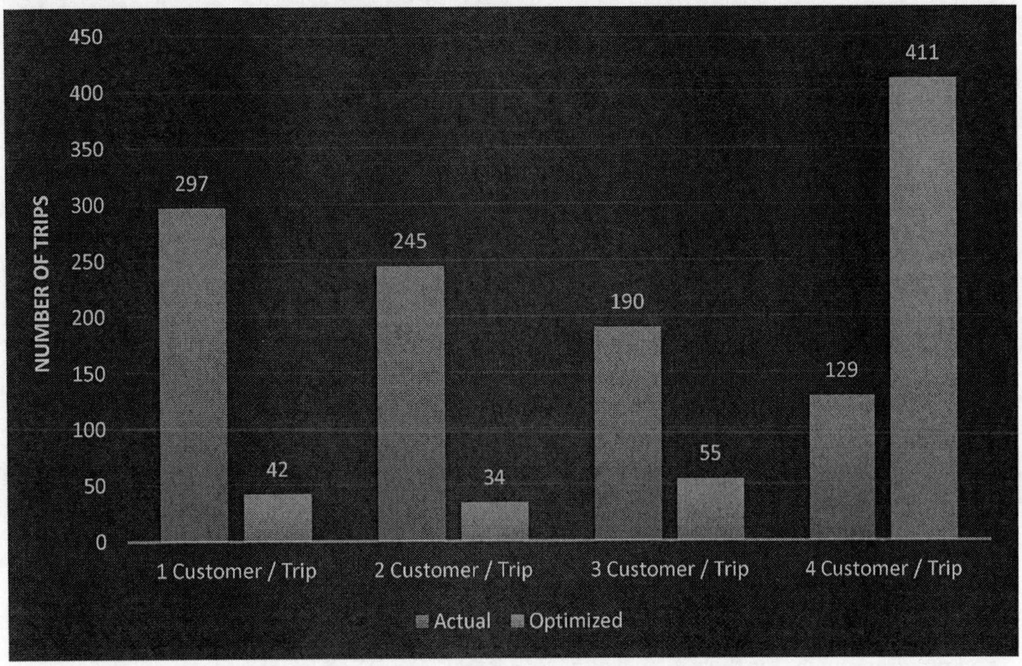

Figure 8. Number of customers per trip

5. CONCLUSION

Collection of fecal sludge in urban environment is complex due to multifarious road network. Study conducted by Bill & Melinda Gates Foundation (Chowdhry and Kone, 2012) revealed that 76% of the operating expenses for African businesses are variable costs, while in Asia, fixed costs make up 62% of the operating expenses. This fact is important as small improvement in the collection part can effect a significant saving in the overall cost. Therefore, interest in the analysis of sludge collection systems arises to optimize the operation of existing systems and to develop data and advanced techniques to design and evaluate new systems for urban areas. Implementation of GIS approach in Nonthaburi city has shown reasonable improvement in length of the routes. The results obtained from this study are encouraging to expand the scope to implementing other algorithms for optimization of the routes.

6. REFERENCES

Adamides, E. D., Mitropoulos, P., Giannikos, I., & Mitropoulos, I. (2009). A multi-methodological approach to the development of a regional solid waste management system. *Journal of the Operational Research Society*, 60(6), 758-770.

Bajeh, A. O., & Abolarinwa, K. O. (2011). Optimization: A Comparative Study of Genetic and Tabu Search Algorithms. *International Journal of Computer Applications* (0975 – 8887) Volume 31– No.5, October 2011

Brandão, J., & Mercer, A. (1997). A tabu search algorithm for the multi-trip vehicle routing and scheduling problem. *European Journal of Operational Research*, 100(1), 180-191.

Brimicombe, A. J. (2003). A variable resolution approach to cluster discovery in spatial data mining. *In Computational Science and Its Applications - ICCSA* 2003 (pp. 1-11). Springer Berlin Heidelberg.

Chowdhry, S., & Kone, D. (2012). Business analysis of fecal sludge management: Emptying and transportation services in Africa and Asia. Resource Document. *The Bill & Melinda Gates Foundation.*

Despotakis, V. K., & Economopoulos, A. P. (2007). A GIS model for landfill sitting. *Global NEST Journal*, 9(1), 29-34.

Dodane, P. H., Mbeguere, M., Sow, O., & Strande, L. (2012). Capital and operating costs of full-scale fecal sludge management and wastewater treatment systems in Dakar, Senegal. *Environ Sci Technol*, 46(7), 3705-3711. doi: 10.1021/es2045234

Esmaili, H. (1972). Facility selection and haul optimization model. *Journal of the Sanitary Engineering Division*, 98(6), 1005-1020.

Furlong, C., Templeton, M. R., & Gibson, W. T. (2014). Processing of human feaces by wet vermifiltration for improved on-site sanitation. *Journal of water, sanitation and hygine for development*, 231-239.

Johansson, O.M. (2006). The effect of dynamic scheduling and routing in a solid waste management system. *Waste Management*, Vol. 26, pp. 875-885, ISSN 0956-053X

ICEA Modèle de simulation financière du secteur de l'assainissement en milieu urbain (Financial simulation model for the urban sanitation sector), *ONAS*: Dakar, Senegal, 2007.

Ingallinella, A., Sanguinetti, G., Koottatep, T., Montangero, A., Strauss, M., Jimenez, B., et al. (2002). The challenge of fecal sludge management in urban areas- strategies, regulations and treatment options. *Water Science & Technology*, 46(10), 285-294.

Kim, B.I.; Kim, S. & Sahoo, S. (2006). Waste collection vehicle routing problem with time windows. *Computers and Operations Research*, Vol. 33, pp. 3624-3642, ISSN 0305-0548

National Statistical Office, Thailand: *Preliminary Report The 2010 Population and Housing census (Whole Kingdom)*. Bangkok, Thailand

Pham, D., & Karaboga, D. (2012). Intelligent optimisation techniques: genetic algorithms, tabu search, simulated annealing and neural networks. *Springer Science & Business Media.*

Sahoo, S., Kim, S., Kim, B-I., Kraas, B., Popov, A. (2005). Routing Optimization for Waste Management. *Interfaces*, Vol. 35, No. 1, January 2005, pp. 24-36, ISSN: 0092-2102

Strande, L., Ronteltap, M., & Brdjanovic, D. (2014). Fecal Sludge Management Systems Approach for Implementation and Operation. *IWA Publishing.*

Tavares, G., Zsigraiova, Z., Semiao, V., Carvalho, M. (2008). A case study of fuel savings through optimization of MSW transportation routes. *Management of Environmental Quality*, Vol.19, No.4, 2008, pp. 444-454, ISSN: 1477-7835

USAID (2010). A Rapid Assessment of Septage Management in Asia: USAID.

Zsigraiová, Z., Tavares, G., Semiao, V., & de Graça Carvalho, M. (2009). Integrated waste-to-energy conversion and waste transportation within island communities. *Energy*, 34(5), 623-635.

Free and Open Source Software for Geospatial (FOSS4G) Conference Proceedings

Volume 15 *Seoul, South Korea* Article 9

2015

Distributed Agents For Contextual Online Searches

Elizabeth-Kate Gulland
Cooperative Research Centre for Spatial Information (CRCSI)

Simon Moncrieff
Department of Spatial Sciences, Curtin University

Geoff West
Department of Spatial Sciences, Curtin University

PO-14

DISTRIBUTED AGENTS FOR CONTEXTUAL ONLINE SEARCHES

Elizabeth-Kate Gulland[1], and Simon Moncrieff[2], and Geoff West[3]

Department of Spatial Sciences, Curtin University
Kent Street, Bentley WA 6102, Australia
and Cooperative Research Centre for Spatial Information (CRCSI)
[1]Email: e.gulland@curtin.edu.au
[2]Email: s.moncrieff@curtin.edu.au
[3]Email: g.west@curtin.edu.au

ABSTRACT

As semantic web use and research blossoms, automated online searches - whether to answer a simple question, seek specific sensor readings, or investigate research in a particular domain - has raised a number of issues. Simple search tools cannot handle context-specific search problems, but specialist search tools have a narrow domain and applicability. Some online tools circumvent these problems by putting more filter controls into the hands of users, but this leads to more complex interfaces which can raise usability barriers. A distributed approach, where specialised search agents act autonomously to find contextualised information, can provide a useful compromise between a simple, general search interface and specialist searches. This paper outlines work in progress on design and use of specialist search agents, with a case study to find public transportation bus stops within a spatial region. The approach is demonstrated with a case-study web interface, developed to interpret a text query to find and show bus stop locations within a named boundary by coordinating multiple online search agents. Search agents were designed to follow a common model to allow for future development of agent types, including specialist agents used in the case study to search standard open web services and extract spatial features.

1. INTRODUCTION

Users searching for data are accustomed to simple text inputs, such as implemented by Google search, which are straight-forward to use but raise processing complications for flexible searches including: 1) how to interpret the user's context, aims and expectations; 2) what filters are appropriate and applicable; 3) how and where the data itself can be accessed; 4) how to format results; and 5) how to rank the relevance of results.

In this paper, we outline an approach for using distributed online search agents, each capable of applying specialist operations and/or accessing specific data sources, to manage complex problems such as these. We demonstrate the approach with a case study web application to find and display bus stop locations by suburb name.

The case study example shows how agents can be used to extract individual spatial features rather than complete datasets or documents, such as shapefiles or web pages. It is tested on data conforming to the Open Geospatial Consortium (OGC) Web Feature Service (WFS) standard[1], although the common design and interoperability of agents mean that this can be expanded in the future to allow for other data formats and search operations. This

[1] http://www.opengeospatial.org/standards/wfs

example coordinates agents, based on a common model, to find: WFS features by name and/or location; regions by name; and bus stop point locations. From the user's perspective, the process from a natural language text input to an output map and list of results is hidden.

In section 2 we introduce some background to the search problem. We discuss the approach for our case study example in section 3 and results in section 4. Discussion and conclusions about our findings are in sections 5 and 6.

2. BACKGROUND

Current online information retrieval (IR) tools typically incorporate flexibility by adding specific filters to the interface, such as choice of spatial location, temporal range, codes or identifiers, or problem-specific categories like author name or government department. This can limit applicability to a narrower purpose or group of users and lead to more complex search tools, but simplifying to a single text input requires more complex processing to interpret natural language text queries and manage potential disparity between what the user asks for and what the data or metadata provides.

Our framework for contextual online searches makes use of multiple software *agents* which can be distributed across different machines. Agents can be designed to take advantage of the *semantic web*, with its definitions of relationships between resources and terminology. They can also be designed to carry out specialised tasks, such as *spatial searches* for geographic features within datasets with differing formats.

2.1 Agents

Software agents are computer programs that can communicate with other agents, machines and/or users to solve a task. Web services are one way to facilitate communication between *distributed* agents, which can be housed on different computers. An agent can use communication standards such as the RESTful framework (Representational state transfer) as an access point to a web service; for example, by sending a *DescribeFeatureType* request to a WFS service with parameters specifying the feature type (i.e. spatial dataset) to be described.

Individual agents can be designed to solve problems such as translating between a user's text query and messages to send to web service(s) to answer that query, accounting for implementation details such as syntax differences between service versions (Huang and Webster 2004, Zhao *et al.* 2008). Complex tasks typically require the interaction of multiple agents and, as a result, systems allowing interaction between agents and services have been developed in a number of contexts, including geospatial (Yue *et al.* 2007, Zhao *et al.* 2012, Tian and Huang 2012).

2.2 Semantic Web

The semantic web also facilitates machine-to-machine communication, by defining relationships between entities, examples of which can include related terminology or a sequence of web services to use in a knowledge discovery process. Entities and relationships are commonly defined in ontologies, using triples such as ["highway", "*is-a-type-of*", "road"]. Linking resources and terms together facilitates the expansion of simple text-based search interfaces by allowing agents to take advantage of more complex processes, without requiring the user to manually enter extra search parameters.

Relating terms together can assist with the processing of natural language queries such as "bus stops in North Perth", and this can be extended to cater for cases where context affects the interpretation of a query: consider the term "K12", which could refer to a geographic feature (mountain in South Asia); the ICD10[2] code for "stomatitis and related lesions" in health, or "kindergarten to year 12" in education. The most appropriate data sources to search would depend on the intended context and, in some cases, the type of data returned would also depend upon context - for instance, a polygon or point object could be more appropriate for a geographic feature, where a textual description or statistical plot could be more useful in the health context. Related terminology is also relevant to spatial queries - a semantic triple like ["North Perth", "*is-in-the-state-of*", "Western Australia"] could inform the search example without requiring spatial operators or features. Where an agent has no ability to check spatial boundaries, a dataset described as "stations in Western Australia" could be marked as potentially relevant, even though it shares no terms with the original query.

Methods for automatic extraction of semantic links have been investigated for a number of information sources, including Wikipedia (Gabrilovich and Markovitch 2009), Google (Cilibrasi and Vitanyi 2007), and context-specific resources such as in the biomedical field (Rybinski and Aldana-Montes 2014). Semantic search tools have been developed to make use of different resources, including ontologies, XML, and images, and some search tools combine semantic web and offline search techniques (Dong *et al.* 2008).

For a flexible search tool to be applicable to a range of contexts and user types, it should be able to take advantage of multiple strategies. It would need to distinguish between contexts, search strategies, data sources, and types of results. Using multiple search agents, where each can apply a more limited range of strategies and search a specific subset of data, is an approach to include this flexibility whilst managing the complexity involved.

2.3 Spatial searches

Spatial service standards such as WFS describe syntax for defining query parameters such as bounding regions, often making use of other standards such as Geographic Markup Language (GML). Each standard has numerous options and syntax requirements can differ across versions and implementations. A level of expert knowledge is necessary to manipulate these settings directly, so it is more feasible to produce requests to a data service programmatically via a user interface or software agent. To enable reuse of standard services without coupling them to specific use-cases, extra details about the services are required. One strategy to manage this approach is to record semantic information about the services themselves (Tian and Huang 2012).

Geographic search engines are designed to search for web pages that are relevant to a text query, both in text content and spatial location. Tools such as the IR-Tree are used to rank the likely relevance of web pages for both aspects of content (Li *et al.* 2011).

The case study described in this paper combines aspects of both approaches, by including textual information with spatial searches and also returning features rather than entire datasets or documents. Encapsulating technology-specific requirements such as syntax into agents allows for communication between agents that uses more general parameters.

[2] International Classification of Diseases and Related Health Problems, http://apps.who.int/classifications/icd10/

A benefit of using multiple agents for different data sources and contexts is that it provides the opportunity to process and search data defined in alternative formats by using standard request parameters. For example, SIRF[3] is an online infrastructure developed by the Australian CSIRO (Commonwealth Scientific and Industrial Research Organisation) for linking data that includes spatial content. It aims to use consistent identifiers for locations that may have multiple IDs and names over different government departments. It has its own unique APIs that could be accessed by SIRF-aware agents. SIRF aims to be an intersection point between government data in various formats, including OGC standards.

3. APPROACH

A *DataAgent* model was designed as a generic search agent, with each specialised agent inheriting from this model. As a minimal requirement, every agent model defines its type (such as "WFS" or "boundary"), name, and optionally service URI and read-only list of data sources. Each agent defines actions: *process* (accepting a dictionary of query parameters), *getCapabilities*, and *addSource*. The process action can be passed on to any data source agents, with or without alteration of the initial parameters.

Subclasses of DataAgent were created to search for WFS features, boundary regions, and public transportation locations respectively. Each implemented agent can be accessed programmatically or via a RESTful web interface for access. Request parameters include at least *query* (the initial query text) and other parameters depending on the agent's purpose and context, for example *bbox*, *boundary*, and *name*. An example RESTful request to an agent searching for boundaries, in this case for regions with a name that includes the word "Perth", is **`http://localhost:8000/spatial/boundary?query=bus stops in Perth&name=Perth`**. Compare this to a WFS request for the same data, as produced by the agent, shown in **Error! Reference source not found.**. The WFS syntax produced could differ from that shown, depending upon implemented WFS capabilities or versions, whilst the request to the agent would remain the same.

```
http://localhost:8080/geoserver/region/ows?service=WFS
&request=GetFeature&version=1.1.0&maxFeatures=50
&outputFormat=application/json&typeName=region:PerthSuburbs
&filter=<Filter>
 <PropertyIsLike wildCard="*"
    singleChar="." escape="!">
 <PropertyName>SSC_NAME</PropertyName>
 <Literal>*Perth*</Literal>
 </PropertyIsLike></Filter>
```

Listing 1. WFS (version 1.1) request to find Perth suburbs by name

Adding a spatial filter to a WFS request increases its complexity and requires details that an agent can discover, such as geometry attribute name, and spatial reference system. An extract from an agent-produced WFS request extract is shown in Listing 2.

```
http://localhost:8080/geoserver/transport/ows?service=WFS
&request=GetFeature&version=1.0.0&maxFeatures=200
&outputFormat=application/json&typeName=transport:Stops
&filter=<Filter
```

[3] Spatial Identifier Reference Framework, http://portal.sirf.net/about-sirf/

```
  xmlns:gml="http://www.opengis.org/gml">
<Within><PropertyName>the_geom</PropertyName>
<gml:MultiPolygon srsName="EPSG:4283">
<gml:polygonMember><gml:Polygon>
<gml:outerBoundaryIs><gml:LinearRing>
<gml:coordinates>115.867820512,-31.9533984945
  115.867814048,-31.95334971
  115.867764352,-31.9529839465
  ... 115.867820512,-31.9533984945
</gml:coordinates>
</gml:LinearRing></gml:outerBoundaryIs>
</gml:Polygon></gml:polygonMember>
</gml:MultiPolygon></Within></Filter>>
```

Listing 2. WFS (version 1.0) request to find point features within a region

All agents return information about the service itself - its type and the date it was invoked - and results, if any are found. Results are returned in a list, with each entry containing metadata about the source and, where available, a list of individual records. Each result is named and each record (where available) is labelled: see the example in Listing 3 for results from a *PublicTransportAgent*, which also shows how geometry information for each record, where available, is returned.

```
{"date":" 30/06/2015 ",
 "request": {"query":"bus stops in Perth"},
 "results": [
   {"name":"transport:Stops",
    "url":"http://...",
    "records": [
      {"label": "Charles St After Gill St",
       "geometry": {"type":"Point",
       "coordinates" :[...]} , ...
      }, ...]
   }, ... ]
}
```

Listing 3. Response (extract) from a PublicTransportAgent

3.1 Case Study Design

The aim of the case study was to reduce a multi-step manual workflow to a single user action: entering a text query. An example manual workflow, assuming spatial data is accessed via WFS, is:

1. Find a data service that contains information about suburbs.
2. Enter a WFS query to find suburbs with names that match the target name.
3. Extract polygon geometry information from returned records.
4. Find a second data source holding bus stop information.
5. If necessary, transform the polygon feature(s) to the second source's a compatible spatial reference system.
6. Convert polygon(s) into a GML filter that the bus stop service can interpret.
7. Enter a WFS query with the new filter to retrieve bus stop features.
8. Extract desired property value(s) and geometries from any returned records and display them.

A local GeoServer instance was set up to host WFS layers created from public spatial data layers: Western Australian bus stops from Transperth[4], and 2011 state suburbs from the Australian Bureau of Statistics[5]. A project was developed using the Django web application framework connected to a PostgreSQL database and tested locally with development settings. Specialised search agents were designed as Django models:

- *DataAgent*: the template for all other search agents to build upon.
- *DataAgentSource*: a link between an agent and a data source agent.
- *WFSAgent*: to handle a Web Feature Service layer.
- *BoundaryAgent*: to retrieve boundary region records from one or more sources.
- *PublicTransportAgent*: to extract a subset of bus stop features from one or more sources.

Three instances of WFSAgent were created, whose sources were the local GeoServer WFS layers and a public online WFS for Australian waste management point sites[6] respectively. The latter was selected as a proxy for bus stop locations as it contained point data in the same location as the suburb data. An instance of BoundaryAgent was created with the suburb WFSAgent as a source, and an instance of the PublicTransportAgent was created with the other two WFSAgents as source. Although the data sources were manually pre-set in this case, consistent parameter and output formats mean that agents with service discovery capabilities could be designed at a later stage and used as alternative data sources.

Internally, the WFSAgent adapts the query parameters it creates to send to its WFS source by interrogating its WFS version, capabilities, and details about its feature type (dataset) via WFS requests such as *GetCapabilities* and *DescribeFeatureType.* This allows it to check for capabilities such as spatial operators before attempting to apply them. The agent also looks for names of attributes likely to contain a geometry field or a label. In the latter case, partial matches were sought - in the suburb data, for instance, a match was found to an attribute called "SSC_NAME".

A web application was designed with a textbox for the user's initial text query. Python code was added to extract region names and target feature types (e.g. bus stop or station) from simple text patterns such as "bus stops in Subiaco" or "Subiaco bus stops" in a user's text query. If the region name was ambiguous, a selection of possible names was extracted - for example, "Mount Lawley stations" becomes ["Mount", "Mount Lawley", "Mount Lawley stations"]. The web application coordinated the use of the specialised agents, as shown in Figure 1: requested suburb name(s) were passed as a parameter to the BoundaryAgent and, if a polygon feature was retrieved by the agent, this was sent as a boundary parameter to the PublicTransportAgent which used it to filter out point features. As it cannot be assumed that all types of sources return geometries or even individual records, any results it returned to the coordinating web application were ignored if they contained no records or no geometries. All other results were combined and processed for display on the user interface.

---

[4] http://www.transperth.wa.gov.au/About/Spatial-Data-Access
[5] http://www.abs.gov.au/AUSSTATS/abs@.nsf/DetailsPage/1270.0.55.003July\%202011?OpenDocument
[6] http://www.ga.gov.au/gis/services/topography/National_Waste_Management_Facilities/MapServer/WFSServer

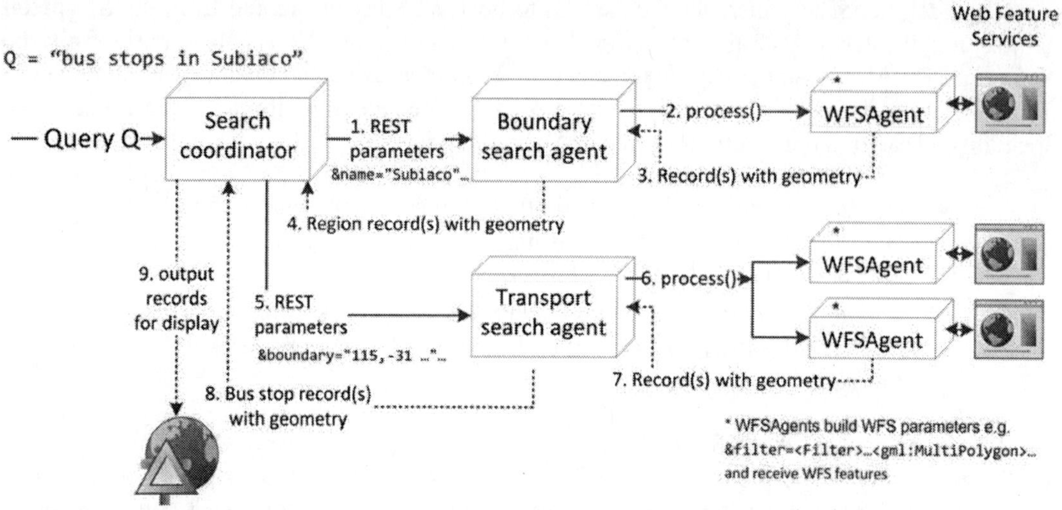

Figure 1. Processing steps between search agents in the case study implementation

If no region is found, or the query's target feature type (such as bus stop or station) is not recognised, an error message is shown (Figure 2) and a plain map with no added features is displayed.

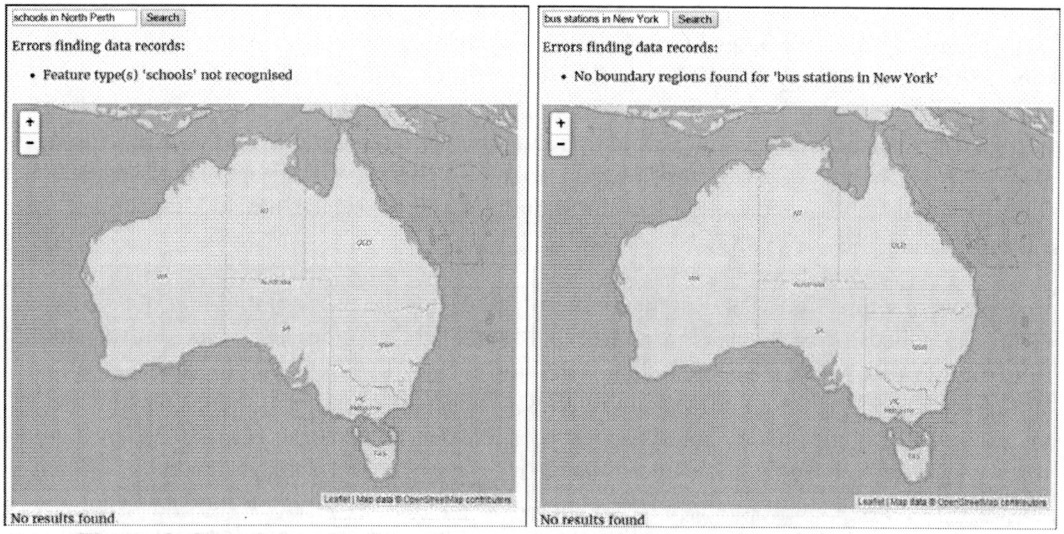

Figure 2. Search results for unknown target (left) or unknown region (right)

Where point features were retrieved, a map was produced using the Leaflet[7] JavaScript library with marker clusters[8], and a scrollable text area was added to list selected attributes from the returned features, as seen on Figure 3. Attributes are also displayed on the map when the mouse rolls over a point feature, as shown in the zoomed-in area on the left of Figure 3.

[7] http://leafletjs.com/
[8] https://github.com/Leaflet/Leaflet.markercluster

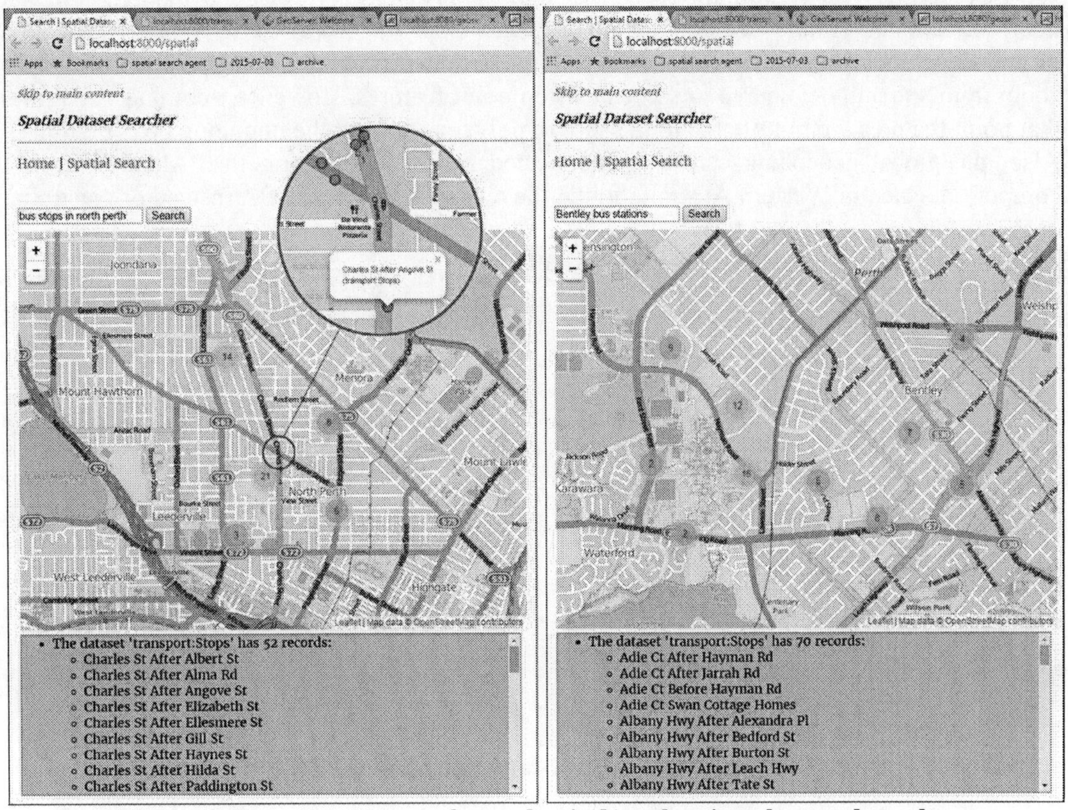

Figure 3. Screen captures of search window showing clustered results.

4. RESULTS

Manual workflows were tested to extract spatial features from the local GeoServer host using queries for WFS versions 1.0.0, 1.1.0 and 2.0.0. Filters were tested for matches to exact, partial, and non-existent suburb and bus stop names. The public waste management service, described with WFSAgent instances in section 3.1 was also tested manually. The case study tool was tested with queries that included exact, partial, case-insensitive or misspelled suburb names, or that missed a suburb name entirely. It was also tested with known feature types (such as bus stops and stations) and unrecognised feature types.

Testing of the case study web application showed that it could find points within suburbs after a partial or case-insensitive name match. In contrast, a WFS request for "subiaco" found no matching polygon features, even when the *PropertyIsLike* filter was used, because the name attribute expected a capital letter: "Subiaco".

Where an agent was not flexible enough to interpret a request, it returned no results. For instance, when the WFSAgent requested JSON output from the waste management service WFS, which can only output in XML format, no points were returned. However, the coordinating PublicTransportAgent still returned features from its other source - the local Transperth WFS. This demonstrated the robustness of the overall design. All the returned points had labels that could be used for display, although there was no attribute called "label" in the original WFS data records.

Comparing results from the case study tool to the Google browser showed that both can interpret similar, simple semantic queries ("bus stops in North Perth") but with some differences. Google can interrogate a data source to find a geocoded suburb bounding box to zoom in to and another source to show bus stop point features. However, zooming out shows that point features are not filtered to the suburb polygon, or even the bounding box (Figure 4). Also, the global geocoding can give unexpected results - searching for "Albany" from a computer located in Western Australia found the city in New York, USA instead of Australia.

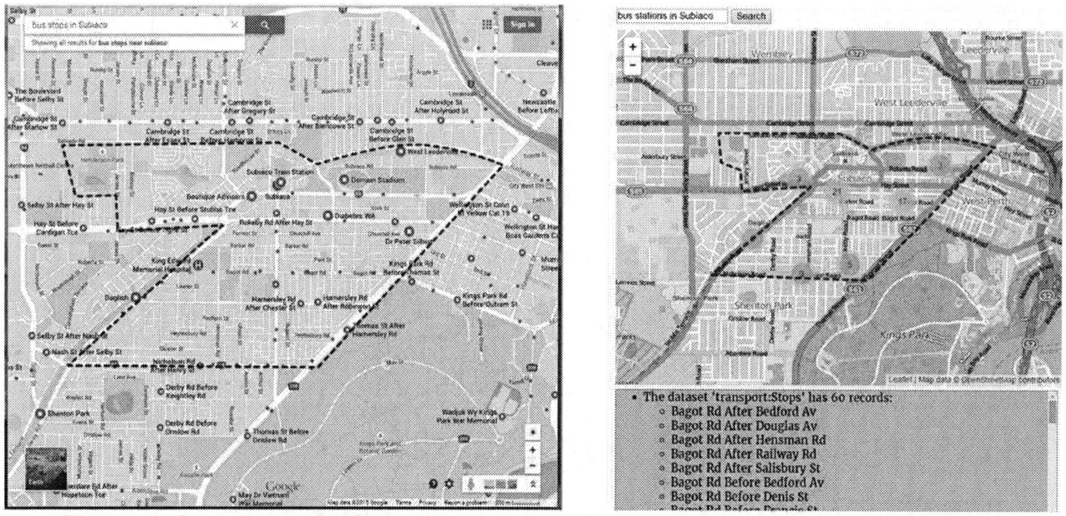

Figure 4. Search results from Google Maps (left) and the case study tool (right)

5. DISCUSSION

Automating the creation of WFS parameters from generic query parameters within WFSAgent was complicated by differences between versions and implementations, but returning empty datasets where problems arose allowed other agents to continue searching any other sources they had available. Encapsulating syntax requirements into a WFSAgent allowed for specialised search agents like the BoundaryAgent and PublicTransportAgent to focus on relevant, contextual actions without needing to consider quirks of different data sources. More agents could be added as sources to these agents to allow access to additional data sources without needing to alter the agents, even if the sources have different usage requirements (such as SIRF).

In the next stage, additional contextualised agent(s) will be implemented to further test coordination of distributed agents. The coordinating agent will need to be extended to allow it to await responses from distributed agents that are concurrently processing initial search parameters. As agents are designed to return an empty set upon failure or lack of results, the coordinator can send out responses to all known agents, rather than pre-determining which ones to access. An exception, as demonstrated in the case study, is the BoundaryAgent, which could be called first in order to find boundary parameters to send to any other agents that can make use of it.

Another agent currently under development focusses upon expanding a query with semantically related terms, which may include related terms from different contexts. This will allow for future expansion into ranking of results and contextual display, such as showing results within facets (Adams and McKenzie 2013) defined by the results' or agents' domains.

In combination with ranking of results within and between facets, this would assist users to focus in on datasets from topics they are particularly interested in.

6. CONCLUSIONS

This paper has described a design for distributed agents that can be coordinated to solve complex, contextual search problems based on a textual query. A case study was developed to test its use with a spatial query problem requiring multiple processing stages and data sources. This initial test showed promise for simplifying the task of finding spatial data from text queries by using a combination of search agents. Embedding specialised syntax and contextual requirements within agents was shown to reduce the amount of expert knowledge required by users of a simple query interface, who would otherwise need to manually solve a multi-stage problem based on online spatial data and service formats.

7. ACKNOWLEDGEMENTS

The research reported in this paper is supported by the Australian Primary Health Care Research Institute (APHCRI), which is supported by a grant from the Australian Government Department of Health. The information and opinions contained in it do not necessarily reflect the views or policy of the Australian Primary Health Care Research Institute or the Australian Government Department of Health. The Cooperative Research Centre for Spatial Information has supported this work, whose activities were funded by the Australian Commonwealth Cooperative Research Centres Programme.

8. REFERENCES

Adams, B., and McKenzie, G., 2013. Inferring Thematic Places from Spatially Referenced Natural Language Descriptions. In Sui, D.Z., Elwood, S., and Goodchild, M.F., editors, *Crowdsourcing Geographic Knowledge: Volunteered Geographic Information (VGI) in Theory and Practice*. Springer, Dordrecht. 201-221.

Cilibrasi, R.L., and Vitanyi, P.M.B., 2007. The Google Similarity Distance. *IEEE Transactions on Knowledge and Data Engineering*. 19(3): 370-383.

Dong, H., Hussain, F.K., and Chang, E., 2008. A Survey in Semantic Search Technologies. Second IEEE International Conference on Digital Ecosystems and Technologies, Phitsanulok Thailand, 26-29 February.

Gabrilovich, E., and Markovitch, S., 2009. Wikipedia-based semantic interpretation for natural language processing. *Journal of Artificial Intelligence Research*. 34:443-498.

Huang, W., and Webster, D., 2004. Enabling context-aware agents to understand semantic resources on the WWW and the semantic web. *IEEE/WIC/ACM International Conference on Web Intelligence*, 138-144.

Li, Z., Lee, K. C. K., Zheng, B., Lee, W.-C., Lee, D. L., and Wang, X., 2011. IR-tree: An efficient index for geographic document search. *IEEE Transactions on Knowledge and Data Engineering*, 23(4): 585-599.

Medelyan, O., Milne, D., Legg, C., and Witten, I.H., 2009. Mining meaning from Wikipedia. *International Journal of Human-Computer Studies*. 67(9): 716-754.

Rybinski, M., and Aldana-Montes, J.F., 2014. Calculating semantic relatedness for biomedical use in a knowledge-poor environment. *BMC Bioinformatics*, 15(S2).

Tian, Y., and Huang, M., 2012. Enhance discovery and retrieval of geospatial data using SOA and semantic web technologies. Expert Systems with Applications, 39(16):12522-12535

Yue, P., Di, L., Yang, W., Yu, G., and Zhao, P. Semantics-based automatic composition of geospatial web service chains. *Computers and Geosciences*, 33(5):649-665.

Zhao, P., Foerster, T., and Peng, Y., 2012. The Geoprocessing Web. *Computers and Geosciences*, 47:3-12.

Zhao, T., Zhang, C., Wei, M., and Peng, Z.-R., 2008. Ontology-Based Geospatial Data Query and Integration. *Geographic Information Science: 5th International Conference, Park City UT USA, September 23-26.* 370-392.

Zhang, Z., Gentile, A.L., and Ciravegna, F., 2012. Recent advances in methods of lexical semantic relatedness - a survey. *Natural Language Engineering*, 19(4):411-479.

Free and Open Source Software for Geospatial (FOSS4G) Conference Proceedings

Volume 15 *Seoul, South Korea* Article 10

2015

Introduction To A New Geo-Referenced Street Level Content: 3D GEOPANO

Boram Yoon
Department of Geoinformatics, The University of Seoul

Impyeong Lee
Department of Geoinformatics, The University of Seoul

Recommended Citation

Yoon, Boram and Lee, Impyeong (2015) "Introduction To A New Geo-Referenced Street Level Content: 3D GEOPANO," *Free and Open Source Software for Geospatial (FOSS4G) Conference Proceedings*: Vol. 15 , Article 10.
DOI: https://doi.org/10.7275/R5JD4TZN
Available at: https://scholarworks.umass.edu/foss4g/vol15/iss1/10

PO-15

INTRODUCTION TO A NEW GEO-REFERENCED STREET LEVEL CONTENT: 3D GEOPANO

Boram Yoon[1] and Impyeong Lee[2]

[1]Department of Geoinformatics, The University of Seoul
163, Seoulsiripdae-ro, Dongdaemun-gu, Seoul 02504, Rep. of Korea
Email: hoyohoo@uos.ac.kr
[2] Department of Geoinformatics, The University of Seoul
163, Seoulsiripdae-ro, Dongdaemun-gu, Seoul 02504, Rep. of Korea
Email: iplee@ uos.ac.kr

ABSTRACT

Geospatial mash-up is to produce a new worthy value by combining geo-referenced contents and different kinds of service. Despite of its high popularity in aerial view contents such as Google Earth, it is hardly achieved especially in street level due to the poor georeferencing quality of the underlining street level contents. In this study we propose a new street-level georeferenced panorama image content, '3D GeoPano'. While the most existing contents provide only the location and orientation for each entire image not for each pixel, this new content is accurately georeferenced pixel by pixel so that one can derive 3D absolute coordinates for every pixel. Based on the precisely determined 3D ground coordinates per every pixel, GIS database linkage can be more close, flexible and organic. This can improve the quality of geospatial mash-up service in street level. Such '3D GeoPano' will be developed in open source community. By doing so, we expect to activate the geospatial mash-up in street-level as well as to spread the geospatial technology.

1. INTRODUCTION

Geospatial mash-up is to produce a new value by combining geo-referenced contents and existing resources at a low cost. The most successful of the mash-up case is Google Earth. Google Earth have produced a lot of map based services. The key of the success is the high quality of Google Earth in terms of its user interface and a detailed description of real world etc. This means that the mash-up is highly depending on the base resources like Google Earth. Mash-up service is influenced by what the base resources to provide and how to provide.

Despite of the success of aerial view geospatial contents represented by Google Earth, it is still depressed to mash up with geospatial contents in street level. In this study, we try to understand in aforementioned context of mash-up. By analyzing the current geo-referenced contents especially in street level, we draw some limitations to mash-up. Furthermore, we suggest a solution to overcome the depression in the street view contents. We call our proposed content 3D GeoPano. This new geo-referenced street level content is worthy of notice for its pixel by pixel geo-referencing technique. Based on the geo-referencing result per every pixel, GIS database linkage can be more close, flexible and organic. We expect 3D GeoPano to improve the quality of geospatial mash-up service in street level. This paper describes the details of analysis of current street view contents and our proposed solution 3D GeoPano.

2. CURRENT STREET VIEW CONTENTS

2.1 2D image based street view contents

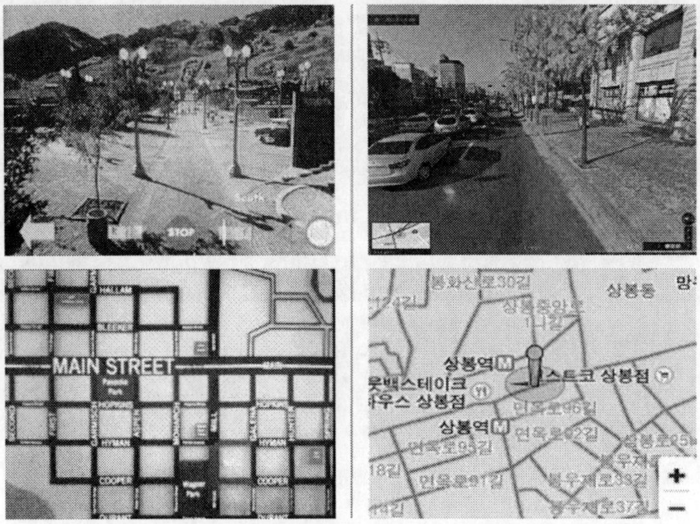

Figure 1. Movie-Map (left) and Google Street View

Although a map of aerial view, a long-standing representative of geospatial contents, has high quality in terms of position accuracy and visibility etc., it is not able to depict the real world in street level due to its limitative field of view. This cause appearance of a map in street level such as Google Street View. There was an early version of map at the viewpoint of street in the late 1970s, which is called Movie-Map. Movie-map is an image based hypermedia system which provides street image sequences of Aspen city in the place of a driver for military purpose (Lippman, 1980). When troops get into strange area, the map can help them quickly and intuitively become familiar with that area. Surprisingly, as you see Figure 1, it gives several functions to handle the map even in the late 1970s as Google Street View does the graphic user interface on the content today.

Map services in the street level, such as Google Street View and Daum Road View, represent the real world as the scenes come to eyes of a walking man or a car driver. They mainly reflect the side of structures on the ground, for example building façade. This representation can intuitively give massive information based on the way a user recognizes the real world. That's the main reason why the Movie-map was developed for military purpose and the street view services are widely used today. Such street view maps are applied to not only the outside but also the interior of buildings. Especially assimilating with the augmented reality technology, they are used for navigating a complex building, advertising a store and guiding a tour of museum.

The state of the art of the street view map is Street View of Google. Its reputation is based on that Google Street View provides high quality of virtual user experiences. It gives image based spatial information over almost the whole world and well organizes them so that a user can easily navigate the virtual world like car driver. To do that, various kinds of data are collected by cameras, laser scanners and navigation sensors. Such sensors are mounted on several kinds of platforms depending on where data is acquired, for example a passenger car, a snow mobile and even a kite. Acquired data mainly generates panorama images with high resolution and position information. After that those products are integrated with probabilistic

Figure 2. 'Click-to-go' function of Google Street View

Figure 3. Microsoft Street Slide (Vincent, 2007)

graphical model representing the road networks and they are linked with Google Maps. Where we can easily find the effort to provide high quality of virtual user experience in Google Street View is the function called click-to-go. Click-to-go is the function helps a user see a specific region of interest at desired location and direction with a floating white window. In Figure 2, you can see different building façade at different viewpoints.

Despite of the fact that Google Street View is state of the art and provides outstanding user interface and virtual experience, it has limitation on augmenting attribute information of GIS as well as user-defined information over the map. Furthermore it also cannot gives an overview of one side of a street. When a user want to scan a street, user should manually change the view point following the road. To overcome those limitation, Microsoft studied Street Slide as Figure 3. Street Slide is based on multi-perspective strip panorama image which is orthographic projection image of one side of street. This representation can provide intuition for wide area. Moreover related attribute information can be augmented over the content more intuitive than Google Street View like map of aerial view. However it is lessen the user satisfaction of virtual experience.

2.2 3D model based street view contents

There have been a lot of efforts and needs for 3D model based geospatial information. The high quality of 3D geospatial model is required to have complete geometrical figures, high resolution of texture information and user-interactive functions regardless of the indoor and the outdoor (Na *et al*, 2013). It is quite cost-consuming and labor-intensive to achieve those requirements.

The most successful case of 3D geospatial service, Google Earth, also have undergone several trial and error. When constructing a 3D model of a building using aerial images or satellite images, the quality and the accuracy of the model are not good enough. To improve

Figure 4. Improved 3D building layer of Google Earth (Anguelov, 2010)

them, Google launched a user participation pipeline, SketchUp, to encourage user to manually build 3D model and to open in the Google Earth. However, this cannot meet the large demand for 3D models. Google's another approach to improve the low quality of 3D model is to exploit the data produced by Google Street View project (Anguelov, 2010). Point cloud data from laser scanner is used to build geometric model framework, and optical image from street view camera and aerial/satellite image are used for texture information of the façade and the roof of the model. The accuracy and the resolution of 3D model can be improved by this approach like Figure 4.

Building indoor 3D model approaches can be divided into two parts. One is based on the laser scanning data, the other is based on the blueprint of a building. Each methods has own strengths and weaknesses. Laser scanning data based 3D model can achieve its high position accuracy and resolution and can be built even in the absence of the blueprint of a building. However it takes large amount of time and cost to acquire and process the primitive laser scanning data. What is more it require cost-consuming and labor-intensive procedure which each components of the model is discretized so that each of them has own attributes (Ministry of Land, Transport and Maritime, 2013). Unfortunately, this step is too complicated and difficult to be fully automated by a computer. On the other hand, the blueprint based 3D model is easy to assign the attribute information to each components of it. Besides, the blind spot of data acquisition can be easily modeled. Nevertheless, there is the disagreement between the blueprints and the constructed objects. Above all, geo-referencing procedure which locates the model in the real world coordinate system is so complicated and result of it is not reliable enough (Seoul, 2014).

3. 3D GEOPANO

To assess aforementioned current 2D/3D street view contents, user's virtual experience and visual satisfaction are considered as being the most important. As Bleisch (2012) mentioned, the way and the effect to express various information on the virtual space are less discussed. This is the limitation of current geospatial contents to active and in-depth application. When various kinds of information and data can be appropriately represented on the geospatial contents, powerful strengths of the geospatial contents, spatial intuition and effective visibility, can be meaningfully maximized and valuable enough.

In an effort to activate the geospatial contents, we propose new type of geospatial content in street level. Designing such geospatial content, we set three directing points of our proposal as follows: (1) the real world should be projected to texture information of our proposal as it

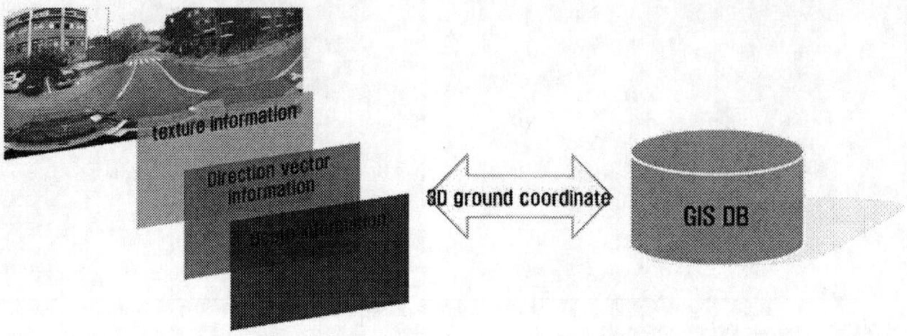

Figure 5. The conceptual structure of 3D GeoPano

is; (2) our proposal should provide 3D geometry of each discretized polyhedral model; (3) attribute information of each ground structures can be flexibly and organically linked to our proposal. Existing geospatial contents are also developing toward these points. However most of them have difficulty building the high quality of 3D model. Because they mainly focus on the visual experience of user not the way to represent other types of information, their 3D model is just aimless 3D contents for only 3D as Shepherd (2008) mentioned. We look at the problem with the opinion that a complete 3D model is not essential and fully textured 2D image is enough to augment other various information on the contents. By separating the texture information and 3D geometry, the linkage between the contents and other attribute information can be more flexible.

Our solution is omnidirectional panorama image called 3D GeoPano, which can be accurately georeferenced per every pixel. This can represent the real world as it is, and implicitly contains accurate 3D geometry. The difference compared to current Google Street View and Daum Road View is it is possible to determine the 3D ground coordinates corresponding to each pixel on the content. The 3D ground coordinates derived from 3D GeoPano play a role as a key of reference in the GIS database which makes flexible connection with database possible. In addition, such 3D GeoPano is able to economically construct and maintain the 3D virtual model and GIS database.

Figure 5 shows the conceptual structure of 3D GeoPano. The 3D GeoPano consists of three kinds of information, texture, direction vector and depth. Texture is an image the real world is projected. Direction vector and depth information participate in estimating the 3D ground coordinates. Direction vector is the unit vector heading from the center of projection of the texture to the point of real world corresponding each pixels of texture, and depth is the distance between the projection center and the ground point. These three kinds of information make up each layers of the 3D GeoPano.

Table 1 is the functional comparison of the 3D GeoPano and Google Street View. A user is passively provided with image contents corresponding given ground coordinates and orientations using Google Street View. However a user can access to the 3D coordinates of real objects in the world as well as the high resolution of image contents using 3D GeoPano. This characteristic of our proposal enables following functions. As previously mentioned, a variety of information can be easily augmented on the content based on the 3D ground coordinates as a key of reference. Furthermore, building 3D geometric model is possible from the content and existing 3D model is able to be augmented on the content. Lastly, it is also possible to measure the geometric size of given real object from the content.

Figure 6 is a designed software architecture for 3D GeoPano. There are two main modules for 3D GeoPano generation and database interaction and other supporting modules.

In the 3D GeoPano generation module, three main components of 3D GeoPano are produced and combined. On the other hand, there are four sub-modules in the database interaction module. They respectively are in charge of 3D world coordinate determination, augmentation, measurement and 3D model construction.

Table 1. The Functional comparison of 3D GeoPano and Google Street View

	3D GeoPano	Google Street View
3D ground coordinate	Possible to determine the 3D ground coordinates of spatial object in the content as well as the content itself	- Possible to render the content corresponding given ground coordinates and orientations - Impossible to acquire the 3D coordinates of given content and spatial object in the content
Measurement (length, area etc.)	Possible to measure the geometric size of spatial object in the content	impossible
3D modeling	- Possible to build the 3D geometric model from the content - Possible to augment the 3D model on the content - Possible to separate the each object model	- Mesh model of building façade from point cloud data - Geometric model information is given as lookup table of 3D planar equation in runtime(user cannot directly access and utilize the geometric model information) - Not separated model object by object
Content augmentation	possible to augment various information based on the 3D ground coordinates as a reference key	- Simple attribute marker is augmented, but its location accuracy is low. - User cannot augment various information on the content

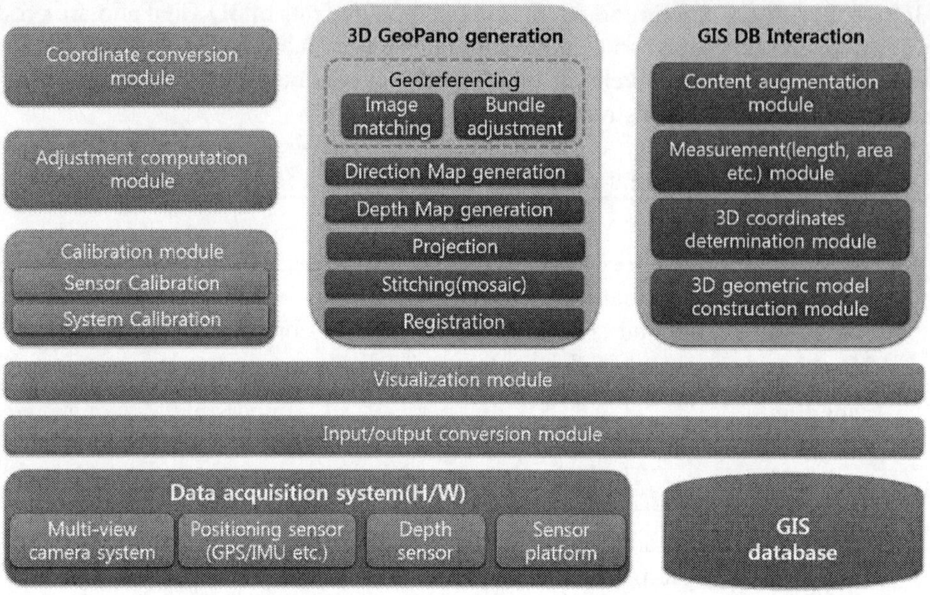

Figure 6. Software architecture of 3D GeoPano

4. CONCLUSION

The high popularity of geo-referenced mash-up improves the quality and availability of geo-referenced contents. Such contents which accurately reflect the real world in street level as well as in aerial level have appeared. Despite the advent of geo-referenced contents of street level, it is still hardly used for mash-up. In this study, we thus find out its limitations to geospatial mash-up and propose solution to overcome them. By analyzing the current geo-referenced street view contents with respect to their method and the way of representation, we draw the limitations of current street level contents. Furthermore, we propose the 3D GeoPano as a solution to overcome those limitations. With 3D GeoPano, it is able to determine the 3D coordinates of given object in real world. Using those coordinates, database linkage can be much closer, flexible and organic. Such 3D GeoPano will be developed in the open source community Github. By doing so, we hope contribute to activation of geo-referenced contents in street level and its mash-up.

5. REFERENCES

Anguelov, D; Dulong, C; Filip, D; Frueh, C; Lafon, S; Lyon, R; Ogale, A; Vincent, L; Weaver, J. 2010, Google street view: Capturing the world at street level, *Computer*, 43(6):32-38.

Bleisch, S. 2012, 3D Geovisualization-Definition and Structures for the Assessment of Usefulness, *the ISPRS Annals of Photogrammetry, Remote Sensing and Spatial Information Sciences*, Melbourne, August 25 - September 1.

Lippman, A. 1980, Movie-maps: An application of the optical videodisc to computer graphics, *ACM SIGGRAPH Computer Graphics*, 14(3):32-42.

Ministry of Land, Transport and Maritime, 2013, Research on integrated management system for indoor spatial information and location search.

Na, J. Y; Hong, C. H. 2013, A Study on the Weight Lightening Algorithm of 3-Dimensional Large Object based on Spatial Data LOD, *Journal of Korea Spatial Information Society*, 21(6):1-9.

Seoul, 2014, Research on 3D indoor space modeling construction and application method in Seoul.

Shepherd, I. D. 2008, Travails in the third dimension: a critical evaluation of three-dimensional geographical visualization. *Geographic Visualization: Concepts, Tools and Applications*, Wiley, pp. 199-222.

Vincent, L. 2007, Taking online maps down to street level. *Computer*, 40(12):118-120.

Free and Open Source Software for Geospatial (FOSS4G) Conference Proceedings

Volume 15 *Seoul, South Korea* Article 11

2015

Neo-geography and Geography Education in the Japanese Universities

Yuichiro Nishimura
Nara Women's University, Japan

PO-18

Neo-geography and Geography Education in the Japanese Universities

Yuichiro Nishimura*

Nara Women';;;s University, Japan

Corresponding Author : Yuichiro Nishimura (nissy_yu@mac.com)

Abstract

The author would like to discuss the state of the geographic education and the neo-geographical practices in Japan. Most of the geography education in universities are still old-fashioned and aimed for educating professionals, the new type of mapping practices like neo-geography are dissociated from it (Dodge and Perkins, 2008)

In recent years, such neo-geographical practices also became popular in Japanese society. People come to recognize the importance of geospatial data especially when they facing at the several heavy disaster situations in Japan. For example, Japanese registered users of OpenStreetMap are over 3,700 in September 2013 (http://wiki.openstreetmap.org/wiki/User:Techstrom/JapanData) and the users still increasing now. Open data/government policies also interested Japanese ordinary people in the neo-geographical practices, because they need the geographic viewpoints and analysis using such open data/government for solving the local issues. Such neo-geographical practices could be related to Japanese Neo-liberalism. Recent Japanese 'New Public Commons' could be associated with these practices. The openness, transparency and participatory natures of these practices attract attention by the Japanese local/national government.

On the other hand, geography education in most of Japanese universities remains conservative style. The GIS education is biased toward the methods of using desktop, proprietary software like ArcGIS desktop. Although Japanese GIScience BoK(Body of Knowledge) has small mention about neo-geography, public participatory GIS(PPGIS) and internet GIS in the chapter of GIS and society, there is no mention related to these words in the chapter of GISc education(http://curricula.csis.u-tokyo.ac.jp).

However, such situation is now changing. The student have new chance to using various different open source softwares, web based platforms, and data collection initiatives (Cowan and Hinton 2014). The several universities in Japan now hire the new curriculum for learning GIS by neo-geographical tools. The students also learn the possibility and the limitation of neo-geographical mapping. Some university students now doing crisis mapping by OpenStreetMap and making the disaster map by FOSS4G software.

There are some issues for excersize the new type of curriculum. The first problem is about keeping student's motivation at neo-geographical practices. Some times the data conflict caused by collaborative mapping would lose the motivation of student. At longer time scale, Sustainable neo-geographical practices will face difficulty because most of students did neo-geographical practices when only they take the class. The new type of geography education needs the collaborative process with local neo-geographers for materialize sustainable practices.

Academic Discipline and Sub-Disciplines : geography; GIS; education

Keywords : neo-geography; Geography Education; Japanese Universities; critical GIS

Agencies : Yuichiro Nishimura, Nara Women's University, Department of History, Sociology and Geography

Volume 15 *Seoul, South Korea* Article 12

2015

A Study of the Development and Distribution of Open Geospatial Data in Japanese Local Governments

Toshikazu Seto
Center for Spatial Information Science, the University of Tokyo

Yoshihide Sekimoto
IIS of the University of Tokyo, Japan

Shu Higashi
Open Knowledge Japan, Japan

PO-17

A Study of the Development and Distribution of Open Geospatial Data in Japanese Local Governments

Toshikazu Seto[1*], Yoshihide Sekimoto[2], Shu Higashi[3]

[1]*CSIS of the University of Tokyo, Japan*
[2]*IIS of the University of Tokyo, Japan*
[3]*Open Knowledge Japan, Japan*

Corresponding Author : Toshikazu Seto (tosseto@csis.u-tokyo.ac.jp)

Abstract

Since the end of the last decade, the use of open data (secondary use and machine-readable formats) has emerged as a political and cultural movement for the realization of citizen participation. Open government, citizen participation, transparency in government affairs, and cooperation of public and private entities were established as goals by the Obama administration in the U.S. in 2009. In the "G8 Open Data Charter," which was declared at the G8 Lough Erne Summit in June 2013, geospatial information data was recognized as an area of high value. In addition to open data policy, data flow is a necessity; for example, the CKAN platform with data catalogs have been developed as open source with the provision for the flow of information. Various policies and government strategies on open data have been enforced since 2012 in Japan including the introduction of various guidelines and standard government terms and conditions

Visualization being an important aspect of geographic information, the use of various tools, such as FOSS4G, is required. On the other hand, since the formats of open data currently vary, a cross-evaluation is necessary to determine the usability of the available data, especially in the case of geographical information comprising of latitudes and longitudes, as well as readable mechanical data. The format in which governments use or distribute data—in addition to desktop GIS and web GIS—is particularly important, as general-purpose tools are also important requirements for using open data.

Based on Japanese trends on open data policies/datasets/activities in recent years, the purpose of this study is: first, to compare the extent and circumstances surrounding the openness of geospatial information in Japan, and; second, to analyze especially open source platforms and applications for using open data.

The classified data categories of Japanese open data are, population and statistical data account for more than 20% of total open data, while the next most common category, public relations, accounts for about 16% of government information. The geographic data format is primarily distributed for disaster prevention, education, and tourism sectors, as much of the original data and urban planning diagrams contain positional information regarding facilities.

In recent years, the Code for Japan, a civic tech community in Japan, has focused on the context of the FOSS4G. Consequently, the Japanese have published open data in more than 100 local governments; this data is simply provided as a file on the website of the local governments.

However, the staff in the technology department of Japanese local governments introducing platforms such as CKAN and the information policy issues is insufficient. The other hands, CKAN and NetCommons (Japanese open source CMS) have been readily adopted in some local governments, such as Fukuoka City and Shizuoka Prefecture. In addition, some local governments provide open geospatial data using the OSM platform. Therefore, an increase in programs that combine enhancements (more provide and use case) and platforms that offer easy access to open geospatial data is necessary.

Academic Discipline and Sub-Disciplines : GIScience; Social Geography

Keywords : Open Data; Open Government; CKAN

Agencies : the University of Tokyo, JSPS KAKENHI Grant Numbers 25870907, 25244042, 15K03008.

Free and Open Source Software for Geospatial (FOSS4G) Conference Proceedings

Volume 15 *Seoul, South Korea* Article 13

2015

Vulnerability Assessment Using Free and Open Source Software (FOSS) and Open Source Data: Talomo District, Davao City, Philippines

Glenn Depra
Ateneo de Davao University, Philippines

PO-20

Vulnerability Assessment Using Free and Open Source Software (FOSS) and Open Source Data: Talomo District , Davao City, Philippines

Glenn Depra*

Ateneo de Davao University, Philippines

Corresponding Author : Glenn Depra (gddepra@addu.edu.ph)

Abstract

In recent years, the city of Davao has been experiencing floods all throughout the year. Things like flooded streets, overflowing of rivers, stranded commuters, and residents fleeing to evacuation centres are now considered normal especially if an heavy rain will continue for more than an hour. The positive economic growth of the city is attributed to its strategic location and favourable climate since it is considered to be typhoon free unlike the national capital region Manila and other major cities in the Visayas region. But these changing climate patterns can affect these upward economic growth if the local government units will not act on these issues promptly. An up- to- date spatial information is needed to support authorities for a timely evidence-based decisions in disaster management. This study will assess the vulnerability of the exposed population and other entities of Talomo District in Davao City, Philippines. Emphasis on the vulnerability component of the United Nations framework on Risk which is Hazard x Vulnerability and the factors affecting vulnerability are exposure, susceptibility and capacity. An open source software (QGIS and MapWindow) where used to process digital elevation models (DEMs) in combination with other spatial data (OpenStreetMap) can be an important database for topography-related analyses.

Keywords : FOSS, DEM, Vulnerability

Free and Open Source Software for Geospatial (FOSS4G) Conference Proceedings

Volume 15 *Seoul, South Korea* Article 14

2015

Building a Time Server with RaspberryPI & GPS, Connecting with a Free Web GIS Software

Orkut Murat YILMAZ
Bogazici University, Turkey

Kerem Halicioglu
Bogazici University, Turkey

Osman Bal
Bogazici University, Turkey

Ayca Eraslan
Bogazici University, Turkey

Ahmet Alper YILMAZ
Bogazici University, Turkey

See next page for additional authors

Recommended Citation

YILMAZ, Orkut Murat; Halicioglu, Kerem; Bal, Osman; Eraslan, Ayca; YILMAZ, Ahmet Alper; Julaiti, Wumiti; and LiBassi, Nicholas Paul (2015) "Building a Time Server with RaspberryPI & GPS, Connecting with a Free Web GIS Software," *Free and Open Source Software for Geospatial (FOSS4G) Conference Proceedings*: Vol. 15 , Article 14.
DOI: https://doi.org/10.7275/R52F7KNV
Available at: https://scholarworks.umass.edu/foss4g/vol15/iss1/14

Building a Time Server with RaspberryPI & GPS, Connecting with a Free Web GIS Software

Authors
Orkut Murat YILMAZ, Kerem Halicioglu, Osman Bal, Ayca Eraslan, Ahmet Alper YILMAZ, Wumiti Julaiti, and Nicholas Paul LiBassi

This paper is available in Free and Open Source Software for Geospatial (FOSS4G) Conference Proceedings: https://scholarworks.umass.edu/foss4g/vol15/iss1/14

PO-21

Building a Time Server with RaspberryPI & GPS, Connecting with a Free Web GIS Software

Orkut Murat YILMAZ*, Kerem Halicioglu, Osman Bal, Ayca Eraslan, Ahmet Alper YILMAZ, Wumiti Julaiti, Nicholas Paul LiBassi

Bogazici University, Turkey

Corresponding Author : Orkut Murat YILMAZ (orkutmurat.yilmaz@boun.edu.tr)

Abstract

"Orkut" Murat Yilmaz1, Kerem Halicioglu1, Osman Bal1, Ayca Eraslan1, Ahmet Alper Yılmaz1, Wumiti Julaiti1, Nicholas Paul LiBassi1

1 Bogazici University Kandilli Observatory and Earthquake Research Institute, Geodesy Department, Istanbul, Turkey

20 years ago, Kandilli Observatory and Earthquake Research Institute Geodesy Department Building was official time measurement and synchronization laboratory for institutions of Istanbul. Geodesy Department's calculations and mechanical clocks were the way of synchronization.

In this study, we have organized a revival project for Time Synchronization and used a Raspberry PI & GPS Module to capture The GPS Signal and parsed the Signal for building NTP Server.

After building NTP Server, we are going to make a GPS handset by adding touchscreens to our Raspberry Pis and we're going to use this handset for locating the trees at the Kandilli Observatory and Earthquake Research Institute Campus. At the end, collected the data will be used at the free and open source web gis software for our project.

Academic Discipline and Sub-Disciplines : Geodesy, GPS, FOSS4G

Keywords : Raspberry PI, Open Hardware, FOSS4G, Web GIS, Free Software, NTP, Time

Free and Open Source Software for Geospatial (FOSS4G) Conference Proceedings

Volume 15 *Seoul, South Korea* Article 15

2015

Development of Opensource-based Photogrammetric UAV System Using Smart Camera

Jinwoo Park
pukyong national university, Korea

Hohyun Jung
pukyong national university, Korea

DongYoon Shin
pukyong national university, Korea

Chuluong Choi
pukyong national university, Korea

Recommended Citation

Park, Jinwoo; Jung, Hohyun; Shin, DongYoon; and Choi, Chuluong (2015) "Development of Opensource-based Photogrammetric UAV System Using Smart Camera," *Free and Open Source Software for Geospatial (FOSS4G) Conference Proceedings*: Vol. 15 , Article 15.
DOI: https://doi.org/10.7275/R5668BCK
Available at: https://scholarworks.umass.edu/foss4g/vol15/iss1/15

PO-23

Development of Opensource-based Photogrammetric UAV System Using Smart Camera

Jinwoo Park, Hohyun Jung, DongYoon Shin, Chuluong Choi*

pukyong national university, Korea

Corresponding Author : Chuluong Choi (cuchoi@pknu.ac.kr)

Abstract

Normally, aero photography using UAV uses about 430MHz bandwidth RF (radio frequency) modem and navigates and remotely controls through the connection between UAV and ground control system. When using the exhausting method, it has communication range of 1-2km with frequent cross line and since wireless communication sends information using radio wave as a carrier, it has 10mW of signal strength limitation which gave restraints on life my distance communication.

The purpose of research is to use communication technologies such as LTE (long-term evolution) of smart camera, Bluetooth, Wi-Fi and other communication modules and cameras that can transfer data to design and develop automatic shooting system that acquires images to UAV at the necessary locations. We conclude that the Photogrammetric UAV system using Smart Camera can not only film images with just one smart camera but also connects UAV system and ground control system together and also able to obtain real-time 3D location information and 3D position information using UAV system, GPS, a gyroscope, an accelerometer, and magnetic measuring sensor which will allow us to use real-time position of the UAV and correction work through non-datum aero triangulation.

Academic Discipline and Sub-Disciplines : Photogrammetric; UAV; Smart Camera; Open Source

Keywords : Photogrammetric; UAV; Smart Camera; Open Source

Free and Open Source Software for Geospatial (FOSS4G) Conference Proceedings

Volume 15 *Seoul, South Korea* Article 16

2015

Case Study: Mapping Mangroves and Coastal Wetlands

Christopher Dubia
MapWorks Learning

Kent Lewis
MapWorks Learning

Andy Long
MapWorks Learning

PO-24

Case Study: Mapping Mangroves and Coastal Wetlands

Christopher Dubia*, Kent Lewis, Andy Long

MapWorks Learning

Corresponding Author : Christopher Dubia (chris@mapworkslearning.org)

Abstract

Mapping the Mangroves (MTM), a project of MapWorks Learning, provides formal and informal education, and gives citizen scientists and the larger scientific community the ability to engage with and explore mangroves and their ecosystems. Mangroves are a cornerstone species and play important roles in habitat formation, stabilization of coastal environments, and carbon sequestration. The MTM open curricula and GIS tool provide opportunities for anyone to learn about authentic applications of GIS in the field, explore mangroves and their ecosystems, and share their findings. Learners develop an understanding and appreciation for the role mangroves play in a healthy environment and how GIS can aid in conservation.

Keywords : GIS, Coastal Conservation, Open Data, OER (Open Educational Resources)

Free and Open Source Software for Geospatial (FOSS4G) Conference Proceedings

Volume 15 *Seoul, South Korea* Article 17

2015

Land Suitability Assessment for Cash Crops Using Geospatial Techniques

Albert B. Jubilo
Chemical Engineering Department, Ateneo de Davao University

Glenn D. Depra
Mathematics Department, Ateneo de Davao University

Aurecel L. Alejandro
Humanities and Arts Cluster, Ateneo de Davao University

PO-05

Land Suitability Assessment for Cash Crops Using Geospatial Techniques

Albert B. Jubilo[1], Glenn D. Depra[2], and Aurecel L. Alejandro[3]
1 Chemical Engineering Department, Ateneo de Davao University, Davao City
ajubilo@rocketmail.com
2 Mathematics Department, Ateneo de Davao University, Davao City
depraglenn@gmail.com
3 Humanities and Arts Cluster, Ateneo de Davao University, Davao City
aurecelalejandro@gmail.com

Abstract

Land suitability assessment studies for cash crop production in certain areas provide opportunities and limitations to decision makers and farmers. The objective of this study was to assess the suitability of land for banana, corn and cacao cultivation and production in Davao Region, Philippines. The methodology adopted combines most aspects of climate, temperature, slope and soil type that affect crop suitability at large. Suitability maps of the region for cash crop cultivation were established by scanning and digitizing the soil maps using geospatial techniques. Bananas are best suited for Davao del Norte. Corn are good for Davao del Norte and Compostela Valley Province. While cacao is best for Davao City, Davao del Norte and Davao del Sur.

Keywords: Land Suitability Assessment, GIS, Soil, Agriculture

I. INTRODUCTION

The Davao Region of the Island of Mindanao, Philippines is considered an agricultural area of the country. It is located in the southeastern part of the Philippines. It is consist of the City of Davao, as the regional capital, and four provinces – Davao del Norte, Davao del Sur, Davao Oriental and Compostela Valley Province. The Regional Development Council of Davao Region is now intensifying its socio-economic programs. These include the area of agriculture.

The Department of Agriculture of Davao Region is leveling up agricultural aspects in order to improve the food sufficiency of the area. Hence, there is a need to identify the potential improvement of the production of cash crops – namely, banana, corn and cacao. Thereby, it will increase the socio-economic condition of the region.

Geospatial study is needed to determine the best site for those cash crops with regards to climate condition, temperature, slope, elevation, soil type and other parameters. The outputs of this study will help many governmental agencies and non-governmental organizations in their agricultural developmental and strategic planning. This will also aid the farmers and investors in the different localities.

As state in the report of the Department of Agriculture, Republic of South Africa (2008), "bananas (scientific name: *Musa acuminata*) are best grown in areas with warm, humid, and frost-free climate; temperature ranging 22 to 31°C; slope zero to one degree; and up to 300 meters above sea level of elevation. It is for any type of soil but well-drained with pH of 5.5 to 7.0; forest loam, rocky sand, marl, red laterite, volcanic ash, sandy clay, or even heavy clay."

Figure 1. Banana Plants

Corn (scientific name: *Zea mays*) is best planted in areas with warm weather and long sun-filled days; temperature ranging from 29 to 32°C; up to 12% slope; and up to 1800 m above sea level elevation. Corn likes rich soil with good drainage. Ideal soil for corn is sandy loam that stays moist, without being too wet.

Figure 2. Corn Plants

Cacao (scientific name: *Theobroma cacao*) is best planted in areas warm weather; temperature ranging from 18 to 32°C; up to 15 degrees slope; and up to 1200 m above sea level. It is a shade tolerant crop. Cacao is suitable in well drained loamy, clayey (with good structure).

Figure 3. Cacao Plant

Objectives of the Study

The main objective of the study is to assess land suitable for banana, corn and cacao production. Specifically, the study is aimed to:

(1) Determine the profile of the land in terms of:
   1.1 Climate
   1.2 Temperature
   1.3 Elevation
   1.4 Soil type
(2) Create a map showing land areas suitable for bananas, corn and cacao production.

II. METHODOLGY

Research Design

This study is a GIS – based type research using Free Open Source Software (FOSS) Geospatial technique. The data collected are processed and analyzed using Quantum Global Information System (QGIS) software.

.

Research Locale

The study was conducted in Davao Region which includes Davao City, Davao del Norte, Davao del Sur, Davao Oriental, and Compostela Valley Province. Figure 4 shows the Philippine map with Davao Region as the study area.

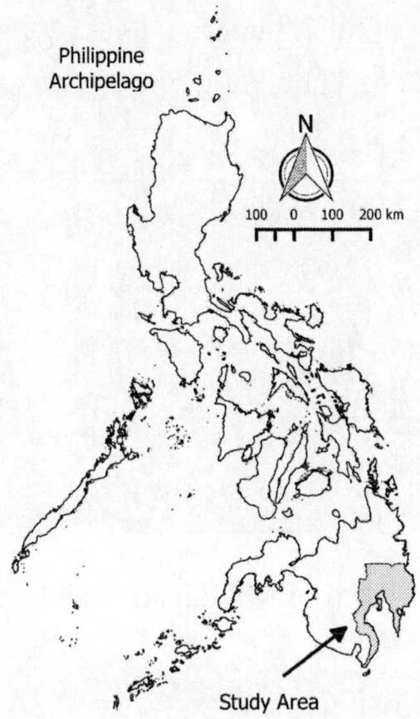

Figure 4. Study Area (Davao Region)

Research Procedure

The following is the procedure used by the proponents:

(1) Gathering of information and data
(2) Tabulation of data
(3) Mapping using geospatial technique
(4) Analysis and interpretation of data

The data gathering was conducted from April to July, 2015. The data gathered were tabulated, analyzed and interpreted. QGIS software was used to develop map. The free software to be used shall be QGIS (known as "Quantum Global Information System"). It is an application using cross-platform free and open-source desktop geographic information system (GIS). It offers data presentation and viewing, editing, and analysis capabilities (www.qgis.org, 2015).

III. RESULTS AND DISCUSSION

Climate and Temperature

The region has a generally uniform distribution of rainfall through the year. It lies outside the typhoon belt. PAGASA reported that Davao has a tropical rainforest climate (Köppen climate classification Af), with little seasonal variation in temperature. Average monthly temperatures are always above 26 °C (78.8 °F), and average monthly precipitation is above 77 millimetres (3.03 in). Davao Region climate is very suitable for cultivating banana, corn and cacao. Figure 5 shows the temperature map of Mindanao which includes Davao Region

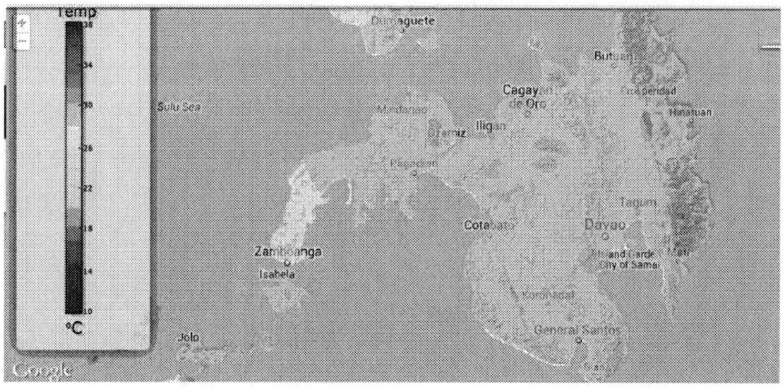

Figure 5. Temperature Map of Mindanao Island (with Davao Region)

Topography - Elevation

The Davao Region topography is dominated by peninsular and island topography characterized by extensive mountain ranges extending along the western border, in the northern central area and in the northwestern area leading to the peninsula in the southeast with uneven distribution of plateaus and lowlands. It has a coastline of 1,600 square kilometers (davaocitybybattad.blogspot.com, 2015).

About 5,021.48 square kilometers, or 26% of the total land area of Davao Region has a slope range of 0-8% which is primarily for agriculture, industries, and settlement. Among the provinces in the region, Compostela Valley has the largest area with 1,300.9 square kilometers with a slope range of 0-8%, followed by Davao del Norte with 1,219.8 square kilometers.

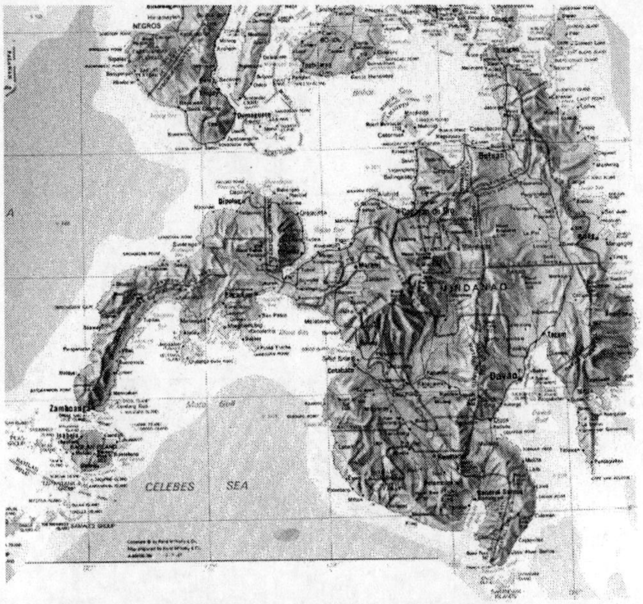

Figure 6. Topographic Map of Mindanao (with Davao Region)

Soil Type

The soil of Davao Region can be classified into the following groups: inceptisols, alfisols and ultisols. These are generally loam and clay types.

Land Areas suitable for Bananas

As shown in figure 7 (*red colors*), bananas are suitable for cultivation mostly in the southern part of Davao del Norte which comprises the cities of Tagum and Panabo, Municipalities of Asuncion and Sto. Tomas; near coastal areas of Davao City in the southeastern part facing Davao Gulf; western coastal areas of Davao Oriental facing Davao Gulf; northeastern coastal part of Davao Oriental facing Pacific Ocean, and northern and western part of Davao del Sur.

Land Areas suitable for Corn

As shown if figure 8 (*red colors*), planting of corn is best suited mostly in the central areas of Davao del Norte and Compostela Valley Province. There are areas in Davao del Sur and Davao Oriental for corn production.

Land Areas suitable for Cacao

As shown in Figure 8 (red colors), cacao is suitable in mostly in Davao City, Davao del Norte and Davao del Sur; and in some parts of Davao Oriental.

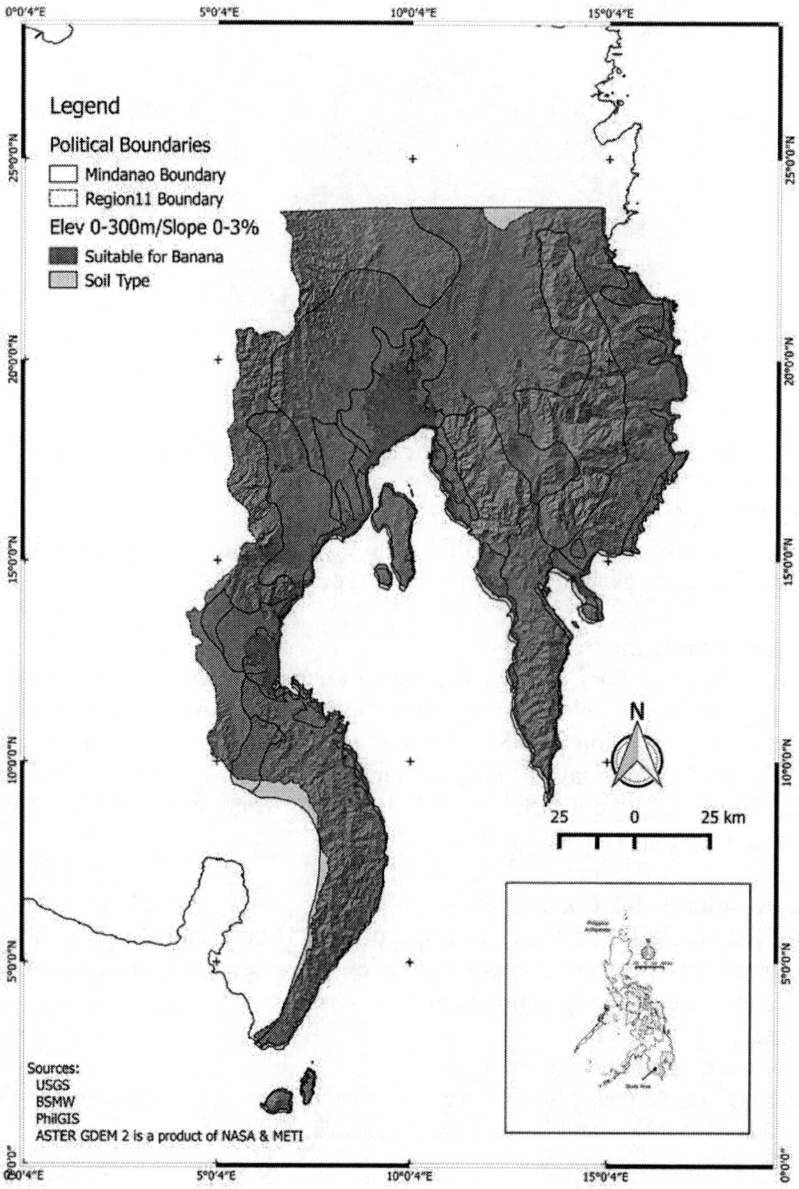

Figure 7. Land Areas suitable for Bananas

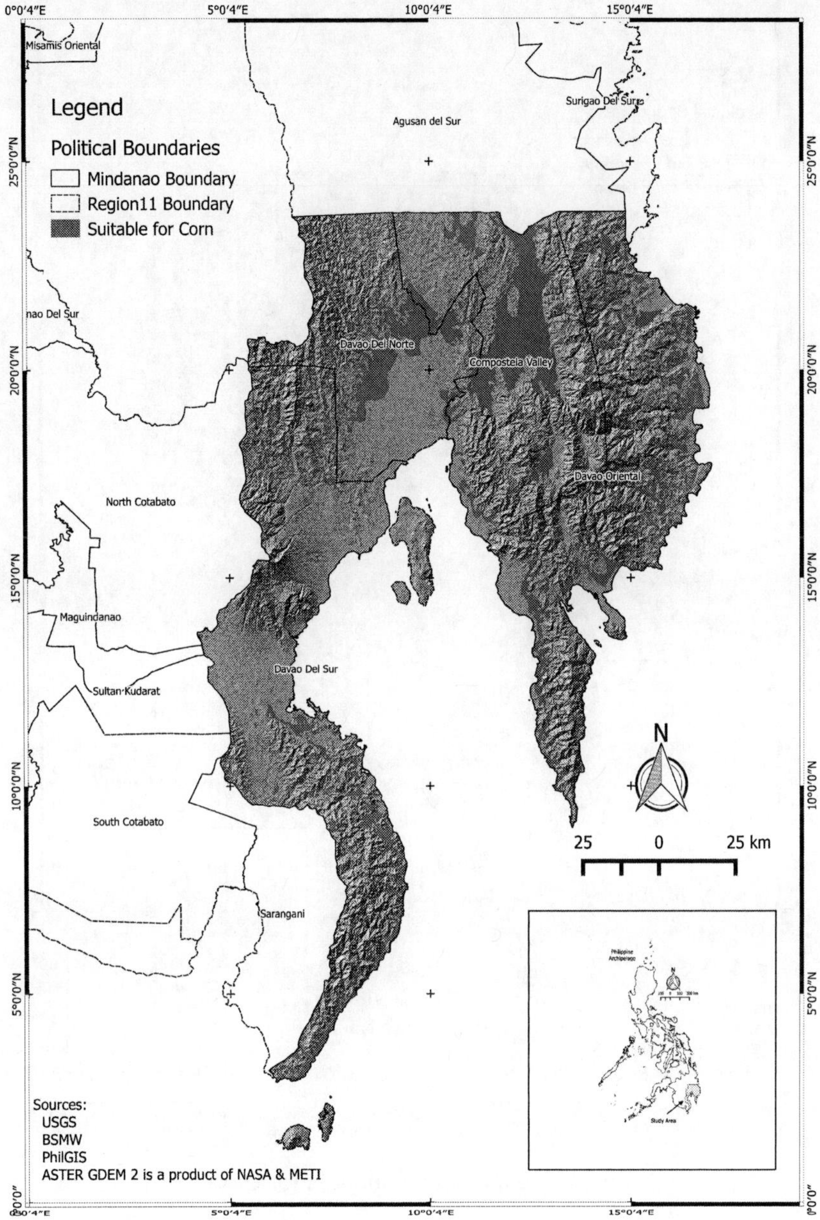

Figure 8. Land Areas suitable for Corn

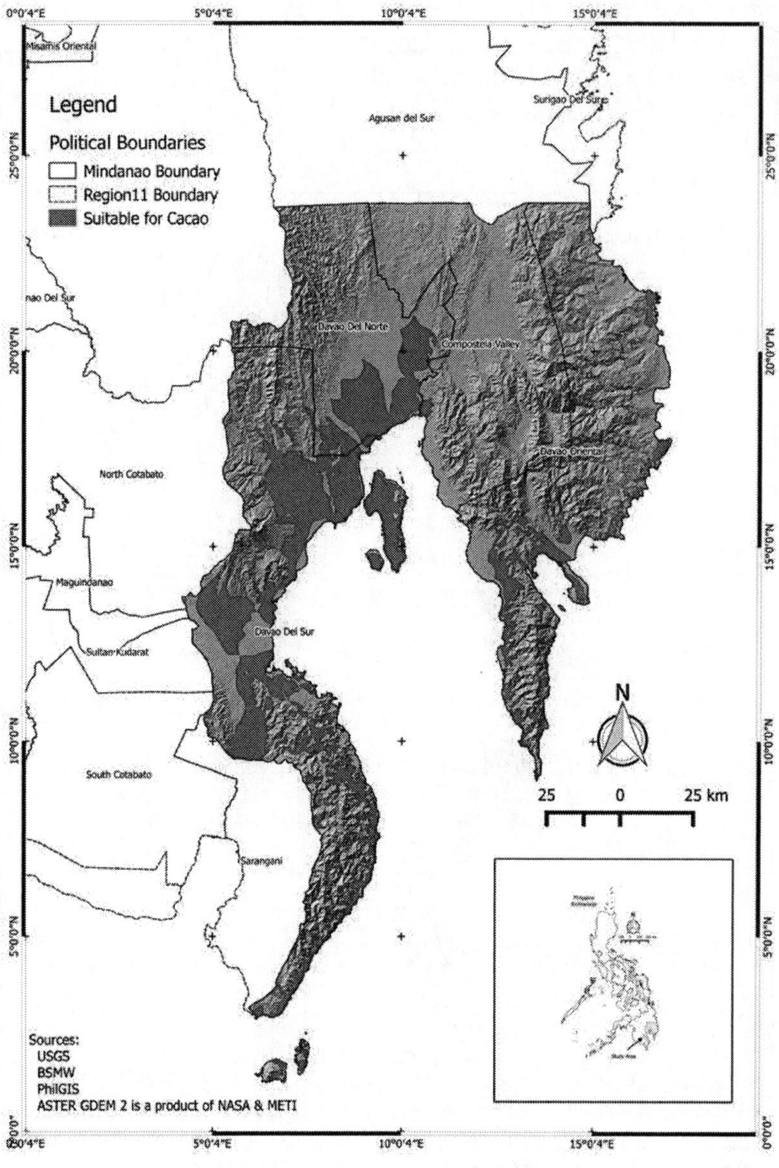

Figure 9. Land Areas suitable for Cacao

IV. CONCLUSIONS AND RECOMMENDATIONS

The following conclusions were obtained from the analysis and findings of this study:

(1) Davao Region climate and temperature are best for cash crop production; slopes vary in different areas while the soils are generally of loam and clay types.

(2) The created map shows bananas are best suited for Davao del Norte; corns are good for Davao del Norte and Compostela Valley Province; and cacao is best for Davao City, Davao del Norte and Davao del Sur.

It is respectfully recommended that:

(1) A study on land use relating to cash crop production shall be conducted.
(2) A study on the implication of climate change in the cash crop production shall be accomplished.

REFERENCES

"Climatic Normals of the Philippines" (PDF). The Naval Research Laboratory accessdate=2013-01-12.

Department of Agriculture, Republic of South Africa (2008). *Banana (Musa acuminate).* Page 2

http://davaocitybybattad.blogspot.com/2011/10/davao-region.html. Retrieved August 14, 2015

http://www.qgis.org/. Retrieved August 4, 2015

Free and Open Source Software for Geospatial (FOSS4G) Conference Proceedings

Volume 15 *Seoul, South Korea* Article 18

2015

Mobile Application Of Open Source Stack To Geo-Based Data Visualisation On E-Government Web Framework

Kwangseob Kim
Graduate Student, Department of Information Systems Engineering, Hansung University

Hansaem Park
Graduate Student, Department of Information Systems Engineering, Hansung University

Kiwon Lee
Department of Information Systems Engineering, Hansung University

Recommended Citation

Kim, Kwangseob; Park, Hansaem; and Lee, Kiwon (2015) "Mobile Application Of Open Source Stack To Geo-Based Data Visualisation On E-Government Web Framework," *Free and Open Source Software for Geospatial (FOSS4G) Conference Proceedings*: Vol. 15 , Article 18.
DOI: https://doi.org/10.7275/R5Q52MTR
Available at: https://scholarworks.umass.edu/foss4g/vol15/iss1/18

PO-01

MOBILE APPLICATION OF OPEN SOURCE STACK TO GEO-BASED DATA VISUALIZATION ON E-GOVERNMENT WEB FRAMEWORK

Kwangseob Kim[1], Hansaem Park[1], and KiwonLee[2]

[1]Graduate Student, Department of Information Systems Engineering, Hansung University,Samseon-dong2-ga, Seongbuk-gu, Seoul, Korea,136-792
Email: lovekph@nate.com, park_hs91@naver.com
[2]Professor, Department of Information Systems Engineering, Hansung University, Samseon-dong 2-ga, Seongbuk-gu, Seoul, Korea, 136-792
Email: kilee@hansung.ac.kr

ABSTRACT

Mobile web technologies by diverse open sources are expanding and developing so fast. Furthermore, open data policy of public organizations and governments in many countries prevails. In Korea, e-government standard framework has been developed based on Spring Framework and distributed to improve implementation efficiency of web-based information systems for public domains. For wide applications and adoption of this webframework, e-government standard framework for mobile environment was also released. These trends evoke new type of geo-based applications, and practical cases are being implemented. This open source project is for data visualization on mobile, which reflects these aspects. For implementation, open source stack including GeoServer, jQuery, OpenLayers and D3.js was applied, and e-government mobile standards framework was used for mobile UX/UI components, as well as HTML5. It is expected that this approach provides practical reference and technological basis to develop target-oriented mobile application business models using open data and e-government mobile web framework.

1. INTRODUCTION

Popularity of mobile devices evokes mobile web service. W3C(World Wide WebConsortium) developed and released several standardization technologies for mobile web implementation such as graphics and layout, device adaptation, forms, or data storage (W3C, 2015).While, open data policies of country administration, local government or public organization are globally the significant another global trend. In Korea, Government 3.0 policy is for government operation to deliver customized public services by opening and sharing data which were produced or owned by government, to make service-oriented, competent, and transparent ones. As of July 2015, approximately 13,000 open data from about 790 sites are available in the form of web-accessible open API or downloadable data sets (http://data.go.kr). Seoul metropolitan city also provides open data policy in the Seoul open data plaza (http://data.seoul.go.kr). Among them, lots of geo-based data are also included. Kwakkel*et al.*(2014) emphasized on spatial data visualization and its value linked with regional characteristics. However, spatial data is more complicated data structure than text or image data sets, and general users with less experiences or understandings regarding geo-based data structure and main features with it might have some difficulties to utilize their applications in any fields.

From 2009 in Korea, there is an important open source policy of web application framework named the e-government, original notation of eGovernment, standard frameworkwhich covers an infrastructure environment for implementing application software, supporting basic functions in the application runtime. It is toward increase the quality of government-based public services, the efficiency of IT investing cost and the standardized reusability of application.

Geo-spatial web running on desktop web browser is moving to mobile web environment. Mobile web needs user experience and interface, different from desktop web. Moreover, uses of easily accessible open data greatly expands owing to the mobile portability. Also open data is good for web-based application based on open source. Components for mobile web implementation of the e-government standard framework are regarded as the practical needs from business side for enterprise geo-spatial web. This work is to develop an application model of the e-government standard framework for geo-based mobile web with open data visualization, based on full open source stack including GeoServer, PostgrSQL/PostGIS, OpenLayers and D3(Data-driven Document).

2. E-GOVERNMENT WEB FRAMEWORK IN KOREA

The main features with the e-government standard framework, Apache license version 2.0, in Korea can be summarized: compliance with open standards on the national scale, integration with commercial solutions, flexibility to cope with the newest technologies, and easiness to use and function-rich environments. It minimizes redundant developments by providing common essential functions and allows developers to concentrate on the business logic by defining an infrastructure, and increases the standardization of the development framework allows components alreadydeveloped on this framework to be reused in other application systems, and the interoperability of application systems by using standard inter-system integration interface. This framework, composed of the runtime environment with the multi-layer such as the screen processing, the business logic processing, the data processing, the integration processing and the common foundation, the development environment, the management environment, the operation environment and the common components which is a collection of reusable application programs to be used in the application software development of many on-going e-government projects.

The runtime version of this standard framework is 3.1 as of July 2015, based on Spring Framework 3.2.9. For mobile web service and hybrid app development, the additional version of this standard framework was released in 2011. Mobile version 2.7 under MIT license is composed of mobile runtime environment and mobile common component. The former uses jQuery and jQuery mobile for dynamic manipulating of HTML document and JavaScript. Kim and Lee (2015) developed a mobile application using this mobile runtime environment to present simultaneous visualization functionality for public data and geo-based data.The eGovFrame portal (http://www.egovframe.go.kr) is to present lots of information for both public uses and business promotion of the e-government standard framework. The National Information Society Agency,one of government-supported agencies for information technology promotion, has carried out administrative tasks forofficial certificationfor interoperability and compatibility confirmation service of this standard framework. Currently, about 45 software products have passed this certification test in the field of DBMS, WAS, portal solution, CMS solution, UX solution, and security solution. But no geo-based solution have yet been received this certification.

3. GEO-BASED VISUALIZATION BASED ON OPEN SOURCE

Open source stack has been utilized to build the visualization technology of geo-spatial information derived from open data on the mobile web. Tomcat was used for the web server and the container. Web frameworks were utilized in the e-government standards based on the Spring Framework. In particular, lots of open source for geo-spatial information were applied. GeoServer and PostgreSQL/PostGIS have been used to store and manage geo-spatial information. In the integrated system, GeoServerwas used to easily publishweb-recognizable standards as WMS (Web Map Service) and WFS (Web Feature Service) with files or databases in PostgreSQL/PostGIS, a relational database management system to store and edit spatial information. As an integrated system based on theseopen source environments, OpenGDS MAS (Mobile Application Server) was implemented in this work.

Mobile runtime environment in the e-government standard framework was applied for an optimized configuration for UX(User Experience)/UI(User Interface)in the client panel for a mobile web. Mobile runtime environment consists of jQuery and jQuery Mobile, and additional eGov mobile is extending the jQuery Mobile library.For the web mapping, OpenLayers 3 supporting mobile web were used. For on-line visualization of open data, D3.js which is JavaScript library set generating interactive charts and graphswas utilized. It usesSVG (Scalable Vector Graphics) of HTML5.

As the integrated open source projectlinked with D3.js and OpenLayers, OpenGDS MCL(Mobile Client Library) was implemented in this work. Table 1 summarizes open sources and the development environments used in this study.

Table 1. Open source environment of geo-based data visualization

Environment		Name / Version
Application Server	Web Server / Container	Tomcat / 7.0.56
	Web Framework	E-GovFramework / 3.1 - Spring Framework / 3.2.9
	Geo-based Data Server	GeoServer / 2.6.0
	Geo-based Database	PostgreSQL / 9.3 PostGIS / 2.1 (Extension)
	Geo-based&Public Data Management / Request	OpenGDS Mobile Application Server / 1.0
Mobile Web Client	Mobile UX/UI	E-GovFramework Mobile UX, UI / 2.7 - jQuery / 1.9.1 - jQuery Mobile / 1.3.2 - eGovMobile / 1.3.2
	Geo-based DataVisualization	OpenLayers / 3.5
	Data Visualization	D3.js / 3.5.1
	Coordinate Translate	Proj4js / 2.2.2
	Integrated Visualization	OpenGDS Mobile Client Library/ 1.3 (using jQuery Mobile, OpenLayers, D3.js)

The architecture of the application system running on the e-government standard framework is shown in Figure 1. Application server is composed of data access, service, and controller object. According to the request of the mobile client, the controller is charging the responses fordata visualizing and processing. Attribute request function is to retrieve the table, connecting to PostgreSQL. In this time, Mybatisin conjunction with the Spring Framework was used for SQL mapping framework to manage requesting database in Java.

Open data is mainly used in the form of open API capable of functioning real-time data requests and responses from public data portal in Korea government or open data portal in Seoul metropolitan city.

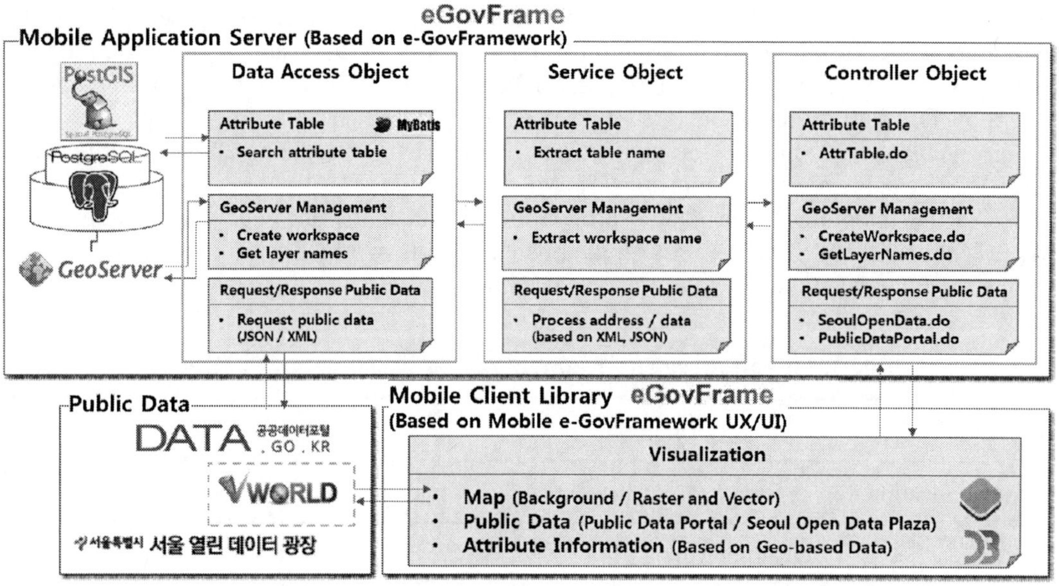

Figure 1. An architecture of mobile visualization web for geo-based data and open data.

4. MOBILE WEB APPLICATION IMPLEMENTATION

Mobile web applications for managing and visualizing of multi-typed open data and geo-based data was implemented using OpenGDS MAS and MCL. Geo-based open data in this work were stored and managed in PostgreSQL, and integrated visualization was carried out by using GeoServer and OpenLayers 3 on mobile web browser.

Figure 2 shows the result of visualization of multi-typed geo-based data. The Seoul district boundary of the polygon type was provided by the Ministry of Land, Infrastructure, and Transport in Korea. Point type data for air quality measurement was collected by downloading from the Seoul open data plaza. This mobile web application provides client functions such as transparency controlling of overlaid layers and layer visibility, listing, ordering, and deletion.

Figure 3 shows the results of attribute visualization in table. Figure 3(a) and (b) are is visualization of the spatial properties in Seoul and Busan, respectively. Figure 4 is a public open data visualization results by open API, not downloading open data. Figure 4 (a) is on-line mapping for area characterization. Other Figure 4 (b) shows bar chart by data visualization library of D3.js.

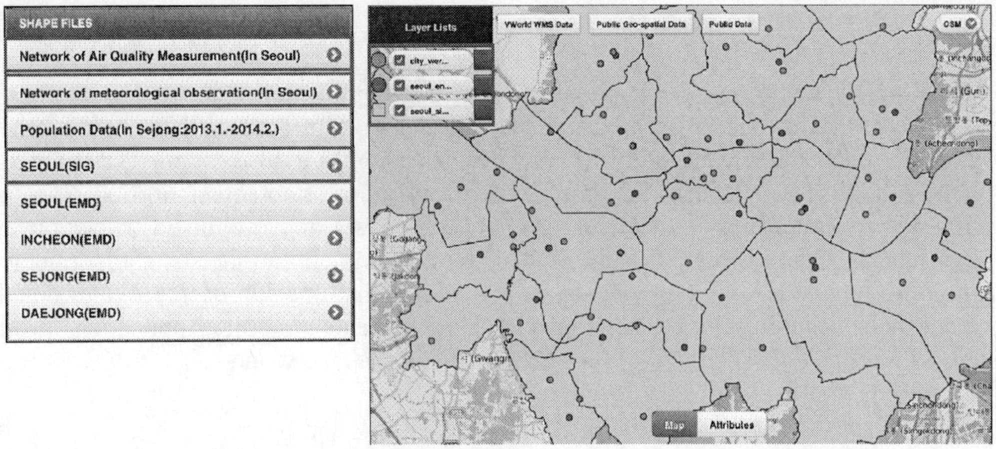

Figure 2. Visualization of multiple typed geo-based data using mobile UX/UI on the e-government web framework.

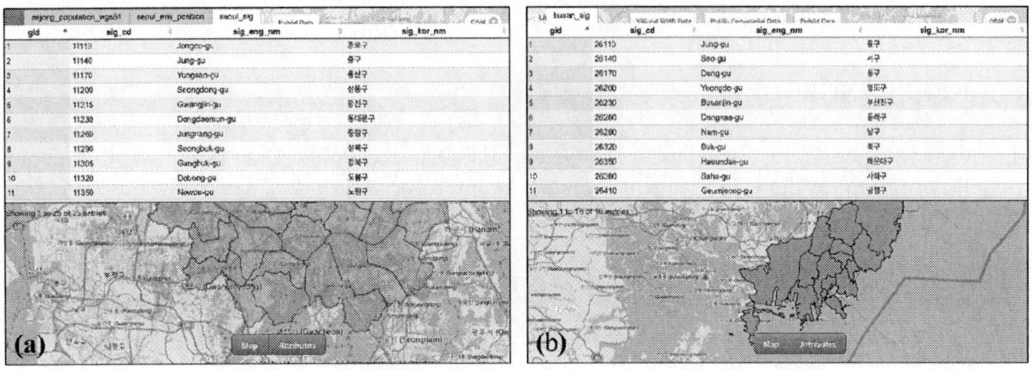

Figure 3. Attribute visualizationof public geo-based data: (a) Seoul and (b) Busan.

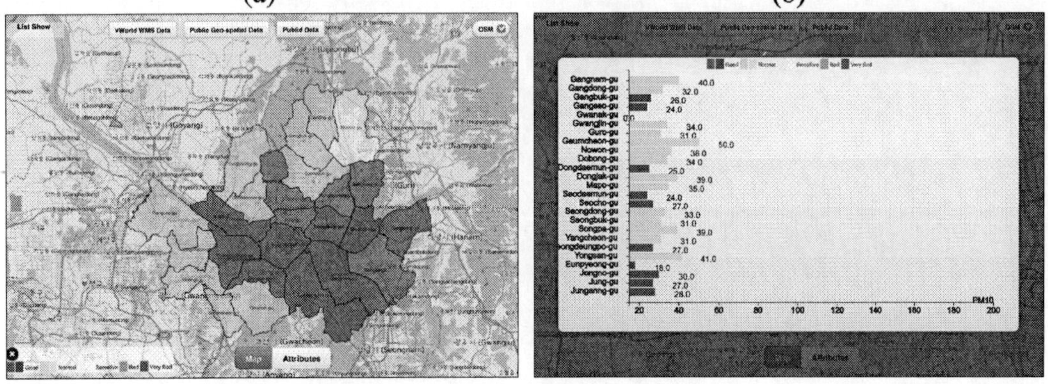

Figure 4. Visualization results using open API on the e-government web framework: (a) on-line mapping and (b) bar chart.

5. CONCLUDING REMARKS

Open source stack was developed for open data application. This project focuses on mobile web visualization of geo-based data and other typed open data. While, the e-government web framework and the mobile extension are also applied as important implementation specifications. For the actual application implementation, geo-spatially specific open sources such as GeoServer and OpenLayers and mobile visualization open sources such as jQuery and D3.js were used. The e-government mobile standards framework was used for mobile UX/UI components. It is expected that this approach provides practical reference and technological basis to develop target-oriented mobile application business models using open data and the e-government mobile web framework.

6. ACKNOWLEDGMENTS

This research was supported by National Land Space Information Research Program by Ministry of Land, Infrastructure and Transport of Korean government (No. 14NSIP-B080144-01).

7. REFERENCES

Kim, K. and K. Lee, 2015. Visualization of Geo-spatial Data and Public Data using Mobile Operating Environment in the eGovernment Standard Framework, *Journal of Korea Spatial Information Society* 23, 9-17.

Kwakkel, J. H.,C. Stephen, C. John, and C. Scott W., 2014. Visualizing Geo-spatial Data in Science, Technology and Innovation, *Technological Forecasting & Social Change*81, 67-81.

W3C, 2015.*Standards for Web Applications on Mobile: Current State and Roadmap*, http://www.w3.org/Mobile/mobile-web-app-state (Accessed July 22, 2015).

Free and Open Source Software for Geospatial (FOSS4G) Conference Proceedings

Volume 15 *Seoul, South Korea* Article 19

2015

Semi-Automated Building Footprint Extraction, Delineation And 3D Visualisation Of The University Of The Philippines Main Campus From LIDAR Data Using GIS-Based Open Source Software

Jara Kaye Villanueva
Laboratory Melchor Hall, College of Engineering University of the Philippines Diliman

Ma. Rosario Concepcion Ang
Laboratory Melchor Hall, College of Engineering University of the Philippines Diliman

Loureal Camille Inocencio
Laboratory Melchor Hall, College of Engineering University of the Philippines Diliman

Ma. Victoria Rejuso
Laboratory Melchor Hall, College of Engineering University of the Philippines Diliman

PO-09

SEMI-AUTOMATED BUILDING FOOTPRINT EXTRACTION, DELINEATION AND 3D VISUALIZATION OF THE UNIVERSITY OF THE PHILIPPINES MAIN CAMPUS FROM LIDAR DATA USING GIS-BASED OPEN SOURCE SOFTWARE

Jara Kaye Villanueva[1], Ma. Rosario Concepcion Ang[2], Loureal Camille Inocencio[3], Ma. Victoria Rejuso[4]

Rm. 406 AGST Laboratory Melchor Hall, College of Engineering
University of the Philippines Diliman, Quezon City 1101 +6329818500 loc 3147
Email: [1]jarakayevillanueva@gmail.com, [2] moang@up.edu.ph,
[3]lcvinocencio@gmail.com, [4] victoria.rejuso@gmail.com

ABSTRACT

The advent of three-dimensional building footprint extraction and visualization has recently been explored due to its applications in urban planning, transportation, environmental monitoring, and modelling. Through the use of LiDAR (Light Detection and Ranging) data products such as Digital Terrain Model (DTM) and Digital Surface Model (DSM), useful information such as elevation, size, and shape can be obtained from the processed point clouds. Different studies regarding building extraction procedures have been conducted and explored. Algorithms have also been developed using LiDAR and GIS-based software that assume a rectangular form of polygons as building structures. As such, this study discusses an approach in developing a semi-automated building footprint extraction and three-dimensional visualization of the University of the Philippines (UP) main campus using solely LiDAR data and GIS-based open source software - QGIS Chugiak Version and GRASS 7.0. The UP-Diliman Campus situated in Quezon City, Philippines consists of various polygonal structured buildings which makes it a good subject area for this study. The proposed scheme composed of three major parts: LiDAR pre-processing, building footprint extraction and delineation, and three-dimensional building visualization. Normalized digital surface model (nDSM) can be generated using DTM and DSM. Under building footprint extraction and delineation, four parameters were set: height threshold, area scope, topographic modelling, and smoothing tolerance. For the classification by height, a threshold is set to remove structures with height less than the predetermined value of elevation. Classification by area removes the objects with areas that do not fall within the predetermined area scope. Topographic modelling is used mainly to separate building from other entities. Smoothing tolerance simplifies building outlines. The accuracy assessment for the extraction and delineation is quantified using ground truth data in comparison with the extracted polygons. As compared to manual digitization of building polygons, this semi-automation can be more efficient in extracting and delineating building footprint in areas with large scope. Experimental results indicate that the proposed scheme provides a promising solution for 3D building extraction and delineation using LiDAR data processed in QGIS Chugiak Version and GRASS 7.0. This methodology is a valuable tool for urban planning and modelling.

1. INTRODUCTION

1.1 Background of the Study

Three-dimensional modelling can be useful in urban management, planning, and monitoring as compared to two-dimensional platform.For building delineation and extraction and the generation in three dimensional space, its automation reduces the time and cost it entails as compared to manual process. Several studies have been conducted with its focus on both the automation and semi-automation of this process using high-resolution image data.

In recent years, Light Detection and Ranging or LiDAR Technology has recently been explored due to its promising applications in environmental planning (Perén*et al* 2015), urban management(Coirier*et al* 2006), and energy prediction (Kalmikov*et al* 2010). This high-resolution data contain useful information for building extraction such as height, shape, and size. Using spatial tools in Geographic Information System or GIS, the analysis and visualization of data can be accomplished. Free and open source GIS software such as QGIS and GRASS has been widely utilized due to its increasing security against malware and decreasing cost for software acquisition.

At an appropriate level of detail, this paper discusses a semi-automated methodology for building footprint extraction, delineation, and 3D visualization of the University of the Philippines Main Campus from LiDAR data using QGIS 2.4 and GRASS 7.0 as the GIS-based open source software.For urban settings with a considerable large area, the semi-automation of extraction and visualization methods can greatly reduce the processing time and cost.

1.2 Objectives of the Study:

This research aims to develop a semi- building footprint extraction, delineation and three-dimensional visualization of the University of the Philippines Main Campus from LiDAR data using GIS-based open source software through the utilization of LiDAR data products such as Digital Terrain Model (DTM) and Digital Surface Model (DSM).

1.3 Scope of Study:

This study covers a portion of University of the Philippines Diliman Campus in Quezon City with an area of .7782 square kilometer. The area includes buildings with complex geometries and tree canopies which make it an ideal test site for building extraction methodology.The GIS software used is QGIS version 2.4 with the help of GRASS 7.

Figure 1. The image shows the portion of UP Main Campus used in this study

For the 3D visualization, a QGIS plug-in created by Minoru Akagi was used. Three2js is a QGIS plug-in that uses 3D library repository and WebGL-enabled browser for visualization. The parameters and their corresponding threshold used is dependent on the study area as a result of trial and error and the researchers' judgement and skill.

1.4 Overview of Related Works

Several algorithms for building extraction and 3D visualization using different methods have already been developed. This can be manual building detection, semi-automated/automated building detection using Orthoimages or LiDAR data, and semi-automated/automated delineation using both data. A brief overview of related works is discussed in this chapter.

Previous studies developed the use of solely orthoimages for urban feature extraction (Ming *et al* 2005, Wei *et al* 2005, and Varshney*et al* 2011). Detection of building patterns are done through parcel units. Parameters did not include height information which explains why LiDAR image was not used. High resolution LiDAR images are usually accompanied by RGB data in the extraction of urban features. Studies by Jiang (*et al* 2008) and Syed (*et al* 2005) used RGB data to get the shape, spectral, and semantic information of the objects through multi-resolution segmentation. The height information contained in DSM was used to distinguish buildings from ground. Both adopted the concept of object-oriented classification.

However, high resolution images like LiDAR and RGB data are not readily accessible and propriety GIS software for spatial processing costs a lot. For this study, the use of solely LiDAR data with the use of spatial tools embedded in Free and Open Source Software (FOSS) to delineate, extract, and visualize urban geometries in three-dimensional space is highlighted.

2. METHODOLOGY

2.1 Process Workflow

The workflow as presented in Figure 2 is the proposed scheme and is composed of three main processes. The first step is the pre-processing of LiDAR data products DTM and DSM. Using nDSM as the result from the first step, building delineation and extraction is done by using height, slope, and area thresholds. This building footprint is then modelled into three-dimensional space for visualization. Further discussions on each process can be found in the succeeding parts of methodology.

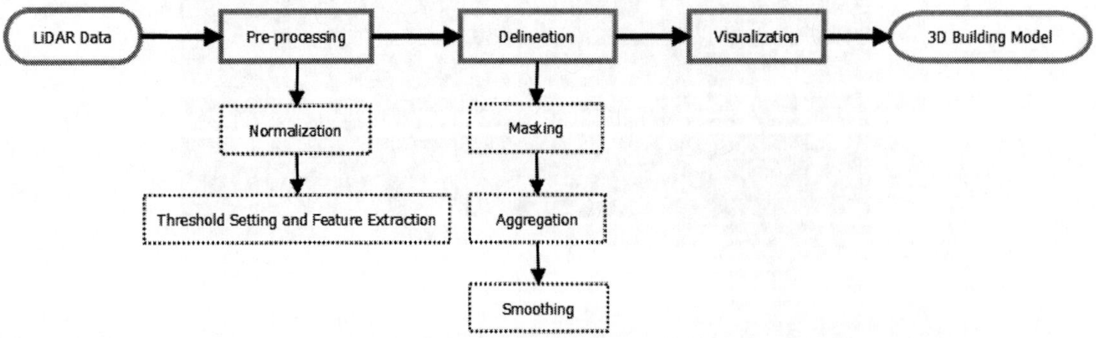

Figure 2. Workflow of the semi-automated building extraction and visualization

2.2 Test Area, and Datasets and GIS Tools Used

The data used are derived LiDAR products which include DTM and DSM. Both are acquired from National Mapping and Resource Information Authority's (NAMRIA) project, Collective Strengthening of Community Awareness on Natural Disasters (CSCAND) agencies. Both LiDAR data products have an image resolution of 1m x 1m with an uncompressed size of 3.03 MB. The area covers a portion of University of the Philippines Diliman Campus in Quezon City, measuring .7762 square kilometer.

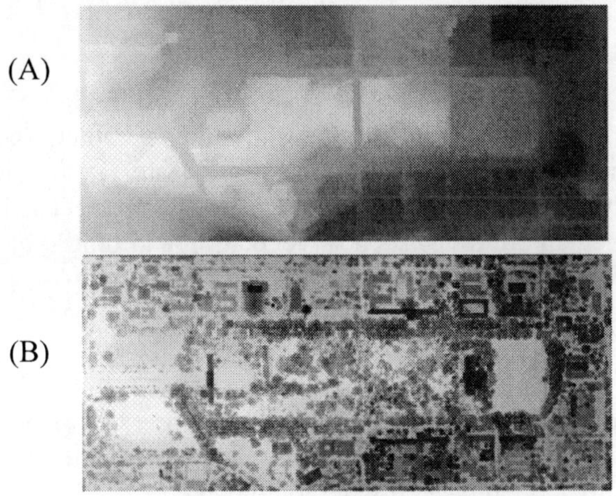

Figure 3.LiDAR data products used include (A) Digital Terrain Model (DTM) and (B) Digital Surface Model (DSM)

The semi-automation can be achieved through Geographic Information System (GIS) tools. GIS is a set of tools that analyze, store, interpret, process, and visualize all types of geospatial information. A couple of free and open source GIS software have recently been developed. Two of these free and open source GIS tool used in the study are QGIS 2.4 and GRASS 7.0. Quantum GIS or QGIS and GRASS are both free and open source software that can create, edit, analyze, and visualize geographic data. They are operational in various operating systems such as Windows, Mac, and Linux.

2.3. Processing

2.3.1. Pre-processing

The initial process consists of two processes: normalizing the Digital Elevation Model and threshold setting through trial and error.

2.3.1.1. Normalization

In LiDAR pre-processing, the concept of normalized digital surface model or nDSM is used. It can be obtained from Digital Terrain Model from Digital Surface Model. DTM delineates the elevation information of ground data or the terrain, while DSM contains the elevation information of surface features. Thus,by subtracting DTM from DTM, we will obtain the nDSM which represents the z data of surface features.

2.3.1.2. Threshold Setting and Feature Extraction

Feature extraction involves the defining of certain thresholds and parameters to differentiate and separate one feature from another. For delineation and extraction of building footprint, certain parameters are set. These include height, slope, and area. For each parameter, the optimal threshold is obtained. The values for the threshold are entirely dependent on the area. Thus, there is no constant threshold values applicable to all test sites. Finding the parameter values can be done through trial and error. The following parameters are used to delineate building from other features:

2.3.1.2.a. Height

To separate high features from ground, the initial parameter considered is height. In an urban setting, buildings have generally higher Z information than other classes. By applying the height threshold, initial building mask can be made. The high features extracted may still classes other than buildings such as vegetation and other objects. Through trial and error, the height threshold set is 2.5 meters.

2.3.1.2.b. Slope

Slope is the change in elevation and can be expressed in percent or in grade. The slope of the building is generally lower since it is planar, as compared to vegetation features such as trees.For this area, the threshold is set to 50 percent.

2.3.1.3.c. Terrain Ruggedness Index (TRI)

Terrain ruggedness index (TRI) describes the shape or roughness of surface which are derived from land cover. TRI can be determined through different methods. It can also be done through computing the standard deviation of slope or elevation. For this particular algorithm, the mean difference between eightneighbouring cells and center pixel is computed. Urban geometries have lower ruggedness index as compared to vegetation. Thus, this can be a variable threshold between buildings and trees.

2.3.1.2.d. Area

The area parameter setting is done after aggregation of delineated building polygons. Polygons with least areas are separated. This is to delete the small ungrouped polygons. The polygon area values used to segregate buildings from other polygons may vary from the largest to smallest area depending on the sizes of features. Visual judgment is used in this particular procedure.

2.3.2.Delineation

The entire extraction process consists of masking, aggregation, and smoothing of polygons. Through spatial tools in QGIS 2.4 and GRASS 7.0,of values not within the threshold set. GRASS 7.0 is used in getting the normalized DSM and the slope. In order to extract the designated attributes as previously defined in threshold setting, functions in QGIS such as Polygonize tool and Extract by Attribute tool is used. The Terrain Ruggedness Index (TRI) can be computed from GDAL plug-in.

2.3.3. Visualization

QGIS has a website for a repository of different plug-ins. For visualization of delineated and extracted buildings in three-dimensional space, Qgis2threejs plug-in was used. Qgis2threejs was created by Minoru Akagi. It is used to export map layers and building geometries into a WebGL-capable browser. This plug-in uses Three.js library is a lightweight 3D library that handles 2D graphics and renders it into three-dimensional space.

2.3.3.1. Data Overlaying

The user can use more than one layer of image data. The plug in reads and allows both raster and vector data types. Two data overlays were used in urban visualization. A screenshot of UP Diliman map from Google Maps is georeferenced in accordance with the projection of the building shapefiles. The second layer is the vector file of the building footprint containing height information.

2.3.3.2. 3D Modeling

Using the height information in the delineated building polygons, extrusion is done using the Three2jsplugin. In addition to that, the building polygons were also classified according to their Z information for visualization purposes in two-dimensional space.

3. RESULT AND DISCUSSION

The results from the semi-automated building footprint extraction, delineation, and 3D visualization of UP Diliman using QGIS and GRASS spatial toolswere presented below.The result of the difference between Digital Terrain Model and Digital Surface Model is shown below. Normalized Digital Surface Model (nDSM) contains the height information of features. Darker areas correspond to higher elevation values.

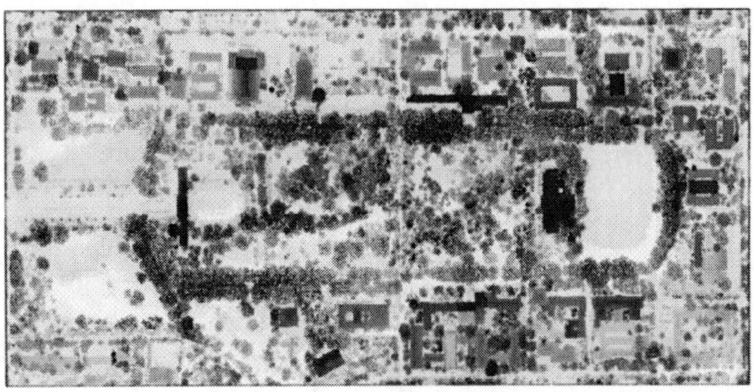

Figure 4. The image shows the normalized DSM or nDSM

Height threshold parameter is set to 2.5 meters such that high features and ground will be separated. These high features, however, still include buildings and trees.

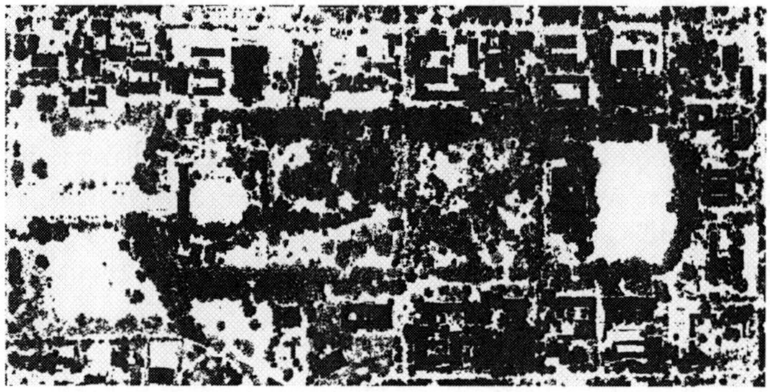

Figure 5. Image shows the result of the application of height threshold

To delineate urban geometries from vegetation, the slope and terrain ruggedness index parameters are used. Darker areas have higher values of slope and tri. As seen from the figures, the trees and edge of buildings have darker tones which indicate these have greater slope and TRI.

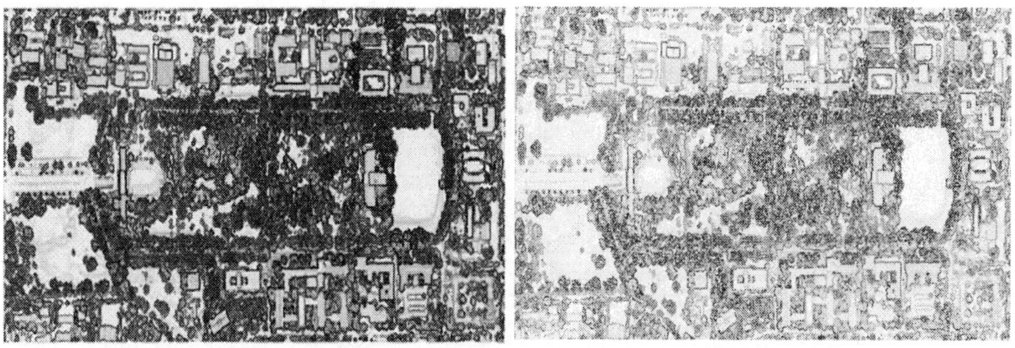

Figure 6. Left image shows the slope image and right corresponds to the TRI of the area

Using these thresholds, masking features other than urban geometries are done. Buildings are more delineated in this process. Some small polygons were eliminated through creating a threshold for area.

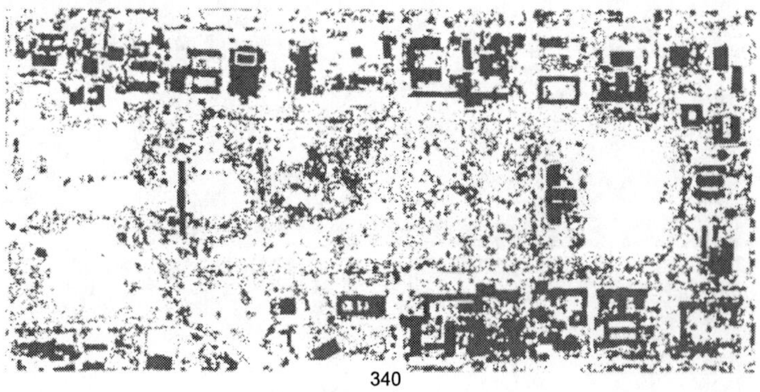

Figure 7. Result of the application of threshold parameters and masking

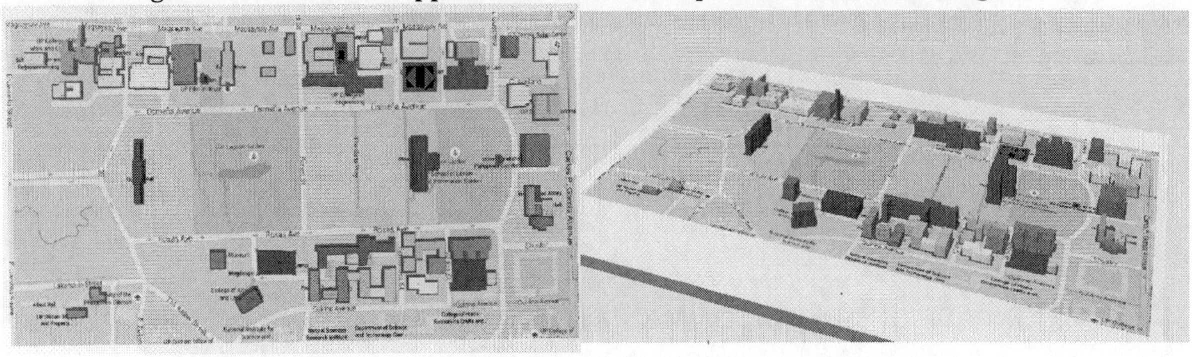

Figure 8. Visualization of building polygons using Qgis2threejs plug-in

3.2 Validation and Accuracy Assessment

In quantifying the quality of the result, comparison between the output of the semi-automated and manual detection was done.The process for validation is adopted from a quantitative evaluation of building detection as devised by Chungan Lin and RamakanNevatia.

The data from Open Street Map (OSM) were used for validation. These OSM shape files can be downloaded from their website and are readily available for public's use.

Figure 9. Open Street Map shapefiles in comparison to result of the study

One of the parameters used is detection percentage. An urban feature is considered detected even if only a fractional part of it is delineated. The corresponding formula is shown below:

$$\text{Detection Percentage} = \frac{(100\ X\ TP)}{(TP+TN)} \quad (1)$$

Detection percentage quantifies the proportion of delineated building in comparison to the actual data or the data that can be obtained through manual digitization. Where:

TP = True Positive, urban feature detected by both manual and semi-automatic method

TN = True negative, urban feature detected by manual and not by semi-automatic

Table 1. Quantitative evaluation of the extraction result

True Positive	55 polygon buildings
True Negative	19 polygon buildings
Detection Percentage	74.324%

4. CONCLUSION AND RECOMMENDATION

This study utilized the use of Free and Open Source Software in the semi-automation of building footprint delineation, extraction, and visualization in UP Diliman. Particularly, the scheme made use of GIS spatial tools embedded in QGIS and GRASS for the analysis and modelling.

All the parameters and threshold used are based from the researchers' judgment and skill. Since these values are not constant, they are entirely dependent on the area of study. These can be determined through trial and error. These variables may differ depending on the type of urban setting. For locations with fairly less trees, cleaning up of unclassified polygons would be less, considering the height threshold that was initially set. However, for urban areas with buildings almost covered with trees, extraction of building footprint is more difficult.

This semi-automated process of analysis and visualization, however, is limited to urban features with flat roof. The current algorithm does not take into account the aspect and slope of the roof. Thus, for future studies and application which requires high accuracy, computation and modeling of shape and slope of roof is recommended. Semi-automation in batch process can be further developed in FOSS through python scripting or creation of a QGIS plug-in. Model builders can also be useful for simultaneous utilization of spatial tools.

In conclusion, for considerably large areas, this method can be less time consuming and more cost efficient as compared to manual digitization of building shapefiles and using proprietary software. Free and open source GIS software like QGIS and GRASS have a significantly great potential for the analysis, manipulation, and modelling of microscale regions particularly the urban areas.

5. References

Teo, T., Rau, J., Chen, L., Liu, J., & Hsu, W. (n.d.). Reconstruction of Complex Buildings using LIDAR and 2D Maps. *Lecture Notes in Geoinformation and Cartography Innovations in 3D Geo Information Systems,* 345-354.

Jiang, N., & Lin, X. G. (n.d.). Object-Oriented Buinding Extraction ByDsm and Very High-Resolution Orthoimages. *Archives.*

Ming, D. M. D., Luo, J. L. J., Shen, Z. S. Z., & Li, J. L. J. (2005). Features based parcel unit extraction from high resolution image. *Proceedings. 2005 IEEE International Geoscience and Remote Sensing Symposium, 2005. IGARSS '05.,3*(section 3), 5–8. http://doi.org/10.1109/IGARSS.2005.1526374

Syed, S., Dare, P., & Jones, S. (2005). Semi-automatic 3D building model generation from lidar and high resolution imagery. *Proceedings of SSC Spatial Intelligence*, 1421–1428. Retrieved from http://scholar.google.com/scholar?hl=en&btnG=Search&q=intitle:SEMI-AUTOMATIC+3D+BUILDING+MODEL+GENERATION+FROM+LIDAR+AND+HIGH+RESOLUTION+IMAGERY#0

Wei, W. W. W., Chen, X. C. X., & Ma, A. M. A. (2005). Object-oriented information extraction and application in high-resolution remote sensing image. *Proceedings. 2005 IEEE International Geoscience and Remote Sensing Symposium, 2005. IGARSS '05.,6*(C), 5–8. http://doi.org/10.1109/IGARSS.2005.1525737

Varshney, A., Rajesh, E., Sarabandi, P., & Solutions, R. M. (1962). Advanced region – based pattern recognition algorithms for building footprint extraction- an automatic approach to generate building inventory.

GIS GPS Geodesy Bookmark: Free Open Street Map data (i.e.shapefiles etc.) for the Philippines. (n.d.). Retrieved December 15, 2014, from http://gisgpsgeodesy.blogspot.com/2011/04/free-road-data-shapefiles-for.html

GIS 4 Geomorphology. (n.d.). Retrieved August 15, 2015, from http://gis4geo morphology.com/roughness-topographic-position/

Shridhar J., Panditrao S., and Luis A., "Airborne LiDAR and High Resolution Satellite Data for Rapid 3D Feature Extraction." National Remote Sensing Centre Indian Space Research Organisation. Web. 4 May 2015.

Shen, W. (2008) Building Boundary Extraction Based on LiDAR Point Clouds Data. The /International Archives of the Photogrammetry, Remote Sensing and Spatial Information Sciences. Vol. XXXVII. Part B3b. Beijing

"Automated Building Extraction and Reconstruction from LIDAR Data." (2000) Geographic Resources Center. Web. 14 May 2015.

Dy, E. (2013) Automatic Generation of 3D Buildings and Trees from DSM and DEM Data using GIS. Print.

Vosselman G., and Dijkman, S., (2001) 3D Building Model Reconstruction from Point Clouds and Ground Plans. IAPRS, Vol. XXXIV, Part 3/W4, pp.37-43

Demir, N., and Baltsavias E., (2012) Automated Modeling of 3D Building Roofs Using Image and LiDAR Data. ISPRS Annals of Photogrammetry, Remote Sensing and Spatial Information Sciences.

Shiravi S., Zhong M., Beykaei S., (2012) Accuracy Assessment of Building Extraction using LiDAR Data for Urban Planning/Transportartion Applications. 2012 Conference of the Transportation Association of Canada Fredericton, New Brunswick.

Coirier, W. J., & Kim, S. (2006, February). CFD modeling for urban area contaminant transport and dispersion: Model description and data requirements. In *Sixth Symposium of the Urban Environment, Atlanta, GA.*

Kalmikov, A., Dupont, G., Dykes, K., & Chan, C. (2010, May). Wind power resource assessment in complex urban environments: MIT campus case-study using CFD Analysis. In *AWEA 2010 WINDPOWER Conference. Dallas, USA.*

Perén, J. I., van Hooff, T., Leite, B. C. C., &Blocken, B. (2015). CFD analysis of cross-ventilation of a generic isolated building with asymmetric opening positions: impact of roof angle and opening location. *Building and Environment,85*, 263-276.

Free and Open Source Software for Geospatial (FOSS4G) Conference Proceedings

Volume 15 *Seoul, South Korea* Article 20

2015

Geo-Based Image Analysis Service In Open Source Cloud Computing Environment

Sanggoo Kang
Department of Information Systems Engineering, Hansung University

Guseon Yoon
Department of Information Systems Engineering, Hansung University

Kiwon Lee
Department of Information Systems Engineering, Hansung University

Recommended Citation

Kang, Sanggoo; Yoon, Guseon; and Lee, Kiwon (2015) "Geo-Based Image Analysis Service In Open Source Cloud Computing Environment," *Free and Open Source Software for Geospatial (FOSS4G) Conference Proceedings*: Vol. 15 , Article 20.
DOI: https://doi.org/10.7275/R5TX3CKM
Available at: https://scholarworks.umass.edu/foss4g/vol15/iss1/20

PO-02

GEO-BASED IMAGE ANALYSIS SERVICE IN OPEN SOURCE CLOUD COMPUTING ENVIRONMENT

Sanggoo Kang[1], Guseon Yoon[1], and Kiwon Lee[2]

[1]Graduate Student, Department of Information Systems Engineering, Hansung University, Samseon-dong 2-ga, Seongbuk-gu, Seoul, Korea, 136-792
Email: lainer23@naver.com,gxexe3@naver.com
[2]Professor, Department of Information Systems Engineering, Hansung University, Samseon-dong 2-ga, Seongbuk-gu, Seoul, Korea, 136-792
Email: kilee@hansung.ac.kr

ABSTRACT

Globally, cloud computing is one of huge trends in ICT communities, exerting its influence on the most information application fields including geo-spatial domain. In general, cloud computing services are being regarded as commercial or public internet data center or infrastructure of computing resources. However, there is no limitation for applications of cloud computing. Especially, it is on the early developing stage in geo-based applications. This project is to present a practical application for geo-based image processing and analysis on open source cloud environment. OpenStack Juno version was applied for open source cloud computing environment. On this cloud environment, PostgreSQL and Django were used as open source for metadata server and web framework, respectively. For geo-based image processing server, OTB was used with GDAL for image manipulating. Some image processing algorithms were implemented in this cloud environment, and provides as a web service for public users. Web-accessible users in this cloud service do not need any software installation and downloading data sets. This full open source approach is expected to be an element geo-spatial information processing linked with cloud computing service.

1. INTRODUCTION

In globally scaled information communication technology (ICT) application, cloud computing leads new paradigm for computing resources uses. The most acceptable one among numerous literal definitions with respect to cloud computing is pay-as-you-go web-accessible computing environment by on-demand typed processing including server hardware, network infrastructure, or software service. Since the early 2010s, cloud computing is widely applied in many application domains, and practical cloud application cases have been reported.In general, cloud-based system shows many advantageous points of economic cost reduction, stability, and scalability (Chou, 2015).

Despite of these huge trend, development of geo-based application in cloud environment is rather slow, it is an early stage. Evangelidis*et al.* (2014) expected that linkage of cloud computing and standards of Open Geo-spatial Consortium, Inc.(OGC) would expand. Thus it is necessary that experiments for linkage of geo-spatial applications and cloud computing technologies and cloud-based application cases.

The purpose of this work is to demonstrate a cloud-based geo-spatial system which was newly designed and implemented for satellite image processing by open sources covering geo-processing engine and even cloud environments. This system by open source stack and integration supports cross-browser functionalities and does not need any installation processes by users. Computing resources such as data server, web server, geo-data and geo-based image processing engine or client user interface for web mapping are included into cloud environment, so that end-users with accessing account, basically open accessing, just use web browser. But internal processing in the cloud environment is hidden on the stable mode and quite differ from the non-cloud system.

As the previous works, Lee and Kang (2013) implemented a real-time processing system of satellite image processing for tablet devices by using Amazon elastic compute cloud (EC2) in Amazon Web Service (AWS). Kang and Lee (2014) performed a test case for geo-data fusion application in AWS and OpenStack cloud, one of open source cloud platform. Lee *et al.* (2015) developed a prototype for geo-based image analysis system on OpenStack cloud. This work used more advanced development environments and user interface components, and proposes a new system architecture, different from those previous works.

OpenStack, open source cloud computing platform for public and private clouds, was used for building cloud computing environment, so that generation of instance servers with various performance capability was possible. It means fast adoption and customization of cloud environment according to application purposes.As for OpenStack, it reduce the risks of lock-in associated with proprietary platforms, and offers target-based flexibility and numerous communities and developers choice (OpenStack, 2015). The OpenStack technologies have been built by interrelated projects which control processing, storage, and networking computing resources, capable of being handled by a web-based dashboard or command-line tools.

In many areas of information system business, use cases of OpenStack are increasing every year. But in the geo-spatial field, OpenStack application is a somewhat beginning stage. In spite of current status, integration or linkage of open source cloud computing and geo-based application is a prospective approach. Cloud computing environment helps management and manipulation of large volume of various typed geo data sets and geo-based images including satellite image data.

2. SYSTEM IMPLEMENTATION

The work was fully implemented by open source basis from operating system (OS) to application service level in Table 1. The operating environment is divided intoserverand client. The server was built with OpenStack cloud computing environment and used the operating system, Ubuntu 12.04. Django was used for web framework of Python-based implementation, and it improves security and reliability of fast implemented products. PostgreSQL/PostGIS, open source database management system, was used for metadata cataloging and query process of geo-based image sets. As for image processing engine, Orfeo ToolBox (OTB) providing a variety of open source satellite image processing algorithms was used. In the client side, Bootstrap was used for user interface framework, taking advantage of optimization for multi-screen devices and consistency for displaying contents. As well, JavaScript libraries such as iCheck and selectric based on jQuery were applied for additional user interfacing components. While, OpenLayerswas used for web mapping library, to on-line visualize base map, index image and the processed images.

Figure 1 shows an architecture of the implemented system, with work flow and system modules. Instance server is divided into 4 operating environments. Processing system and user interface system are basically operated on Apache web server and Django. There are many cloud-based user requesting and responding modules. In the geo-based image management module, an important preprocessing for metadata query and display regarding index image archived in the storage units is performed before the image processing or analysis request management module. For actual image processing, OTB and GDAL are used.

Among lots of image processing algorithms available from OTB, Image fusion and thresholding scheme are provided for the cloud-based and real-time processing service. While, attached volume environment means the block storage constructed from the cloud-based processing system, and it works storing and managing space of data sets mounted in the instance server. Web client environment is also applied with many open source libraries.

Table 1. Open source list applied in this work

	Development Environments	Name / Version
Instance Server-side	Operating System	Ubuntu Server 64bit / 12.04
	Web Framework	Django / 1.8.2
	DBMS and Extension	PostgreSQL / 9.3 PostGIS / 2.1
	Geo-based Image Processing	Orfeo ToolBox / 4.4.0
	Geo-based Image Manipulation	GDAL / 1.11.1
	Geo-based Coordinates Manipulation	Proj / 4.8.0
Client-side	HTML, CSS, and JS Framework	HTML / 5, CSS / 3 Bootstrap / 3.3.4 jQuery / 2.1.3
	Web Mapping Library	OpenLayers / 3.4
	User Interface Libraries	iCheck / 1.0.2 selectric / 1.9.3 noUiSlider / 7.0.10

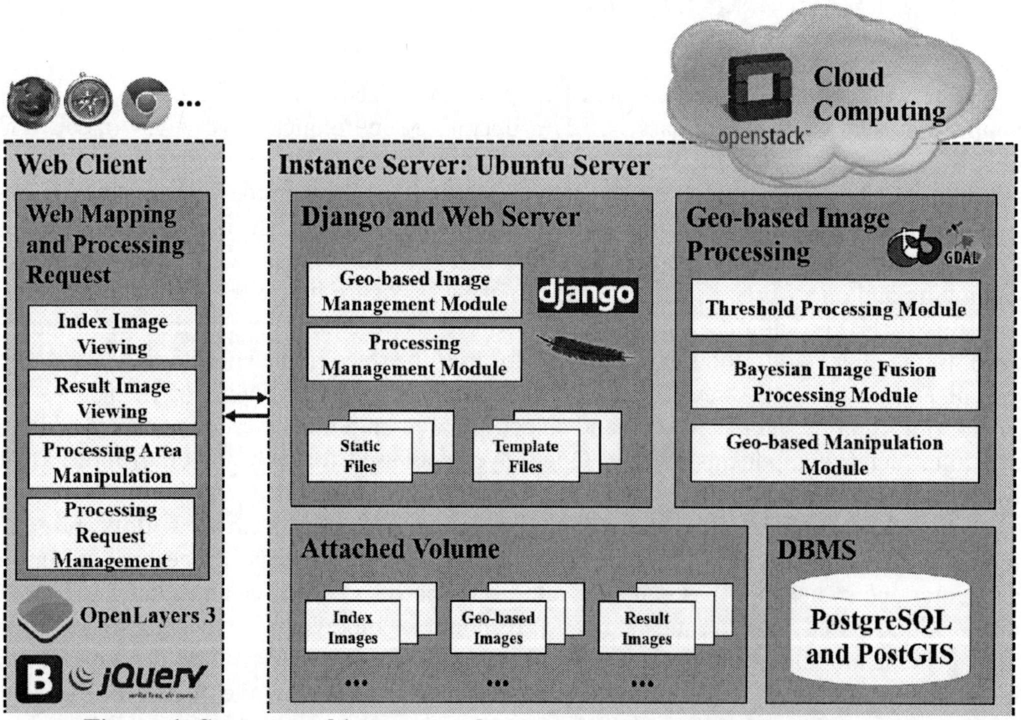

Figure 1. System architecture and components proposed in this work.

3. IMPLEMENTATION RESULTS

Figure 2 represents user interface for end-user, accessing cloud-based web system. In one page, generic image processing can be carried out by web users: identification index image regarding geo-based image which can be actually processed in the OTB, selection of processing algorithm and concerned parameters choice, and visualization of the final processed result. Figure 2(a) is the menu system for on-line base mapping. Open street map (OSM) is the default, and V-world, open platform for geo-spatial data in Korea, is also provided. Index image is the reduced image by approximately 20% re-sampling compared to size of the original image data. It is similar to thumbnail image, a term in the digital image processing. Figure 2(b) and (c) are the processing algorithm selection menu and parameter choice menu system from user inputs, respectively. Figure 2(d) is a simple click event button for requesting the cloud-based server processing after several user inputs.

In this works, KOrea Multi-purpose SATellite image (KOMPSAT 3) sets were used. Figure 3 shows binary image generation process by thresholding image processing scheme, and Figure 4 isthe produced result by Bayesian image fusion algorithm.

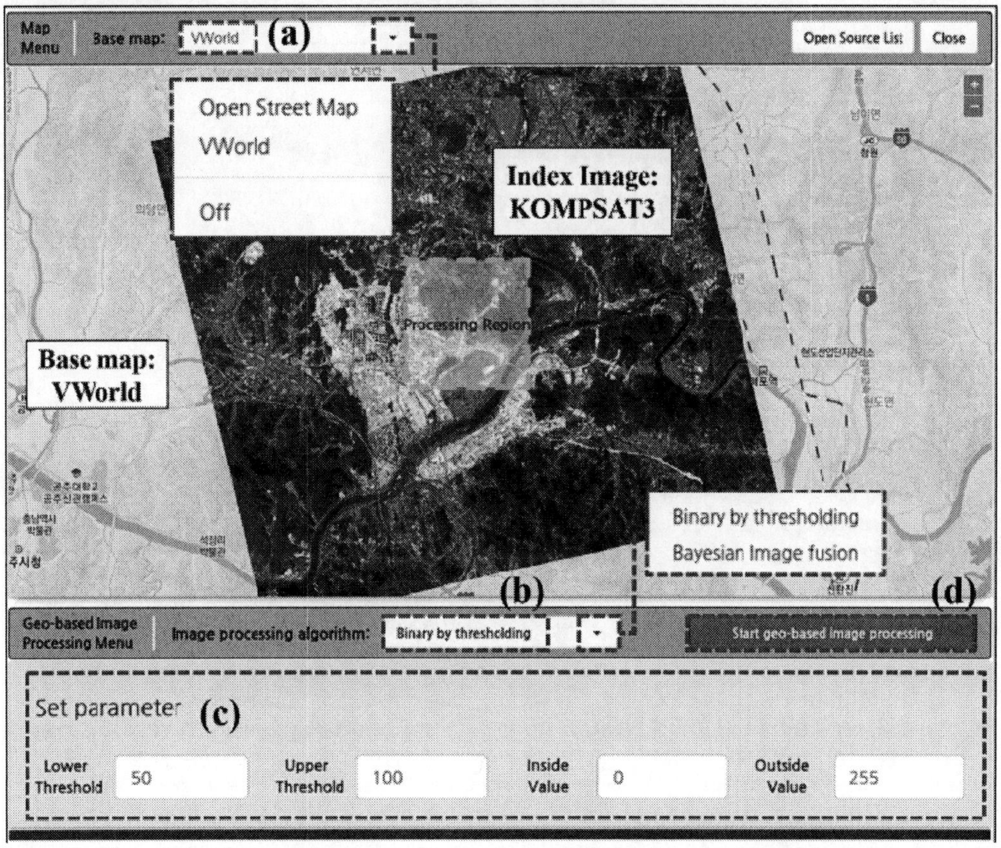

Figure 2. User interface and processing components in the geo-based image processing system on OpenStack cloud computing environment.

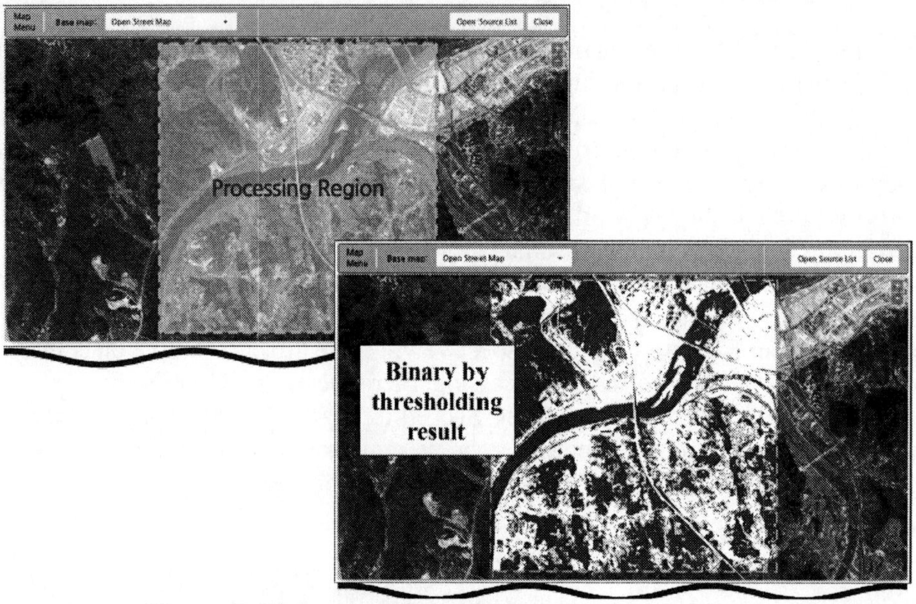

Figure 3. Binary image generation by thresholding.

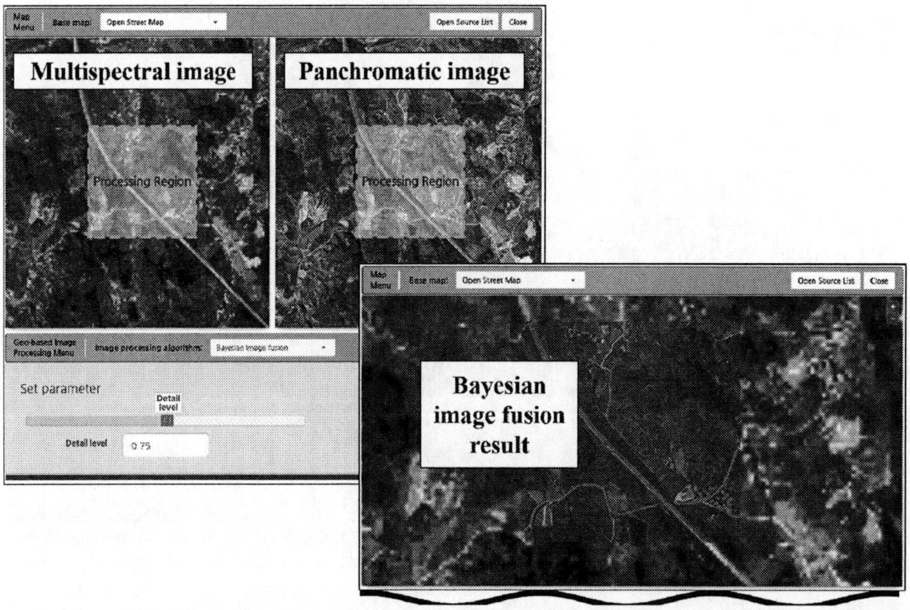

Figure 4. User interface for Bayesian image fusion and the processed result example.

4. CONCLUDING REMARK

Cloud computing and its services are still on the early maturing stage in the geo-based data manipulation and processing application. There are many ways and platforms to build a cloud-based application. This study employed OpenStack cloud computing platform. This study based on full open source presents a test case with geo-based image processing functions. Bayesian image fusion and image thresholding as geo-based image processing algorithm are based on the open source algorithms of OTB. For the further study, performance test with multiple source data sets needs for practical uses of mobile cloud in geo-science application. Also the on-the-fly image matching or image registration on mobile devices for heterogeneous image sets from different sensor are also necessary.

5. ACKNOWLEDGMENTS

This research was partly supported by National Land Space Information Research Program by Ministry of Land, Infrastructure and Transport of Korean government (No. 14NSIP-B080144-01).

6. Reference

Chou, D. C., 2015. Cloud computing: A value creation model, *Computer Standards & Interfaces* 38, 72-77.

Evangelidis, K., K. Ntouros, S. Makridis, and C. Papatheodorou, 2014. Geospatial services in theCloud, *Computers & Geosciences*63, 116-122.

Kang, S. and K. Lee, 2014. A Performance Test of Mobile Cloud Service for Bayesian Image Fusion, *Korean Journal of Remote Sensing* 30, 445-454.

Lee, K. and S. Kang, 2013. Mobile cloud service of geo-based image processing functions: a test iPad implementation, *Remote Sensing Letters* 4, 910-919.

Lee, K., S. Kang, T.Chae, and Y. Kim, 2015. Satellite Image Processing Application System onOpenStack Cloud Computing Environment, *2015 ICEO&SI and ICLEI Resilience Forum*, Kaohsiung.

OpenStack, 2015.*OpenStack User Stories*, http://www.openstack.org/user-stories (Accessed July 22, 2015).

Free and Open Source Software for Geospatial (FOSS4G) Conference Proceedings

Volume 15 *Seoul, South Korea* Article 21

2015

WebGIS-based Natural Disaster Information System of China

Jia Du
Institute of Geographic Sciences and Natural Resources Research, CAS, China

Yaping Yang
Institute of Geographic Sciences and Natural Resources Research, CAS, China

Xiafang Yue
Institute of Geographic Sciences and Natural Resources Research, CAS, China

Wenlong Jing
Institute of Geographic Sciences and Natural Resources Research, CAS, China

PO-03

WebGIS-based Natural Disaster Information System of China

Jia Du*, Yaping Yang, Xiafang Yue, Wenlong Jing

Institute of Geographic Sciences and Natural Resources Research, CAS, China

Corresponding Author : Jia Du (jossmail@126.com)

Abstract

We have constructed China natural disaster information system utilizing open source WebGIS framework. Through comparisons and researches on different WebGIS system structures, implementation patterns, spatial database management patterns and several popular software, open source WebGIS platform MapServer, PostgreSQL+PostGIS (spatial database and spatial data engine), client-side browser framework openLayer were used. This system have basic function of web information publishing, spatial data storage and management, and it collects natural disaster information that have been publishing on the internet through grabing internet information, and based on the analysis to the text information, different kinds of natural disaster information databases were constructed. Additionally, locations of the disaster occurred were combined to provide map services, and the system is able to meet multi-user access issues, and have functions of thematic mapping, query and statistics, charting and output, et.al.

Keywords : Natural Disaster;WebGIS;Visualization;Geography e-science

Free and Open Source Software for Geospatial (FOSS4G) Conference Proceedings

Volume 15 *Seoul, South Korea* Article 22

2015

GeoDjango-Framwork-based Population and Social Economic Information System For China

Wenlong Jing
Institute of Geographic Sciences and Natural Resources Research, CAS, China

Xiaodan Zhao
Institute of Geographic Sciences and Natural Resources Research, CAS, China

Yaping Yang
Institute of Geographic Sciences and Natural Resources Research, CAS, China

Jia Du
Institute of Geographic Sciences and Natural Resources Research, CAS, China

Xiafang Yue
Institute of Geographic Sciences and Natural Resources Research, CAS, China

Recommended Citation

Jing, Wenlong; Zhao, Xiaodan; Yang, Yaping; Du, Jia; and Yue, Xiafang (2015) "GeoDjango-Framwork-based Population and Social Economic Information System For China," *Free and Open Source Software for Geospatial (FOSS4G) Conference Proceedings*: Vol. 15 , Article 22.
DOI: https://doi.org/10.7275/R53F4MTP
Available at: https://scholarworks.umass.edu/foss4g/vol15/iss1/22

PO-04

GeoDjango-Framwork-based Population and Social Economic Information System For China

Wenlong Jing*, **Xiaodan Zhao**, Yaping Yang, Jia Du, Xiafang Yue

Institute of Geographic Sciences and Natural Resources Research, CAS, China

Corresponding Author : Wenlong Jing (jingwl@lreis.ac.cn)

Abstract

We have constructed China population and social economic information system based on open source geographic Web framework GeoDjango. By comparing different WebGIS system structures, implementation patterns, spatial database management patterns and mainstream popular open source APIs, open source WebGIS framework GeoDjango, MVT pattern, Python and MySQL database, client-side browser framework openLayer were implemented. The database system utilized a Top-Down design method — DATAID-D to establish distributed database, and construct a uniform encoding standard, through normalized processes of the statistic yearbook, we constructed population and social economic database. This system has basic functions of web information publishing, spatial data storage, distributed management and spatial query. Additionally, the system is able to meet multi-user access issues, and have functions of population and economy data visualization, thematic mapping, query and statistics, charting and output, et.al.

Keywords : Population and social economy;GeoDjango;WebGIS

Free and Open Source Software for Geospatial (FOSS4G) Conference Proceedings

Volume 15 *Seoul, South Korea* Article 23

2015

Research on Voyage Navigation S/W Development Architecture Using Open Source Base GIS Modules

Lee Junho

Kim Jiyoon

PO-06

Research on Voyage Navigation S/W Development Architecture Using Open Source Base GIS Modules

Mr.Lee junho , Dr. Kim Jiyoon

Abstract

Recently, as the number of voyage accident is increasing, massive damage of human life and property is following depends on scale of accident. ECDIS is navigation equipment essential for safe sailing and clean marine environment. However, ECDIS S/W is relatively expensive, and ocean-related international organizations such as IMO and IHO are designing the programs mainly for medium and large ships. OPEN CPN, SeaClear are existing as product based on Open Source Project, however they are lack of functions to use in actual voyage. ECDIS development by making combination with modules based on open source could be an alternative to follow the standards and develop efficiently

KeyWord : Open Source; Marine GIS; eNavigation;ECDIS; ENC

1. Introduction

Recently, as the number of voyage accident is increasing, massive damage of human life and property is following depends on scale of accident. Early, IALA suggest to install ECDIS (Eletronic Chart Display Information System) for safe voyage, and appoint time of installation of ECDIS depends on scale of ship. An **Electronic Chart Display & Information System** (ECDIS) is a computer-based navigation information system that complies with International Maritime Organization (IMO) regulations and can be used as an alternative to paper nautical charts. IMO refers to similar systems not meeting the regulations as Electronic Chart Systems (ECS)[1].

ECDIS is a system which shows geographical features of sea space information for voyage, weather on screen. ECDIS is basically loaded ENC format electronic navigation chart, and has a function of interlocking information from all sorts of sensor and telecommunications equipment to electronic navigation chart.

Nowadays, ECDIS is mainly developed by companies of voyage telecommunication equipment such as PRUNO.JRC, TRANSAS, and most of them are for common use. Although OPEN CPN, SeaClear are existing as product based on OPENSOURCE, however they are lack of functions to use in actual voyage. Also, existing voyage S/W based on open source doesn't guarantee interlock with various voyage communication equipment, which is used in field, such as AIS, GPS, RADAR ,ECHO Sounder. Also, they must follow standards of voyage communication equipment such as IEC 6174 and should be installed necessary function in S/W for this to use open source base voyage S/W. Because of these reasons, there are difficulties on developing ECDIS in one unit of open source project.

Therefore, ECDIS S/W, based on open source, needs combination with open source project, not a unit of project based on open source. ECDIS development by making combination with modules based on open source could be an alternative to follow the standards and develop efficiently. This research will suggest the way to develop

ECDIS through introduction of Open Source and S/W structure to achieve the goal

2. Discussion

1) Analysis of ECDIS standards and ECDIS-related open source modules

There are basic S/W and H/W standards required for building ECDIS (Electronic Chart Display Information System). Only when these standards are applied and tested, the system is usable for actual voyages. One of the major characteristics of ECDIS standards is that they are modularized into the functional units of ECDIS.

The most representative ECDIS standard is IEC 61174, The performance requirements for ECDIS are defined by IMO and the consequent test standards have been developed by the International Electrotechnical Commission (IEC) in International Standard IEC 61174.[2]

individual functions of ECDIS are tested and verified through the IEC 61174 standard. Figure 1 defines ECDIS-related standards and roles played in the functional parts of ECDIS.

ECDIS S/W and H/W and related modules can be mounted on a ship as radio communication and navigation equipment only when they are designed and developed to satisfy such modularized standards and finally to meet the IEC61174 standard. Because ECIDS functions accommodate various IT technologies including GIS and network, ECDIS producers have so far developed their products using components of ECDIS functions, and such components have been called ECDIS kernels. Some of them were commercialized and provided to ECDIS developers (SEVENCS, CMAP, etc.), and sometimes ECDIS producers developed and used their own ECDIS kernels. The basic functions of ECDIS kernels are defined as follows.

ECDIS Standard	Role	Current Version
IEC 61174	Operational and performance requirements, methods of testing and required test results	edition 3.0
ISO/IEC 61162	Maritime navigation and radiocommunication equipment and systems – Digital interfaces –	ISO/IEC 61162-1,2,3
S-57	IHO Transfer Standard for Digital Hydrographic Data,	IHO S-57 Edition 3.0
S-52	Specifications for Chart Content and Display Aspects of ECDIS	IHO S-52 Edition 4.0
S-63	standard for International Hydrographic Organization (IHO) standard for encrypting, securing and compressing electronic navigational chart (ENC) data.	IHO S-63 Edition 1.1
S-58	Recommended ENC Validation Checks	IHO S-58 Edition 5.0
S-66	Facts about Electronic Charts and Carriage Requirements	IHO S-63 Edition 1.1

Table 1. ECDIS-related standards and definitions

Electronic chart database visualization and management - Sea chart security processing and update - Sea chart visualization (apply the S-52 visualization standard) - Sea chart object information service - Sea chart screen manipulation (zoom in/out, change)	Navigation planning and waterway surveillance - Navigation plan plotting - Navigation plan information management and validity check - Navigation monitoring and information service
Interoperation with navigation equipment - Interoperation with radio communication and navigation equipment such as AIS, DGPS, and Echo Sounder through communication - Radar image presentation and interoperation - Interoperation on the electronic chart screen	Navigation recording and other functions - Navigation record storage and display - Seafaring victim information management - Other information services including weather information and interoperation

Figure 1. Classification of ECDIS kernel functions

Figure 1 classified and restructured the functions defined in IEC 61174. If a system is equipped with all of these functions, it is regarded to have all of the basic functions required of ECDIS, and if the functions meet the ECDIS system test standards of IEC 61174, the system is usable in actual voyages.

In order to satisfy these conditions, most of ECDIS producers or engineers develop ECDIS using ECDIS kernel module components compatible with the IEC 61174 standard. As mentioned above, however, most of ECDIS kernels are commercial products to be purchased. In addition, because they support only limited development platforms and languages, they impose restrictions on the composition of manpower and the setting of development environment. Table 2 below shows the results of surveying the development environment of commercial ECDIS kernels.

ECDIS kernel	Development platform	Operating environment
SEVEN CS	QT, MS(C++,C#).opengl	Standalone
Geomaris	Microsoft .net, direct X	Standalone
CMap	Microsoft .net, direct X	Standalone
Cher Soft	Microsoft.net	Standalone

Table2. The development environment of major commercial kernels

The development environments and platforms listed above have the problem that they are far different from the latest IT development environments such as Cloud and smartphone. In order to solve this problem, it is desirable to adopt development using open sources, but it is difficult for open sources related to ECDIS and marine GIS to reflect recent IT trends while providing the IEC 61174 functions. Moreover, there are representative open sources related to marine systems such as OPEN CPN[4] and SeaClear[5], but commercial products have the problem mentioned above and they support only part of the functions. In addition, they have been found to be poor in supporting the standards.

Table 3. The support of open sources to navigation-related S/W ECDIS standards

Open source	Running environment	S-57	S-63	S-52	ISO/IEC61162	IEC61174
OPENCPN	Standalone	Support	Not support	Support	Partially support	Partially applied
SeaClear	Standalone	Support	Not support	Support	Partially support	Partially applied
OpenSeaMap	Web	Not support	Not support	Not support	Not support	Not support

Because, like ECDIS kernels of commercial environment, current projects based on navigation-related open sources use running environment far different from the latest IT operating environment, it is difficult to develop ECIDS using current open source projects.

2) Development of ECDIS S/W using open source S/W and data standards

Open source-based projects related to marine navigation are poor in supporting ECDIS standards, and in interoperability with radio communication and navigation equipment. Because there is not an established community of developers, it is not easy to patch up and upgrade.

Furthermore, time and human resources need to be considered in implementing a large number and variety of ECDIS functions within a single project. In order to develop ECDIS as an open source, therefore, the development architecture should be in the form of mesh-up integrating multiple open source projects, and the structure of database used in ECDIS must guarantee interchangeability to overcome problems raised by differences among development languages and development platforms.

Service data used most often in ECDIS are largely electronic charts, navigation records, and sensor data generated from radio communication and navigation equipment. For electronic charts, the database embedded in ECDIS is used instead of the S-57-based original copies of sea charts for reduced system load and high performance.

The embedded database, which is called System ENC, has a structure unique to each ECDIS system. Because of the uniqueness of the structure, most of ECDIS systems cannot interchange and share data with one another.

In order to solve this problem, OGC and general-purpose data standards may substitute for the use of database embedded in ECDIS. The application of data standards makes it possible to integrate many open source projects.

OGC(R) standards are technical documents that detail interfaces or encodings. Software developers use these documents to build open interfaces and encodings into their products and services.[6]

Existing data	Alternative open data structure	Related open source projects
System ENC	-GML, -GEOJSON - ST_GEOMETRY - XML	- GeoAPI - GeoServer - DotSpatial - MapServer - FDO
Navigation records	GML XML GEOJSON	- GeoAPI - Map Window - FDO
Sensor information (AIS, GPS, etc)	Sensor ML	- Apache - GeoAPI

Table 4. Application of open data structure to ECDIS data

By applying standards such as OGC and W3C as data structure used in ECDIS, we can diversify ECDIS development platforms, and this enables interoperation with various data including weather information. Moreover, SLD, one of the map style methods of Geoserver, an open source Java server, is utilized in order to observe S-52 electronic chart standards, which is most essential in geographic information services of ECDIS. The CSP algorithm should be used to implement the S-52 standards. Constraint satisfaction problems (CSPs) are mathematical problems defined as a set of objects whose state must satisfy a number of constraints or limitations. CSPs represent the entities in a problem as a homogeneous collection of finite constraints over variables, which is solved by constraint satisfaction methods.[7] This is the core algorithm for presenting electronic charts in ECDIS using different object expression methods depending on the condition of data. The method of utilizing SLD of Geoserver defines the display method according to the data value of the corresponding object, so it is possible to observe standards in development without directly implementing the CSP algorithm.

Besides, communication among H/W units, one of major functions of ECDIS, becomes possible through the application of open sources.

Communication in ECDIS usually adopts RS-232 and RS-422, but with the advance of communication technologies, Internet-based communication is being studied for ships. At present, terminal communication is being used in various open source S/W programs including IFTOOLS(), and it is also used often in actual development.

3. Conclusion

ECDIS is navigation equipment essential for safe sailing and clean marine environment. However, ECDIS S/W is relatively expensive, and ocean-related international organizations such as IMO and IHO are designing the programs mainly for medium and large ships. Because ECDIS S/W is essential navigation equipment for preventing marine accidents, it should also be available to economically stricken ship owners or small vessels.

Therefore, the development of ECDIS based on open sources or free ware is demanded socially. There are a number of applicable open source S/W programs, but they do not contain all the functions required for sailing. The range of ECDIS functions is too broad to be covered by a single open source project. Accordingly, we need to develop necessary functions through integrating open source projects and utilizing standardized data structures.

This study opened the possibility for developing efficiently the extensive functions of ECDIS S/W by suggesting a development method utilizing open source S/W and OGC standards. Based on this method, future research will develop and test ECDIS compatible with the IEC61174 standard.

Reference

[1] Introduction to Electronic Chart Systems and ECDIS. International Hydrographic Organization. Retrieved 2010-07-13

[2] Maritime navigation and radiocommunication equipment and systems - Electronic chart display and information system (ECDIS) - Operational and performance requirements, methods of testing and required test results". International Electrotechnical Commission. Retrieved 2007-09-01.

[3] IEC 61162. https://en.wikipedia.org/wiki/IEC_61162#Sections_of_IEC_61162

[4] OPENCPN . http://opencpn.org/

[5] SeaClear. http://www.sping.com/seaclear/

[6] OGC Standard. http://www.opengeospatial.org/standards

[7] Constraint satisfaction problem. https://en.wikipedia.org/wiki/Constraint_satisfaction_problem

Free and Open Source Software for Geospatial (FOSS4G) Conference Proceedings

Volume 15 *Seoul, South Korea* Article 24

2015

Experiments on Long-term Geodetic Monitoring by Low-cost GNSS Receivers and GoGPS Positioning Engine

Daisuke Yoshida
Osaka City University, Japan

Eugenio Realini
Geomatics Research & Development (GReD) srl, Italy

Stefano Caldera
Geomatics Research & Development (GReD) srl, Italy

Masayuki Uchida
NIKON-TRIMBLE CO., LTD., Japan

Recommended Citation

Yoshida, Daisuke; Realini, Eugenio; Caldera, Stefano; and Uchida, Masayuki (2015) "Experiments on Long-term Geodetic Monitoring by Low-cost GNSS Receivers and GoGPS Positioning Engine," *Free and Open Source Software for Geospatial (FOSS4G) Conference Proceedings*: Vol. 15 , Article 24.
DOI: https://doi.org/10.7275/R5BV7DS6
Available at: https://scholarworks.umass.edu/foss4g/vol15/iss1/24

PO-07

Experiments on Long-term Geodetic Monitoring by Low-cost GNSS Receivers and GoGPS Positioning Engine

Daisuke Yoshida[1], Eugenio Realini[2*], Stefano Caldera[2], Masayuki Uchida[3]

[1]*Osaka City University, Japan*
[2]*Geomatics Research & Development (GReD) srl, Italy*
[3]*NIKON-TRIMBLE CO., LTD., Japan*

Corresponding Author : Eugenio Realini (eugenio.realini@g-red.eu)

Abstract

This paper describes experiments of GNSS-based monitoring by means of different models and brands of low-cost receivers. Single-frequency observations were processed by Open Source goGPS Positioning Engine, in relative positioning mode with respect to a virtual reference station and three physical permanent stations. The experiments were carried out in Osaka, Japan and the results here described cover a timespan of eight months. The daily solutions show standard deviations ranging from about 1mm to 1cm, mainly depending on the base station observations quality.

Academic Discipline and Sub-Disciplines : Geodesy

Keywords : GNSS; low-cost receivers; monitoring; goGPS; Open Source

Agencies : JENOBA Co., Ltd; Grants-in-Aid for Scientific Research of JSPS

Free and Open Source Software for Geospatial (FOSS4G) Conference Proceedings

Volume 15 *Seoul, South Korea* Article 25

2015

Development of an Agent Based Urban Simulation Model

Donghan Kim
Division of National Territorial Planning and Research Korea Research Institute for Human Settlements

Recommended Citation
Kim, Donghan (2015) "Development of an Agent Based Urban Simulation Model," *Free and Open Source Software for Geospatial (FOSS4G) Conference Proceedings*: Vol. 15 , Article 25.
DOI: https://doi.org/10.7275/R5GM85HX
Available at: https://scholarworks.umass.edu/foss4g/vol15/iss1/25

PO-08

Development of an Agent Based Urban Simulation Model

Donghan Kim*

Korea Research Institute for Human Settlements, Korea

Corresponding Author : Donghan Kim (dhkim@krihs.re.kr)

Abstract

Since the pioneering work of Lowry in the 1960s, many urban models has developed to understand how urban systems work and change. Early urban models tend to focus on the relationship and interaction between land use and transportation. Yet, recent urban models pay more attention to urban morphology with an interest on self-organising nature of urban systems.

Agent based urban modelling approach is gaining popularity as a new means to study an urban systems in this context. The methodology has roots in the natural science field such as the complexity science, but it also provides meaningful insights for social systems as well as urban systems. The key strength is that it offers a way to understand urban systems through the interaction among individual members in the systems such as households and developers.

This research aims to develop an agent based urban model as a new scientific tool to understand changes in urban systems and to support planning policy making. It starts with reviewing history and issue in urban modeling and then move on the trend of urban development in general and in Korea. The research then defines the key behaviour and algorithms of the model before developing it as a functional computer program. We developed the model by using an open source programming platform, Repast Simphony. Then the model was applied to a case study area of Busan Metropolitan Area(Busan, Gimhae, Yangsan) to see if it properly works and generates valid simulation outcomes. We has confirmed that the model offers new and meaningful ways to understand urban future and to support planning policy.

Yet, the model developed in this study is subject to further development. It is necessary to analyse more empirical data on agents' type and behaviour. At the same time, it is desirable to improve technical interfaces to make is as a planning support system. These require continued research and development.

Academic Discipline and Sub-Disciplines : Urban model

Keywords : Agent based model; Urban growth; Urban model

Free and Open Source Software for Geospatial (FOSS4G) Conference Proceedings

Volume 15 *Seoul, South Korea* Article 26

2015

3D Visualization of City GML Building Data on World Wind

Vishal Tiwari
Lab for Spatial Informatics, IIIT-Hyderabad, India

KS Rajan
International Institute of Information Technology Hyderabad Gachibowli, Hyderabad, India

Recommended Citation
Tiwari, Vishal and Rajan, KS (2015) "3D Visualization of City GML Building Data on World Wind," *Free and Open Source Software for Geospatial (FOSS4G) Conference Proceedings*: Vol. 15 , Article 26.
DOI: https://doi.org/10.7275/R5MC8X78
Available at: https://scholarworks.umass.edu/foss4g/vol15/iss1/26

PO-10

3D Visualization of City GML Building Data on World Wind

Vishal Tiwari*, K.S. Rajan

Lab for Spatial Informatics, IIIT-Hyderabad, India

Corresponding Author : Vishal Tiwari (vishal.tiwari@research.iiit.ac.in)

Abstract

Large area cityGML visualization currently deals with its conversion to another model such as VRML, X3D or KML data format which can be visualized using various visualization tools like Google Earth, VTP, Cesium, NASAs World Wind, etc. But these conversion to VRML or X3D, leads to a loss of the geographical information, geometries, which restricts the aspect of geospatial analysis, in addition to adding an extra preprocessing step. One of the tool that supports 3D visualization of cityGML data is Aristoteles developed by the Institute for Cartography and Geoinformation. But it lacks features like move around camera controls, projecting the coordinates from the cityGML CRS, and underneath terrain rendering. On the other hand virtual globes like NASAs world wind or cesium globe, are powerful enough to harness the power of the systems GPU to render large amount of geometries and DEM data to generate terrains, give scene controls, etc, but doesn't give support for visualization of cityGML data in its current form. Thus there is a need to integrate the virtual globe technology and the visualization functionality of tools like Aristoteles, so as to render buildings on top of terrain and give a more realistic view of the environment.

In this paper we present an opensource application build on NASAs WorldWind technology, for the visualization of cityGML levelofdetail2 building data. The application enables support for importing multiple cityGML files as layers, which have the building class and have the GroundSurface, Wallsurface, Roofsurface and Solids features. The first step includes extraction of geometry data for each of the features mentioned in a data model defined by us. This is achieved by using the citygml4j library's visitor functions, which iterates over each of the elements in the gml file with the passed feature class. In a similar manner the CRS of the features is also extracted. These geometries are in the local coordinates system, and needs to be transformed to the WGS84 coordinate system. For this conversion we use GDAL-OGR library, which takes the EPSG code from extracted CRS. To support multiple cityGML files a list of data models are created and are passed for rendering. In the rendering stage, a layer is created for each cityGML file. This layer will store all the features of the buildings. For each feature a set of basic shape-attributes are set, like its texture, color, opacity, etc. These shape-attributes, along with the transformed coordinates , are passed to the WorldWind polygon rendering function, which uses the underlying openGL array buffers to render these features.

One of the limitations we find in this method is when the number of features and their sizes increases, for which we plan to add vector tile support soon for rendering such large cityGML files. Depending on the camera location and its height, only those buildings in the current and surrounding tiles will be rendered, rather than the entire dataset, leading to a reduced computaitonal load for the GPU, and an increase in the rendering speed, thus achieving better overall performance.

Academic Discipline and Sub-Disciplines : GIS; Data Visualization; 3D Modeling

Keywords : CityGML; WorldWind; Open-Source Visualization

Free and Open Source Software for Geospatial (FOSS4G) Conference Proceedings

Volume 15 *Seoul, South Korea* Article 27

2015

Analysis of Spatial Density Utilizing The Big Data of Floating Population of Seoul City

Hailin Kim
Hankuk Academy of Foreign Studies

AT7-01

ANALYSIS OF SPATIAL DENSITY UTILIZING THE BIG DATA OF FLOATING POPULATION OF SEOUL CITY

Hailin Kim

Hankuk Academy of Foreign Studies
50, Oedae-ro 54beon-gil, Mohyeon-myeon, Cheoin-gu, Yongin-si, Gyeonggi-do, Korea
Email: kwla1024@gmail.com

ABSTRACT

The goal of this research is to support spatial decision that deals with foreign tourists' inconveniences in Seoul City—associated with place—by using data mining technology. The analysis of domestic mobile phone records of foreign tourists not only produces the number of floating population demonstrated in the aspect of its nationality and time period, but also conducts the spatiotemporal localization of foreign visitors. R-Studio, a statistical computing software, is used to analyze the spatial density from the call traffic; QGIS, an open-source Geographic Information System(GIS) software, is useful in displaying the data on a map. As a result, fourteen areas, where foreign floating population's mobility patterns are noticeable, are selected: Gyeongbok Palace, Namsan Hanok Village, Deoksu Palace, Insa-dong, Myeong-dong, Namdaemun Market, Dongdaemun Market, Garosu-gil, Apgujeong-dong, Itaewon, Hongik University, Gangnam Station, COEX, and Lotte World.

Further this research introduces the application to alleviate the inconveniences regarding the signposts, bus route, and insufficient tourist information centers; it suggests to add more foreign languages on the signposts at certain locations, modify the current route of the city tour bus, and position additional tourist information centers at the area where floating population of foreign tourists turns out to be dense. Likewise, the connection between big data and GIS technology in this research poses a significant impact on enhancing the efficiency of spatial decision.

1. INTRODUCTION

Increased demand for tourism has brought expansion of the tourism industry and emphasis on the tourist policies. The developed countries are strengthening their competitiveness in the tourism industry by establishing proper policies which match their national characteristics. Despite its tremendous number of foreign tourists, Korea's productivity in the tourism industry still remains lower than that of developed countries. This phenomenon derives from the shortcoming of the industry which has relied upon an unscientific method of analysis. Although the increase in the use of digital media—smartphones or Social Network Service (SNS)—has accumulated large amount of data on domestic tourism, they are not being utilized effectively.

This thesis differs from current tourists policies of Korea in its approach to analysis: whereas current tourist policies of Korea depend on surveys or simplistic statistics, this thesis analyzes mobile phone records of foreign tourists by using data mining technology. The development of Information & Communication Technology(ICT) sheds light on the ignored data and enables the scientific support for the tourist policies by integrating a myriad of information. Hence, big data analysis would be an innovative marketing strategy of the tourism industry. Since it produces more significant information than the offline survey and immediately provides constantly-changing current issues, it would promote the tourism industry as well as tourism-related researches.

This research aims to provide a diverse, scientific spatial decision by connecting Big Data to Geographic Information System(GIS) and establishing tourist policies associated with location. The mobile phone records of foreign tourists are utilized to analyze floating population and its mobility pattern—in other words, spatial density, and a variety of applications is suggested. In particular, the spatial analysis of the distribution of foreign floating population sorted by nation and time is employed for tourist policies. Thus, instead of previous incomprehensive methods of analysis such as offline surveys, the data mining of spatial density of foreign tourists will result in customized services—additional tourist information centers at apt location, additional foreign languages on the signposts, and effective city tour bus route.

2. LITERATURE REVIEW

2.1 Seoul Metropolitan Government's Owl Bus

SeoulMetropolitan Government is operating 'Owl Bus' for its late-night passengers. Its night bus service focuses on drawing optimum bus route from the scientific grounds for appropriate locations—bus stops— and time interval between buses.

Wooyong Kim notes in his report about the late-night bus service that the whole city is divided on a map into 1,250 hex-cells with a radius of 1km; the amount of calls during the late-night(12AM~5AM) is analyzed so that the floating population density of each area is evaluated. Then, the results are visualized in the unit of hex-cell. Also, the pattern for each time period of the current route is analyzed; the floating population near the route is weighted, and the bus route is optimized. For more efficient operation of the bus, the traffic is measured in the unit of a bus stop. The amount of traffic for each stop is differentiated by the thickness of the line and visualized on a map; thus, apt time interval between buses for each day of the week is determined.(Kim, 2013)

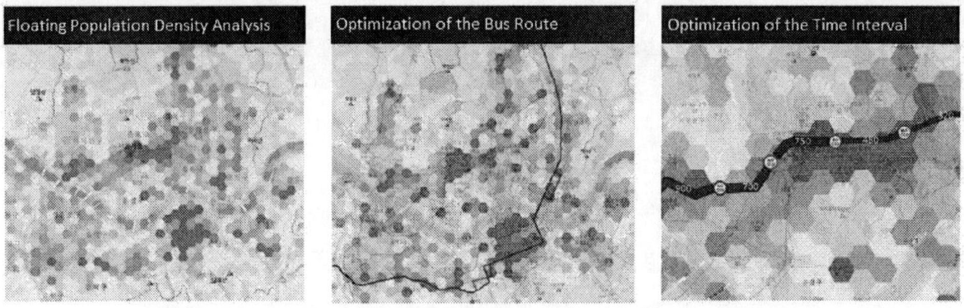

Figure 1. Optimizing the route of Seoul City's Owl Bus

2.2 Crime Prediction

According to Mark Ward in his report about big data used to fight against crime, publicly shared information combined with data from local authorities, social services and intelligence gathered by beat officers is helping police forces around the world spot trouble before it starts.(Ward, 2014)While previous researches have used either background historical knowledge or offenders' profiling, Andrey Bogmolov presents a novel approach to predict crime in a geographic space from multiple data sources, in particular mobile phone and demographic data. He also states that aggregated human behavioral data captured from the mobile network infrastructure, in combination with basic demographic information, can be used to predict crime.(Bogmolov, 2014) Likewise, previous researches on crime prediction using big data can contribute to insuring security for the late-night travelers in the urban area.

2.3 Significance of the Research

Although Seoul City's *Owl Bus*is in operation, the inefficiency has been revealed in several parts of the current route. Therefore, using mobile phone records of foreign tourists, here I modify the route into more populated area during night time based on their mobility pattern. Moreover, researches on the crime prediction based on human behavioral data will be utilized to guarantee the safety of the foreign visitors.

3. RESEARCH METHODS

3.1 Data

The mobile phone recordsused in this paper is provided by Tourism Big Data Contest hosted by Korea Culture & Tourism Institute; it is a part of domestic Roaming Call Data Record of foreigners which can be geographically analyzed in the unit of cell tower. The dataset contains 1,774,618 domestic call records of foreign tourists from ten countries (Code N01~N10) from December 1st, 2013 to February 28th, 2014. The location is nearby cell towers which received the signals from mobile telephones; thus, there is a disparity between the actual location of devices and the location in the set.

As Figure 2 shows, the dataset consists of ID—a code that differentiates encrypted subscribers, DATE—the date on which a call took place(e.g.20131201), TIME—the time at which a call took place, COUNTRY_CODE, ZIP—the location of the cell tower which received the signals from mobile telephone, LATITUDE—the location of the cell tower which received the signals from mobile telephone, LONGITUDE—the location of the cell tower which received the signals from mobile telephone.

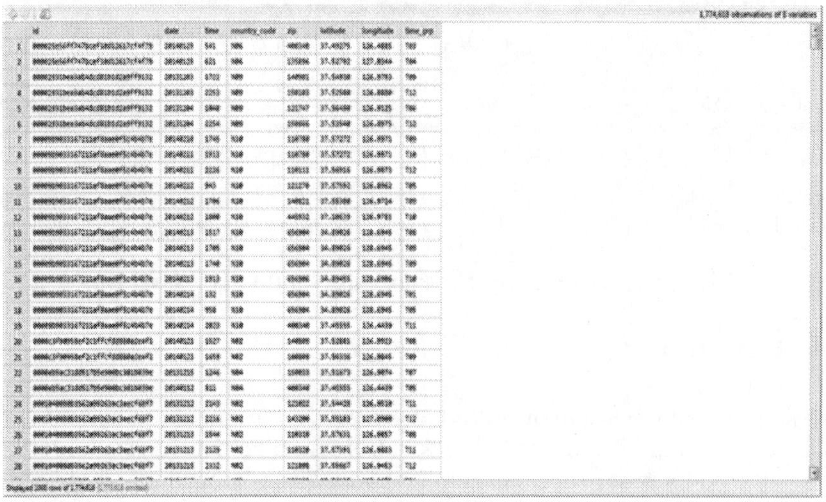

Figure 2. Dataset

In particular, 1,774,618 calls associated with 1,560 cell towers are analyzed in order to evaluate floating population density of foreign tourists as in Figure 3.

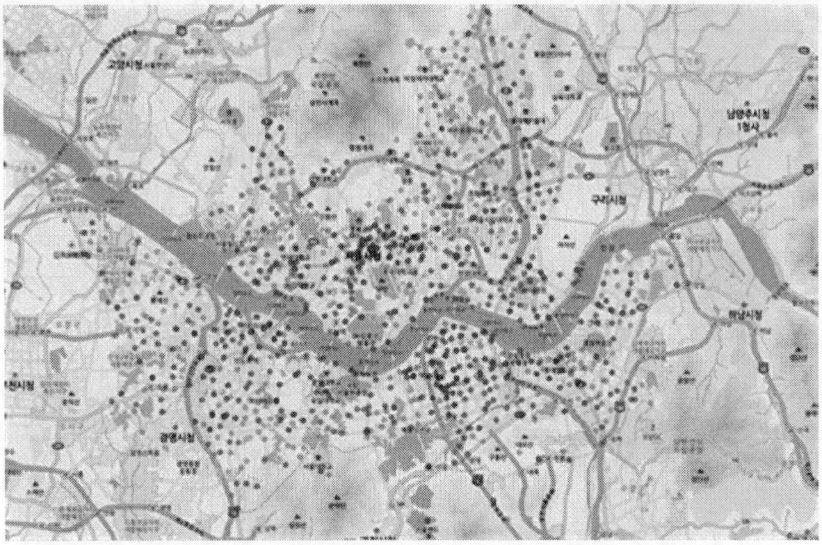

Figure 3. 1,774,618 call records registered to 1,560 cell towers.

3.2 Analysis Software

R-Studio, an open-source statistical computing software, is used to sample, process, and analyze big data of cellular phone subscribers. Also, an open-source Geographic Information System(GIS) application called QGIS is used to spatially visualize and analyze the data. QGIS converts statistical data into spatial data, displays the location, and analyzes spatial density.

3.3 Procedure

In order to analyze the spatial density of the floating population, the workspace in R-Studio should be set: the working directory must be set, and R packages must be installed. As CSV format dataset is read in R-Studio, eight columns—ID, DATE, TIME, COUNTRY_CODE, ZIP, LATITUDE, LONGITUDE—should be confirmed. Then, the time, location, and country variables must be created. Since the analysis of the single variable does not produce any meaningful results, the dataset associated with combination of variables should be analyzed: Table 1 shows various combinations of three variables. Using the temperature map on QGIS, the floating population density of 1,560 foreign tourists is displayed: darker spots indicate denser floating population. All points—coordinates(LATITUDE, LONGITUDE)—within a given area are closer to its corresponding tower than to any other tower in the map. With this sampling method, top fourteen places in which the floating population is dense—Gyeongbok Palace, Insadong, Namsan Hanok Village, Deoksu Palace, Myeongdong, Namdaemun Market, Dongdaemun MArket, Garosugil, COEX, Lotte World, Gangnam Station, Hongik University, Apgujeong Station, and Itaewon—are selected.

Table 1. The combination of COUNTRY, TIME, and LOCATION variables.

The number of tourists sorted by LATITUDE, LONGITUDE	The number of tourists sorted by LATITUDE, LONGITUDE/COUNTRY
The number of tourists sorted by TIME/LATITUDE, LONGITUDE	The number of tourists sorted by LATITUDE, LONGITUDE/TIME/COUNTRY
The number of tourists sorted by TIME/COUNTRY/LATITUDE, LONGITUDE	The number of tourists sorted by TIME(20PM~2AM)/LATITUDE, LONGITUDE
The number of tourists sorted by LATITUDE, LONGITUDE/TIME	The number of tourists sorted by TIME(22PM~2AM)/LATITUDE, LONGITUDE

4. STUDY RESULTS

Amongst 1,774,618 foreign subscribers in the dataset, majority of them locates in Gangbuk Area—Myeongdong and Gyeongbok Palace, and Gangnam Area—Apgujeongdong and Gangnam Station. (See Figure 4)

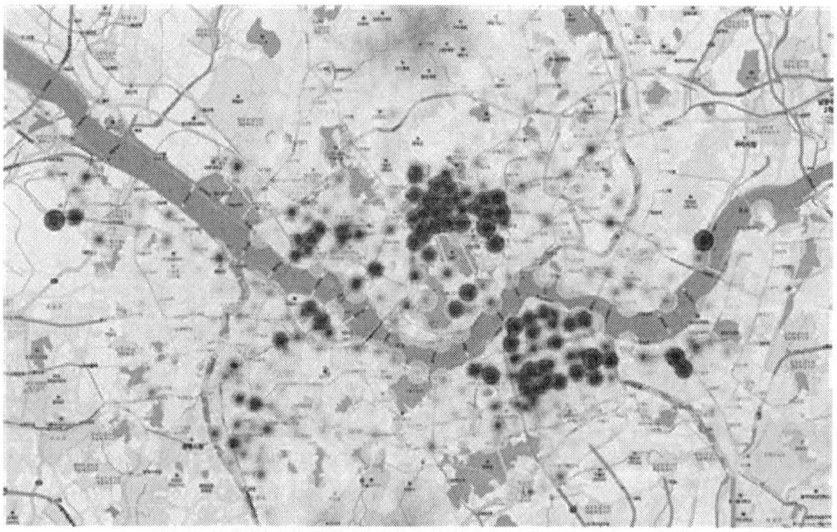

Figure 4. The spatial density of foreign tourists

4.1 Call Patterns Sorted by Country, Time

The analysis produces the number of foreign visitors—number of calls—from country N01 to N10—which is ranked in order of N01>N02>N10>N04>N06>N03>N07>N09>N08>N05. Table 2 demonstrates that the visitors from N01 and N02 account for 86% of the overall distribution.

Table 2. The number of foreign tourists sorted by country

Country Code	Number of tourists	Percentage(%)	Rank
N01	800,937	45.13	1
N02	555,390	31.3	2
N03	70,457	3.97	6
N04	76,887	4.33	4
N05	5,765	0.32	10
N06	76,736	4.32	5
N07	34,522	1.95	7
N08	17,327	0.98	9
N09	32,605	1.84	8
N10	103,992	5.86	3
TOTAL	1,774,618	100	

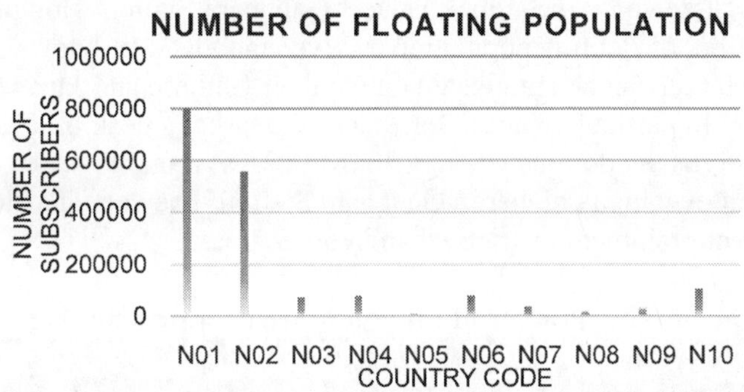

Figure 5. The number of foreign tourists sorted by country

Each particular time of the day shows distinct activity patterns. Call patterns are most active from 16PM to 18PM and evenly distributed from 10AM to 10PM. (See Table 3)

Table 3. The number of foreign tourists sorted by time of the day

Time	Number of Tourists	Percentage(%)	Rank
00~02	37,410	2.11	9
02~04	12,109	0.68	11
04~06	10,955	0.62	12
06~08	37,094	2.09	10
08~10	136,131	7.67	7
10~12	243,752	13.74	4
12~14	250,734	14.13	3
14~16	255,606	14.40	2
16~18	276,330	15.57	1
18~20	231,408	13.04	5
20~22	169,602	9.56	6
22~24	113,487	6.40	8
Total	1,774,618	100	

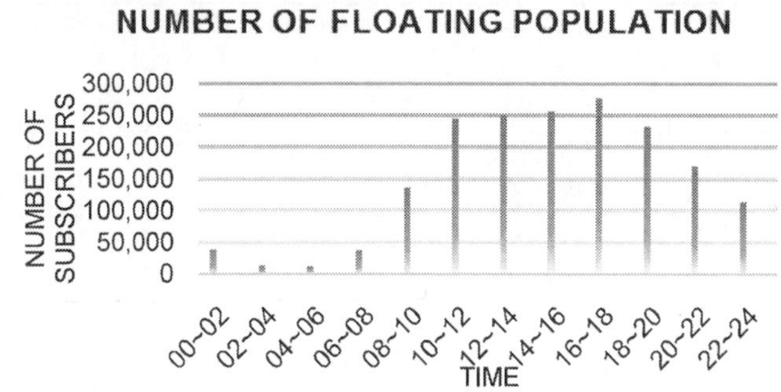

Figure 6. The number of foreign tourists sorted by time of the day

Moreover, call traffic patterns vary depending on areas. As Table 4 and Figure 7 show, the number of calls is ranked in order of Myeongdong>Dongdaemun Market>Deoksu Palace>COEX>Lotte World>Insadong>Garosugil>Sungryemoon>Namsan Hanok Village>Apgujeong>Itaewon>Gyeongbok Palace>Gangnam Station>Hongik University.

The great number of floating population in Myeongdong(well-known shopping district) and Dongdaemun Market can be anticipated due to their Chinese and Japanese customers who enjoy shopping. In particular, since Dongdaemun Market's peak time is known to be late at night, the call patterns are dominated from 20PM to 2AM around its area. Also, the aggregated pattern at evening is higher in Gangnam Station, Itaewon, and Hongik University(busiest entertainment districts) than in other areas.

Table 4. The number of foreign tourists in principal areas

Area	Number of Subscribers	Country Code										Time											
		N01	N02	N03	N04	N05	N06	N07	N08	N09	N10	0~2	2~4	4~6	6~8	8~10	10~12	12~14	14~16	16~18	18~20	20~22	22~24
Myeong-dong	**74,764**	44,175	19,215	1,599	241	72	3,276	1,264	304	1,030	3,588	719	128	114	5,392	2,675	8,013	10,423	12,976	15,665	13,235	7,418	2,859
Dongdaemun market	**40,759**	25,176	7,772	3,809	157	11	1,560	650	67	360	1,197	3,853	1,830	806	462	1,222	2,920	3,484	3,975	4,788	4,779	6,072	6,568
Deoksu Palace	**19,763**	5,152	6,556	431	864	180	2,627	387	249	261	3,056	643	171	160	613	2,178	2,598	2,070	2,176	2,811	2,495	2,095	1,753
COEX Mall	**17,475**	5,651	5,318	814	919	165	1,670	372	285	213	2,068	582	212	159	525	1,555	2,497	2,246	2,078	2,493	2,031	1,603	1,494
Lotte world	**16,226**	11,598	1,102	679	189	28	486	544	73	131	1,396	56	16	20	73	427	1,835	2,376	3,140	3,884	2,954	1,200	245
Insa-dong	**10,851**	3,805	3,524	539	240	118	592	372	139	323	1,199	264	97	51	185	816	1,782	1,590	1,602	1,498	1,206	1,019	741
Garosu-gil	**8,697**	3,944	2,560	329	193	23	759	99	121	148	521	180	52	35	68	353	857	1,182	1,529	1,785	1,359	845	452
Namdaemun market	**8,518**	2,151	4,148	452	38	4	589	260	70	100	706	170	53	19	70	499	1,246	1,579	1,602	1,412	896	526	446
Namsan Hanok Village	**8,459**	3,697	2,688	357	157	17	599	297	76	76	495	289	76	57	209	1,164	903	903	1,043	1,251	946	881	782
Apgujeong-dong	**6,654**	3,939	1,337	89	427	26	341	63	49	44	339	99	49	29	111	407	1,020	1,018	1,077	1,069	897	545	333
Itaewon	**6,653**	682	1,643	301	420	68	1,654	240	141	109	1,395	321	101	58	202	669	725	635	525	774	972	857	814
Gyeongbok Palace	**6,353**	4,138	784	420	61	3	261	181	14	167	324	21	6	5	16	460	1,677	1,594	1,250	584	513	190	37
Gangnam Station	**5,835**	2,277	1,697	183	173	17	427	129	113	150	669	195	57	36	89	370	742	802	740	901	785	626	492
Hongik University	**4,674**	1,873	1,021	428	92	3	345	153	40	368	351	168	42	28	42	184	463	632	604	734	732	619	426

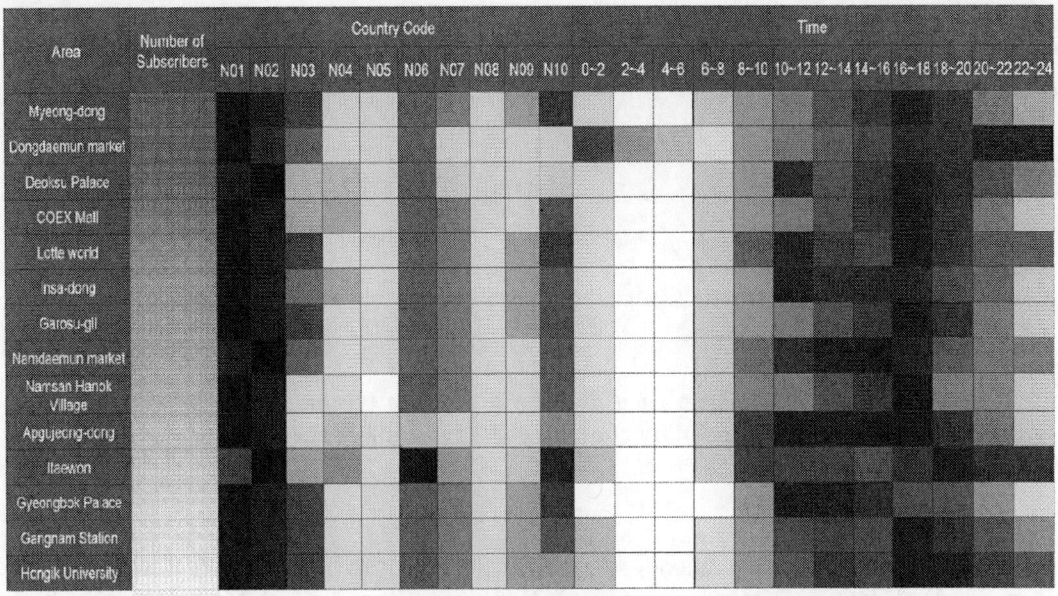

Figure 7. Overview diagram of the floating population in principal areas

4.2 Gyeongbok Palace

The number of foreign tourists in Gyeongbok Palacefrom N01 is overwhelmingly high. Since it is a historical site, the call volume around this area is busiest during the daytime—from 10AM to 16PM.(See Figure 8 and 9)

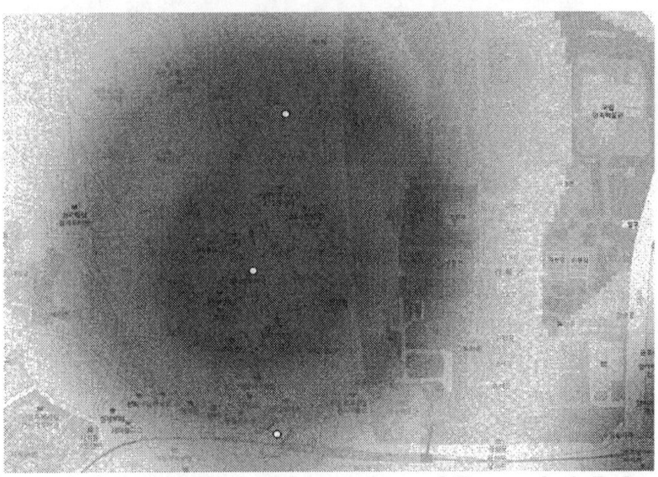

Figure 8. Spatial call activity pattern of Gyeongbok Palace displayed on a map with QGIS

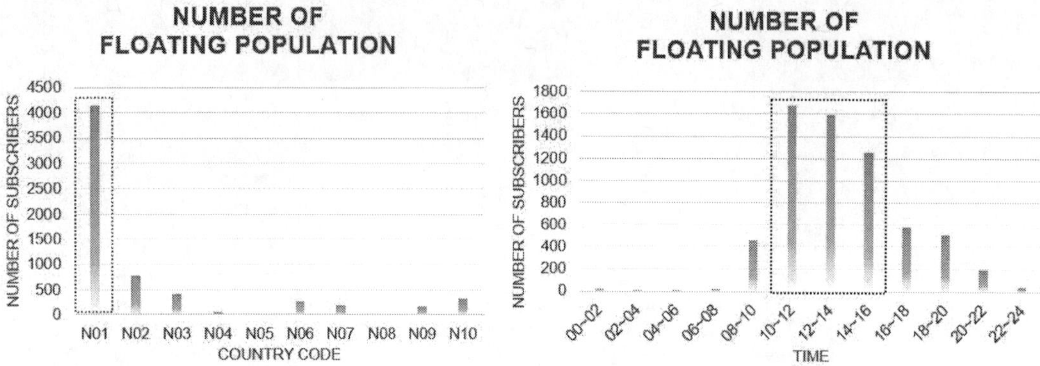

Figure 9. The number of foreign tourists in Gyeongbok Palace sorted by country code and time of the day

4.3 Dongdaemun Market

The primary cellular phone subscribers reported to nearby cell towers around Dongdaemun Market are those from country N01. Since Dongdaemun Market's peak time is known to be late at night, the call patterns are dominated from 20PM to 2AM around its area, and active call patters in the early morning can be inferred.

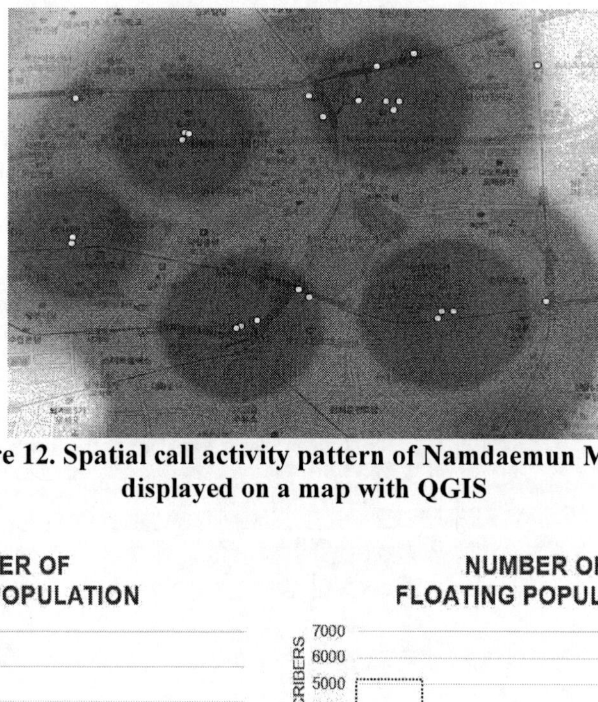

Figure 12. Spatial call activity pattern of Namdaemun Market displayed on a map with QGIS

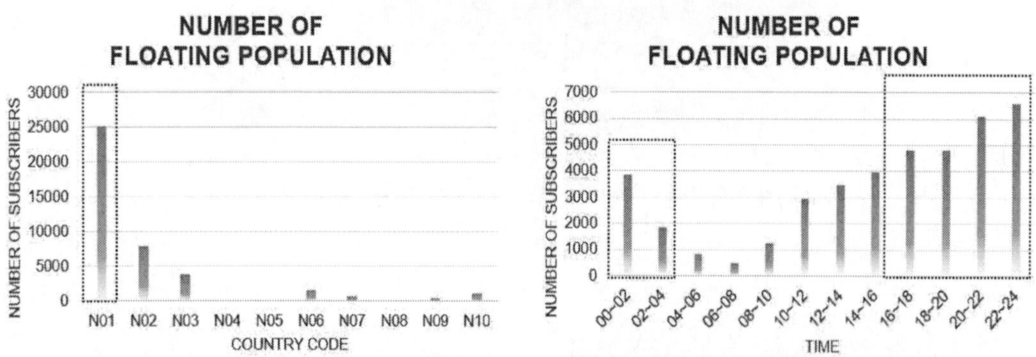

Figure 13. The number of foreign tourists in Dongdaemun Market sorted by country code and time of the day

4.4 Apgujeong-dong

The primary cellular phone subscribers registered to the cell towers around Apgujeong-dong are those from country N01. Numerous hospitals for plastic surgery in Apgujeong-donghave promoted medical tourism in Korea. Hence, the call records of foreign visitors for medical purposes are evenly distributed from 8AM to 20PM.

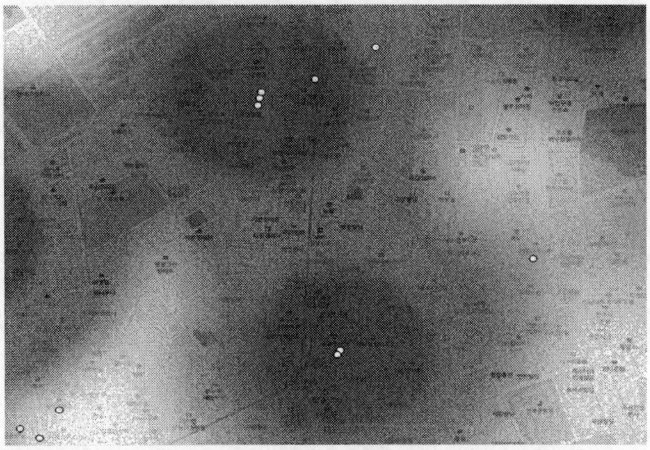

Figure 14. Spatial call activity pattern of Apgujeong-dong displayed on a map with QGIS

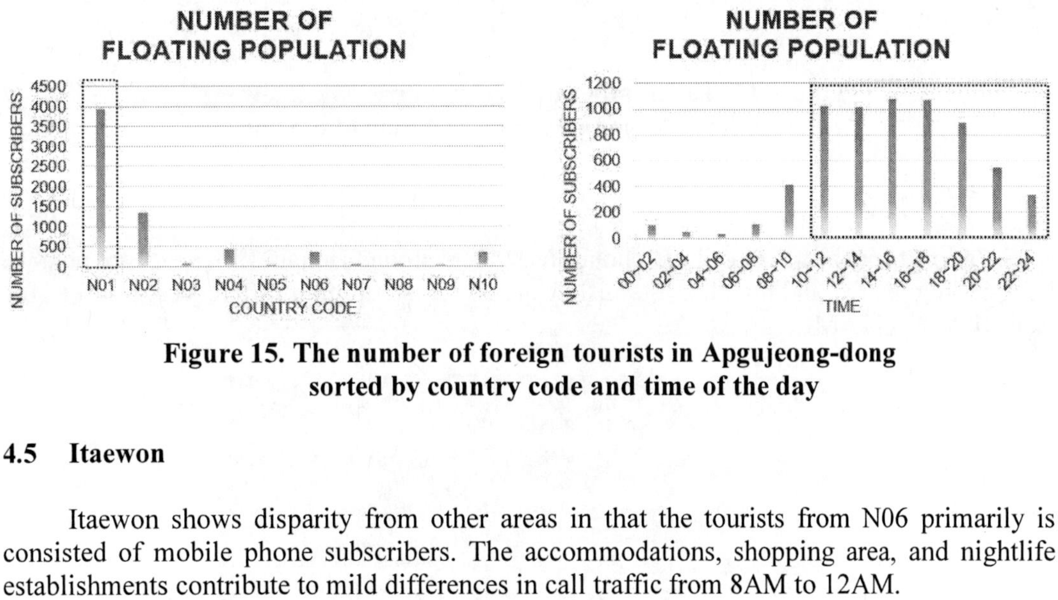

Figure 15. The number of foreign tourists in Apgujeong-dong sorted by country code and time of the day

4.5 Itaewon

Itaewon shows disparity from other areas in that the tourists from N06 primarily is consisted of mobile phone subscribers. The accommodations, shopping area, and nightlife establishments contribute to mild differences in call traffic from 8AM to 12AM.

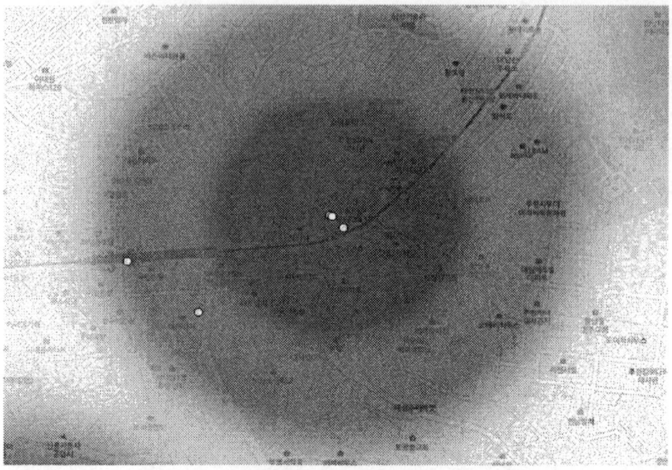

Figure 16. Spatial call activity pattern of Itaewon displayed on a map with QGIS

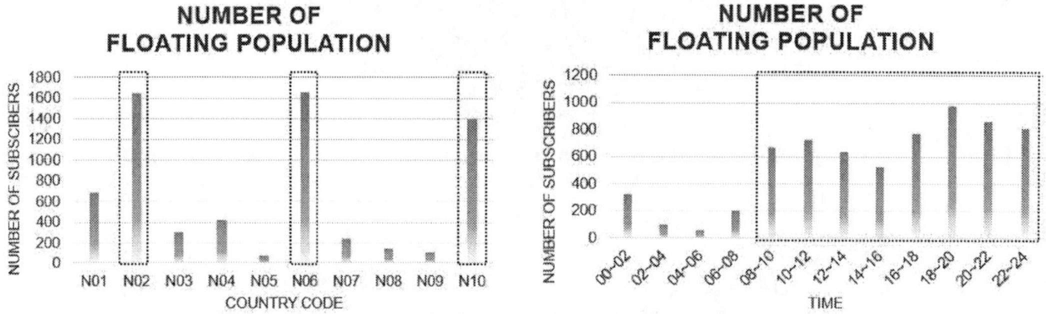

Figure 17. The number of foreign tourists in Itaewon sorted by country code and time of the day

4.6 Hongik University

Variousnationalities of cellular phone subscribers around Hongik University are notable. Its shopping district and nightlife establishments contribute to mild differences in call traffic from 10AM to 12AM.

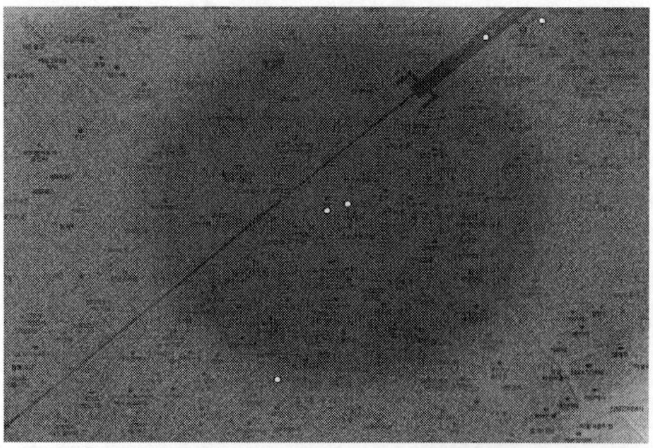

Figure 18. Spatial call activity pattern of Hongik University displayed on a map with QGIS

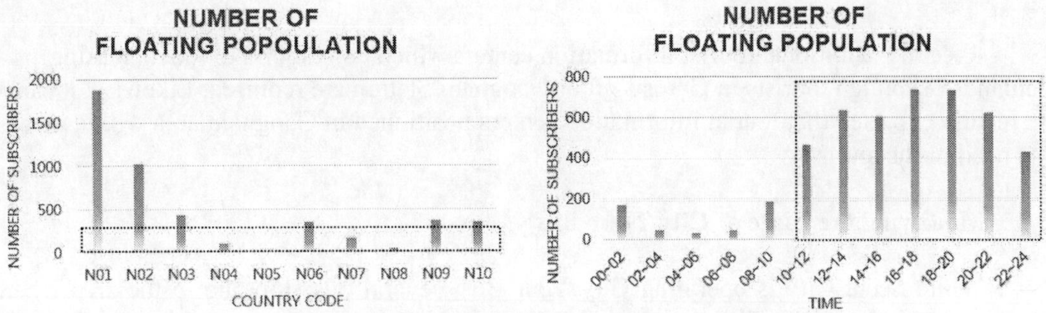

Figure 19. The number of foreign tourists in Hongik University sorted by country code and time of the day

5. DISCUSSION

The inconveniences foreign visitors in Seoul City are frequently experiencing are language problem, lack of tourist information, or signposts only written in Korean. The analysis of spatiotemporal mobility patterns of foreign floating population by means of mobile phone big data can be used to alleviate the discomfort of which foreign tourists often complain; using the trajectories of 1,774,618 foreign mobile phone users, here I make suggestions of adding more foreign languages on the signposts at certain locations, modifying the current route of the city tour bus, and positioning additional tourist information centers at apt location.

5.1 Positioning additional tourist information centers.

Call traffic of foreign tourists in Seoul City is evenly distributed across Gangbuk and Gangnam area. However, among twenty-five tourist information centers in Seoul City, twenty-two locate in Gangbuk area.

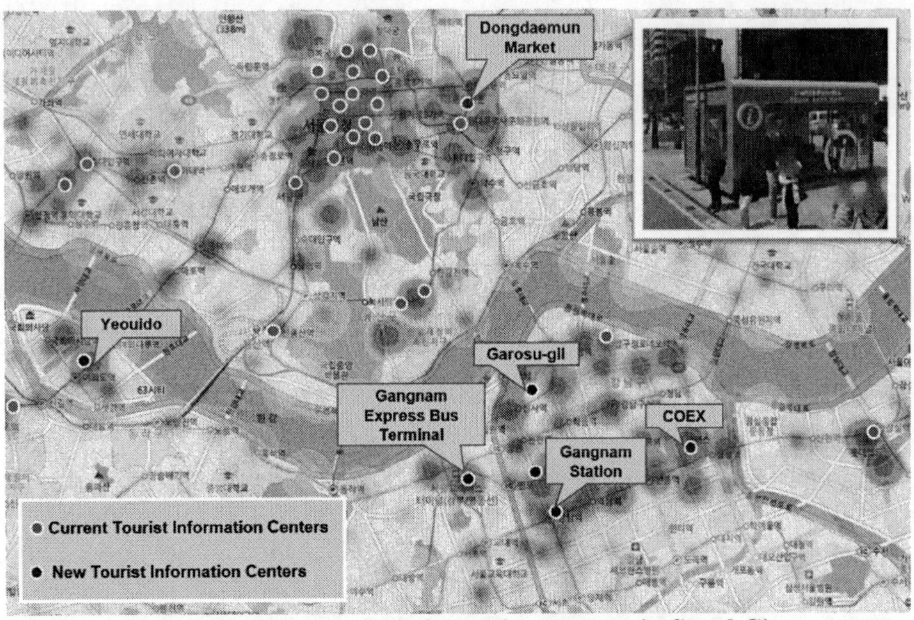

Figure 20. The tourist information centers in Seoul City

Recently, additional tourist information centers which correspond to the increasing demands of foreign tourists in Garosu-gil or Gangnam Station are required. Likewise, it can be readily expected that tourist information centers positioned in Gangnam area will match the need of the tourists.

5.2 Modifying the route of City Tour Bus

While Seoul City is operating City Tour Bus, several bus stops lie in the area where foreign population's mobility pattern is hardly found. Therefore, more foreign tourists will benefit from the efficient bus route modified into the area whose call traffic is active.(See Figure 21)

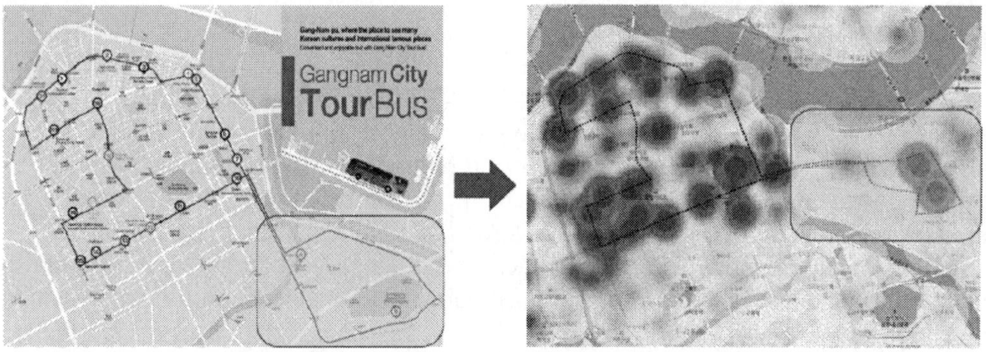

Figure 21. Modifying the city tour bus route

Moreover, the late-night bus route connecting the areas where call patterns of foreign mobile phone users—Itaewon, Gangnam Staion, COEX mall, Hongik University, and etc.—are dominated by the hectic activity will guarantee safety of the tourists during night-time.(Blue spots on Figure 22represent busy spatial activity pattern.)

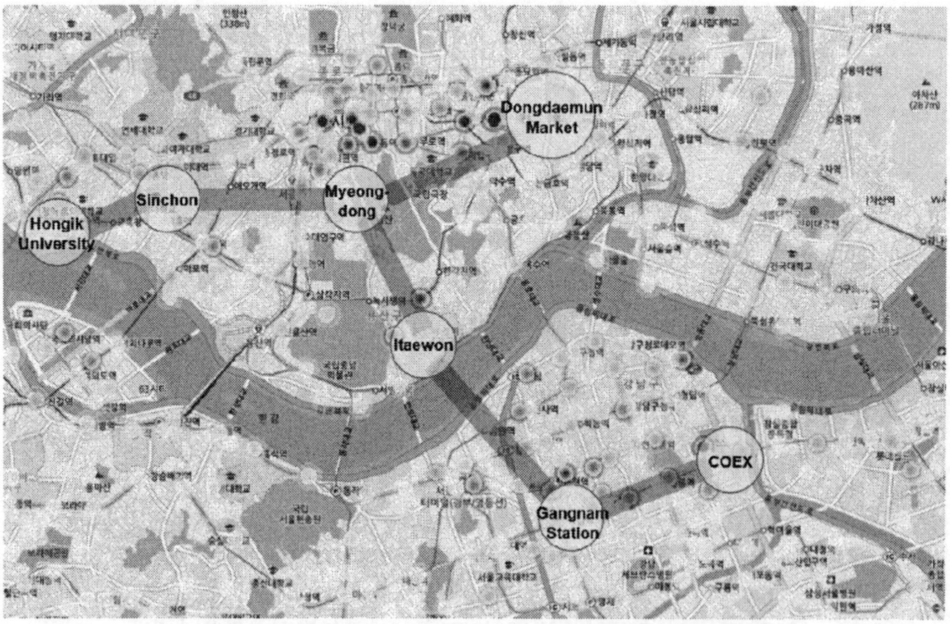

Figure 22. Suggesting new late-night bus route

5.3 Positioning the signposts written in Korean, English, Chinese, and Japanese

Numerous foreign tourists complain of language problems which can also be found on the signposts. The traffic signs only written in Korean and English are confusing the tourists from China and Japan; therefore, installing signposts written in four languages first in the area where large number of floating population is found will solve the language problems.(See Figure 23)

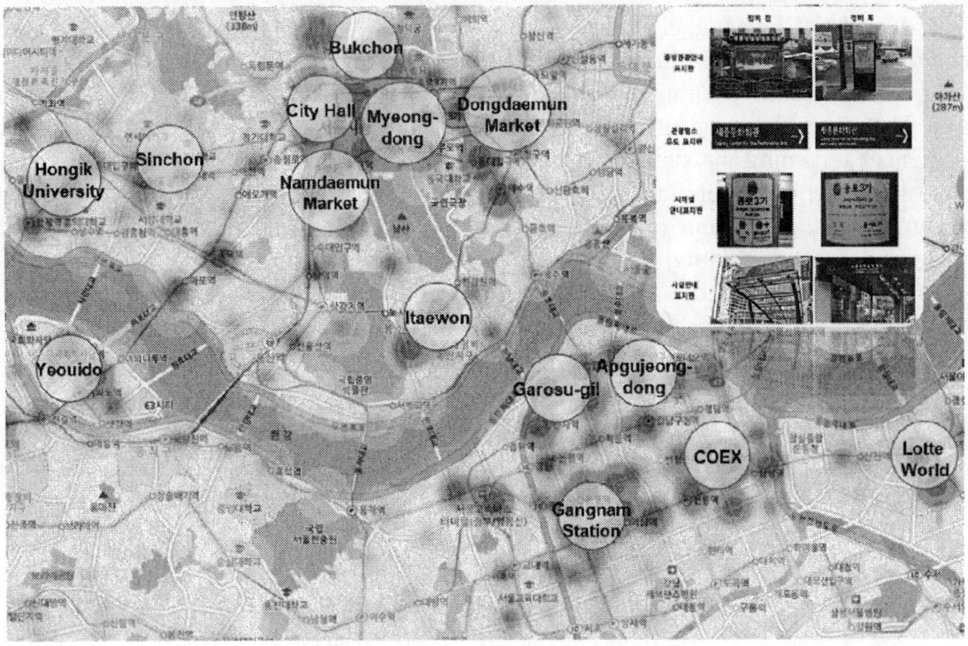

Figure 23. Areas where signposts with four languages are recommended

6. CONCLUSION

Since the dataset used is mobile phone records reported to nearby cell towers, the location in the set did not mark the actual location of foreign cellular phone subscribers precisely.

6.1 Future Study

Although spatiotemporal localization of foreign tourists provided various applications in this research, with more data on individual duration of stay or the distinction between individual tour and group tour, more tourist policies can be suggested.

6.2 Closing Comments

This paper aimed to deal with the inconveniences foreign tourists may confront—associated with place—by utilizing the big data technology. It analyzes the spatial density of floating population of foreign tourists and suggests a variety of applications based on the results.

This thesis associates the roaming big data of foreign tourists with Geographic Information System(GIS) to analyze the floating population density by location, time, and country. It uses R-Studio, a statistical computing software, and QGIS, an open source GIS software to visualize the data on a map.

The first step to conduct the research is to open CSV format data and set variables for analysis. Then it analyzes the spatial density of floating population of foreign tourists sorted by time, location, and country, and the number of visitors is ranked for each area. Based on the analysis results, the data are displayed on a map through QGIS. Lastly, several adjacent points—cell towers— are matched to a given area.

This paper selects fourteen locations in which floating population density of foreign tourists is notable: Gyeongbok Palace, Namsan Hanok Village, Deoksu Palace, Insadong, Myeongdong, Namdaemun Market, Dongdaemun Market, Garosu-gil, Apgujeongdong, Itaewon, Hongik University Entrance, Gangnam Station, COEX, and Lotte World.

This research paper introduces four pragmatic proposals in order to improve the inconveniences regarding the signposts, bus route, and insufficient tourist information centers; it suggests to add more foreign languages on the signposts at certain locations, modify the current route of the city tour bus, and position additional tourist information centers at the area where floating population of foreign tourists turns out to be dense. Thus, a link between big data and GIS in this thesis results in customized services based on spatially distinct mobility patterns, thereby enhancing the efficiency of spatial decision in the tourism industry.

7. REFERENCES

Bogomolov, Andrey, Bruno Lepri, Jacopo Staiano, Nuria Oliver, Fabio Pianesi, and Alex Pentland. "Once Upon a Crime: Towards Crime Prediction from Demographics and Mobile Data." *Proceedings of the 16th International Conference on Multimodal Interaction - ICMI '14* (2014): n. pag. Web.

Byun, Miree, and U-Seok Seo. "How to Measure Daytime Population in Urban Streets?: Case of Seoul Pedestrian Flow Survey." *Survey Research* 12.2 (2011): 27-50. Web.(Korean)

Candia, Julian, Marta C. Gonzalez, Pu Wang, Timothy Schoenharl, Greg Madey, and Albert-Laszl ´ O Barabasi. "Uncovering Individual and Collective Human Dynamics from Mobile Phone Records." *JOURNAL OF PHYSICS A: MATHEMATICAL AND THEORETICAL* 41.22 (2008): n. pag. Web.

GIS United. *Big Data Spatial Analysis Public Policy*. N.p.: Soup, 2014. Print.(Korean)

Keum, Kiyong. *An Analysis of the Effectiveness of Seoul's Inbound Tourism Policy*. Thesis. 2009. N.p.: Seoul Development Institute, n.d. Print.(Korean)

Keum, Kiyong, and Yunjoung Park. *A Study for Qualitative Improvement of Seoul Tourism*. Thesis. 2013. N.p.: Seoul Institute, n.d. Print. (Korean)

Kim, Wooyong. "Seoul City's Operating Late-Night Bus." *ZDNet Korea* 2 July 2013: n. pag. Web.(Korean)

Lee, Manjai. "Big Data and the Utilization of Public Data." *Internet and Information Security* 2.2 (2011): 47-64. Web.(Korean)

Maeil Business Newspaper. *Big Data World.* N.p.: Maeil Book, 2014. Print.(Korean)

Pan, Jeonghwa, Junghyun Lee, and Misun Gang. *Strategies to Vitalize the Utilization of Seoul Industrial Tourism Resources*. Thesis. 2011. N.p.: Seoul Development Institute, n.d. Print.(Korean)

Shim, Won Seob. *A Case Study on the Tourism Policy of Foreign Countries and Future Directions.* Thesis. Korea Culture & Tourism Institute, 2011. N.p.: n.p., n.d. Print.(Korean)

"Use of Big Data in Seoul City's Owl Bus." *ZERONOVA*. N.p., 07 Aug. 2013. Web. 02 Aug. 2015.

Ward, Mark. "Crime Fighting with Big Data Weapons." *BBC* 18 Mar. 2014: n. pag. Web.

Free and Open Source Software for Geospatial (FOSS4G) Conference Proceedings

Volume 15 *Seoul, South Korea* Article 28

2015

Evaluating Flood Hazard Potential in Danang City, Vietnam Using FOSS4G

An Tran Thi
Graduate School for Creative Cities, Osaka City University

Venkatesh Raghavan
Graduate School for Creative Cities, Osaka City University

Shinji Masumoto
Graduate School of Science, Osaka City University

Susumu Nonogaki
Geological Survey of Japan, National Institute of Advanced Industrial Science and Technology

Tatsuya Nemoto
Graduate School of Science, Osaka City University

See next page for additional authors

Evaluating Flood Hazard Potential in Danang City, Vietnam Using FOSS4G

Authors
An Tran Thi, Venkatesh Raghavan, Shinji Masumoto, Susumu Nonogaki, Tatsuya Nemoto, Vinayaraj Poliyapram, and Go Yonezawa

This paper is available in Free and Open Source Software for Geospatial (FOSS4G) Conference Proceedings:
https://scholarworks.umass.edu/foss4g/vol15/iss1/28

AT6-05

EVALUATING FLOOD HAZARD POTENTIAL IN DANANG CITY, VIETNAM USING FOSS4G

An Tran Thi[1], Venkatesh Raghavan[1], Shinji Masumoto[2], Susumu Nonogaki[3], Tatsuya Nemoto[2], Vinayaraj Poliyapram[1], Go Yonezawa[1]

[1] Graduate School for Creative Cities, Osaka City University
3-3-138 Sugimoto, Sumiyoshi-ku, Osaka 558-8585, Japan.
Email: tranthian.gis@gmail.com
[2] Graduate School of Science, Osaka City University
3-3-138 Sugimoto, Sumiyoshi-ku, Osaka 558-8585, Japan.
[3]Geological Survey of Japan, National Institute of Advanced Industrial Science and Technology
Central 7, 1-1-1 Higashi, Tsukuba, Ibaraki 305-8567, Japan.

ABSTRACT

Floodplain in the alluvial lowland coastal area in the south of Danang City is a part of Vu Gia - Thu Bon river basin where is affected by serious and repetitive flood hazard. The topography and geomorphology which directly related to flood hazard but such relationship are still poorly understood. This study aims to measure flood inundation during a flood event in 2007 and estimate the potential flood hazard areas. ALOS PALSAR imageries were used for observing flood inundation dynamics. A 5m resolution DEM of study area were generated using field survey elevation points data and BS-Horizon interpolation method (Nonogaki et al., 2012). The generated DEM and flood inundation map were used as input data to extract the geomorphologic features of the study area and investigate relations on flood hazards. The flood potential mapping based on the investigation of ALOS PALSAR interpretation, topographical analysis and the land cover classification was carried out. As the result, flood inundation map shows good correlation to field survey flood map in 2007 and potential flood hazard map also coincides with flood scenario assessment by the Danang City government.

1. INTRODUCTION

Flooding is one of the most frequent and damage causing natural disaster in Vietnam and more so in Danang City. Several researchers have tried to characterize the flood potential in Danang area, based on hydro-geomorphological methods integrated with remote sensed data (Ho *et al.*, 2012; Do *et al.*, 2014). However, in the previous studies, lack of high resolution digital elevation model (DEM) and near-flood events satellite data, the results may not be sufficient for reliable analysis.

This study aims to build potential flood hazard map for a lowland area in the south Danang City, Vietnam and estimate the change of flood hazard potential. ALOS PALSAR imageries before and during flood event in 2007 were used to characterize flood inundation. Land cover map in 2007 extracted from Landsat TM data was applied for the landform classification process. The potential flood hazard map was generated based on the probability of submergence of each landform unit. In addition, optical RapidEye remotely sensed data acquired in 2014 was used to evaluate the land cover change between 2007 to 2014. The potential impact of recent land cover change as a result of future flood event has been discussed. The result was compared with field survey flood pillar data, flood map of the past, flood scenario provided by Danang City government to assess future flood inundation potential in the study area.

2. METHODOLOGY

2.1 Study area and data used

Study area comprises of 98 km^2 of lowland area located in the center of Vietnam, covered by Cam Le river in the north-west and Vinh Dien river in center (Figure 1). The study area includes semi-urban area in the south part of Danang City and a northern parts of Dien Ban district, Quang Nam province. This area currently undergoing rapid urbanization under the development policy of governments.

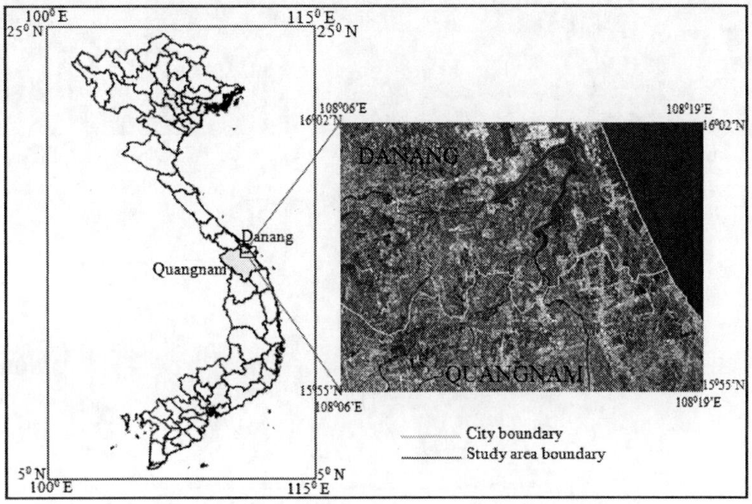

Figure 1. Location of study area in Vietnam. (PALSAR data on 15/09/2007).

The *in-situ* spot height data surveyed by Department of Natural Resource and Environment (DONRE), Danang City in 2009 was used for DEM generation process. In order to characterize flood inundation pattern, major flood event from 28/10/2007 to 09/11/2007 was investigated. A pair of ALOS PALSAR image before and close to this flood event were acquired and analyzed. In addition, optical remote sensed data include Landsat TM in 16/03/2007 and high resolution RapidEye satellite image in 02/03/2014 were used to classify the land cover of study area that relate to flood hazard potential. Flood pillars data collected by a field survey in March 2015 was used as reference data to assess the accuracy of flood map and potential flood hazard map.

Firstly, 5m resolution DEM was generated from field survey elevation point data using BS-Horizon method (Nonogaki *et al.,* 2012). BS-Horizon is a bi-cubic spline interpolation method implemented as Fortran program which takes advantage of equality-inequality elevation constraints. The DEM generation methodology was already mentioned in previous study of Tran *et al.* (2015). This DEM was used as input data for flood inundation mapping and assessment of the flood hazard potential.

2.2. Flood inundation mapping

A pair of ALOS PALSAR HH polarized image (Figure 2a and 2b) for time before and during flood event in 2007 were used for flood inundation mapping. Firstly, Lee filter was applied to reduce speckles in PALSAR data. Subsequently these data were converted from Digital Number (*DN*) value to Normalized Backscattering Coefficient (σ^0), expressed in decibels (dB) as shown in equation (1) below;

$$\sigma^0 = 10*\log_{10}(DN^2) + CF \qquad (1)$$

Where σ^0 is the Normalized Backscattering Coefficient, *DN* is the radar amplitude expressed as a digital number and CF is the calibration coefficient for PALSAR standard products, and equals -83 dB.

a. ALOS PALSAR 15/09/2007 b. ALOS PALSAR 31/10/2007

c. Flood Inundation map extracted from ALOS PALSAR

Figure 2. ALOS PALSAR HH Polarization (a, b) and flood inundation map (c).

Flood inundated areas were extracted using the difference in σ^0 between 15/09/2007 and 31/10/2007 images and DEM data. Firstly, the difference in σ^0 between two PALSAR data was calculated, then a threshold of 7dB was used to separate water and non-water objects. The resulting flood inundation map still have too many noisy pixels which were filtered using the 5m DEM. Comparing with field survey data, all of flood pillar points have elevation lower than 5m. Therefore a threshold of 5m elevation was used to remove all the noisy pixels in the higher elevation area. Subsequently a 5*5 smoothing filter was applied. All of the data analysis was carried out using GRASS GIS.

As a result, comparing to the original PALSAR image in 31/10/2007, this extracted flood inundation map matched well with the flooded regions in SAR data (Figure 2(c)). Using this method, flood inundated areas were not only separated from non-water areas, but also demarcated from permanent water such as river channels and lakes.

2.3. Characterizing the topographic features and the relation to flood inundation

Geomorphological analysis is recognized as an essential part of understanding of floodplains, based on a simple principle: the floodplain boundaries of a stream correspond to the envelope curve of its past flood. Firstly, contour lines with 2m interval were extracted from 5m DEM to detect the low-lying area (Figure 3a). The 2m contour line was also corresponded to the flood inundation boundary. It can be seen that the areas from 0m-2m is under the high vulnerability of submerging when floods occur. Also, it is observed that land cover has significant effect on flood hazard. Land cover map extracted from Landsat TM in

2007 reveals that agricultures are submerged areas during flood time. As a result, it is easy to determine the flood areas based on surface elevation and land cover class.

Statistical evaluation of DEM and land cover shows that the average elevation of each land cover is related to flood condition. Table 1 shows the calculation of mean elevation based on land cover classification. It is observed that water and agriculture usually locate in the lowland area where can be easily submerged if flood occurs. Built-up and bare land are generally located in the upper area. Based on this average elevation, integrated with flood inundation map and land cover classification, a method for landform classification of this study area was proposed.

Table 1. Average elevation of each land cover (Calculate from 5m DEM).

Land cover	Average Elevation
Agriculture	1.97
Built-up	3.67
Water	1.42
Bare land	5.06

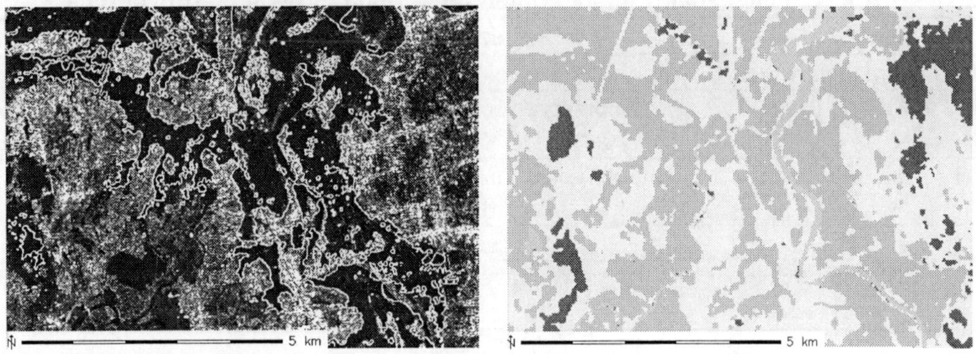

Figure 3. Contour data and land cover classification map.
(a) 2m contour lines overlay with PALSAR in 31th October. (b) Land cover classification: Green: Agriculture; Yellow: Built-up; Cyan: water; Brown: Bare land.

Figure 4 shows a cross section across the floodplain with different landform units. The elevation and distance to river channel are criteria to separate landform types. Based on the rule-based landform classification proposed by Ho *et al.* (2012), integrated with land cover (LC) map in 2007, 5m resolution DEM and flood inundation map extracted from ALOS PALSAR data, a new landform classification method was implemented as below:

- Permanent water: if LC = water
- Flood basin: if LC = agriculture or Flood map = flood or DEM <= 2m
- Natural Levee: if LC = non-water and Distance from channels <= 500m and DEM <= 3m
- Lower terrace: if LC = non-water and DEM <= 4m
- Middle terrace: if LC = non-water and 4m < DEM <= 5m
- Higher terrace: if LC = non-water and DEM > 5m.

Using the landform map, potential flood hazard was analyzed based on the potential of submergence of each landform unit. Geomorphological landform classification map is intended to estimate the nature and extent of flood. The potential flood hazard map that reflects relationship between the landforms and flood condition in this alluvial lowland area is shown in Figure 5 and summarized in Table 2.

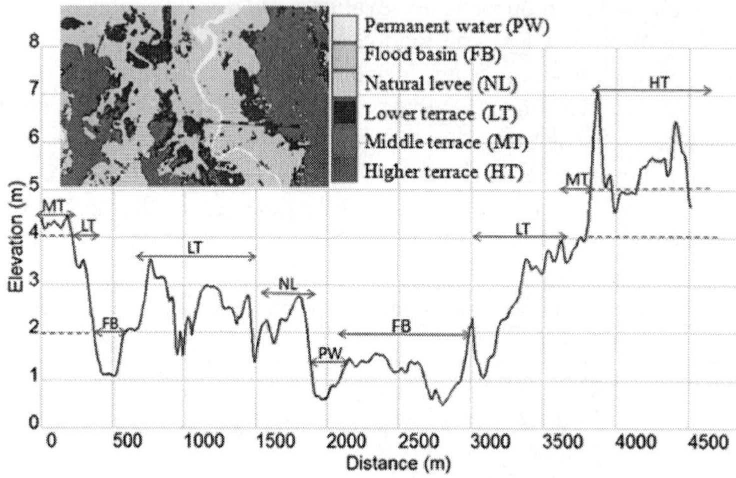

Figure 4. The geomorphological features from a cross session in a floodplain.

Table 2. Flood condition and potential of flood base on landform types.

Landform type	Flood condition	Potential of flood
Permanent water	Water level rise, but no damage	Non-flood hazard
Flood basin	Submerged for long time with deep water	High
Natural levee	Submerged in an extra-ordinary flood but drain well	Moderate
Lower terrace	Submerged in major flood	Low
Middle terrace Higher terrace	Never submerged	Very low

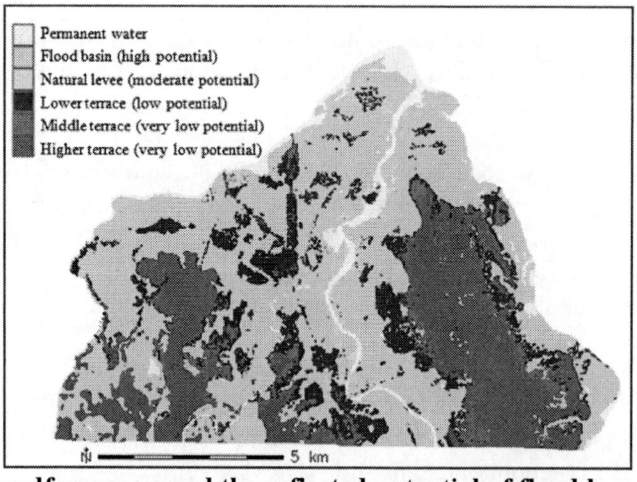

Figure 5. Landform map and the reflected potential of flood hazard in 2007.

2.4. Estimating the change of flood hazard potential based on land cover change

The study area is currently undergoing the rapid change of land cover. Landsat TM image in 2007 and high resolution RapidEye data in 2014 were applied for land cover classification of study area. Both data were acquired in March, so it is more reliable to detect the land cover changes.

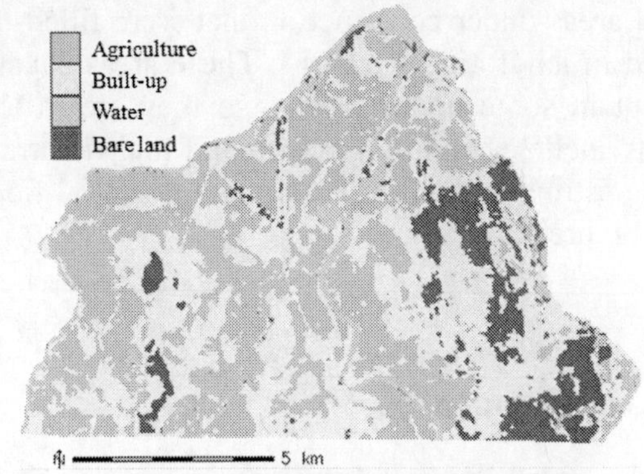

a. Land cover 2007 (Landsat TM 16/03/2007)

b. Land cover 2014 (RapidEye 02/03/2014)

Figure 6 : Land cover maps of study area in 2007 and 2014.

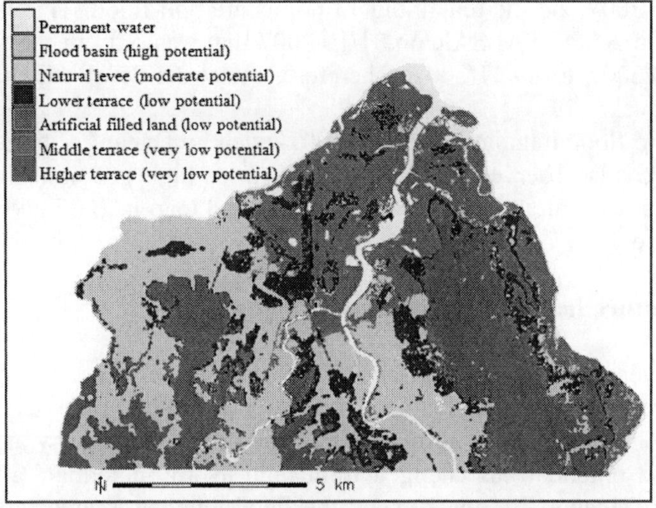

Figure 7. Landform map and the reflected potential of flood hazard in 2014

Comparison of land cover maps for 2007 to 2014 indicates that there is a significant change especially in built-up area and agriculture. The agriculture area in the northern was totally replaced by built-up and bare land. These bare land areas are mainly under construction for developing of new township. The urban development policy of Danang City has planned the conversion of these areas from rural communes to urban areas. A large area of bare land observed in the eastern of study area in 2007 was also replaced by built-up. The changes in this eastern included new residential areas in Ngu Hanh Son district (Danang city) and Dien Ban district (Quang Nam province), the development of series of universities (Danang Universities Zone - Ngu Hanh Son district) and Dien Nam-Dien Ngoc urbanized zone in Dien Ban district.

The change in land cover leads to the change in topography (valley-fills) as well as flood hazard potential. In this consideration, it is difficult to obtain DEM data in 2014 and also there is no flood event in this year, hence flood inundation map in 2014 is also not available. Land cover in 2014 and the previous landform map were used to estimate potential flood hazard in future. Agriculture was considered as flood basin in 2014, due to the close relation between this land cover type and flood inundation. Permanent water was extracted from water class in land cover map in 2014. Natural levee usually does not change in several years, therefore it was kept as same in landform map 2007. The significant areas of land cover change are the new residential areas and areas under construction that were filled-up. These areas were named as artificial filled land in landform map 2014. These areas usually are not easily affected by annual floods except in some major floods, so they are in low potential of flood hazard. Other landform units include, middle terraces and higher terrace were maintained from landform map 2007. The estimated landform map and potential flood hazard map for the year of 2014 can be seen in Figure 7.

4. RESULTS AND DISCUSSION

4.1 Flood inundation mapping and landform classification

Field work was conducted in March 2015 to collect 36 flood pillar data which contains all of the historical flood depth during past flood events. This survey point data then was imported to flood inundation map extracted from ALOS PALSAR and landform map in 2007 to evaluate the accuracy of these maps. There are 23 flood pillar points belongs to inundated area in flood map 2007 and the remaining 13 points are non-flood. This is because of flood map extracted from ALOS PALSAR on 31/10/2007 that was only in the first stage of flood event from 28/10/2007 to 09/11/2007. Therefore, the water level still has not reached to maximum flood level.

Comparing to flood inundation map in 2007 given by Danang City government (Phong Tran *et al.,* 2013), it is observed that the flood basin or high potential flood areas in 2007 extracted from this study are mostly coincided with flood map in 2007 given by Danang City government (Figure 8).

4.2. Evaluating future flood hazard potential

The land cover change as well as land filling has made change in flood condition in the corresponding areas. The estimated potential flood hazard map in 2014 shows the replacement of flood basin in the north and east parts of study area by artificial filled land, therefore the flood potential has changed from high to low potential. Comparing to flood scenario given by Danang City government, the flood inundation areas are coincide to high potential flood areas in estimated map 2014, and the new residential areas becomes non-flood areas (Figure 9).

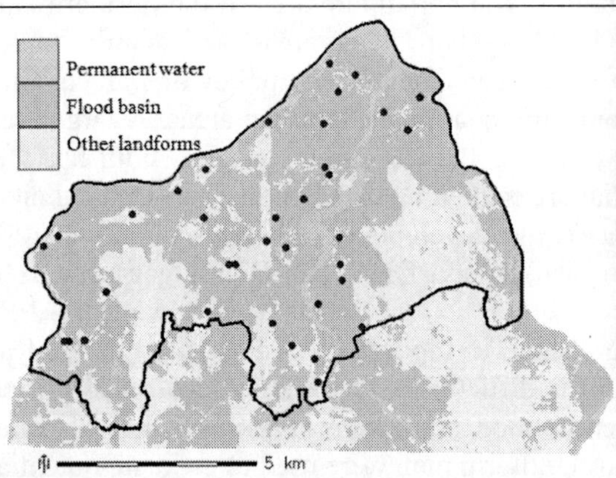

Flood basin extracted from landform 2007 and field survey flood pillar points data.

Flood map in 2007 at the maximum flood level (DOC, Danang City)

Figure 8. Accuracy assessment for flood inundation mapping

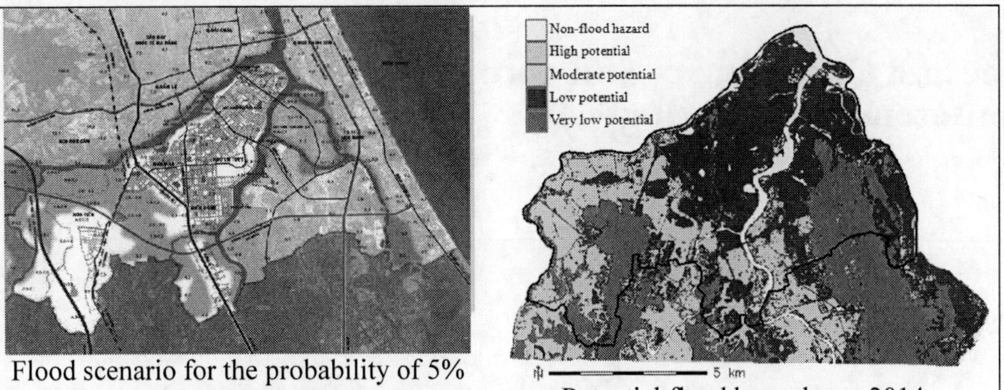

Flood scenario for the probability of 5% (20 years flood)

Potential flood hazard map 2014

Figure 9. Accuracy assessment for potential flood hazard mapping.

CONSCLUSION

ALOS PALSAR data supports for creating flood inundation map rapidly and accurately. The availability of PALSAR data taken during flood time is effective for characterizing flood inundation in this study. Moreover, high resolution DEM integrated with flood pillar data have contributed to evaluate the accuracy of flood map.

In this study flood basins are the lowland agriculture areas situated along either side of Cam Le river and Vinh Dien river. The land cover as well as topography has significantly changed from 2007 to 2014. The flood basin in 2007 were replaced by artificial filled land in 2014 in a large area in north and east parts of study area. The potential flood hazard in these areas also change accordingly from high potential to low potential.

Landforms units derived from rule-based classification of land cover map, 5m DEM data and flood inundation map not only facilitates the understanding of the nature of flood but also in flood hazard potential evaluation. The methodology developed in this study would be useful in coastal lowlands in Vietnam and other parts of the world.

REFERENCES

1. Do, T.V.H. and Nagasawa, R., 2014., Potential flood hazard assessment by integration of ALOS PALSAR and ASTER GDEM: a case study for the Hoa Chau commune, Hoa Vang district, in central Vietnam, *Journal of Applied Remote Sensing*, 8, 1-12.
2. Ho, L.T.K., Yamaguchi, Y. and Umitsu, M., 2012. Rule-based landform classification by combining multi-spectral/temporal satellite data and the SRTM DEM, *International Journal of Geoinformatics*, 8, 27-38.
3. Nonogaki, S., Masumoto, S. and Shiono, K., 2012. Gridding of geological surfaces based on equality-inequality constraints from elevation data and trend data, *International Journal of Geoinformatics*, 8, 49 - 60.
4. Phong Tran, Tran Tuan Anh, Tran Huu Tuan, 2013, *The economics of alternative development pathways: Preliminary scenarios case study Danang City, Vietnam*, Institute for Social and Environmental Transition (ISET) - International, September 2013.
5. Tran T.A., Raghavan V., Masumoto S., Yonezawa, G., Nonogaki S., Nemoto T. (2015) Using FOSS4G to Study Flood Inundation around Danang City, Vietnam, *Proceeding of FOSS4G India 2015*, 8-10 June, 2015.

Free and Open Source Software for Geospatial (FOSS4G) Conference Proceedings

Volume 15 *Seoul, South Korea* Article 29

2015

An Evaluation of Open Source Geographic Information Systems Routing Tools in Vaccine Delivery in Kano State, Northern Nigeria

Kehinde A. Adewara
eHealth Africa, Kano, Nigeria

AT9-02

An Evaluation of Open Source Geographic Information Systems Routing Tools inVaccine Delivery in Kano State, Northern Nigeria.

KEHINDE A. ADEWARA,

eHealth Africa, Kano, Nigeria.

ABSTRACT

The recent proliferation of online/desktop/mobileopen source geographic information systems (GIS)routing tools such as qgis road graph plugin (QRG), open street routing machine (OSRM), google maps engine (GME), graphhopper (GH), and OsmAnd has led to the need to provide a method forcomparatively evaluating the strength and weakness of these routing tools. This is crucial in view of its implication on the prospect and otherwise of routing related projects such as supply chain logistics, supply/delivery operations, and emergency services, among others. In this paper, comparative evaluation of these tools has been carried out using drive test survey and desktop routing estimation with respect to routine vaccine delivery in Kano, Nigeria. Kano state being one of the states in Nigeria with huge burden of health challenges with records of 3062 maternal death between 2005 – 2010(Ibrahim, 2014). *Thus vaccine delivery is one of such healthcare delivery programmes used to addressing some of these health challenges. The primary objective of this paper is to demonstrate comparative advantage of using open sourceGIS routing tools to optimize vaccine delivery process such that there would be significant reduction in logistics, manpower and cost associated with routine vaccine delivery. The capacity of the selected open source GIS routing tools was evaluated against this backdrop. Hence drive test survey was used to define the benchmark for determine the best rank among these desktop routing tools. The drive test survey was carried on selected number of delivery routes and the results were compared with values derived from desktop routing estimation using these tools. Two rounds of drive test survey were carried out for the delivery routes and an average was considered in order to minimum possible error associated with possible inconsistent in driving behavior. Significant discrepancies were observed in the outputs derived between desktop (QRG), online (OSRM, GME, GH) and mobile (OsmAnd) routing platforms. OSM vector base map was used across all the routing tools except GME.The overall outcome indicated QRG had the highest cumulative error margin of 67.52km while the lowest was reported for GraphHoper (46.17km) using same OSM base map. This is an indication that the routing algorithm used is not the same. When compared with GME that uses different base map, the cumulative error margin is very close (QRG – 67.52, GME – 55.99), an indication that similar routing algorithm has been used. Drive test outcome may not be sufficient to determine best or otherwise routing tool, it may be appropriate to consider other valuable criteria for the purpose of ranking these tools. Hence, those criteria were not limited to drive test/routing output error margin, others include capacity for multiple routing, base map completeness/content, support for traffic input, routing platform, and alternative routing option. With these considerations, QRG was ranked 1$^{st.}$ while OsmAnd (5) was least ranked. GME and GH had same ranking (2). QRG was ranked above other OSM based routing tools because it uses desktop platform and a capacity to integrate traffic input. It was ranked above GME majorly because of its robust OSM base map compared to google base map.*

1. INTRODUCTION

The need to be objective in the choice of appropriate open source GIS routing tools cannot be over-emphasized in view of its decision making consequence. This may have adverse economic implication on critical service deliveryand supply chain systems. However the need to embrace open sourceGIS routing tools is necessary and crucial in view of high cost associated with proprietary software yet must be embraced with caution in order to avoiddamaging consequent it may have on the decision making as well as the integrity of few other credible ones(Graser, Straub, & Dragaschnig, 2015).

open source GIS routing tools are some of the resources provided by the open source communities (such as open source geospatial foundation - OSGeo) through crowd sourcing initiative to support international efforts in addressing contemporary global challenges in agriculture (climate change, food security, drought, flooding etc.), health care (response to disease outbreak and eradication, wellbeing etc.), emergency (flooding, earth quakes etc.) among others. These efforts have been applauded worldwide most especially the kind of humanitarian support provided to the developing countries in Africa sub-Sahara(MapAction, 2011).

Hence this article took advantage of the enormous benefits of open source resources to addressing some of the basic problems being faced in Northern Nigeria by government in providing avoidable healthcare service delivery to its citizen through routine or periodic immunization. In view of limited fund to address some of these problems, the need to optimize vaccine delivery activities (which is key to the immunization programme) using reliable and credible cost effective open sourcerouting tools cannot be over-emphasized.Bimonthly routine vaccine delivery is a healthcare programme designed to ease access to vaccine supply in order to eliminate vaccine preventable disease. Thus, it is desirable to use open source GIS routing tools to carry out the necessary planning involving budgeting, manpower allocation, delivery schedules among others.

Due to issues associated with recent proliferation of these routing tools in recent time, it is imperative to conduct a comparative evaluation of strength and weakness of prospective open source GIS routing tools before use in order to avoid possible errors in the critical aspects of delivery planning. This effort would undeniably reduce possible revenue loss in the entire process of vaccine delivery system in terms of efficient deployment of manpower and logistics(Yekta, Yalçın, & and Akçay, 2008).

2. DATA AND METHOD

The approach adopted in solving the problem involves comparing outputs from both drive test survey and desktop routing estimation. Drive test output is used to define the benchmark to determine strength and weakness of the open source GIS routing tools as well as providing suitable multi criteria ranking among them.

2.1. Drive Test Survey

Drive test approach is the technique adopted in defining the benchmark for determining the most reliable, credible and efficient GIS routing tools. Such definition is based on the levels of deviation of the GIS routing tools from the drive test output.

Drive test technique seems to be a reliable means of measuring accuracy in this respect because comparing routing outputs from these tools with drive test results is a strategy to avoid conceptual controversies associated with different routing algorithms used in these tools (since any outputs is a manifestation of strength or weakness of its algorithms). Drive test analysis is a technique that has been used in different studies as a control measure depending on study objective and scope. It is widely used in telecommunication studies(Sanders, Linder, Pratt, Dickherber, Floyd, & Pickering, 2004; Boxberger, Lawver, & Smithey, 2010). However it was used in this case as a technique for acquiring baseline information which serves as a benchmark for determining accuracy among different routing outputs.

In view of the fact that it may be practically impossible to conduct drive test survey for all the facilities in the state due to logistic challenge, hence a representative sample size was considered with minimum error of margin. To ensure that that the smallest possible sample size is considered and to ensure it is representative of the entire sample population, proximity/distance coverage (between the state cold store and the farthest facility in each zone) was considered as the sample population.

The sample size was determined at 95% confidence level and narrow confidence interval (CI) of 5. The CI of +/-5 represents the possible narrow margin of error in the sampling(Myles, Douglas, & Eric, 2013). Thus the sample population of 437 km was considered which was the cumulative distances between farthest facilities in each zones and a sample size of 326 km was determined using online calculator (see table 1). The actual distant covered was 411.74 km for 10 facilities (see table 2), which was 85.74 km in excess of the sample size. Drive test exercise and routing estimation were then conducted for the 10 health facilities.

Table 1 - Sample Distribution

Zone	Farthest Distance (km)	Sample Size (km)	Number of facilities served	Distant Covered (km)	Driver
Rano	196	130	2	115.06	A

Bichi	90	73	3	130.4	B
Nassarawa /Wudil	151	123	5	166.28	C
Total	**437**	**326**	**10**	**411.74**	

Table 2–Kano State Primary Health Facilities targeted for Routing Destinations

S/N	PrimaryName	Settlement	Local Government Name
1.	Gwarzo General Hospital	Tudun Burtu	Gwarzo
2.	Shanono Comprehensive Health Center	Unguwar Hakimi	Shanono
3.	Tsanyawa Comprehensive Health Center	Yan Amar	Tsanyawa
4.	Gezawa General Hospital	Kuka	Gezawa
5.	Kunya Primary Health Center	Kunya	Minjibir
6.	Panda Model Primary Health Center	Wudilawa	Albasu
7.	Hungu Primary Health Center	Gidan Tudu Yamma	Albasu
8.	Kachako Primary Health Center	Unguwar Tsamiya	Takai
9.	Falgore Health Post	Tudun Ningi	Doguwa
10.	Dadin Kowa Health Post	Tasha	Doguwa

Hence, drive test was conducted to determine actual travel distant measurement between the state cold store and the 10 selected health facilities. The travel distance and time during the drive test survey was determined using mobile mapping GPS enabled solution called OsmAnd tablet device. The device is a wide screen android based digital tablet preinstalled with OsmAnd app. The device is made up of a detailed base map, navigation tool and a plugin called trip recording. The plugin is the app used for recording distances covered during the drive test survey.

Multiple drive test rounds were conducted for the purpose of improving the quality of the drive test outputs and to adequately compensate for inconsistency in driver behaviors. In order to manage off route driving behavior, the drive test survey considered average of two drive test rounds. Each of the drive test rounds considered optimum delivery routes between the state cold store and the 10 health facilities (fig.1). The delivery routes are the same for the two rounds. During the survey, each driver is supported by a field coordinator who operates the device. The device is used to record the beginning and end of each trip and save accordingly. This is repeated for all the trips, the outputs are saved as gpx files and shared for later download and processing.

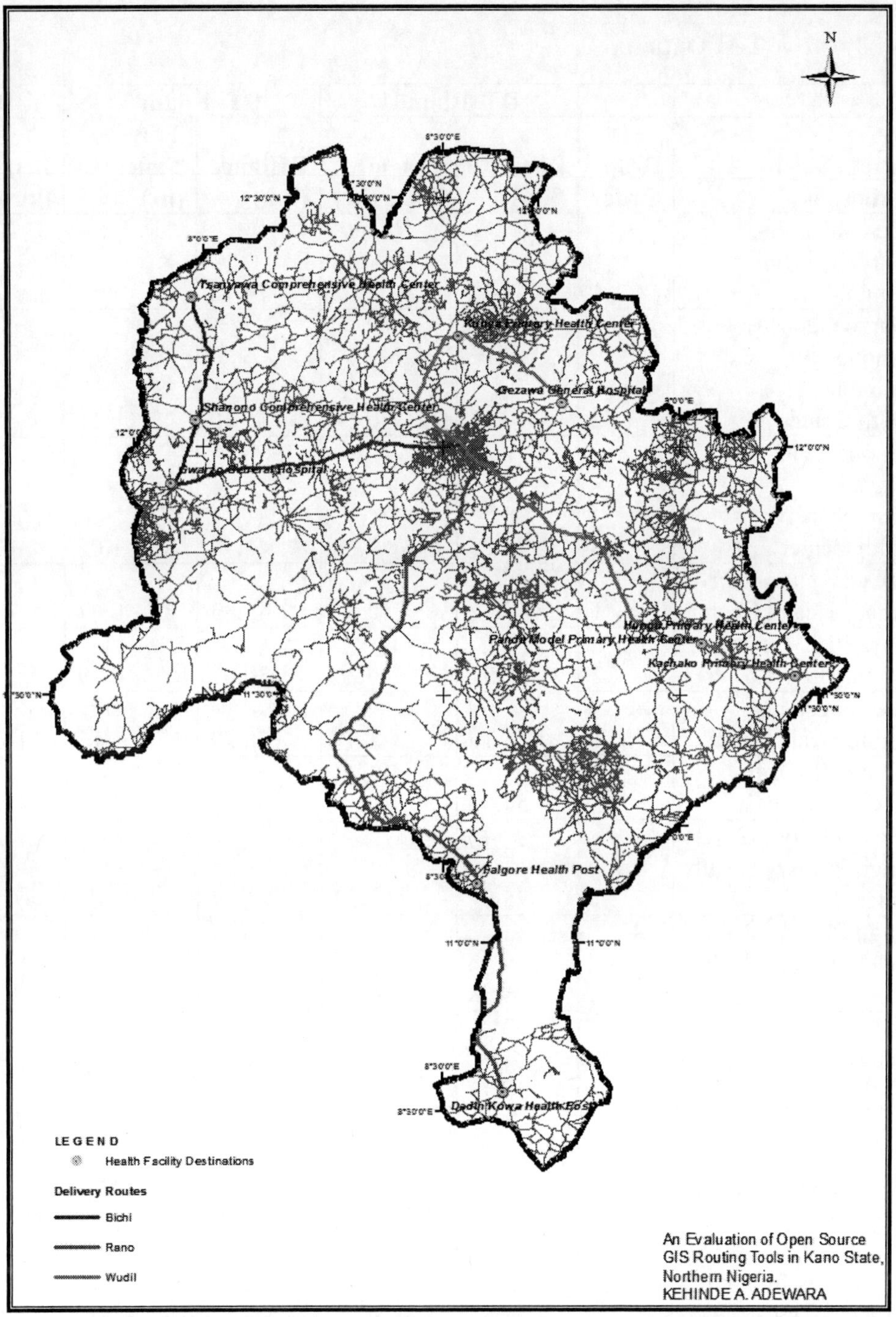

Figure 1 - Health Facility Delivery Routes in Kano State, Nigeria.

Outputs from drive test survey is a control measurement for determining the accuracy of the desktop routing tools because it is the direct measurement of actual distance travelled compared to estimation being conducted during desktop routing exercise. The drive test outputs were derived as gpx file which was later downloaded, converted and integrated with OSM base map to determine distances and travel time (see table 3).

Table 3 - Drive Test Outputs

		DT_Round1		DT_Round2		DT_Average	
Routes (Origin-Destination)	**Drop Order**	**Distance (km)**	**Travel Time (hr)**	**Distance (km)**	**Travel Time (hr)**	**Distance (km)**	**Travel Time (hr)**
State cold store-Gwarzo General Hospital	1	74.34	1.12	74.78	1.55	74.56	1.34
Gwarzo Gen Hosp-Shanono Comprehensive Health Center	2	15.53	0.14	15.45	0.13	15.49	0.14
Shanono CHC-Tsanyawa Comprehensive Health Center	3	79.35	0.58	45.30	1.02	62.33	0.80
State cold store-Kunya Primary Health Center	1	49.12	0.54	45.33	0.47	47.23	0.51
Kunya PHC-Gezawa General Hospital	2	32.91	0.26	30.07	0.19	31.49	0.23
State cold store-Falgore Health Post	1	146.19	1.49	138.79	1.16	142.49	1.33
Falgore HP-Dadin Kowa Health Post	2	54.07	0.36	53.99	0.49	54.03	0.43
State cold store-Hungu Primary Health Center	1	75.81	1.00	75.94	1.02	75.88	1.01
Hungu PHC-Panda Model Primary Health Center	2	5.58	0.05	5.53	0.04	5.56	0.05
Panda MPHC-Kachako Primary Health Center	3	25.73	0.16	25.76	0.18	25.75	0.17

2.2. Desktop RoutingEstimation

The desktop routing estimation was the technique used to compare routing outputs from multiple routing tools(Haklay, 2010). It was conducted for all the delivery routes for the 10 facilities considered during the drive test survey. A number of open source GIS routing tools were used for this purpose, which were primarily classified as online, mobile and desktop routing tools (see table 4).

Table 4 - GIS Routing Tools

S/N	GIS Routing Tool	Platform Type
1	QGIS Road Graph plugin (QRG)	Desktop
2	Open Street Routing Machine (OS RM)	Online
3	Google Maps Engine (GME)	Online
4	GraphHopper (GH)	Online
5	OsmAnd	Mobile

These routing tools use different routing protocols to determine distance and travel time. Speed limit defined for each road class is used as impedance factor in travel time estimation. Hence, speed limit profile for some of these tools are outlined below –

Table 5 - Speed Limit Consideration

Road Class	os rm	GH	QRG
motorway	90	#N/A	#N/A
motorway_link	45	#N/A	#N/A
trunk	85	80	100
trunk_link	40	75	5
primary	65	65	100
primary_link	30	60	5
secondary	55	60	80
secondary_link	25	50	5
tertiary	40	50	60
tertiary_link	20	40	5
unclassified	25	30	30
residential	25	30	30
living_street	10	5	5
service	15	20	5
track	5	15	5
ferry	5	#N/A	#N/A
movable	5	#N/A	#N/A
shuttle_train	10	#N/A	#N/A
default	10	#N/A	#N/A

In carrying out routing estimation, a number of assumptions were made and certain limitations were unresolved. It is assumed that all routing tools consider travel by vehicle (not pedestrian/walking, bicycle) and shortest path option. Limitations include inability to factor in traffic conditions on these delivery routes, as well as inability to consider same routes among all the routing tools due to insufficient base map. Deficiency in base map was as a result of derived error in the base map contents (include digitizing error, geo-referencing errors among others). Hence these errors varied among various routing tools even with those that use same base map. These limitations and other factors such as geo positioning accuracy as well as limitations of inbuilt routing/network model were the reasons for the observed discrepancies in the routing outputs using same base map (see figure 2).

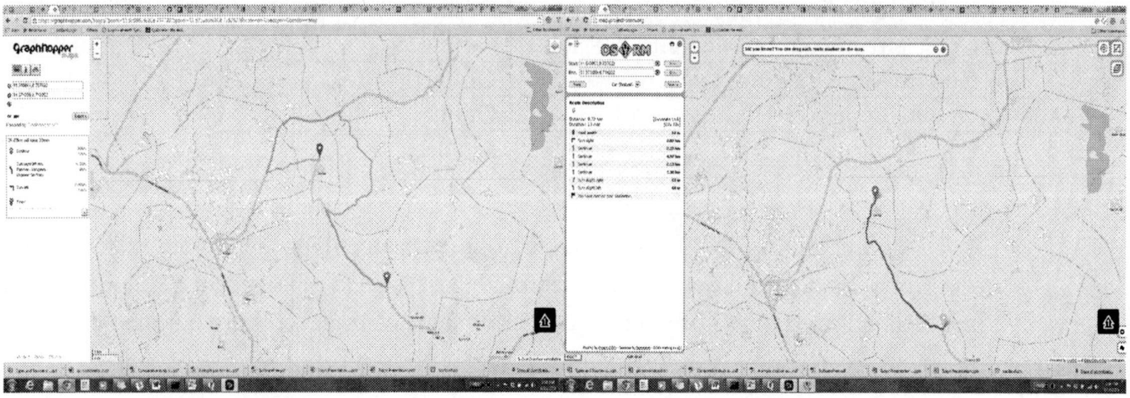

Figure 2 - Different Outputs from Routing Tools using same base map

Hence, each of the routing tools was used to carry out the routing estimation for all the delivery routes considered serving the 10 health facilities. The delivery route is an optimum route (with shortest possible distance) connecting the state cold store to the 10 health facilities across the state (table 6).

Table 6 - Desktop Routing Outputs

			QRG	GME	GH	OS RM	OsmAnd
Routes	**Drop Order**	**Zone**	**Distance (km)**	**Distance (km)**	**Distance (km)**	**Distance (km)**	**Distance (km)**
State cold store-Gwarzo General Hospital	1	Bichi	69.71	71.00	71.19	71.20	73.70
Gwarzo Gen Hosp-Shanono Comprehensive Health Center	2	Bichi	14.97	15.40	15.63	15.60	15.60
Shanono CHC-Tsanyawa Comprehensive Health Center	3	Bichi	30.41	32.00	46.96	30.60	23.90
State cold store-Kunya Primary Health Center	1	Nassarawa	35.90	36.00	36.08	36.10	36.00
Kunya PHC-Gezawa General Hospital	2	Nassarawa	29.51	29.40	29.56	29.60	29.50
State cold store-Falgore Health Post	1	Rano	132.51	135.00	135.00	143.00	143.00
Falgore HP-Dadin Kowa Health Post	2	Rano	52.77	58.80	52.62	52.60	53.00

State cold store-Hungu Primary Health Center	1	Wudil	70.54	70.60	70.65	70.70	70.30
Hungu PHC-Panda Model Primary Health Center	2	Wudil	5.55	5.40	5.54	5.54	5.53
Panda MPHC-Kachako Primary Health Center	3	Wudil	25.39	25.20	25.39	25.40	25.30
Total			**467.26**	**478.80**	**488.62**	**480.34**	**475.83**

The outcomes of both drive test survey and desktop routing estimation were analyzed to identify peculiar trend for further investigation. Thus, the nature of discrepancies or inconsistency in the outcomes was reported prior to multi-criteria ranking.

There were significant discrepancies in both desktop routing outputs and the drive test outcome. These discrepancies were based on the variation between the drive test average and routing outputs and were reported as error margin (table 7).

Table 7–Comparative Discrepancy Outcomes

			QRG	**GME**	**GH**	**OS RM**	**OsmAnd**
Routes (Origin-Destination)	**Drop Order**	**Zone**	Error Margin (km)	Error Margin (km)	Error Margin (km)	Error Margin (km)	Error Margin (km)
State cold store-Gwarzo General Hospital	1	Bichi	4.85	3.56	3.37	3.36	0.86
Gwarzo Gen Hosp-Shanono Comprehensive Health Center	2	Bichi	0.52	0.09	-0.14	-0.11	-0.11
Shanono CHC-Tsanyawa Comprehensive Health Center	3	Bichi	31.92	30.33	15.37	31.73	38.43
State cold store-Kunya Primary Health Center	1	Nassarawa	11.33	11.23	11.15	11.13	11.23
Kunya PHC-Gezawa General Hospital	2	Nassarawa	1.98	2.09	1.93	1.89	1.99
State cold store-Falgore Health Post	1	Rano	9.98	7.49	7.49	-0.51	-0.51
Falgore HP-Dadin Kowa Health Post	2	Rano	1.26	-4.77	1.41	1.43	1.03
State cold store-Hungu Primary Health Center	1	Wudil	5.33	5.28	5.22	5.18	5.58
Hungu PHC-Panda Model Primary Health Center	2	Wudil	0.00	0.15	0.01	0.01	0.02

Panda MPHC-Kachako Primary Health Center	3	Wudil	0.36	0.55	0.36	0.35	0.45
Sum Total			**67.52**	**55.99**	**46.17**	**54.45**	**58.96**

The error margins reported in table 7 and figure 3 were largely as a result of road quality in the rural area and traffic condition in the urban area, which affected drivers driving behaviors with respect to choice of routes. The cumulative margin of error ranges between 67.52 km to 46.17km. The maximum error margin was reported for QRG routing tool while the minimum was reported for GH both using OSM base map. In this case, choice of different routes must have being considered as a result of different routing algorithms used (a case for future investigation). When the cumulative error margin of a routing tool using different base map (GME) is considered with QRG, the difference is very close (QRG – 67.52, GME – 55.99), an indication that similar algorithm has been used.

In view of the degree of error margin reported, it is not enough to draw conclusion on the best or otherwise routing tools based on this evident. Hence it is imperative to consider other valuable parameters which will be used as criteria for subsequent ranking of these routing tools.

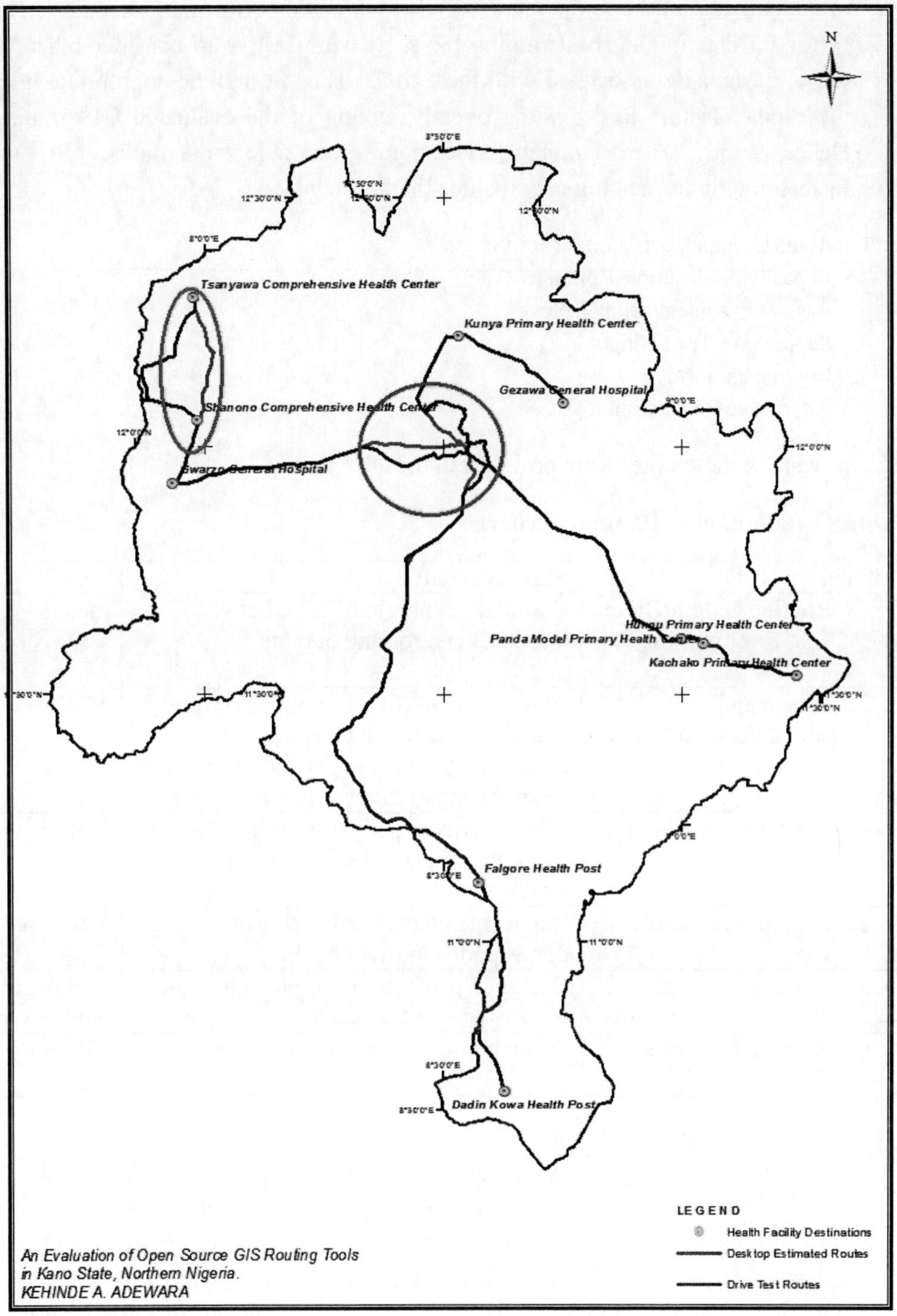

Figure 3 - Comparing Discrepancy in Drive Test Survey and Desktop Routing Estimation

2.3. MultiCriteria Ranking

In view of the limitation associated with using error margin condition as a sole factor for determining best routing tools, it is imperative to consider other 'win and loose' advantages associated with these tools. Thus, it may be appropriate to consider these other factors in providing overall ranking of the evaluated GIS routing tools. Hence multi criteria ranking technique was adopted(Tofallis, 2014). These factors/criteria for ranking may include the following –

1. Routing Output/Drive Test Error margin
2. Capacity For Multiple Routing
3. Base-Map content/completeness
4. Support For Traffic Input
5. Routing Platform
6. Alternative Route Option

The measure of these criteria are provided in the table below -

Table 8 - Definition of Ranking Criteria

Criteria	Measurement	Unit
1. Routing Output/Drive Test Error margin	Cumulative margin of error between drive test and routing outputs.	meters
2. Base map content/completeness	Analysis of % coverage of vector base map compared with imagery, considering average of five clusters each in urban and rural areas (Haklay, 2010).	%
3. Capacity For Multiple Routing	The capacity to support multiple routing was determine by the options provided by the app designed feature.	Yes or no
4. Support For Traffic Input	Determine by onboard designed feature that support traffic modeling	Yes or no
5. Routing Platform	Designed platform for implementation	Mobile, desktop, online/web
6. Alternative Route Option	Provision for choice of alternatives	Yes or no

Hence, a multi-criteria ranking technique was adopted to provide suitable ranking to these routing tools. There are several methods of conducting such ranking well documented in the literature,most common is the additiveweighing factor approach(Tofallis, 2014). This involves assigning certain weights to these criteria and normalizes values reported in different units. This approach is often constrained by non-uniform way of normalizing data which often results in different outcomes(Sagar & Najam, 1998; Tofallis, 2014). But the multiplicative weighing score was able to overcome this limitation and it is commonly used in the decision making hierarchy of world leading organizations such as United Nations Development Programme (UNDP) in developing annual human development index (HDI), an instrument used for ranking nations based on human per capita income, life expectancy and education(UNDP, 2010).

2.3.1. Threshold Definition and Weighing

The range of values that were derived for these criteria were given in different units of measurement, hence the thresholds for each of the criteria were defined and ordinal weights assigned (table 9).The threshold for the drive test/routing error margin was determined using a median (56.62) of the cumulative error margin (46.17 – 67.52) as the threshold. Thus any error margin greater than 58.75 is assigned 0 and less than is assigned 1. Threshold for base map completeness was derived from average consensus option expressed in percentage for both five sample clusters in rural and urban. The threshold values for rural and urban clusters were determined as 50% and 70% respectively by consensus. Hence the average (60%) of the two was used for assigning weight of 1 for greater than 60% and 0 for less than. Consensus opinion was used to assign threshold of 1 for desktop routing platform and 0 for others. For other criteria, ordinal values of 1 and 0 were assigned for yes and no responses accordingly.

Table 9 - Normalized Thresholds

Criteria	Value Range	Threshold condition	Assigned Weight s
1. Routing Output/Drive Test Cumulative Error margin	46.17 – 67.52 km	>56.62 <56.62	0 1
2. Base map content/completenes s	0 - 100%	≤ 60 ≥ 60	0 1
3. Capacity For Multiple Routing	Yes/No	No Yes	0 1
4. Support For Traffic Input	Yes/No	No Yes	0 1
5. Routing Platform	Mobile, desktop, online/web	Mobile/Online Desktop	0 1
6. Alternative Route Option	Yes/No	No Yes	0 1

2.3.2. Scoring/Ranking

In view of this outcome (table 10), scoring was derived based on frequency of 1 occurrence while the highest frequency of 4 was ranked 1st and least score of 1 was ranked 4th.

Table 10 - Normalization Outcome

Ranking Criteria	QRG	GME	GH	OS RM	OsmAnd

1. Routing Output/Drive Test Error margin	0	1	1	1	0
2. Capacity For Multiple Routing	1	1	1	0	0
3. Base map content/completeness	1	0	1	1	1
4. Support For Traffic Input	1	0	0	0	0
5. Routing Platform	1	0	0	0	0
6. Alternative Route Option	0	1	0	0	0
Score	**4**	**3**	**3**	**2**	**1**

Table 11- Ranking Outcome

Routing Tools	**Score**	**Rank**
QRG	4	1
GME	3	2
GH	3	2
OS RM	2	3
OsmAnd	1	4

3. RESEARCH FINDINGS, CONCLUSION AND RECOMMENDATIONS

It is not surprising to find distant variation in the routing outputs of routing tools using different base maps (OSM and google) because different choice of routes are chosen. But to discover a significant variation in routing outputs of routing tools using same OSM base map is worrisome. It is an indication that there is an inconsistency in the routing algorithm used. The outcome of drive test clearly shown that GH tool has a better routing algorithm with lowest cumulative error margin. Future research is hereby encouraged to investigate how road graph plugin of QRG would integrate GH routing algorithm for better performance.

It is important to emphasize that the GME better drive test performance over QRG was just because the selected delivery routes were largely within urban area where GME base map content is relatively good. It is thus recommended that outstanding GME features such as alternative routing option should be considered for QRG integration.

Scheduling delivery and routing within Kano metropolis was largely constrained by lack of traffic details, hence the routing outputs which considered speed limit as an input for travel time estimation was not consistent with travel time output derived from drive test survey. Thus it is anticipated that future research would consider investigating metropolitan traffic. Speed limit consideration is still valid for interstate and rural areas.

QRG is at the moment constrained by inability to handle batch routing. All the routing tools considered are equally unable to do batch routing. Hence it is expected that future research would focus of developing batch routing component and to integrate other features such as alternative route options.

4. ACKNOWLEDGMENT AND CONTRIBUTIONS

I strongly acknowledge the contributions and selfless support I have received from the Management and staff of eHealth Africa (eHA), such contributions are hereby outlined below (not necessarily in order of important) –

eHA Management	-	For providing the platform and financing the trip
GIS Team	-	For the valuable criticisms
Human Resources	-	For assisting with necessary documents
Monitoring, Research& Evaluation	-	For their immense support from researchperspective.
Health Camps	-	For providing personnel support during the drive test survey
Operations (Reservations)	-	For assisting hotel reservation and flight booking
Operations (Logistics) survey	-	For providing drivers that carried out the drive test
Finance	-	For assisting with necessary documents and fund disbursement

REFERENCES

Boxberger, S. D., Lawver, M. B., & Smithey, P. S. (2010). Method and system for automatic analysis and management of drive test routes . *Sprint Spectrum L.P.*

Graser, A., Straub, M., & Dragaschnig, M. (2015). Is OSM Good Enough for Vehicle Routing? A Study Comparing Street Networks in Vienna. In G. H. Gartner, *Progress in Location-Based Services* (pp. pp 3-17). Switzerland: Springer International Publishing.

Haklay, M. (2010). How good is the volunteered geographic information? A comparative study of OpenStreetMap and Ordainance Survey datasets. *Environment & Planning B: Planning and Design* , Vol. 37 pp 682 - 703.

Ibrahim, Y. S. (2014). Temporal Analysis of Maternal Mortality in Kano State, Northern Nigeria: A Six-Year Review. *American Journal of Public Health Research*, 2(2), 62-67.

MapAction. (2011). *Field Guide to Humanitarian Mapping.* Buchinghamshire, U.K: MapAction.

Myles, H., Douglas, A. W., & Eric, C. (2013). *Nonparametric Statistical Methods, 3rd Edition.* John Wiley & Sons.

Sagar, A., & Najam, A. (1998). The human development index: A critical ical review. *Ecological Econom*, 25:249–264.

Sanders, A. D., Linder, P. S., Pratt, T. G., Dickherber, A. J., Floyd, T. S., & Pickering, L. W. (2004). Radio network test analysis system. *Telecom Network Optimization, Inc.*

Tofallis, C. (2014). Add or Multiply? A Tutorial on Ranking and Choosing with Multiple Criteria. *Institute for Operations Research and the Management Sciences (INFORMS)*, 14(3):109-119.

UNDP. (2010). *United Nations Human Development Report 2010: The Real Wealth of Nations.* New York: Palgrave Macmillan.

Yekta, Y., Yalçın, Y., & and Akçay, A. E. (2008). *An Open Source Java Code For Visualizing Supply Chain Problems.* Jönköping, Sweeden: CELS.

Free and Open Source Software for Geospatial (FOSS4G) Conference Proceedings

Volume 15 *Seoul, South Korea* Article 30

2015

Modifications To Web Processing Service Standard For Client-Side Geoprocessing

Evgeny Panidi
Department of Cartography and Geoinformatics, Saint-Petersburg State University

Eduard Kazakov
Department of Cartography and Geoinformatics, Saint-Petersburg State University

Anton Terekhov
Department of Cartography and Geoinformatics, Saint-Petersburg State University

Evgeny Kapralov
Department of Cartography and Geoinformatics, Saint-Petersburg State University

AT8-04

MODIFICATIONS TO WEB PROCESSING SERVICE STANDARD FOR CLIENT-SIDE GEOPROCESSING

Evgeny Panidi, Eduard Kazakov, Anton Terekhov and Evgeny Kapralov

Department of Cartography and Geoinformatics, Saint-Petersburg State University
33 10-th Line V.O., Saint-Petersburg 199178, Russia
Email: panidi@yandex.ru

ABSTRACT

Nowadays we see the rapid growth of solutions number for geospatial data processing in the Web (i.e. geoprocessing). One of the main trends of Web geotechnologies evolution is the transition from Web map applications to the Web GIS applications, which are supplement the maps delivery with the analytic tools providing to the end user through Web interface. In fact, the only general open standard describes implementation rules for Web geoprocessing services. This is the Open Geospatial Consortium Web Processing Service standard, which is fully server-oriented. Moreover, the vast majority of currently used solutions (both open source and proprietary) are server-oriented, i.e. assume the server resources only as the computational resource. However, some researchers underline that it is possible way to transmit the executable code to the client for client-side computations and geoprocessing. Also, some general Web architecture concepts assume the effectiveness of client-side computations, e.g. Fog Computing concept. Our practical experience also shows that in some cases it is useful to have ability of client-side geoprocessing, which is not opposite but complement technology to the server-side processing technologies. In addition, we believe that it is more useful to have the ability to run the same processing tool by choice on server or client side. We name such double-sided services as Hybrid Geoprocessing Web Services.

We study and discuss the approaches to gap filling in client-side geoprocessing general schema. For this purpose, we implemented previously the getProcess request as addition to the WPS protocol. Additionally at the previous steps of our study, we proposed a possible structure of getProcess request and draft XML file structure for its response, which describes the list of executable resources and their dependencies. Currently we working on detailed methodology of processing tools implementation and testing. We use the Python programming language as primary development tool, because of its applicability to build both server- and client-side cross-platform processing tools using single core program code. We use Python also for implementation of needed infrastructure components, such as HGWS server that supports the getProcess request/response performing, and client-side Runtime Environment that provides executable code orchestration on the client. Achieved results need to be discussed widely and carefully. However, main conclusion of our current work is that client-side geoprocessing schema in general could be relatively simple and compatible backward with current standards. The HGWS concept is applicable when implementing client-side geoprocessing Web services in small-scale projects and could be the entering point for study of distributed geoprocessing systems implementation.

1. INTRODUCTION

Only a few years ago the primary use cases of the Web, when operating with geospatial data (geodata), were the data delivery to the end user and the graphical representation in the browser or in the desktop application, such as the well-known Google Earth geobrowser (*http://www.google.com/earth/*), and desktop software of Geographic Information Systems (desktop GISs). Nowadays we may observe the rapidly growing amount of different Web-geospatial applications. Geospatial technologies implementation in the Web became the basis for the Geoweb (Scharl A. *at al.*, 2007) knowledge area formation, which integrates the geospatial aspect into data structuring and data search technologies. Also the use of Web technologies in Geoinformation Science and Technology (GIS&T) made possible the emergence of such disciplines as Web Mapping and Web Cartography (Kraak M.-J. *at al.*, 2001; Fu P. *at al.*, 2010), which are incorporate the techniques and methods of graphical map-like data visualization in the Web. This way of common use of the Web-based tools and geodata was quite natural at the early evolution stages, taking into account that the Web infrastructure in General was formed primarily as a mean of universalized data representation. However, some new trends are formed at this cross-scientific and multidisciplinary knowledge area.

One of the most important trends is the shift from use of passive operations with geodata (i.e. storage, transmission and visualization) to active operations (i.e. data mining and pattern recognition, data processing and different area specific types of data analysis). One of the sources for this shift is the paradigm change in Web technologies, namely the implementation of the Web 2.0 concept (O'Reilly T., 2005), that implies (despite the presence of different interpretations) the user interaction with the Web environment and the transition from static Web sites to Web(-based) applications. On the other hand, the evolution of Web mapping/Web cartography included not only the implementation of Web mapping applications and interactive geodata representation, but repeated the way of the GIS&T in whole and transited from presentation to processing. In this way, the term "Web GIS" has become widely used in recent years. This term indicates the Web-based systems and applications, targeted on solving analytical problems using geodata and geodata manipulation problems in general (Fu P. *at al.*, 2010).

These evolutionary processes are confirmed with the Open Geospatial Consortium (OGC) standards appearance chronology (*http://www.opengeospatial.org/*):

- Web Map Service (WMS) – 2000 (Doyle A. (ed.), 2000).
- Web Feature Service (WFS) – 2002 (Vretanos P.A., (ed.), 2002).
- Web Coverage Service (WCS) – 2003 (Evans J. (ed.), 2003).
- Web Processing Service (WPS) – 2007 (Schut P. (ed.), 2007).
- Web Coverage Processing Service (WCPS) – 2009 (Baumann P. (ed.), 2009).

Additionally, it should be noted that the paradigm of Cloud Computing (Buyya R. *at al.*, 2011) and the use of thin clients are the most used basis for current Web-based geotechnologies. The services, which are implemented accordingly to the last two of above-mentioned standards, transmit only the result data to the user and look like dynamically generated data services for the user, despite the standards' names. This is due to that these standards are focused on the server-side data processing. The processing tools in this case are hidden deeply from the user at the server side. Thus, current standards describe the implementation rules for geoprocessing Web services, which could be fitted well only within the architecture of cloud solutions or computing clusters where the data processing is performed on the server side under centralized control. This centralized-control approach is used in other current trend areas such as development of the technologies for integration of distributed geospatial data arrays and technologies for big geospatial data.

Finally, another one current trend of automation of the Internet and Web makes the growing impact onto GIS&T. This trend related with such concepts as Web 3.0 (Calacanis J., 2007), Semantic Web (Berners-Lee T. *at al.*, 2001), Semantic Web Services (SWS) (McIlraith S.A. *at al.*, 2001), and Internet of Things (IoT) (Ashton K., 2009). These concepts together form the opportunities for implementation of machine-readable services in addition to machine-readable data. The growing impact of these concepts is confirmed by appearance of OGC standards related to the design and use of sensor grids, which are included into the OGC Sensor Web Enablement framework (Bröring A. *at al.*, 2011).

It is quite remarkable in this context that the gap remains in the area of client resources use. The use of decentralized systems (Grid Computing systems for example) has a small reflection in geoinformation technologies, except of some case projects (Aktas M.S. *at al.*, 2005; Coene Y. *at al.*, 2007). However, the consideration of modern concept of Fog Computing (Fernando N. *at al.*, 2013; Hong K. *at al.*, 2013; Abdelshkour M., 2015) in conjunction with the concept of Semantic Web Services, allows to assume that the question of client-side computing in Web-based geodata processing services will be relevant in the near future.

Thus, taking into account these concepts, when implementing the Web-based geodata processing services, we could make next conclusions:

- We should not neglect the computational resources available on the client side, which are sufficient for solving many computational tasks and data processing tasks.
- When implementing the Web services for data processing (including geodata processing) on the client side, we need to automate the search and use of such services.
- Therefore, standardization of the processing execution rules on the client side, is needed additionally.

The use of Web services for client-side geodata processing involves the transmitting to the client of the executable files of some process or processes, which contain the source code of some processing algorithms. The idea of executable code transmitting is quite obvious and is not new in general. In relation to the geodata processing, it was mentioned by other authors also (Coene Y. *at al.*, 2007; Keens S. (ed.), 2007). However, widely-known and standardized solution of this idea is not known for us. In our project, we study the ways and possibilities of unification (and standardization in the future) of the executable code transmitting methods and means for the Web geoprocessing services (Kazakov, 2013; Kazakov *at al.*, 2015; Panidi, 2013; Panidi, 2014; Panidi *at al.*, 2015). We working on implementations of such services, which imply the ability to perform processing on both server and client side. These services we name as the Hybrid Geoprocessing Web Services (HGWS).

2. PROJECT SCOPE

The main subjective reason of the project in our case is the presence of a number of thematic projects related to the geodata processing. Some thematic specialists involved in these projects, who do not have advanced skills in the use of GIS software for data analysis. In this context, being the GIS professionals, we face with the task of design, implementation, configuring, and providing some software tools for data processing and analysis for the end user (some thematic specialist in other words).

It is obvious that the end user could explore new software tools easier if these tools are in a familiar interface. It is also obvious that the most intuitive desktop interface for many users today is the Web interface. Such background became the main prerequisite for the desire to use the Web services for geodata processing in our thematic projects.

However, many times we faced a situation in which the use of currently standardized technologies becomes irrelevant in small and medium-size projects, due to focusing of these technologies on the server-side processing. The server-oriented feature of geospatial Web standards leads to the need of building of the server or cloud-based computing infrastructure, and additionally involves the time spending when transmitting initial data to the server for processing.

Thus, the main goal of our project is the technologies and tools development for complementation of existing Web-based approaches to geodata processing with the client-side processing techniques. Moving to this goal, we assume:

- To minimize requirements (software and hardware) for the server-side part of system by transmitting only compact executable processing application components to the user.
- To ensure backward compatibility with existing standards, especially by providing backward compatibility with the OGC WPS standard and possibility to perform processing on the client or server side by user's choice, accordingly to available resources and context of the project.
- To give the user more flexible control of geodata that stored at the client side when processing with Web services by providing a client side processing for applicable cases.

2.1 Development tools

As the base software platform, we use the Apache HTTP Server (*http://httpd.apache.org*) and PyWPS WPS server (*http://pywps.wald.intevation.org/*). The choice of HTTP server is due only to the Apache administering experience and do not imply any special requirements to HTTP server performance. The absence of these requirements concerned with the fact that currently we perform the project as a case study and a prototype solution, but not as an industrial solution. The choice of PyWPS is due to that this WPS server was the only stand-alone solution (not integrated into any geospatial software complex for servers) at the time of the project launch. PyWPS was used for WPS functionality implementation.

As the primary program code development tool, we use the Python programming language (*https://www.python.org/*). All HGWS functionality that differs from WPS functionality (Mueller M. *at al.*, 2015; Schut P. (ed.), 2007) were implemented using Python as separate HGWS software tools (server-side or client-side).

However, when implementing the project we pay special attention to the exclusion of fixed bindings to the programming tools and try to implement independent architecture that suitable for the design of abstract specification.

2.2 Web interface

For purposes of HGWSs publishing and testing, we created a prototype of geoportal interface (*http://195.70.211.131/*). The Web interface that implemented in this prototype is very simple. It and includes several descriptive Web pages (such as home page, page with a list of publications, etc.) and two main pages for user interactions. These are the personal user account page and process execution page (Fig. 1).

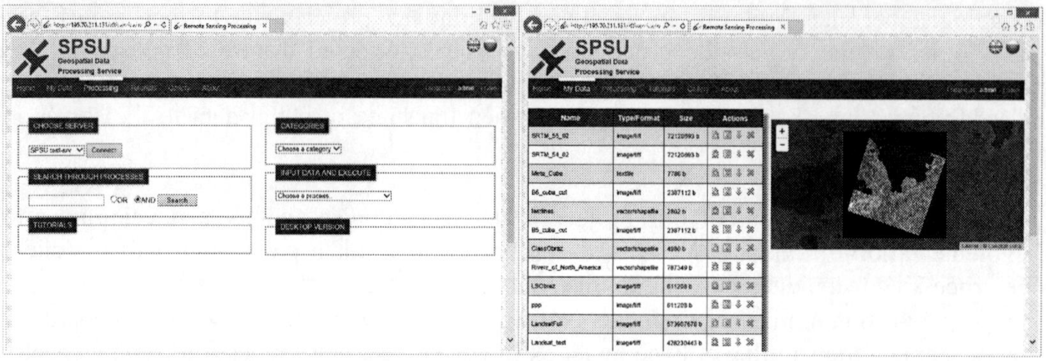

Figure 1. Process execution page (on the left) and personal user account page (on the right) at the prototype of HGWS geoportal.

Besides the possibilities of data processing on the server and downloading client-side processing tools, the resource allows to store amount of datasets on the server. This possibility eliminates the need of repeated uploading and downloading the same datasets, for example in the case of running multiple operations step by step.

Additionally, it is possible to explore the results of processing on the server using, previews directly in the browser window. This feature allows to reject incorrect processing results, without downloading on client computer. The personal user account could be used after user's registration on the Web site, which is verified by site administrator in manual mode.

Process execution page is arranged as follows. A WPS getCapabilities request is generated automatically, when the user selects some of available WPS/HGWS servers in the first dropdown list on the page (Fig. 1). After receiving of response, the response content is used for generating of list of the processes available at the selected server. This list appears in another one dropdown list on the page. Also, the text descriptions are loaded on page in the case of availability. User can do the search in these descriptions by entering any keywords in search box. After selection of a desired process, the WPS describeProcess request is generated and the page is updated with the input boxes for process inputs using response content. In the case of presenting of some geodata files in the process inputs list, the corresponding input boxes include dropdown lists of files, which are stored under user's personal account and have required format.

After filling all required input boxes of process inputs, user could execute the process. In the case of server-side execution, process output files appear in the user's personal account and could be observed online in preview mode or downloaded.

3. IMPLEMENTATION

The implemented prototype system of publishing and execution of HGWSs includes two base subsystems. Server subsystem incorporates a stack of the HTTP, WPS and HGWS servers and implements the access to software tools for geodata processing. Client subsystem incorporates a set of applications required for graphical user interface providing, downloading of the process software components and executing process on the client computer. These subsystems are presented on schematic diagram (Fig. 2).

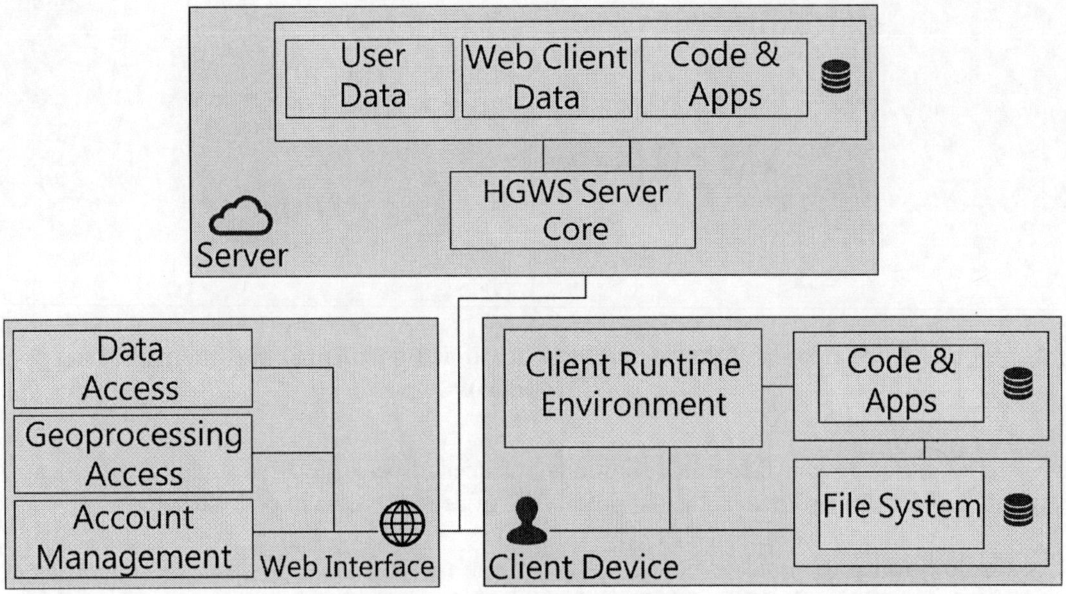

Figure 2. Server (upper block) and client (lower blocks) subsystems of implemented prototype of the HGWSs publishing and execution system.

In our implementation, we have separated WPS and HGWS servers. The last one provides only the client-side processing functionality. However, in general case (as it presented on the diagram) the HGWS server should provide both the client-side and the server-side processing functionality. Server subsystem accepts client requests and controls the flows of data files and executable files, as well as server databases, which store all of these files. The user can interact with stack of the servers in two modes. The first mode is via the Web interface, by uploading data to the server and runing server-side processing. Either the second mode is via the desktop application (client Runtime Environment or RE), by receiving on a local client device all of necessary software components of the process.

3.1 Server subsystem

The stack of servers performs the task of interaction with user via the Web interface or RE. OGC WPS standard implements requests, which are needed for server-side processing. For handling of the request(s), which is(are) needed for client-side processing, we need to define the structure of transmitted data (i.e. transmitted software components).

Here we need to introduce two new concepts concerning the types of downloadable software components. The concept of Functional Module (fModule) assumes discrete file, which serves the program code of some processing algorithms (it can be a Python script file, DLL, Java bytecode file, etc.). The Extra Application (extraApp) is the standalone program (usually compiled binary file) that implements some processing algorithms and can be executed directly for processing of some specific geodata files. In this way, we may reproduce some open source GISs, such as SAGA (*http://www.saga-gis.org*) or GRASS (*http://grass.osgeo.org*) for HGWS implementation (Fig. 3). This approach assumes the separate transmitting of the application core (extraApp) and its libraries (fModules).

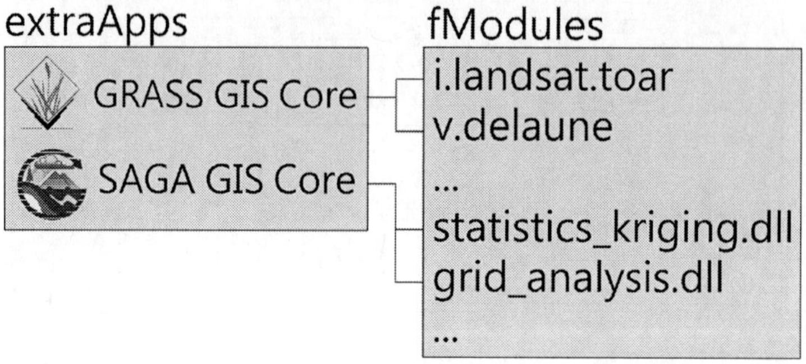

Figure 3. Separation of the cores of applications (extraApps) and their libraries (fModules).

The extraApp and fModule definitions are needed because of potentially complicated structure of process program code and possibility of dependencies in code structure.

Additionally we need to define and store the metadata, which describe properties of process as well as properties and dependences of the software components of both types. Each process could be associated with an unlimited number of fModules and extraApps. The only limitation is needed to store the software component files. This is the uniqueness of file names within the system. Other metadata parameters for software components (except name) are the file type, description, operation system compatibility record. Metadata parameters for process include the process name, abstract (description), inputs and outputs lists, desktop (client-side) version availability flag and lists of fModules and extraApps (Fig. 4).

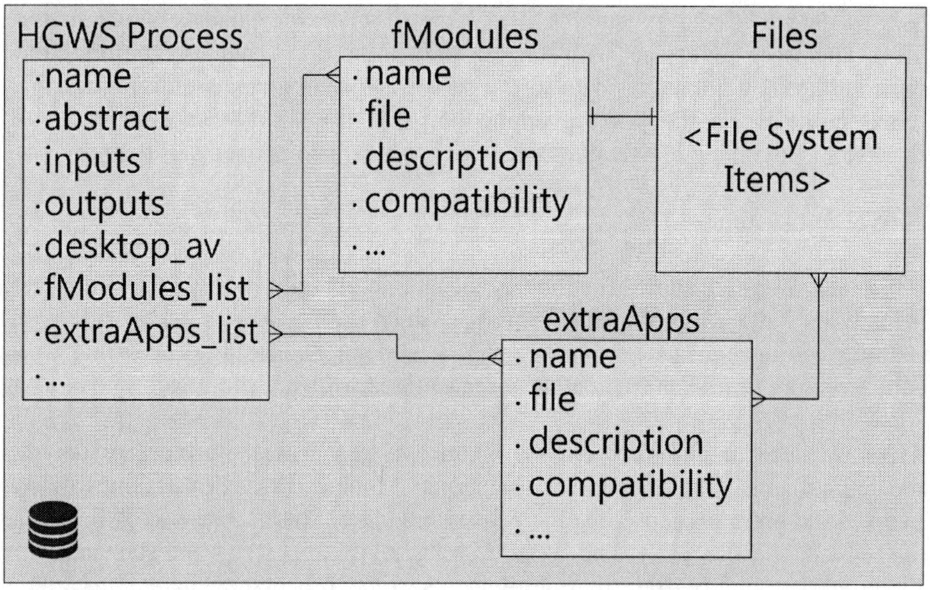

Figure 4. Structure of software components of the HGWS process.

3.1.1 *Requests*

The extension of the WPS standard with the possibilities of client-side processing and software components transmitting involves the implementation of a new request type that will be used for getting software components on the client side. We named this request as getProcess request. The parameters of this new request type are listed in Table 1:

Table 1. getProcess request parameters.

Name	Presence	Description
Request	Required	Identifies service request. Must be "getProcess".
Service	Required	Identifies service type. Must be "WPS".
Version	Required	Identifies service version.
Identifier	Mandatory	Process identifier as listed in the Capabilities document.
Platform	Required	Identifies user's desktop platform. Could be "Win32", "Win64", "Mac", "Linux Debian" etc.

The getProcess request could be executed using GET request method with above-mentioned parameters:

http://some.host/server?
Request=getProcess&
Service=WPS&
Version=x.0.0&
Platform=Win32&
Identifier=index-NDVI

This request takes its place in the chain of interactions between client and server (Fig. 5). The user chooses the server and connects to it. After that, client executes the WPS getCapabilities request to the selected server, and gets the response, which is extended with some value of the "desktop-available" attribute. After this, client could parse the list of available processes as well as the list of the processes available for client-side processing.

At the next step, the user selects some process, and the client executes the WPS describeProcess request. Then client parses the lists of the process inputs and process outputs from the describeProcess response, and generates the Web or desktop graphical user interface, depending on use of the browser or RE. Additionally, in the case of client-side processing, the entry point or the program base interface is generated that connects to the user interface and receives all process inputs from the user interface as arguments.

Next, the HGWS getProcess request executed in the case of client-side processing. Client parses the response and gets the lists of fModules and extraApps. Now client becomes able to download the needed software components. In our implementation, we store all downloaded software components into some standard directory, so previously downloaded software components could be excluded from downloading list.

All inputs that were received from user interface can be transmitted then to one of downloaded fModules, which has the name equal to the process name and provides input function with unified name.

Finally, the Execute request can be executed to start the processing.

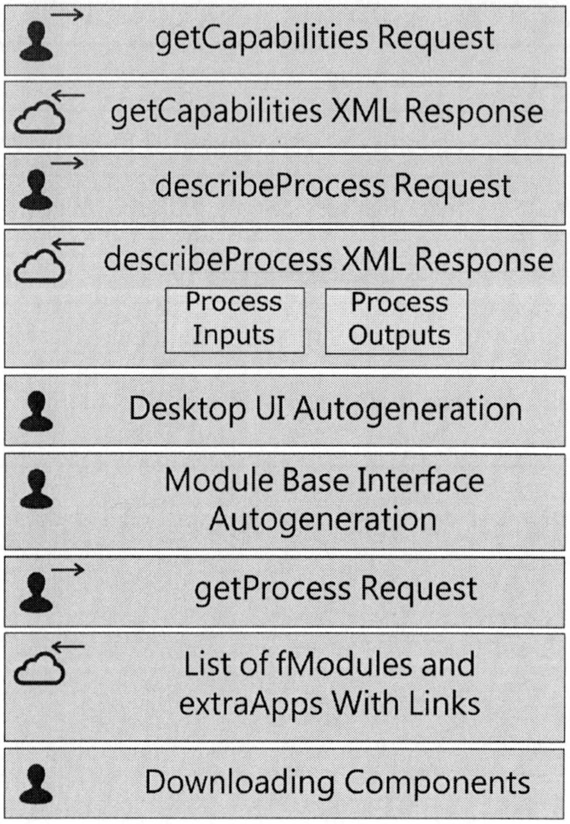

Figure 5. Chain of the requests to server in the case of HGWS processing.

3.1.2 *getProcess response structure*

The getProcess response is XML document that contains all necessary information about the software components needed for client-side processing using particular process. This document could include unlimited number of descriptions of the fModules and extraApps with metadata and download links. Here is an example fragment for such XML document:

```
<!--Module parameters-->
<complexType name="FModule">
    <complexContent>
        <name>sampleFModule</name>
        <annotation>sampleFModule annotation</annotation>
        <compability>
            <!-- Description of platform compatibility for module-->
            Win32-common
        </compability>
        <type>
            <!--Type of software component-->
            DLL
        </type>
        <downloadLink>URL</downloadLink>

        <usedExtraApps>
```

```
                <complexContent>
                    <name>SampleExtraApp</name>
                    <annotation>sampleExtraApp annotation</annotation>
                    <compability>Win32-common</compability>
                    <type>Binary</type>
                    <downloadLink>URL</downloadLink>
                </complexContent>
            </usedExtraApps>
        </complexContent>
    </complexType>
```

3.2 Client subsystem

The RE is needed on a client computer to provide the process execution. In the simplest case, a software client (desktop application) or a Web browser that displays a graphical user interface are able to play the role of the RE. However, the software client is more usable, due to security limitations of running executable files directly from browser interface.

In our solution, the RE is a desktop application, which displays the user interface; implements the tasks of loading, storage and removal of the process executable files; verifies are the client computer parameters applicable to run geoprocessing; and controls the starting and completing of geoprocessing. More broadly, the RE could include some external software, which are necessary to run process executable code, in particular the Python interpreter, in our case.

We assume two scenarios. The first is downloading of the settings file using our Web interface, and automatic opening of this file in the RE application with automatic initialization of the processing (i.e. connection to the server, loading the needed software components, and user interface generating). The second is working in manual mode, when the user enters the HGWS server URL and other parameters, and performs step by step all of the operations needed to select and parametrize some of the available processes.

In case of the first scenario, the settings file could be a JSON file, which incorporates all necessary values of process parameters:

```
{
  "server": {
              "host" : 195.70.211.131/hgws/hgws?
              "port": 8080
              }
  "HGWSversion": "0.1v",
  "Request": "getProcess",
  "ProcessName": "imagery-landsat8-to-radiance",
}
```

Using this file, the client RE may automatically execute needed requests, analyze responses, and alerting the user with errors or correct processing finalization.

The RE downloads software components of the process using getProcess response content and, organizes the storage of downloaded executable files and metadata.

Additionally, RE generates graphical user interface using describeProcess response content, as well as the base program interface that connects the user interface with process executable files.

The RE prototype is developed using Python PyQt (*https://riverbankcomputing.com/software/pyqt/intro*) tools to provide cross-platform interoperability, and Python PycURL library to implement HTTP requests. The main menu (Fig. 6) allows opening the JSON settings file manually, configuring the server connection settings to the server, and observing the list of already downloaded processes (Fig. 7) with the ability to run them. The user interface is generated in a simple manner. Each process input parameter is mapped to interface element that was rendered by PyQt (Fig. 8).

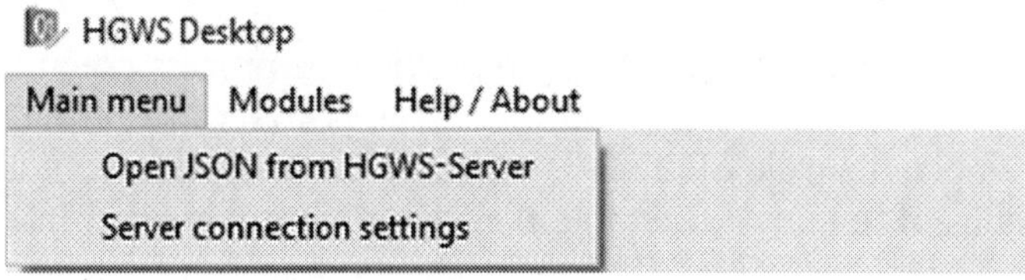

Figure 6. RE main menu.

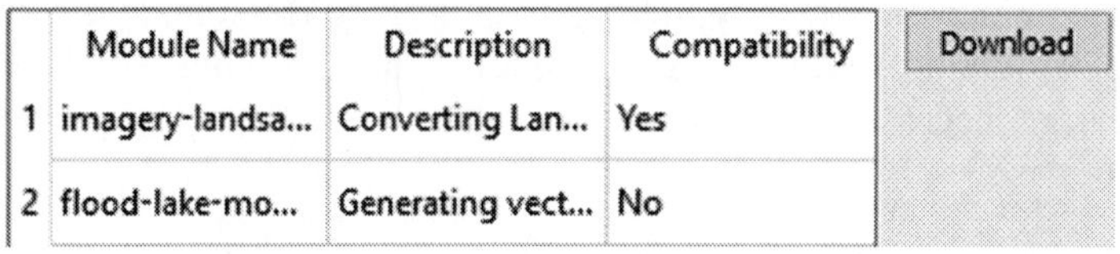

Figure 7. List of available processes.

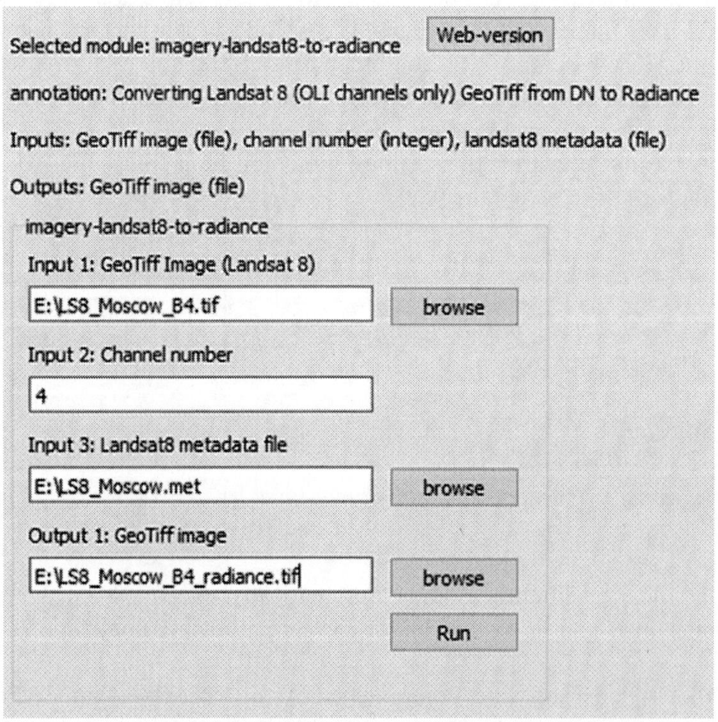

Figure 8. User interface of selected process.

4. RESULTS

Summing up, we should note that selected approach is workable in general and allows to solve the task of providing client-side computations when using Web services for geodata processing. Of course, some issues require further investigations still. The most interesting of these are the security issues of running of the executable code received from a third party server, and issues of the processes orchestration both in case of processing under user control and in case of machine-to-machine interface implementations.

However, we have formulated approach and created prototypes of software tools. These tools we can already use when performing small-scale projects for implementation of traditional advantages of the Software as a Service architecture (Software & Information Industry Association, 2001), such as universal and user-friendly front-end interface and low-cost back-end and processing program code administration.

In conclusion, we may point the basic methodology regulations, which we formulated as a result of working on the project. These regulations formalize the process of designing and implementing of HGWSs:

- Providing the backward compatibility of the HGWS processes and HGWS servers with WPS clients.
- Providing the maximal atomization of software components of the process, i.e. decomposition of the complex libraries into the amount of separated libraries.
- Providing the ability of choice of the side (client/server) for processing when executing processes.
- Providing the metadata with descriptions of the dependencies between the software components of the processes.
- Providing the metadata with descriptions of the client resources and functionality, which are needed for the process execution.

5. OUTLOOK

Our research opened a number of theoretical and practical issues, which require additional consideration. Next, we mention briefly some of them.

We have previously identified the need to ensure the safe running of executable code on the client side. Of course, if we execute program code that received from a third party server, the security issue requires a deep study. However, detailed regulation of security issues is hardly possible in abstract specifications or implementation standards for such services. At the current stage, we are of the opinion that the issue of the safe running of executable code must be considered in detail individually in the context of each specific project. Theoretical and conceptual foundations of security should be discussed as part of the theory of Web Services in general.

The question of services orchestration and formation of complex systems for geodata processing was also mentioned. It is an urgent task, but, we believe, has no universal solution, because of the complexity and nonlinearity of many data-processing procedures. Part of the problem of orchestration is to provide a consistent and unified description and structuring of the processing inputs and outputs. This issue can be solved by developing descriptive standards for the valid structure of data and metadata.

Another important area of research is related closely to the Internet of Things and Semantic Web Services concepts. This complex task associated with providing of automatic interaction between data servers, processing servers and tools servers. The, tools servers are the host servers of transmittable data processing services (HGWS servers in our terminology). The implementation of automated communication between the servers can be useful in distributed geodata (including big geodata) processing systems.

In our opinion, the conceptual gap of the resources allocation and redistribution on the server side for server-side processing requires some studies also. This gap is associated closely with WPS standard, which we used in our project as a base standard. We resolved the problem of redistribution of the limited server resources between users through the mechanism of user personal accounts. However, this mechanism is extremely inconvenient from the point of view of administration costs and from the point of view of processing automation abilities. The decision is likely to be dynamic allocation and redistribution of resources and implementation of data life time limit on the server as the data attribute.

Finally, current results of our project itself require detailed discussion between users and developers. Whereas, the fact that the research was carried out as a case study, we assume that our findings may have gaps and failures. Nevertheless, we hope that our work can be served as the basis for further research of other authors in this direction.

6. ACNOWLEGEMENTS

The study was partially supported by Russian Foundation for Basic Research (RFBR), research project No. 13-05-12079 ofi_m.

7. REFERENCES

Abdelshkour M., 2015. IoT, from Cloud to Fog Computing. Available at: http://blogs.cisco.com/perspectives/iot-from-cloud-to-fog-computing

Aktas M.S., Aydin G., Fox G.C., Gadgil H., Pierce M. and Sayar A., 2005. Information Services for Grid/Web Service Oriented Architecture (SOA) Based Geospatial Applications. Community Grids Lab, Indiana University, Technical Report. Bloomington, Indiana, USA

Ashton K., 2009. That "Internet of Things" Thing. *RFID Journal.* Available at: http://www.rfidjournal.com/articles/view?4986

Baumann P. (ed.), 2009. Web Coverage Processing Service (WCPS) Language Interface Standard. OGC 08-068r2. Version: 1.0.0, 2009-03-25

Berners-Lee T., Hendler J., Lassila O., 2001. The Semantic Web. *Scientific American Magazine*. May 2001

Bröring A., Echterhoff J., Jirka S., Simonis I., Everding T., Stasch C., Liang S., and Lemmens R., 2011. New Generation Sensor Web Enablement. *Sensors*, Volume 11(3), pp. 2652-2699, doi:10.3390/s110302652

Buyya R., Broberg J., Goscinski A. (eds.), 2011. *Cloud Computing: Principles and*

Paradigms. John Wiley & Sons Inc., 637 p., DOI: 10.1002/9780470940105.fmatter

Calacanis J., 2007. Web 3.0, the "official" definition. Available at: http://calacanis.com/2007/10/03/web-3-0-the-official-definition/

Coene Y., Marchetti P.G., Smolders S., 2007. Architecture and Standards for a Distributed Digital Library of Geospatial Services. Third Italian Research Conference on Digital Library Systems (IRCDL 2007) Padova 29–30 January 2007 Proceedings pp. 52–60

Doyle A. (ed.), 2000. OpenGIS Web Map Server Interface Implementation Specification. OGC 00-028. Version 1.0.0, 2000-04-19

Evans J. (ed.), 2003. OGC Web Coverage Services (WCS). OGC 03-065r6. Version 1.0.0, 2003-08-27

Fernando N., Loke S.W., Rahayu W., 2013. Mobile cloud computing: A survey. *Future Generation Computer Systems*, Volume 29, Issue 1, pp. 84–106

Fu P., Sun J., 2010. *Web GIS: Principles and Applications*. ESRI Press. ISBN 978-1-58948-245-6

Hong K., Lillethun D., Ramachandran U., Ottenwälder B., Koldehofe B., 2013. Mobile fog: a programming model for large-scale applications on the internet of things. Proceedings of the second ACM SIGCOMM workshop on Mobile cloud computing, 16 August 2013, Hong Kong, China. pp. 15-20, doi:10.1145/2491266.2491270

Kazakov E.E., 2013. Approaches to Implementation of Web Services for Processing and Analysis of Remote Sensing Data. In Proceedings of the Earth Remote Sensing From Space: Algorithms, Technology, Data – Young Scientists Workshops (Altai State University, Barnaul, Russia, October 2-6, 2013). pp. 82-86. In Russian. Available at: http://elibrary.asu.ru/xmlui/bitstream/handle/asu/203/read.7book

Kazakov E., Terekhov A., Kapralov E., and Panidi E., 2015. WPS-based technology for client-side remote sensing data processing. *The International Archives of the Photogrammetry, Remote Sensing and Spatial Information Sciences*, XL-7/W3, pp. 643-649, doi:10.5194/isprsarchives-XL-7-W3-643-2015

Keens S. (ed.), 2007. Discussions, findings, and use of WPS in OWS-4. OGC Discussion Paper. OGC 06-182r1. Version 0.9.1, 2007-05-10

Kraak M.-J., Brown A., 2001. *Web Cartography Developments and Prospects*. London and New York: Taylor & Francis

McIlraith S.A., Son T.C., and Zeng H., 2001. Semantic Web Services. *IEEE Intelligent Systems*, Volume 16, Issue 2, pp. 46-53

Mueller M., Pross B. (eds.), 2015. OGC WPS 2.0 Interface Standard. OGC Implementation Standard. OGC 14-065. Version 2.0, 2015-03-05

O'Reilly T., 2005. What Is Web 2.0 Design Patterns and Business Models for the Next Generation of Software. Available at: http://www.oreilly.com/pub/a/web2/archive/what-is-web-20.html

Panidi E.A., 2013. Geoservices for Online Remote Sensing Data Processing. In Proceedings of the Earth Remote Sensing From Space: Algorithms, Technology, Data – Young Scientists Workshops (Altai State University, Barnaul, Russia, October 2-6, 2013). pp. 74-81. In Russian. Available at: http://elibrary.asu.ru/xmlui/bitstream/handle/asu/203/read.7book

Panidi E.A., 2014. Towards Client-Side Web Processing Services. Proceedings of OSGeo's European Conference on Free and Open Source Software for Geospatial (FOSS4G-Europe 2014) July 15-17 2014 Jacobs University, Bremen, Germany. 4 p. Issued online. Available at: http://europe.foss4g.org/2014/content/toward-client-side-web-processing-services.html

Panidi E., Kazakov E., Kapralov E., Terekhov A., 2015. Hybrid Geoprocessing Web Services. In SGEM2015 Conference Proceedings (15th International Multidisciplinary Scientific GeoConference SGEM, Albena, Bulgaria, June 18-24, 2015), Book 2 Vol. 1, pp. 669-679

Scharl A., Tochtermann K. (Eds.), 2007. *The Geospatial Web: How Geobrowsers, Social Software and the Web 2.0 are Shaping the Network Society*. London: Springer-Verlag.

Schut P. (ed.), 2007. OpenGIS Web Processing Service. OpenGIS Standard. OGC 05-007r7. Version 1.0.0, 2007-06-08

Software & Information Industry Association, 2001. Software As A Service: Strategic Backgrounder. Washington, D.C., USA

Vretanos P.A., (ed.), 2002. Web Feature Service Implementation Specification. OGC 02-058. Version 1.0.0, 2002-05-17

Free and Open Source Software for Geospatial (FOSS4G) Conference Proceedings

Volume 15 *Seoul, South Korea* Article 31

2015

Image Geocoding as a Service

Nuno Amorim
Universidade do Minho, Portugal

Paulo Almeida
Universidade do Minho, Portugal

Jorge Gustavo Rocha
Universidade do Minho

Recommended Citation

Amorim, Nuno; Almeida, Paulo; and Rocha, Jorge Gustavo (2015) "Image Geocoding as a Service," *Free and Open Source Software for Geospatial (FOSS4G) Conference Proceedings*: Vol. 15 , Article 31.
DOI: https://doi.org/10.7275/R5833Q6K
Available at: https://scholarworks.umass.edu/foss4g/vol15/iss1/31

AT8-02

Image Geocoding as a Service

Nuno Amorim, Paulo Almeida, **Jorge Gustavo Rocha***

Universidade do Minho, Portugal

Corresponding Author : Jorge Gustavo Rocha (jgr@di.uminho.pt)

Abstract

Driven by the ambition of a global geocoding solution, in this paper we present the architecture of an image geocoding service. It takes advantage of the ubiquity of cameras, that are present in almost all smartphones. It is an inexpensive sensor yet powerful, that can be used to provide precise location and orientation.

This geocoding service provides an API similar to existing ones for place names and addresses, like Google Geocoding API. Instead of a text based query, images can be submitted to estimate the location and orientation of the user. Developers can use this new API, keeping almost all the existing code already used for other geocoding APIs.

Behind the scenes, image features are extracted from the submitted photograph, and compared against a huge database of georeferenced models. These models were constructed using structure from motion (SFM) techniques, and heavily reduced to a representative set of all information using Synthetic Views. Our preliminary results shows that the pose estimation of the majority of the images submitted to our geocoding was successfully computed (more than 60%) with the mean positional error around 2 meters.

With this service, an inexpensive outdoor/indoor location service can be provided, for example, for urban environments, where GPS fails.

Academic Discipline and Sub-Disciplines : Computer Vision; Geocoding

Keywords : Structure from Motion; Geocoding

Free and Open Source Software for Geospatial (FOSS4G) Conference Proceedings

Volume 15 *Seoul, South Korea* Article 32

2015

Geosocial Big Data Analysis Using Python and FOSS4G with the Case Study of Korean Data

Ilyoung Hong
University at Buffalo, The State University of New York at Buffalo

Recommended Citation

Hong, Ilyoung (2015) "Geosocial Big Data Analysis Using Python and FOSS4G with the Case Study of Korean Data," *Free and Open Source Software for Geospatial (FOSS4G) Conference Proceedings*: Vol. 15 , Article 32.
DOI: https://doi.org/10.7275/R5CV4FXK
Available at: https://scholarworks.umass.edu/foss4g/vol15/iss1/32

AT7-03

Geosocial Big Data Analysis Using Python and FOSS4G with the Case Study of Korean Data

Ilyoung Hong*

Namseoul University, Korea

Corresponding Author : Ilyoung Hong (ilyoung.hong@gmail.com)

Abstract

Nowadays, there are many researches on the analysis of Geosocial big data, such as geotweeet and as foursquare venues and OSS(Open Source Software) has an important role on this. In the analyzing geosocial big data, there are several different steps such as data collection, data parsing, data conversion, statistical analysis, visualizing and database management. So, the integrated system architecture and the compatible analysis environment has a key role to acquire the relevant analysis results. The Python programming support the interoperable analysis environment for the various and different software functions and enable to process for geosocial big data in the integrated platforms. FOSS4G support software environment for geovisualization and data management for the collected data. In this study, the way and process of geosocial big data analysis is introduced with case study of geotweet and foursquare venues and the analysis results are presented with the case study of Korean data. For this study, Python API libraries for tweeter(tweepy) and foursquare(pyforsquare) used to collect the geosocial data, and Pandas and Simplejson are used to parse and extract the valid data, and GDAL and PySAL are used to convert and analyze for GIS data. PyTagCloud and WordCloud are used to visualize the qualitative text. MongoDB is used to store the collected dataset and QGIS are applied for the geovisualization.

Academic Discipline and Sub-Disciplines : geosocial big data; python; FOSS4G

Keywords : geosocial big data; python; FOSS4G

Free and Open Source Software for Geospatial (FOSS4G) Conference Proceedings

Volume 15 *Seoul, South Korea* Article 33

2015

An On-Board Visual-Based Attitude Estimation System For Unmanned Aerial Vehicle Mapping

M. Ridhwan Tamjis
School of Civil and Environmental Engineering, UNSW

Samsung Lim
School of Civil and Environmental Engineering, UNSW,

AT8-03

AN ON-BOARD VISUAL-BASED ATTITUDE ESTIMATION SYSTEM FOR UNMANNED AERIAL VEHICLE MAPPING

M. Ridhwan Tamjis[1] and Samsung Lim[2]

[1]School of Civil and Environmental Engineering, UNSW
Sydney, NSW, 2052, Australia
Email: m.tamjis@student.unsw.edu.au
[2]School of Civil and Environmental Engineering, UNSW
Sydney, NSW, 2052, Australia
Email: s.lim@unsw.edu.au

ABSTRACT

This paper evaluates the performances of several salient feature detectors, namely; Harris detector, Minimum Eigenvalue (MinEig), Scale Invariant Feature Transform (SIFT), Maximally Stable Extremal Region (MSER), Speeded Up Robust Feature (SURF), Features from Accelerated Segment Test (FAST), and Binary Robust Scale Invariant Keypoint (BRISK), in order to assess the suitability in the application of the proposed visual-based attitude estimation system. Throughout the experiment, three main requirements have been investigated which include Time-to-Complete (TTC), detection rate, and matching rate. It was found that SURF fulfills each of the system's requirements. Moreover, it was also found that keypoints detection capabilities affect the processing time, and the clustering patterns in the results may assist in automated inspection of correct and false matching.

1. INTRODUCTION

History of unmanned aerial vehicles (UAVs) can be traced back to as early as 1849 when unmanned balloons loaded with explosives were launched from the Austrian ship to attack the Italian city of Venice (Anonymous, 2012). Since then, the UAV technology is expanding and evolving from a simple balloon to a much sophisticated mechanism. Nowadays, UAVs play important roles in various fields such as engineering, civil, defense, urban planning, recreation, etc. The technology has undergone a series of evolutions in various aspects such as size, material, and control.

The challenges in today's researches on UAVs are to make it more intelligent, robust, and safe during flight, aiming to make an UAV operable on minimal human intrusions (Clough, 2005; Doherty, 2004). One of the approaches is to maximize the vision capabilities equipped in most modern UAVs for attitude estimation. The main idea is to rely on the visual scenes provided by the camera to calculate/estimate attitude. Different types of approaches for visual-based attitude estimation have been proposed in (Garratt & Chahl, 2008; Srinivasan et al., 2004; Srinivasan, 1994).

As an alternative to the conventional sensor-based attitude estimation system, this paper proposes a visual-based attitude estimation system for an UAV and its camera self-calibration, which will use optical flow to measure the egomotion of the on-board camera, and exploit it to estimate the platform's attitude. That is, this study aims to integrate optical flow with a keypoints detector for on-board attitude estimation and camera self-calibration.

This is to minimize the computation load that may be imposed by optical flow. In this paper, the assessment of the possible detectors for the proposed visual-based attitude estimation system is presented.

2. KEYPOINTS DETECTORS

2.1 Visual-based attitude estimation system

The use of visual information for attitude/pose estimation is not new, however, researches are still on-going to discover its potentials. One of the breakthroughs of this approach is on camera self-calibration (Armstrong et al., 1996; Luong & Faugeras, 1997) which has been widely used in modern digital cameras.

The proposed on-board visual-based attitude estimation system is illustrated in Figure 1. The main idea is to use visual information from overlapping images to measure the platform's egomotion, and estimate attitude from the visual motion. Similar approaches have been studied by Dusha et al. (2007) and Srinivasan (1994). Optical flow computation could be expensive depending on the approach (Fleet & Weiss, 2006). For that reason, this research aims to reduce the computation load at the start of the process by limiting the images to regions of upmost important. This requires an integration of optical flow with salient feature detection and matching.

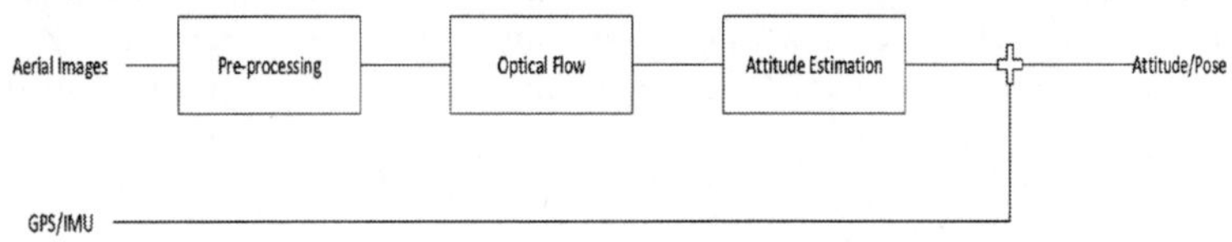

Figure 1. Framework of the proposed system.

2.2 Keypoints Detectors

Several types of detectors are investigated and evaluated in this experiment, namely; Harris detector (Mikolajczyk & Schmid, 2001), Minimum Eigenvalue (MinEig) (Shi & Tomasi, 1994), Scale Invariant Feature Transform (SIFT) (Lowe, 1999), Maximally Stable Extremal Region (MSER) (Matas et al., 2004), Speeded Up Robust Feature (SURF) (Bay et al., 2006), Features From Accelerated Segment Test (FAST) (Rosten & Drummond, 2006) and Binary Robust Scale Invariant Keypoint (BRISK) (Leutenegger et al., 2011). Each of the detectors possesses a different computation algorithm as well as capabilities.

3. APPROACH

A set of aerial images acquired from an UAV is used in the evaluation. The image set consists of 249 overlapping images, taken above Loftus Oval, New South Wales, Australia. The aerial scene shows open yards, vegetation, roads and partial suburbs. The images also show pose variations and distortions which occurred during flight. The selection of detectors for this experiment is based on those attributes, as it was proven that the detectors are invariant to such transformations and distortions in literature.

3.1 Keypoints detection and matching

The main purpose of this experiment is to evaluate the detection rate and matching rate of each nominated keypoints detector and descriptor. For a fair judgement, a standard process framework is required. Figure 2 illustrates the standard detection and matching process that have been used throughout the experiment.

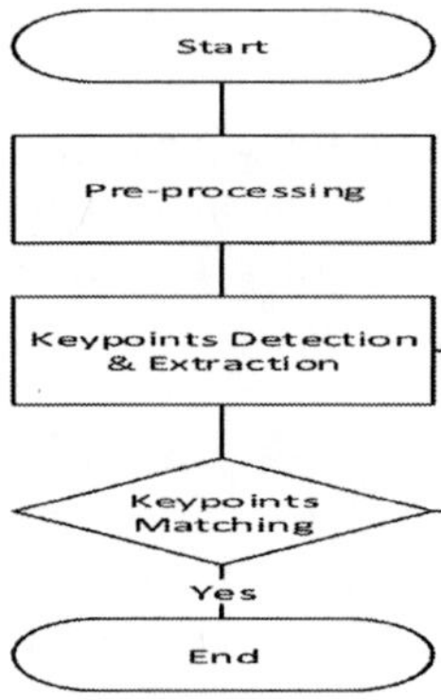

Figure 2 Standard framework for keypoints detection and matching

3.1.1 Image pre-processing

To minimize the complexity and completion time, images are down-sampled to 20% of the original size, an optimal size which does not compromise image features. Furthermore, images are converted to grayscale from its original Red-Green-Blue (RGB) color space to fulfill the processing requirements of keypoints detectors.

3.1.2. Keypoints detection and extraction

The purpose of this process is to detect and extract salient feature points in the image using keypoints detectors. Detection and extraction of the keypoints are conducted in every two consecutive overlapping images in the image sequence. Based on the type of detectors, keypoints are projected to the image, and the locations as well as the descriptors are extracted for the latter matching process.

3.1.3 Keypoints matching

It is the research concern that each detector will work differently in different matching metric. To investigate this issue, two different metrics have been used in the experiment, namely; Sum of Squared Differences (SSD) given in Equation (1), and Sum of Absolute Differences (SAD) given in Equation (2). For two images $f(x,y)$ and $g(x,y)$;

$$SSD(d_1,d_2) = \sum_{i=1}^{n_1} \sum_{j=1}^{n_2} (f(x+i,y+j) - g(x+i-d_1,y+j-d_2))^2 \qquad (1)$$

$$SAD(d_1,d_2) = \sum_{i=1}^{n_1} \sum_{j=1}^{n_2} |(f(x+i,y+j) - g(x+i-d_1,y+j-d_2))| \qquad (2)$$

3.2 EVALUATION CRITERIA

3.2.1. Detection rate

In this experiment, the number of keypoints that can be detected by each detector is investigated. This criterion is chosen to evaluate the capability of each detector to provide a sufficient number of keypoints for the matching process. It is assumed that the probability of obtaining correct matches increases when a higher number of detected keypoints are available, making the matching process robust. Such an advantage has been shown in the SIFT framework, although it will also increase computation time. Therefore, this factor should be considered for on-board applications.

3.2.2. Time-to-Complete

Time-to-Complete (TTC) is measured from the start of keypoints detection to the end of matching, in which matching pairs have been identified. In this evaluation, it is important to select the detector with minimal processing time, e.g. the fastest detector. There are other requirements that are significantly important for the proposed system, however, the highest weight is assigned to TTC.

3.2.3. Match rate

Matching rates of keypoints detectors are determined by the number of correctly matched keypoints over the number of overall matched keypoints. To ensure that the results are statistically significant, an optimal sample size of 151 is chosen based on the sample size calculation, employing 95% confidence interval.

4. EVALUATION OF PERFORMANCE

4.1 Detection capabilities

The experiment results show that BRISK has failed to locate keypoints in at least 60% of the total overlapping images, which is followed by FAST but with a smaller percentage. This is probably due to the capabilities of the detectors to locate keypoints in overlapping images. Furthermore, it is found that there is a relationship between the number of detected keypoints and the number of failed matches, in which the number of failed matches lessens with an increase of the number of keypoints. The same pattern can be observed in the number of matched keypoints.

Table 1. Mean TTC

Keypoints Detector	Mean TTC (seconds)	
	SSD	SAD
BRISK	0.259	0.133
SURF	0.261	0.151
FAST	0.241	0.120
SIFT	1.441	1.410
MSER	0.542	0.403
Harris	0.441	0.324
MinEig	0.811	0.636

4.2 TTC Comparison

The processing time for the detectors to complete the full cycle of the algorithm has been measured in this experiment and the results are shown in Table 1. From Table 1, it can be seen that the processing time varies for each detector while SIFT took the most time to complete the process. This has been expected because of the number of generated keypoints. However, even though MinEig generates a much larger number of keypoints than SIFT, the binary nature of the detector has made the processing time much faster.

4.3 Matching capabilities

In this experiment, it is very difficult to compute the number of correct and false matches in SIFT-processed images due to the density of the matching keypoints, as the lines connecting the points which have been used as an aid for visual inspection rendered the matching points in corresponding images. To tackle the rendering constraints, a different kind of assessment is formulated for SIFT which does not deviate from the comparison goals. For SIFT, correct matches dominate the lines compared to false matches, thus, led to a conclusion that the correct-match percentage for SIFT is particularly high. The other keypoints detectors are assessed straightforwardly, and the results for all detectors (except for SIFT) are tabulated in Table 2.

Table 2. Correct-match results for each keypoints detectors except for SIFT

Keypoints Detector	Correct-Match Percentage (%)	
	SSD	SAD
BRISK	88	82
SURF	61	96
FAST	82	86
MSER	48	94
Harris	91	91
MinEig	93	94

4.4 Assessment and findings

In order to find the best detector for the proposed system, an occurrence-based (best-fit) assessment has been conducted. The results are divided into five selection criteria, namely, (1) TTC, (2) keypoints density, (3) match density, (4) match failure, and (5) percentage of correct matches. Four best detectors are ranked based on the performance in each criterion. The main findings are described below.

4.4.1 Optimal detector

From the assessment, SIFT is ranked first in the performance comparison as it generates the highest number in matched keypoints and correct-match, as well as minimal match failure. However, SIFT is slow in processing, which is a critical criterion in the assessment. The same goes to MinEig. If the processing time is not critical in the assessment, this research highly suggests SIFT for its robustness. In overall, SURF has the optimal performance required by the proposed system, given that SAD metric is employed.

4.4.2 SSD vs SAD

The experiment results have shown that matching metric works differently for different types of detectors. The results of BRISK, FAST, Harris, and MinEig show that the matching metric does not directly affect these binary-type detectors in terms of detection and matching capabilities. On the contrary, integer-based detectors such as SIFT, SURF and MSER are directly affected by it. The SAD metric deteriorates the matching capabilities of the detector, however, increases the number of correct-matches. In terms of computation time, algorithms ran faster when SAD is employed.

4.4.3 Keypoint density vs TTC

It is also found that the detection capabilities affect TTC, in which, a higher number of detected keypoints leads to slower computation time. This is, however, subject to the complexity of the detector's framework.

4.4.4 Clustering pattern

From visual inspection, clustering patterns are detected on the matching points, which may assist in the automated inspection of correct and false matching. The formation of a cluster is dependent on the matching keypoints. Investigation has shown that correct matches can be grouped into one cluster based on the attribute used in the measurement, and made up the large portion of the matched keypoints.

5. CONCLUDING REMARKS

Performances of various keypoints detectors have been evaluated in terms of detection rate, TTC and matching rate. A set of 249 aerial images taken using a fixed-wing UAV's camera have been evaluated, which represents actual flight conditions, translations, rotations, illumination changes, anomaly, and perturbations. Assessments are conducted based on the chosen criteria, which aims to fulfill the UAV's on-board applications requirements.

The assessment results show that the best detector candidate to be integrated to the proposed system is SURF, given that the SAD metric is used to measure similarity between keypoints. It was found that the time taken for SURF to complete the full cycle of the algorithm is relatively small. SURF is also able to provide a sufficient number of salient feature points in each detection without sacrificing the computation time. Additionally, SURF is able to provide a high percentage of correctly matched keypoints, which fulfills the requirements of the proposed visual-based attitude estimation system and camera self-calibration.

REFERENCES

Anonymous. (2012, 15 November 2012). UAV Evolution - How Natural Selection Directed the Drone Revolution. *army-technology.com.* Retrieved 23 April 2015, 2015, from http://www.army-technology.com/features/featureuav-evolution-natural-selection-drone-revolution/

Armstrong, M., Zisserman, A., & Hartley, R. (1996). Self-calibration from image triplets *Computer Vision—ECCV'96* (pp. 1-16): Springer.

Bay, H., Tuytelaars, T., & Van Gool, L. (2006). Surf: Speeded up robust features *Computer vision–ECCV 2006* (pp. 404-417): Springer.

Clough, B. T. (2005). Unmanned aerial vehicles: autonomous control challenges, a researcher's perspective. *Journal of Aerospace Computing, Information, and Communication, 2*(8), 327-347.

Doherty, P. (2004). *Advanced Research with Autonomous Unmanned Aerial Vehicles.* Paper presented at the KR.

Dusha, D., Boles, W., & Walker, R. (2007). *Attitude estimation for a fixed-wing aircraft using horizon detection and optical flow.* Paper presented at the Digital Image Computing Techniques and Applications, 9th Biennial Conference of the Australian Pattern Recognition Society on.

Fleet, D., & Weiss, Y. (2006). Optical flow estimation *Handbook of Mathematical Models in Computer Vision* (pp. 237-257): Springer.

Garratt, M. A., & Chahl, J. S. (2008). Vision - based terrain following for an unmanned rotorcraft. *Journal of Field Robotics, 25*(4 - 5), 284-301.

Leutenegger, S., Chli, M., & Siegwart, R. Y. (2011). *BRISK: Binary robust invariant scalable keypoints.* Paper presented at the Computer Vision (ICCV), 2011 IEEE International Conference on.

Lowe, D. G. (1999). *Object recognition from local scale-invariant features.* Paper presented at the Computer vision, 1999. The proceedings of the seventh IEEE international conference on.

Luong, Q.-T., & Faugeras, O. D. (1997). Self-calibration of a moving camera from point correspondences and fundamental matrices. *International Journal of Computer Vision, 22*(3), 261-289.

Matas, J., Chum, O., Urban, M., & Pajdla, T. (2004). Robust wide-baseline stereo from maximally stable extremal regions. *Image and vision computing, 22*(10), 761-767.

Mikolajczyk, K., & Schmid, C. (2001). *Indexing based on scale invariant interest points.* Paper presented at the Computer Vision, 2001. ICCV 2001. Proceedings. Eighth IEEE International Conference on.

Rosten, E., & Drummond, T. (2006). Machine learning for high-speed corner detection *Computer Vision–ECCV 2006* (pp. 430-443): Springer.

Shi, J., & Tomasi, C. (1994). *Good features to track.* Paper presented at the Computer Vision and Pattern Recognition, 1994. Proceedings CVPR'94., 1994 IEEE Computer Society Conference on.

Srinivasan, M., Zhang, S., Chahl, J., Stange, G., & Garratt, M. (2004). An overview of insect-inspired guidance for application in ground and airborne platforms. *Proceedings of the Institution of Mechanical Engineers, Part G: Journal of Aerospace Engineering, 218*(6), 375-388.

Srinivasan, M. V. (1994). An image-interpolation technique for the computation of optic flow and egomotion. *Biological Cybernetics, 71*(5), 401-415.

Free and Open Source Software for Geospatial (FOSS4G) Conference Proceedings

Volume 15 *Seoul, South Korea* Article 34

2015

FREEWAT: FREE and open source tools for WATer resource management

Massimiliano Cannata
Institute of Earth Sciences, SUPSI-DACD, Switzerland

Iacopo Borsi
TEA SISTEMI Spa, Pisa, Italy

Laura Foglia
Institute of Geosciences Technical University Darmstadt, Germany

Violeta Velasco Mansilla
Institute of Environmental Sciences and Water Research, Spanish Research Council

Rudy Rossetto
Institute of Life Sciences, Scuola Superiore Sant`Anna, Italy

Recommended Citation

Cannata, Massimiliano; Borsi, Iacopo; Foglia, Laura; Mansilla, Violeta Velasco; and Rossetto, Rudy (2015) "FREEWAT: FREE and open source tools for WATer resource management," *Free and Open Source Software for Geospatial (FOSS4G) Conference Proceedings*: Vol. 15 , Article 34.
DOI: https://doi.org/10.7275/R5JM27VK
Available at: https://scholarworks.umass.edu/foss4g/vol15/iss1/34

AT8-05

FREEWAT: FREE and open source tools for WATer resource management

Massimiliano Cannata[1]*, Iacopo Borsi[2], Laura Foglia[3], Violeta Velasco Mansilla[4], Rudy Rossetto[5]

[1]*Institute of Earth Sciences, SUPSI-DACD, Switzerland*
[2]*TEA SISTEMI Spa, Pisa, Italy*
[3]*Institute of Geosciences Technical University Darmstadt, Germany*
[4]*Institute of Environmental Sciences and Water Research, Spanish Research Council*
[5]*Institute of Life Sciences, Scuola Superiore Sant`Anna, Italy*

Corresponding Author : Massimiliano Cannata (massimiliano.cannata@supsi.ch)

Abstract

This talk will present FREEWAT: an HORIZON 2020 project financed by the EU Commission under the call WATER INNOVATION: BOOSTING ITS VALUE FOR EUROPE.
FREEWAT main result will be an open source and public domain GIS integrated modelling environment for the simulation of water quantity and quality in surface water and groundwater with an integrated water management and planning module.
FREEWAT aims at promoting water resource management by simplifying the application of the Water Framework Directive and other EU water related Directives.
Specific objectives of the FREEWAT project are: to coordinate previous EU and national funded research to integrate existing software modules for water management in a single environment into the GIS based FREEWAT and to support the FREEWAT application in an innovative participatory approach gathering technical staff and relevant stakeholders (in primis policy and decision makers) in designing scenarios for the proper application of water policies. The open source characteristics of the platform allow to consider this an initiative "ad includendum" (looking for inclusion of other entities), as further research institutions, private developers etc. may contribute to the platform development.

The core of the FREEWAT platform will be the SID&GRID (Rossetto et al. 2013) framework in its version ported to the QGIS desktop.
SID&GRID will be complemented with solute transport (also density dependent) capabilities in aquifers within the MARSOL (2014) EU FP7 project.
Activities will be mainly carried out on two branches:
(i) integration of modules, so that the software will fit the end-users requirements, including tools for better producing feasibility and management plans;
(ii) a set of activities devoted to fix bugs and to provide a well-integrated interface for the different tools implemented.

Further capabilities to be integrated are:
- a dedicated module for water management and planning that will help to manage and aggregate all the distributed data coming from the simulation scenarios;
- a whole module for calibration, uncertainty and sensitivity analysis;
- a module for solute transport in the unsaturated zone;
- a module for crop growth and water requirements in agriculture;
- tools for dealing with groundwater quality issues;
- tools for the analysis, interpretation and visualization of hydrogeological data.

Through creating a common environment among water research/professionals, policy makers and implementers, FREEWAT main impact will be on enhancing science- and participatory approach and evidence-based decision making in water resource management, hence producing relevant and appropriate outcomes for policy implementation. Synergies with the UNESCO HOPE initiative on free and open source software in water management greatly boost the value of the project. Large stakeholders involvement is thought to guarantee results dissemination and exploitation.

Free and Open Source Software for Geospatial (FOSS4G) Conference Proceedings

Volume 15 *Seoul, South Korea* Article 35

2015

Analysing the Performance of NoSQL vs SQL Databases with Respect to Routing Algorithms

sarthak agarwal
International Institute of Information Technology Hyderabad Gachibowli, Hyderabad, India

KS Rajan
International Institute of Information Technology Hyderabad Gachibowli, Hyderabad, India

Recommended Citation

agarwal, sarthak and Rajan, KS (2015) "Analysing the Performance of NoSQL vs SQL Databases with Respect to Routing Algorithms," *Free and Open Source Software for Geospatial (FOSS4G) Conference Proceedings*: Vol. 15 , Article 35.
DOI: https://doi.org/10.7275/R5PC30K9
Available at: https://scholarworks.umass.edu/foss4g/vol15/iss1/35

AT9-01

Analysing the Performance of NoSQL vs SQL Databases with Respect to Routing Algorithms.

Sarthak Agarwal[1*], KS RAJAN[2]

[1]*Student, India*
[2]*IIIT Hyderabad, India*

Corresponding Author : Sarthak agarwal (sarthak.a@research.iiit.ac.in)

Abstract

With the increased shift towards GeoSpatial Web Services on both the Web and mobile platforms especially in the user-centric services, there is a need to improve the query response time. The traditional routing algorithm requires server to process the query and send the results to a client but here we are focussing on query processing within the client itself. This paper attempts to evaluate the performance of an existing NoSQL database and SQL database with respect to routing algorithm and evaluate whether or not we can deploy the computations on the client system only.

While SQL databases face the challenges of scalability and agility and are unable to take the advantage of the abundant memory and processing power available these days, NoSQL databases are able to use some of these features to their advantage. The nonrelational databases are more suited for handling the dynamic rise in the data storage and the increased frequency of data accessibility.

For this comparative study, MongoDB is the NoSQL engine while the PostgreSQL is the chosen SQL engine. The dataset is a synthetic dataset of road network with several nodes and we find the distance between source and destination using various algorithms. As a part of paper The implementation we are planning on using pgRouting for the analysis which currently uses PostgreSQL at the backend and implements almost all the routing algorithms essential in practical scenarios. We have currently analyzed the performance of NoSQL databases for various spatial queries and have extended that work to routing.

Initial results suggest that MongoDB performs faster by an average factor of 15x which increases exponentially as the path length and network data size increases in both indexed and nonindexed operations. This implies that nonrelational databases are more suited to the multiuser query systems and has the potential to be implemented in servers with limited computational power.

Further studies are required to identify its appropriateness and incorporate a range of spatial algorithms within non-relational databases.

Academic Discipline and Sub-Disciplines : open source ; spatial databases ; mongoDB ; routing ; GIS ;

Keywords : performance analysis ; spatial databases ; mongodb ; pgrouting ; routing problem ; mobile devices

Agencies : International Institute of information technology

Free and Open Source Software for Geospatial (FOSS4G) Conference Proceedings

Volume 15 *Seoul, South Korea* Article 36

2015

A Cross National Comparison on the Awareness of Adopting FOSS4G to NSDI in Developing Countries

Junyoung Choi
Spatial Information Office, Korea Land and Housing Corporation

Jaesung Ahn
Department of Geoinformatics, Kyungil University

Hyungtae Kim
Lx Education Institute, Korea Land and Geospatial Informatix Corporation

Recommended Citation

Choi, Junyoung; Ahn, Jaesung; and Kim, Hyungtae (2015) "A Cross National Comparison on the Awareness of Adopting FOSS4G to NSDI in Developing Countries," *Free and Open Source Software for Geospatial (FOSS4G) Conference Proceedings*: Vol. 15 , Article 36.
DOI: https://doi.org/10.7275/R5T43R86
Available at: https://scholarworks.umass.edu/foss4g/vol15/iss1/36

AT9-03

A Cross National Comparison on the Awareness of Adopting FOSS4G to NSDI in Developing Countries

Junyoung Choi[1], Jaesung Ahn[2] and Hyungtae Kim[3]

[1]Spatial Information Office, Korea Land and Housing Corporation
19 Chungui-ro, Jinju-si, Gyeangsangnam-do, Republic of Korea
Email: junyoung@lh.or.kr
[2]Department of Geoinformatics, Kyungil University
50 Gamasil-gil, Hayang-eup, Gyeongsan-si, Gyeongbuk-do, Republic of Korea
Corresponding Author, Email: jsahn@kiu.kr
[3]Lx Education Institute, Korea Land and Geospatial Informatix Corporation
213-20 Dongbu-ro, Cheoin-gu, Yongin-si, Gyeonggi-do, Republic of Korea
Email: khtkht1@lx.or.kr

ABSTRACT

In this study, we constructed an assessment framework that was consisted of 9 indicators about functional, economic and public value for FOSS4G adoption to NSDI and alternatives such as data sharing, data management, utilization and construction and derived relative weights using AHP method. For the AHP, we conducted a survey to developing countries' 10 respondents from 9 Asian and Latin American countries. Firstly, result of the survey showed that economic value indicator came in the highest weight with 0.425, followed by functional value indicator with 0.345 and public value indicator with 0.230. Secondly, result of the alternatives analysis showed that data sharing alternative came in the highest adoption rate with 0.824, followed by data management with 0.780, data utilization with 0.778. This means that developing countries want to introduce FOSS4G to their NSDI from economic motivation. This study focused on the comprehensive aspect for adopting FOSS4G to NSDI that is different from the previous researches that were focused on the software engineering aspect to the adoption.

1. INTRODUCTION

Spatial data infrastructure (SDI) plays an important role in the sharing and exchange of spatial information of the country. In this respect, it has also played a major role in the sustainable development of the country's economy and society (Rajabifard & Williamson, 2001). In recent years, Free Open Source Software for Geospatial or FOSS4G provides functionalities that are not inferior to commercial software, which lead to the diffusion of that software to the environment management and natural disaster of public and private sectors (Wawer et al. 2008; Jolma et al., 2008; Herold & Sawada, 2012; Moreno - Sanchez, 2012). More and more countries have been interested in adopting FOSS4G in their National Spatial Data Infrastructure (NSDI) and INSPIRE is acknowledged as one of the best practices of integrating FOSS4G technologies into NSDI (Anguix et al., 2008).

Developing countries which have poor information infrastructures, there are increasing discussions about the adoption of FOSS4G to their NSDI in order to utilize the benefits of the foundation. Especially among the benefits of FOSS4G adoption, low introduction cost and interoperability of software that does not depend on the specific software are frequently suggested factors. Developing countries, however, have their own economic and technological development stages and different cultural background and institutions, it can vary from country to country what are the important factors for their countries in adopting the FOSS4G to NSDI. So there are differences on the pros and cons of FOSS4G adoption to their NSDI, but little researches have been done to analyze what are the favorable factors for adopting the FOSS4G. Moreover, many researches for introduction criteria of adopting Open Source Software (OSS) were done but those researches are more focused on the software engineering aspect (Jusoh et al., 2012; Jusoh et al., 2014) and researches didn't deal with political or public aspect to FOSS4 adoption.

In this research, we will develop a framework to compare and evaluate the relationships between indicators on the developing countries' public, functional and economic factors in the introduction of FOSS4Gs to their NSDI and conduct a survey of 12 ex- or current government officials from the developing countries about the FOSS4G adoption to their NSDI. To prioritize and derive relative weight of indicators, we used AHP (Analytical Hierarchical Process) method. In the end, we could draw priorities of adopting FOSS4G to NSDI in developing countries and propose a deployment strategy for overcoming the disadvantages when developing countries consider the introduction of FOSS4G to their NSDI.

2. Literature review

This section highlights the research suggested models and frameworks for preferred list of criteria for software selection. The selection criteria FOSS4G to NSDI will be applied to the adoption framework with modification of software quality models and OSS evaluation models.

Many researchers have suggested software quality models. These models are McCall's Quality Model(McCall et al, 1977), Boehm's Quality Model(Boehm et al., 1978), FURPS Quality Model(Grady, 1992), Dromey's Quality Model(Dromey, 1996), and ISO/IEC 9126(ISO, 2001). To evaluate software quality, the qualitative indicators of system characteristics are listed and the quality models explain the relationship between such characteristics. For example, ISO/IEC 9126, an international standard for software quality evaluation, contains a set of software quality metrics related to each of the six quality characteristics: functionality, reliability, usability, efficiency, maintainability, and portability. The model measures any of the six quality characteristics then covert to a percentage value to represent the corresponding quality characteristics. The main advantage of these types of models is models could be applied to evaluate the quality of every type of software product.

The OSS is software which has following features: source code availability to its users and free. To evaluate the OSS, most of software quality characteristics will be applied to the OSS with appropriate modifications. Confino and Laplante(2010) and Ahmad(2011) cited four OSS evaluation models which are the Open Source Maturity Model(OSMM) created in 2003, Open Source Maturity Model(OSMM) created in 2004, the Qualification and Selection of Open Source software model(QSOS), and the Business Readiness Rating(BRR). They also proposed the nine evaluation criteria: functionality, product evaluation, licensing, longevity/pedigree, community, market penetration, documentation, support and code quality. Jusoh et al.(2014) suggested 12 criteria for OSS selection. The criteria suggested are reliability, usability, performance efficacy, functionality, maintainability, security, tangible, reliability, responsiveness, assurance, empathy, competence.

3. Awareness for adopting FOSS4G to NSDI

3.1 AHP methodology

The AHP is a systematic decision making method which was introduced by Thomas L. Saaty in 1980(Saaty, 1980). It assesses the relative importance of multiple criteria, compares alternatives for each criterion, and determines an overall ranking of the alternatives. The main advantage of the AHP is that it allows users to take into account a variety of multiple criteria, which rating is based on a multiple-value choice. It has been widely used for a long time in many fields which include research on the selection the OSS (Open Source Software) product (Jusoh et al., 2014).

In this study, the processes for selecting the FOSS4G to NSDI follow the basic procedure of the AHP method. The first step is to structure a decision problem and selection of criteria. In this step, all the criteria arranged in a hierarchy and the alternatives are generated. The second step is priority setting of the criteria by pairwise comparing. A numerical weight is derived for each criteria of the hierarchy. All criteria should be compared to one another in a consistent way. In order to verify the consistency of comparison, Consistency Ratio (CR) was proposed. Saaty suggested that CR should satisfy the condition of the value less than or equal 0.1. However, if CR less than or equal 0.2, it can be understood that it has consistency in an acceptable degree. In this study, the condition of CR less than or equal 0.2 is adopted. In the next step, pairwise comparisons of alternatives on each criterion are carried out. This process is to multiply the alternative weight by the criteria weight to score each alternative. Finally, an overall relative score for each alternative is obtained. The result of the final ranking is presented for users to select the FOSS4G to NSDI.

3.2 Selecting indicators

Some researches argue that economic aspect such a low introduction cost (Makanga and Smit, 2010), but others stressed the importance of functionality and opening and sharing of geospatial information (Steiniger and Hunter, 2008). In the research, we constructed assessment frame into 2 tier indicator system according to proceeding researches. First tier is conceptual indicator and second tier is specific sub-indicators that explain the first tier indicators. First tier assessment indicators deduced 3 indicators such as functional, public and economic values (Table 1). Functional values are explained as evaluating how much helpful FOSS4G is to do one's task, public values are explained as evaluating how much helpful FOSS4G is to achieve value about obtaining geospatial information technologies and knowledge and opening and sharing of the data in each countries. Lastly, economic values are explained as evaluating the economic benefits of adoption of FOSS4G.

Second tier consists of 12 specific sub-indicators that mean each indicator has 3 sub-indicators (Table 2). A functional value consists of three sub-indicators that are fit of task, maturity of open source software and usability. A public value consists of achievement of self-reliance of technology, diffusion of spatial information knowledge and openness of spatial information and enhancement of its sharing. Economic value consists of industry ripple effect, total cost of ownership and reusability.

Table 1. First tier assessment indicators for adoption of FOSS4G to NSDI

Indicator	Explanation
Functional value	How much helpful FOSS4G is to do one's task
Public value	How much helpful FOSS4G is to obtain opening and sharing of the data
Economic value	Economic benefits of adoption of FOSS4G

Table 2. Second tier indicators for adoption of FOSS4G to NSDI

Indicator	Sub-indicator	Explanation
Functional value	Fit of task	Achievement of efficiency through the business support
	Maturity of open source software	Completeness of FOSS4G in contrast to commercial software

	Usability	Provision of user-friendly functionality and interface
Public value	Achievement of self-reliance of technology	Contribution to country's technological development
	Diffusion of spatial information knowledge	Acceleration of spatial information knowledge into one's country
	Openness of spatial information and enhancement of its sharing	How many people can be benefited from the FOSS4G based SDI(Spatial Data Infrastructure)
Economic value	Industry ripple effect	Contribution of Geospatial information related industry
	Total cost of ownership	Economic evaluation of Total cost of ownership including construction and maintenance
	Reusability(Interoperability)	Additional cost saving using reusable FOSS4G SW

After determining the weights of individual indicators and sub-indicators, it needs to know which goal is more important than others in adopting the FOSS4G to NSDI. In this respect, we set 4 alternative goals such as data construction, data management, data sharing and data utilization (Figure 1).

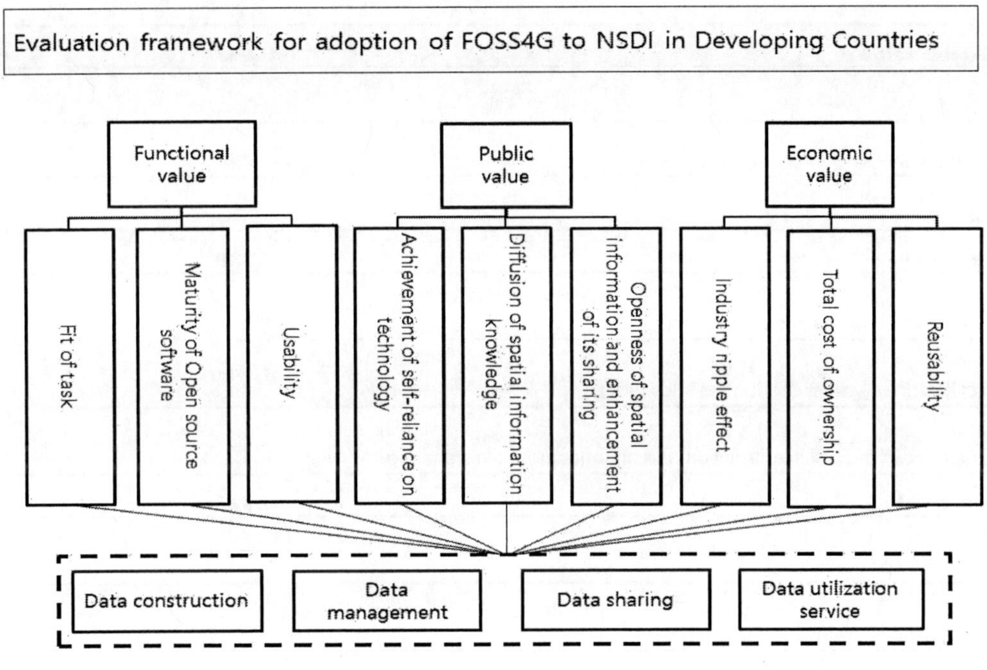

Figure 1. Overall assessment framework.

3.3 Respondents

Survey was conducted to 12 participants, but among the received questionnaires, only 10 were effective for the AHP analysis. That is because we selected questionnaires which are below 0.2 of AHP Consistency Ratio(CR). Participants are from 5 Asian and 3 Latin American countries who are ex or current government officials in those countries. Asian participants are from Vietnam, Sri Lanka, Cambodia, Iraq and Philippines and Latin American participants are from El Salvador, Peru and Colombia. The respondents are now studying in Korea and with the help of International Urban Development Cooperation Center in Korea Land and Housing Corporation which is government company. Participants' ages are from late 20' to 40's and their affiliations of their countries are urban planning, construction, environment department.

3.4 Results

Result of the survey showed that economic value indicator came in the highest weight with 0.425, followed by functional value indicator with 0.345 and public value indicator with 0.230(Table 3). Among the economic value sub-indicators, total cost of ownership has the highest weight with 0.493. That means developing countries have not sufficient budget and they mostly focused on saving of expenses. Among the functional value sub-indicators, fit of task has the highest weight with 0.581. We can presume that developing countries place a high value on business utilization. In the public value sub-indicators, openness of spatial information and diffusion of spatial information get a similar weight.

Table 3. Weight

Criteria	Weight
Functional value	**0.345**
Fit of task	0.581
maturity of open source software	0.138
usability	0.281
Public value	**0.230**
Achieve of self-reliance of technology	0.121
Diffusion of spatial information knowledge	0.435
Openness of spatial information and enhancement of its sharing	0.444
Economic value	**0.425**
Industry ripple effect	0.222
Total cost of ownership	0.493
Reusability	0.284

Finally we assess the goal for adopting FOSS4G to NSDI. Result of the survey showed that Data sharing alternative came in the highest adoption rate with 0.824, followed by data management with 0.780, data utilization with 0.778 (Table 4). That means although data sharing is slightly important than other alternatives, all alternatives have similar importance in adopting FOSS4G to their NSDI.

Table 4. Weight for the adoption of alternatives

	adopt	Not adopt
Data construction	0.774	0.226
Data management	0.780	0.220
Data sharing	0.824	0.176
Data utilization	0.778	0.222

4. Conclusion

In this study, we conducted a survey to find the awareness of FOSS4G introduction of developing countries' 10 respondents from 9 Asian and Latin American countries. We constructed a framework that was consisted of 9 indicators about functional, economic and public values for FOSS4G adoption to NSDI and investigated relative weights between indicators using AHP method. In the result, we identified that economic value is the highest value in developing countries. This means that developing countries want to introduce FOSS4G to their NSDI from economic motivation. Also weight of alternatives which are NSDI alternatives from FOSS4G such as data sharing, management, utilization, etc., were analyzed. As a result, those alternatives have similar result of high adoption rate. This means that those alternatives are also important in developing countries. This study focused on the comprehensive aspect for adopting FOSS4G to NSDI that is different from the previous researches that were focused on the software engineering aspect to the adoption. In the future study, additional survey is needed to find the differentiated adoption factors by country, by economic status and by political condition, etc.

References

Ahmad, N. 2011. A Systematic Approach to Evaluating Open Source Software. *International Journal of Strategic Information Technology and Applications*. 2(1). 48-67.

Anguix, A., Díaz, L., & Carrera, M. A. R. I. O. (2008). gvSIG: A GIS desktop solution for an open SDI. *Journal of Geography and Regional Planning, 1*(3), 041-048.

Boehm, B.W., Brown, J.R., Kaspar, H., Lipow, M., McLeod, G. and Merritt, M. (1978). Charcteristics of Software Quality. North Holland Publishing, Amsterdam.

Budhathoki, N. R., & Nedovic-Budic, Z. (2008). Reconceptualizing the role of the user of spatial data infrastructure. GeoJournal, 72(3-4), 149-160.

Confino, J., Laplante, P. 2010. An Open Source Software Evaluation Model. International Journal of Strategic Information Technology and Applications. 1(1). 60–77.

Dromey, R.G. (1996) Concerning The Chimera [Software Quality]. *IEEE Software*, **13**, 33-43.

Fitzke, J., Greve, K., Müller, M., & Poth, A. (2004, February). Building SDIs with Free Software– the deegree project. InProceedings of GSDI-7, Bangalore, India.

Grady, R.B. (1992) Practical Software Metrics for Project Management and Process Improve ment. Prentice Hall, Englewood Cliffs.

Herold, S., & Sawada, M. C. (2012). A review of geospatial information technology for natural disaster management in developing countries.*Geographic Information Systems: Concepts, Methodologies, Tools, and Applications: Concepts, Methodologies, Tools, and Applications*, 175.

Herold, S., & Sawada, M. C. (2012). A review of geospatial information technology for natural disaster management in developing countries.*Geographic Information Systems: Concepts, Methodologies, Tools, and Applications: Concepts, Methodologies, Tools, and Applications*, 175.

Horvat, Z. Building Spatial Data Infrastructure Using Free and Open Source Software. SDI DAYS.

ISO (2001) ISO/IEC 9126-1: Software Engineering—Product Quality—Part 1: Quality Model. International Organization for Standardization, Geneva.

Jolma, A., Ames, D. P., Horning, N., Mitasova, H., Neteler, M., Racicot, A., & Sutton, T. (2008). Chapter Ten Free and Open Source Geospatial Tools for Environmental Modelling and Management. *Developments in Integrated Environmental Assessment*, *3*, 163-180.

Jusoh, Y. Y., Chamili, K., Che Pa, N., and Yahaya, J. H.(2014). Open source software selection using an analytical hierarchy process. Ameraican Journal of Software Engineering and Application, 3(6), 83-89.

Jusoh, Y. Y., Chamili, K., Yahaya, J. H., & Pa, N. C. (2012). The Selection Criteria of Open Source Software Adoption in Malaysia. *International Journal of Advancements in Computing Technology*, *4*(21).

Moreno‐Sanchez, R., 2012. Free and Open Source Software for Geospatial Applications (FOSS4G): A mature alternative in the geospatial technologies arena. *Transactins in GIS*, 16(2), 81-88.

Makanga, P., & Smit, J. (2010). A review of the status of spatial data infrastructure implementation in Africa. South African Computer Journal, (45), 18-25.

Makanga, P., & Smit, J. (2008). A review of spatial data infrastructure implementation in Africa. FOSS4G 2008.

McCall, J.A., Richards, P.K. and Walters, G.F. (1977) Factors in Software Quality. US Rome Air Development Center Reports, US Department of Commerce, Washington DC

Rajabifard, A., & Williamson, I. P. (2001). Spatial data infrastructures: concept, SDI hierarchy and future directions(https://minerva-access.unimelb.edu.au/handle/11343/33897).

Saaty, T. L. (1980). The Analytical Hierarchy Process. New York, McGraw Hill.

Steiniger, S., & Hunter, A. J. (2012). Free and open source GIS software for building a spatial data infrastructure. Geospatial free and open source software in the 21st century, 247-261.

Wawer, R., Steenberghen, T., Rusztecka, M., Podolczak, A., & Orlitowa, E. (2008). Evaluation of FOSS4G software projects for enviromental aplications. Evaluation of gvSIG.

Free and Open Source Software for Geospatial (FOSS4G) Conference Proceedings

Volume 15 *Seoul, South Korea* Article 37

2015

Results Of and Evaluation Of Augmented Reality Mobile Development Frameworks For Addresses in Augmented Reality

Danie Jooste
Centre for Geoinformation Science, Department of Geography, Geoinformatics and Meteorology, University of Pretoria

Victoria Rautenbach
Centre for Geoinformation Science, Department of Geography, Geoinformatics and Meteorology, University of Pretoria

Serena Coetzee
Centre for Geoinformation Science, Department of Geography, Geoinformatics and Meteorology, University of Pretoria

Recommended Citation

Jooste, Danie; Rautenbach, Victoria; and Coetzee, Serena (2015) "Results Of and Evaluation Of Augmented Reality Mobile Development Frameworks For Addresses in Augmented Reality," *Free and Open Source Software for Geospatial (FOSS4G) Conference Proceedings*: Vol. 15 , Article 37.
DOI: https://doi.org/10.7275/R5XS5SMB
Available at: https://scholarworks.umass.edu/foss4g/vol15/iss1/37

AT4-02

RESULTS OF AN EVALUATION OF AUGMENTED REALITY MOBILE DEVELOPMENT FRAMEWORKS FOR ADDRESSES IN AUGMENTED REALITY

Danie Jooste, Victoria Rautenbach and Serena Coetzee

Centre for Geoinformation Science, Department of Geography, Geoinformatics and Meteorology, University of Pretoria, Hatfield, Pretoria, South Africa
Email: dfjooste@yahoo.com, victoria.rautenbach@up.ac.za, serena.coetzee@up.ac.za

ABSTRACT

Addresses displayed on dwellings and buildings play a key role in society. Amongst others, they are used for deliveries, in household surveys, to navigate, or to find friends. Sometimes, address signs are destroyed, displaced or illegible, for example, as a result of vandalism, disasters, or poor maintenance. In augmented reality, computer-generated information is superimposed onto a live view of the real world. When address signs are not available, displaying the address in augmented reality could be immensely useful. The research presented in this article is part of a larger research endeavour to investigate augmented reality for addressing. This article presents the results of an evaluation of augmented reality mobile development frameworks for the implementation of a mobile application that displays addresses in augmented reality. Firstly, the requirements for addresses in augmented reality were identified. Three use cases informed these requirements: disaster relief, e.g. address signs are destroyed by an earthquake; household surveys, e.g. locating dwellings in informal settlements or rural areas where addresses are not assigned in any specific sequence and signs do not exist; and address data quality management, e.g. validating digital address data against addresses displayed in the physical world. Due to procurement challenges in the use cases, open source licensing and integration with open source products was identified as an important requirement. The internet was searched and a list of augmented reality mobile development frameworks was compiled. Based on the requirements, the list was shortened to seven frameworks, which were evaluated against a set of criteria informed by the requirements. The evaluation results can guide developers in choosing a framework best suitable for their specific needs and/or for integration with open source products.

1. INTRODUCTION

Addresses play a vital role in society. They are used for deliveries, in household surveys, by utility companies, to navigate, or to find friends (Coetzee and Cooper, 2007). Sometimes, address signs, such as street names and house numbers, are destroyed or displaced as a result of vandalism, disasters or poor maintenance. Replacing the address signs takes time and is expensive. In augmented reality, computer-generated information is superimposed onto a live view of the real world. When address signs are not available, superimposing address information onto a live view of the real world could be a viable alternative.

The research presented in this article is part of a larger research endeavour on the display of geocoded address data in augmented reality. As part of this endeavour, a mobile application will be developed. This article presents the results of an evaluation of augmented reality mobile development frameworks. Instead of starting from scratch, the aim was to identify existing development frameworks that could be used for the development of the mobile application.

In augmented reality a live view of the real world is superimposed with computer-generated information, such as text or images. Azuma (1997) defines augmented reality as the real-time combination of the physical world with virtual objects. Augmented reality enhances our understanding and interaction with the physical world (Amin and Govilkar, 2015; Carmigniani et al., 2011). Augmented reality has proven to be useful in a variety of application fields, such as medicine (Azuma, 1997; Kounavis et al., 2012; Dünser et al., 2008), education (Wu et al., 2013; van Krevelen et al., 2010), navigation (Wen et al., 2014; Mulloni et al., 2011) and planning (Anagnostou and Vlamos, 2011; Allbach et al., 2011).

Allbach *et al.* (2011) evaluated augmented reality applications for urban planning and design. They concluded that at the time it was not possible to recommend a single augmented reality browser, but rather commented on the shortcomings of augmented reality applications, such as limited precision of the GPS, size of the mobile device (i.e. information might be too dense to be displayed on a small screen), and the need for internet connectivity which is not always available. However, with the rapid development of augmented reality and mobile technology these shortcomings are fast disappearing (Wen et al., 2014; Wu et al., 2013). Leebmann (2006) suggested augmented reality as a solution for disaster relief. Examples are analysing rescue routes for collapsed buildings and performing analyses from safe distance (Leebmann, 2006; Leebmann, 2004). A drawback is that such applications need a large amount of data (i.e. 3D laser scans and site plans) in order to be useful.

Amin and Govilkar (2015) compared six augmented reality software development kits (SDKs), three of them (Metaio, Wikitude and ARToolkit) are also evaluated in this paper. They compared the license type, platform support, marker generation, tracking functionality and overlaying capability. The choice of the SDKs was not justified. In this paper, mobile development frameworks for augmented reality are evaluated with the specific requirement of superimposing address information on a live view of the real world. At present, case studies of augmented reality applications where the main focus is on augmenting address data could not be found in literature. The remainder of this article is structured as follows: in section 2, three use cases are described and requirements for a mobile application based on the use cases are identified; in section 3, evaluation criteria, derived from the requirements, are described; results are presented in section 4 and discussed in section 5. We conclude in section 6.

2. REQUIREMENTS FOR ADDRESSING IN AUGMENTED REALITY

Three use cases informed the requirements for the display of addresses in augmented reality: 1) disaster relief, e.g. address signs are destroyed by an earthquake; 2) household surveys, e.g. locating dwellings in informal settlements or rural areas without any address infrastructure; and 3) address data quality management, e.g. validating digital address data against addresses displayed in the physical world. In this section, the three use cases are presented and subsequently, requirements for the mobile application, based on these use cases, are described.

2.1 Use case 1: Disaster relief

In the disaster relief use case, dwellings (with the house numbers) and street names signs have been damaged or destroyed. A tsunami, an earthquake or fires could be the cause of such a disaster. Emergency workers are informed that there may be survivors at a specific address. Assuming that the backbone for internet (and mobile) connectivity has been destroyed, how do the emergency workers locate the site? A backup of geocoded address data was recovered from an off-site location. However, address maps are of little use as buildings, streets and signs have been destroyed. Emergency workers are equipped with smartphones connected to a satellite network, but data connectivity via satellites is expensive. Relief efforts are coordinated from a disaster management centre where a server has been set up.

a) Damage caused by the earthquake on 4 September 2010 in Christchurch, New Zealand (Photo: www.foxnews.com)

b) Damage caused by the tsunami on 11 March 2011 in Kesenuma, Japan (Photo: http://www.dailymail.co.uk)

Figure 1. Damaged or destroyed street name signs and house numbers after a disaster

2.2 Use case 2: Household surveys in rural areas

In this use case, a random sample of dwellings has been selected for a survey. Using aerial photography as a backdrop, a unique number was assigned to each dwelling without an address. Subsequently, a random number generator was used to select the sample of dwellings, based on the unique numbers assigned to each dwelling. Enumerators (i.e. people doing the interviews) have to interview the household at each of the dwellings in the sample. Some of the dwellings are in rural areas, others in an informal settlement. In both cases there is no address infrastructure: there are no street signs, no house numbers, and an intricate web of footpaths connects dwellings to each other. Paved roads connect one village or settlement to another; smaller roads beyond are typically nameless dirt roads. Dwellings in the villages or settlement are generally scattered, not necessarily arranged in a fixed pattern. In the rural areas, dwellings are sometimes interspersed with agricultural fields. See Figures 2 and 3. Figures 4 and 5 show dwellings in an informal settlement.

Figure 2. Dwellings in a rural village in the Eastern Cape, South Africa (Photo: Serena Coetzee)

Figure 3. A rural village in the Eastern Cape, South Africa (Image from maps.google.com)

Surveys are planned and coordinated from a head office with access to ample bandwidth and internet connectivity. In the rural villages, internet connectivity is not necessarily available. Survey responses are captured on tables and/or smartphones. Imagine an enumerator had to visit three dwellings on the hill displayed in Figure 2: How does the enumerator find the dwellings without any street signs or house numbers?

Figure 4. Dwellings in an informal settlement in the City of Tshwane, Gauteng, South Africa (Photo: Victoria Rautenbach)

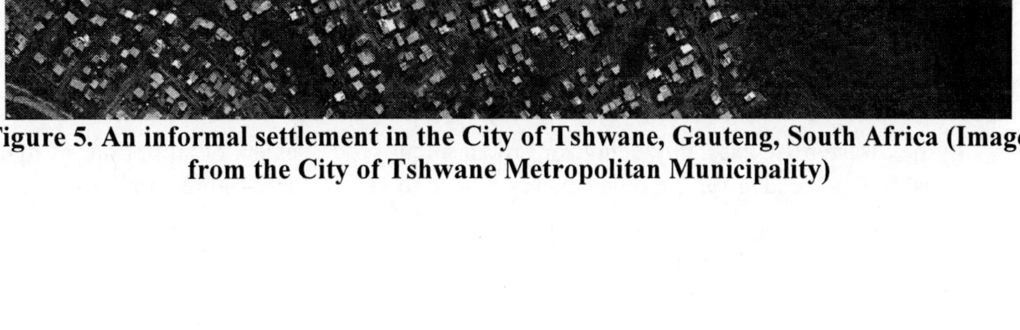

Figure 5. An informal settlement in the City of Tshwane, Gauteng, South Africa (Image from the City of Tshwane Metropolitan Municipality)

2.3 Use case 3: Address data quality management

In this use case, a field worker compares geocoded address data with address signs in the real world. The local authority assigns house numbers to dwellings and buildings when building plans are approved, i.e. before the buildings are constructed. Geocoded house numbers are stored in a geospatial database. House numbers are not verified after the buildings have been erected and when buildings are altered or extended, one does not have to apply for a house number again. As a result, owners and occupants may put up house numbers, which are not reflected in the geospatial database at the local authority. Ultimately, this may lead to returned mail and service delivery interruptions (e.g. when bills by the local authority do not reach the owner). Therefore, from time to time, the local authority needs to compare its digital address database against address signs in the real world in order to harmonize the digital representation with the real world.

2.4 Requirements for address in augmented reality

In all three use cases, superimposing digital address data onto a live view of the real world could solve the problem at hand: to locate the address where survivors need assistance; to visit specific dwellings in a rural village or informal settlement without an address infrastructure; and to compare digital address data with house numbers in the real world. See Figure 6.

a) Use case 1: Disaster relief (Photo: www.citizen.co.za)

b) Use case 2: Household surveys in rural areas

Figure 6. An example of addresses displayed in augmented reality in a rural village setting

In two of the three use cases, internet connectivity is available at a coordinating centre, but not in the field. Therefore, the display of address information in the augmented reality application must be available, even if the device is offline.

In all three cases, geocoded address data is available. In the augmented reality view, the address should be superimposed as close as possible to the actual location of the address and one has to be able to distinguish an address from its neighbouring address. Appropriate precision is therefore important; 'appropriate' because in rural areas dwellings are spaced further apart, requiring less precision; whereas in densely populated informal settlements, better precision is required. It would also be useful to know the distance between the smartphone and the address. For example, in the household survey use case, this would allow the enumerators to plan their route of interviews around the village.

In the disaster use case, any delay for procurement processes is not an option: as many licenses as may be needed for the relief exercise have to be available immediately in order to save lives. The use case of rural villages and informal settlements without an address infrastructure is often found in developing countries with financial constraints and plagued with corrupt and/or lengthy procurement processes. Therefore, ideally, the mobile application should be available free of charge.

Nice-to-have requirements include navigation and/or wayfinding; a map view in addition to the augmented reality view; and the capability to edit or update address data, or any other information linked to the address (e.g. survey responses or notes about the dwelling).

3. METHODOLOGY

In this section, we describe the two-phase approach followed to evaluate the augmented reality mobile development frameworks. They were evaluated in the context of an augmented reality solution for addresses that meets the requirements identified in Section 2. Refer to Figure 7 for an overview of the two-phase evaluation.

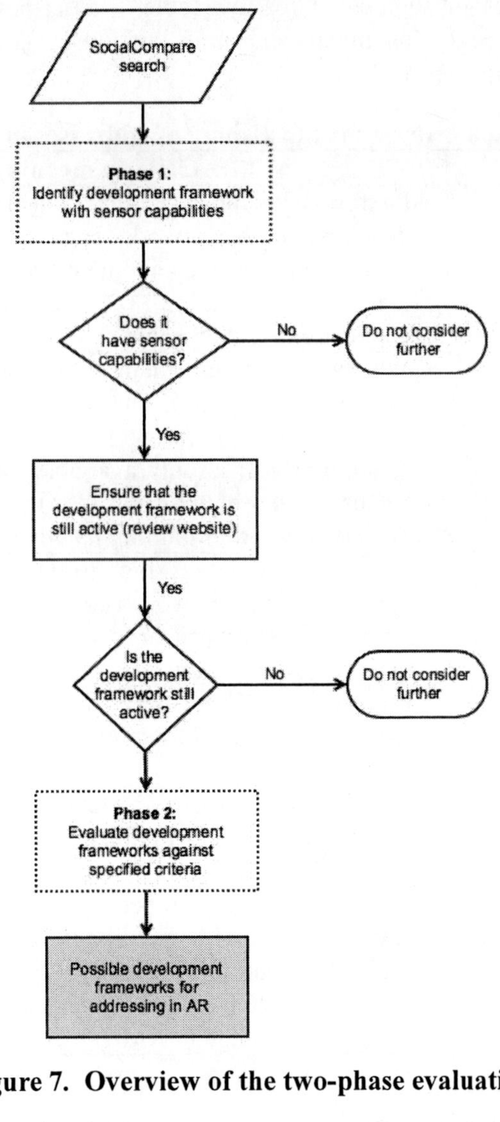

Figure 7. Overview of the two-phase evaluation

We consulted SocialCompare (http://socialcompare.com/en), a collaborative online comparison tool, where a comparative list of augmented reality SDKs and frameworks is actively maintained[1]. Guided by the comments posted on the SocialCompare list, two additional development frameworks were included, namely ARToolkit and Layar (between the time of the evaluation and the writing of this paper, ARToolkit was added to the SocialCompare list). This resulted in a list of 68 SDKs and frameworks.

In the first phase of the evaluation, development frameworks were disqualified if they did not support GPS and inertial measurement unit (IMU) sensors (these are required to locate and correctly display geocoded address data in augmented reality). Thereafter, we reviewed the websites of each of the remaining 12 frameworks. Inactive and irrelevant (e.g. mainly aimed towards gaming) frameworks were disqualified. This left seven frameworks for evaluation in the second phase.

During phase two, each development framework was evaluated in more detail against a set of evaluation criteria and also tested to ensure that it complied with the requirements for the display of addresses in augmented reality. Table 1 describes the three categories of evaluation criteria: general, functional and non-functional. The criteria are based on requirements identified in Section 2.

Table 1. Evaluation criteria for the display of addresses in augmented reality

	Criteria	Criteria meaning
1. General criteria	1.1. Platform	Platforms and/or operating systems supported by the development framework, such as Windows Mobile, Android or iOS. Development frameworks that support the implementation of cross-platform mobile applications are desirable.
	1.2. Programming language	Which programming languages does the development framework supports? The programming language and the supported platform (1.1) are closely related. For example, Swift is only available for iOS but Java can be used for both Android and Windows Mobile. Widely used programming languages are desirable.
	1.3. License	Refers to whether the license under which the development framework is available, e.g. open source, freeware or proprietary. The licensing affects the manner in which the development framework may be used and also how derived implementations may be distributed. As few constraints as possible are desirable.
	1.4. Implemented standards	The implementation of standards contributes to the interoperability and modularity of an implementation. Various standards may contribute to interoperability, for example, data encoded in eXtensible Markup Language (XML) facilitates data exchange from different sources. Recently, the Open Geospatial Consortium (OGC) published the Augmented Reality Markup Language (ARML) that uses XML to describe

[1] http://socialcompare.com/en/comparison/augmented-reality-sdks, last updated 31 July 2015

	Criteria	Criteria meaning
		the location and appearance of augmented visual objects.
	1.5. Offline availability	The connectivity required by the augmented reality application in the field. It is desirable to have all functionality available in the field without any connectivity.
2. Functional criteria	2.1. Data source	The flexibility of the development framework to allow access to various data sources. Least desirable is a dedicated source, e.g. via a quick response (QR) code, hosted by a provider or organization.
	2.2. Data display	The capability to adjust the visual representation of the data being superimposed on the camera feed. For example, can the text size be altered? Flexibility in configuring the visual representation is desirable.
	2.3. Object behaviour	The ability to add behaviour to the object, e.g. by adding triggers or events to the object. For example, if the user clicks on the object, additional information is displayed.
	2.4. Display radius	The flexibility to adjust the amount of content that is displayed in the camera feed based on the distance between the user and objects (e.g. addresses or dwellings). For example, specify that only addresses that are within a 50m radius from the user are displayed.
	2.5. Visual search	The ability to recognize a specific object based on additional information, such as a photo or light detection and ranging (LiDAR) data.
3. Non-functional criteria	3.1. Ease of integration with other applications	Built-in capabilities to facilitate integration with other applications, such as PostGIS, QuantumGIS or ArcGIS. For example, integration with PostGIS will allow seamless access to a database without the need for additional code or third-party products.
	3.2. Ease of extending the framework	Refers to the effort required to add additional functionality to the development framework and to port the solution to different platforms.
	3.3. Usability	Refers to the user friendliness framework and ease of installation of the development framework.
	3.4. Documentation and support available	The documentation and additional avenues available to developers for support, such as forums and mailing lists.

4. EVALUATION RESULTS

4.1. Phase 1

The list of 68 development frameworks compiled from the SocialCompare source was reduced to 12 by disqualifying development frameworks that do not support GPS or IMU sensors. After reviewing the websites of these 12 frameworks, the following were eliminated as they did not meet the requirements for the use cases: ARmedia[2], BeyondAR[3], LibreGeoSocial[4], Total Immersion[5] (previously known as D'Fusion), and Xloudia[6].

LibreGeoSocial is an open source project that looked promising initially. However on closer inspection, it was eliminated as it has not been updated for quite a number of years. The LibreGeoSocial website was also recently removed after being hacked. Total Immersion and Xloudia are more geared towards proprietary solutions and provide little flexibility. For example, these frameworks cannot easily be customized or extended to serve a different purpose than originally intended. Lastly, ARmedia and BeyondAR were eliminated as they focus on superimposing 3D objects. This is useful for gaming or advertising purposes, but not for superimposing textual information as required in the use cases. The remaining seven development frameworks were evaluated further in phase 2.

4.2. Phase 2

The following development frameworks were included in the phase 2 evaluation: ARLab, ARToolkit, DroidAR, Layar, Metaio, PanicAR and Wikitude. A brief overview of each of the evaluated frameworks follows.

1. **ARLab[7]**
   ARLab is developed by a commercial company that focuses on augmented reality solutions. They have divided their products into the following modular units: the AR Browser, image matching, 3D engine, image tracking, object tracking and virtual buttons. The AR Browser is a SDK that allows the development of a location-based augmented reality application for points of interests (POIs) within minutes.

2. **ARToolkit[8]**
   ARToolkit is a GNU GPL v3 open source library developed for Apple iOS and recently also for Android. ARToolkit was not originally included, but was added to the list after being mentioned in various comments on the SocialCompare page. ARToolkit has not been updated since 2013, but there are various blog posts on how to fix errors due to deprecated functions in newer versions of iOS. This shows that ARToolkit still has strong community support. The framework makes use of MapKit and UIKit, two standard Applie iOS libraries.

3. **DroidAR[9]**
   DroidAR is an open source framework (available under the GNU GPL v3 license) that focuses on location and marker based augmented reality. Even though the code has not been updated in the GitHub repository since 2013, there seems to still be an active user community. With small adjustments in the code, DroidAR runs on the latest Android operating systems. It is not clear why the project stagnated, as it averaged about 1500 website hits per month in 2013. The DroidAR website is currently out-dated and the

[2] http://dev.inglobetechnologies.com
[3] http://beyondar.com/platform
[4] http://www.libregeosocial.org
[5] http://www.t-immersion.com/products/dfusion-suite
[6] http://www.xloudia.com/xloudia-imerico/
[7] http://www.arlab.com/arbrowser
[8] https://www.artoolworks.com/products/artoolkit-for-mobile/artoolkit-for-ios/
[9] http://droidar.blogspot.com

development status of the framework is unclear.

4. **Layar**[10]
   Layar was not included in the SocialCompare list. Based on comments posted on this list, it was included in the evaluation because it is one of the leading augmented reality frameworks available today. Layar provides developers with an open platform to publish and discover augmented reality layers (the Layar term for data). Layar employs representational state transfer (REST) services to serve POIs facilitating, the integration of Layar with other Layar applications.

5. **Metaio**[11]
   Metaio is a commercial company providing augmented reality solutions for a variety of application fields. Metaio offers six products that cover all the requirements of the augmented reality value chain, ranging from development tools to out of the box solutions. Metaio was included in the evaluation, but was acquired by Apple Inc. in May 2015 (after the evaluation had been completed) and is no longer offering any subscriptions for purchase.

6. **PanicAR**[12]
   PanicAR is a SDK that is fully customizable and focused on location-based AR. PanicAR is fully integrated with the Apple MapKit, allowing the developer to visualize POIs on a 2D map (in addition to the augmented reality view). PanicAR states that it is fully white label, i.e. it can be completely re-branded.

7. **Wikitude**[13]
   Wikitude was the first openly available location-based augmented reality application, and it has won numerous awards, such as the Android Developers Challenge in 2008. Wikitude provides a multifunctional framework that includes numerous features, for example, location-based augmented reality, 3D model rendering, and image recognition and tracking.

The Apple iOS and Android **platforms** are supported by most of the development frameworks (refer to Table 2). DroidAR is the only framework that does not support Apple iOS. Apple iOS requires high subscription fees for iOS development and implementation of the framework on iOS requires conversion of the code to Objective-C. Wikitude has the widest platform support: apart from Android and iOS, Google Glasses, Blackberry and Windows Mobile are supported. As mentioned in Table 1, the supported platform and the **programming language** are closely related. The primary supported programming language is Java (used on Android), followed closely by Objective-C (used on iOS). The majority of frameworks are available under a proprietary **license**, but offer a free development framework option (freeware) with limited functionalities and a watermark. Among the evaluated frameworks, ARToolkit and DroidAR are the only open source frameworks.

The OGC ARML 2.0 standard (Open Geospatial Consortium 2015) is currently the only augmented reality **standard** published in the geospatial industry. ARML is composed of

[10] https://www.layar.com
[11] http://www.metaio.com
[12] http://panicar.dopanic.com
[13] http://www.wikitude.com

XML (for describing locations and appearance of virtual objects), and ECMAScript (allows dynamic access to the properties of the objects). Layar, Metaio and Wikitude implement ARML. Other standards, such as OGC web services (i.e. web feature service or web processing service) and encoding standards (i.e. XML or JSON), contribute to the modularity and integration of the framework with other applications. At present, no framework implements OGC services. However, all the frameworks rely on standard encodings.

Table 2. Overview the results of the general criteria

			ARLab	ARToolkit	DroidAR	Layar	Metaio	PanicAR	Wikitude
1. General criteria	1.1. Platform	*Android*	X	X	X	X	X	X+	X
		Blackberry				X			
		iOS	X	X		X	X	X	X
		Other					X		X
	1.2. Programming languages	*Java*	X	X	X	X	X	X	X
		Objective-C	X	X		X	X	X	X
		Other				X	X		X
	1.3. License	*Open Source*		X	X				
		Freeware				X	X		X
		Proprietary	X			X	X	X	X
	1.4. Implemented standards	*OGC ARML*				X	X		X
		OGC web services							
		Encoding standards	X	X	X	X	X	X	X
	1.5. Offline availability	*Yes*		X	X				
		No	X*			X*	X*	X*	X*

+ Beta version available.
* Possible, but will require additional programming to circumvent the method of data acquisition

Offline availability is crucial when working in rural areas or in a disaster relief situation where connectivity is limited. Since ARToolkit and DroidAR are open source, users can implement the application in such a way that it reads the data from a local source, for example from a file in comma-separated values (CSV) format or from a JavaScript object notation (JSON) message. With the proprietary frameworks, the default method of accessing information is via a service that requires connectivity. However, if the format of the response is known, the developer can create a file in the same format and the application can then read the data from this file.

In all frameworks, the primary **data source** is a web service. Other options are directly from a database or through native code (refer to Table 3). Layar and Metaio access the data through a proprietary web service that is available as a 'black box' to the user. With Metaio, the user accesses the web service through a channel identifier. The channel is the entry point to the Metaio Cloud backend from where the information is requested. The channel identifier can be distributed in two ways: by QR code or by publishing the channel identifier. When using Layar, the user publishes the data on the Layar service and the data is then seen as a layer (the format of the layer is not know). The user can then only access the layers via this Layar service. All frameworks, except Layer, allow the user to make use of native code to acquire data from custom sources. However, with the proprietary frameworks, this is a challenge as the data request has to be overwritten with custom code. For example, with Metaio the data can only be accessed via a channel identifier. The user has to reproduce the Metaio service data format and inject it into the implementation.

Table 3. Overview the results of the functional criteria

			ARLab	**ARToolkit**	**DroidAR**	**Layar**	**Metaio**	**PanicAR**	**Wikitude**
2. Functional criteria	**2.1. Data source**	*Web service*	X	X	X	X*	X*	X	X*
		QR code					X		
		Database	X	X	X			X	X
		Native code	X	X	X		X	X	X
		Other					X		
	2.2. Data display	*The visual representation can be altered?*	X	X	X	X		X	X
		User can swap between which information is displayed?	X	X	X	X	X	X	X
	2.3. Object events	*Does the framework implement event triggers?*	X	X	X	X	X	X	X
	2.4. Display radius	*The display radius can be altered?*	X	X	X	X	X	X	X
	2.5. Visual search	*Photo*	X+				X		X
		LiDAR	X+				X		X
		Other, e.g. 3D objects	X+		X		X		X

+ Available through a separate application provided by ARLab.
* These frameworks make use of a proprietary web service.

Augmented reality superimposes information (**data display**) on a live feed, for example a camera feed. Various types of data or information can be superimposed, such as text, images or videos. For addressing, the display of text is important. In all frameworks, the fields to be displayed from a table or database, can be configured. However, adjusting the visual representation can be tricky, especially with proprietary frameworks that do not allow rebranding. ARToolkit relies on the UIKit to adjust the visual representation of the objects. The Metaio API specifies the access of POIs and also events (**object events**), but does not document how to change the visual representation of the POIs on the live feed. This code is hidden from the user. In contrast, PanicAR promotes itself as being a white label software, i.e. it can be completely be rebranded by the user. All frameworks implemented object behaviour in the form of events and/or triggers.

The **display radius** is a filtering mechanism based on distance from the user, for example, one can specify that only objects within a 100m radius are displayed. All evaluated frameworks provided the users with a method of adjusting or specifying this radius. However, the display can still get crowded with information if there are many points within close proximity of the user.

Visual search provides added intelligence to the application by not only relying on the location of the object only, but also on additional information, such as a photo, LiDAR data or 3D model. ARLab provides modular solutions, therefore the AR Browser does not implement these functionalities, but they are available in other packages developed by ARLab, such as the AR image matching or 3D engine. Wikitude provides the widest range of functionalities, including support for Google Glasses to recognize objects in the wearer's view based on image recognition. Metaio follows closely behind Wikitude. DroidAR does not support image recognition (i.e. photos) or LiDAR, but does support gesture recognition that can be used to develop virtual reality applications to complement the augmented reality applications. Visual search functionality is less important for the use cases described in this paper.

At present, none of the frameworks have built-in **integration with other products**, such as ArcGIS, QuantumGIS or PostGIS. Although this is not essential, integration with other products would make it easier to access or exchange information. For example, the application could directly access information from a PostGIS database and display it in the augmented reality application.

It is generally not **easy to extend** proprietary frameworks, as the code is not available and licensing constraints prohibit the user from extending the framework. ARToolkit and DroidAR are the only applications that can easily be extended, as they are open source frameworks. Open source frameworks encourage the modification and extension of the code to produce higher quality frameworks, and also to add new functionalities.

All the frameworks were found to be very **usable**. Frameworks either used the Android Studio with additional libraries or a stand-alone SDK that could be installed using a one-click installer. Tools typically provided auto completion and error checking that assisted with fast and effective coding.

All frameworks, except DroidAR, provide a variety of avenues of **support**. They provide extensive, well-structured and up-to-date documentation with numerous examples and code snippets. Additionally, instructional videos, issue trackers, forums and mailing lists are provided. Official documentation for DroidAR is fairly limited: the majority of the documentation is provided by the user community in the form of non-official documentation, instructional videos, and basic examples on Github.

Table 4. Overview the results of the non-functional criteria

			ARLab	**ARToolkit**	**DroidAR**	**Layar**	**Metaio**	**PanicAR**	**Wikitude**
3. Non-functional criteria	**3.1. Ease of integration with other GIS applications**	*Built-in integration with any GIS software product?*							
	3.2. Ease of extending the framework	*Can the framework easily be extended?*		X	X				
	3.3. Usability	*The framework is easy to install?*	X	X	X	X	X	X	X
		The framework is generally easy to use?	X	X	X	X	X	X	X
	3.4. Documentation and support available	*Documentation is clear and up to date?*	X			X	X	X	X
		Forums and mailing lists are available to interact with developers and user community?	X	X	X	X	X	X	X
		Support desk	X			X	X	X	X

5. DISCUSSION OF RESULTS

Our aim was to identify and evaluate existing development frameworks that could be used for the development of a mobile application that displays addresses in augmented reality. The requirements are based on three use cases in disaster relief, household surveys and address data quality management respectively. All the frameworks that were evaluated in phase two are available on the Android platform. Android has various advantages over other operating systems, such as iOS and Windows Mobile, primarily, because Android is an open source Linux-based mobile operating system. Other advantages of developing Android applications include integration with other Java based software, and also wider adoption (Android currently is the operating system used by the majority of mobile phones on the market).

At the moment, only two open source frameworks (ARToolkit and DroidAR) satisfy the requirements for the use cases (see Section 2). Other open source frameworks, such as BeyondAR, LibreGeoSocial and Mixare[14], do not meet the requirements. We also found that the open source frameworks (including ARToolkit and DroidAR) were generally out-dated and the developers had moved on to other things. The need for an open source framework is apparent as proprietary frameworks do not allow extensions and could result in vendor lock-in, forcing users to make use of their web services at additional cost. An open source framework will also allow users to more easily integrate the application with other products. Such integration is important as it facilitates accessing information from a database in PostGIS or the serving of data as a feature service from GeoServer, for example.

ARML 2.0 was published in 2015 by the OGC (Open Geospatial Consortium 2015). It allows users to develop a XML style sheet that specifies the appearance of objects and their anchors (the location or coordinates). Additionally, ARML defines ECMAScript bindings that allow the dynamic modification of the augmented reality scene subject to user input and behaviour. Wikitude originally developed ARML 1.0, and Metaio and Layar also implement ARML 2.0. Currently, none of these implementations conform completely to ARML 2.0. A likely reason is that the standard was only recently published and conformance testing is probably still in progress.

Offline availability of the data and application is critical when working in a rural area or a disaster relief situation where connectivity is limited. A current limitation of all the proprietary frameworks is extensive additional programming required to bypass the data acquisition method, for example, to bypass the data from a commercial web service in order to read from a local file. The limitation implies that the user has to know and understand the data format that is consumed by the application, so that native code can be written to request data in this format from a different source, either locally or online. Publishing data on a commercial server might also raise security or privacy concerns that might discourage the use of the augmented reality application.

A prototype of the augmented reality solution for addresses was implemented in ARToolkit (iOS) and Metaio (Android). ARToolkit was selected as it is an open source application, available on Apple iOS. Metaio provided a customizable solution, namely Junaio that could be used to test the framework. The prototypes were tested on the campus of the University of Pretoria, and successfully displayed addresses in augmented reality. From a developer point of view, both frameworks were user friendly and the prototype was easy to implement. However, some functions in ARToolkit were deprecated and had to be fixed in the code. The ARToolkit community is very active and users will implement fixes if they are reported. However, such small issues might put-off beginners and cause them to rather look at proprietary options. With Metaio, customizing the data source to something other than the channel identifier was challenging as the documentation did not specify the format of the data returned from the service/channel and one could thus not overwrite this from a local source. Configuring the visual representation of addresses in Metaio was not possible as the Metaio branding cannot be changed through its library.

Even though the frameworks provided all the tools to develop an augmented reality application for addressing, the precision of a phone GPS might still cause challenges when the application is used in a densely populated environment. A phone GPS can have precision of approximately 5m horizontally (this level of precision may take about 2-5 minutes to

[14] http://www.mixare.org

achieve), which is sufficient for sparsely populated (rural) areas.

Results of the evaluation show that all the evaluated frameworks could be used to develop an augmented reality application for addresses that meets the requirements identifier for use cases in disaster relief, household surveys and address data quality management. However, each framework has some limitations. The results of our evaluation highlight the respective strengths and weaknesses of the frameworks, which can guide developers to choose the framework best suited for their specific needs and requirement.

6. CONCLUSION

In this paper, seven mobile development frameworks for augmented reality were evaluated. The evaluation was based on requirements for three use cases: 1) disaster relief, e.g. address signs are destroyed by an earthquake; 2) household surveys, e.g. locating dwellings in informal settlements or rural areas without any address infrastructure; and 3) address data quality management, e.g. validating digital address data against addresses displayed in the physical world.

A two-phased evaluation was followed. In the first phase, development frameworks were disqualified if they did not support GPS and IMU sensors. These are required to locate and correctly display geocoded address data in augmented reality. During phase two, seven development frameworks (ARLab, ARToolkit, DroidAR, Metaio, PanicAR and Wikitude) were evaluated against three categories of evaluation criteria: general, functional and non-functional.

Results show that all frameworks are available on Android and all, except DroidAR, are also available on iOS. Java (on Android) and Objective-C (on iOS) are the most widely supported programming languages. Among the evaluated frameworks, only ARToolkit and DroidAR have open source licenses. All frameworks make use of standard encodings (e.g. XML and JSON) and some implement the ARML standard (Layar, Metaio and Wikitude). To date, none of the frameworks implement any standard OGC web services. Built-in integration with GIS products is also not available. Generally, the development frameworks are designed to access data sources via the internet (through a web service). The open source frameworks (ARToolkit and DroidAR) can be customized to access data downloaded to the mobile device.

All frameworks allow the user to specify that only objects within a specified radius from the user are displayed in the augmented reality view. The textual information to be superimposed on the live view of the world can be configured in all frameworks, but configuring the visual presentation (e.g. changing the text size or colour) is only possible in PanicAR and ARToolkit. Visual object behaviour is implemented in all frameworks in the form of triggers and events. The evaluated frameworks were found to be very usable and all of them, except DroidAR, provide a variety of avenues of support. The majority of DroidAR documentation is provided by the user community in the form of non-official documentation, instructional videos, and basic examples on Github.

A prototype of an augmented reality solution for addresses was implemented in ARToolkit (iOS) and Metaio (Android). The prototypes were tested on the campus of the University of Pretoria and successfully displayed addresses in augmented reality. From a developer point of view, both frameworks were user friendly and the prototype was easy to implement. However, some functions in ARToolkit were deprecated and had to be fixed in the code. With Metaio, customizing the data source to something other than the channel identifier was challenging. Configuring the visual representation of addresses in Metaio was not possible because the Metaio branding cannot be changed through its software library.

Based on the evaluation results, ARToolkit and DroidAR are most suitable for the three addressing use cases. Both frameworks seem to have active user communities who implement bug fixes in the case of ARToolkit and develop documentation in the case of DroidAR. However, infrequent updates to the code base are a concern. An alternative to using an existing development framework would be the implementation of augmented reality functionality from scratch. This is possible, for example, with the Android SDK and libraries. However, programming is required and this option does not follow the software engineering good practice of software re-use. Amongst others, a developer will have to (re-)implement functionality for the conversion between coordinate systems, for the calculation of distances between the user and anchor and for optimization the GPS precision.

At present, the Open Source Geospatial Foundation (OSGeo) does not support any augmented reality development framework. The fact that open source frameworks exist but are not updated frequently, could suggest that they are in need of a structured support system, such as OSGeo, that would provide financial, organizational and legal support. An augmented reality framework that displays objects sourced from spatial data layers (e.g. shapefile) or web services (e.g. from a web feature service on MapServer or GeoServer), in augmented reality could provide significant benefits to the free and open source for geospatial community.

The research presented in this article is part of a larger research endeavour on the display of geocoded address data in augmented reality. In future work, we plan to do empirical research to evaluate the use of augmented reality for addresses in each of the three use cases.

REFERENCES

Allbach, B., Memmel, M., Zeile, P., and Streich, B., 2011. Mobile Augmented City – New Methods for Urban Analysis and Urban Design Processes by using Mobile Augmented Reality Services. *Proceedings of RealCORP*, 6, 633–641.

Amin, D. and Govilkar, S., 2015. Comparative Study of Augmented Reality Sdk's. *International Journal on Computational Science & Applications*, 5(1), 11–26.

Anagnostou, K. & Vlamos, P., 2011. Square AR: Using augmented reality for urban planning. In *Proceedings - 2011 3rd International Conferenceon Games and Virtual Worlds for Serious Applications, VS-Games 2011*. 128–131.

Azuma, R., 1997. A survey of augmented reality. *Presence: Teleoperators and Virtual Environments*, 6(4), 355–385.

Carmigniani, J., Furht, B., Anisetti, M., Ceravolo, P., Damiani, E., and Ivkovic, M., 2011. Augmented reality technologies, systems and applications. *Multimedia Tools and Applications*, 51, 341–377.

Coetzee, S. and Cooper, A.K., 2007. What is an address in South Africa? *South African Journal of Science*, 103, 449–458.

Dünser, A., Grasset, R. and Billinghurst, M., 2008. A survey of evaluation techniques used in augmented reality studies. In *ACM SIGGRAPH ASIA 2008 courses on - SIGGRAPH Asia '08*. New York, USA: ACM Press, 1–27.

Kounavis, C.D., Kasimati, A.E. and Zamani, E.D., 2012. Enhancing the tourism experience through mobile augmented reality: Challenges and prospects. *International Journal of Engineering Business Management*, 4, 1–6.

Leebmann, J., 2006. Application of an augmented reality system for disaster relief. In *International Archives of the Photogrammetry, Remote Sensing and Spatial Information Sciences*, XXXIV-5/W10.

Leebmann, J., 2004. an Augmented Reality System for Earthquake Disaster Response. *Virtual Reality*, 35, 909–914.

Mulloni, A., Seichter, H. and Schmalstieg, D., 2011. Handheld augmented reality indoor navigation with activity-based instructions. *Proceedings of the 13th International Conference on Human Computer Interaction with Mobile Devices and Services - MobileHCI '11*, 211–220.

Open Geospatial Consortium, 2015. *OGC Augmented Reality Markup Language 2.0 (ARML 2.0)*.

van Krevelen, D., and Poelman, R., 2010. A survey of augmented reality technologies, applications and limitations. *The International Journal of Virtual Reality*, 9(2), 1–20.

Wen, J., Deneka, A., Helton, W.S., and Billinghurst, M., 2014. Really, it's for your own good...making augmented reality navigation tools harder to use. In *Proceedings of the extended abstracts of the 32nd annual ACM conference on Human factors in computing systems - CHI EA '14*. New York, USA: ACM Press, 1297–1302.

Wu, H., Lee, S.W., Chang, H., and Liang, J., 2013. Current status, opportunities and challenges of augmented reality in education. *Computers & Education*, 62, 41–49.

Free and Open Source Software for Geospatial (FOSS4G) Conference Proceedings

Volume 15 *Seoul, South Korea* Article 38

2015

Integrating Open Source GIS Software in Undergraduate Curriculum, Research, and Outreach - Recent Experiences at Salisbury University

Arthur J. Lembo, Jr.
Department of Geography and Geosciences, Salisbury University

AT5-04

Integrating Open Source GIS Software in Undergraduate Curriculum, Research, and Outreach - Recent Experiences at Salisbury University

Arthur J. Lembo, Jr.[1]

[1]Department of Geography and Geosciences, Salisbury University
1101 Camden Ave, Salisbury, MD, United States
Email: ajlembo@salisbury.edu

ABSTRACT

The Department of Geography and Geosciences at Salisbury University has a long tradition of teaching geographic information science. Until recently, most of the courses and research activities focused on commercial software offerings. However, the Department recently integrated Free and Open Source Software for GIS (FOSSG) into its curriculum, research, and outreach. Curriculum changes included introducing students to FOSSG in traditional GIS courses using QGIS, and creating two entirely new courses in Enterprise GIS and GIS Programming using PostGIS, GDAL, and SpatialLite. Through a competitive National Science Foundation (NSF) Research Experience for Undergraduates grant (REU), students participated in cutting edge research projects in parallel processing with Hadoop and spatialHadoop for cluster computing, and CUDA for GPGPU calculation on embarrassingly parallel processes for raster data. Finally, undergraduate interns working in the Department's Eastern Shore Regional GIS Cooperative (ESRGC) developed geodashboards using node.js, PostGIS, and Leaflet, while a special topics course developed a GIS based iphone and Android application used by 4,000 participants in the annual Sea Gull Century bike ride using GeoJSON, Leaflet, and javascript. In addition to highlighting the successes of these activities, this paper will discuss the process we used to make the necessary changes in our curriculum, secure the necessary funding for external projects, and the training approach we used to get our computer science students proficient in programming with FOSSG tools.

1. INTRODUCTION

The Department of Geography and Geosciences (DoGG) at Salisbury University has a long tradition of teaching geographic information science. As an academic institution, the DoGG focuses on three primary areas: teaching, research, and outreach. Until recently, most of the courses and research activities focused on commercial software offerings. However, the Department has recently integrated Free and Open Source Software for GIS (FOSSG) into its teaching curriculum, research, and outreach.

While many of our students have obtained employment using commercially available GIS software, a growing number of alumni have migrated their work into the open source realm. Additionally, more job postings include a preference for open source software. In response to the increased use of open source GIS software, the DoGG Faculty made a decision to include more offerings of FOSS4G into the student experience. This included introducing FOSS4G in our teaching, research projects, and outreach experiences.

2. TEACHING

In addition to concentrations in Atmospheric Science, Planning, Human Geography and Earth Science, the Department offers numerous courses in GIScience. These courses include *Map and Air Photo Interpretation, Introduction to GIS, Advanced GIS, Remote Sensing, GIS Modelling, GIS Programming, Enterprise GIS,* and *Cartographic Visualization*. Over the last two years, four of our courses were adapted to include FOSSG software including *Advanced GIS, GIS Programming, Enterprise GIS,* and *Cartographic Visualization.*

2.1 Advanced GIS

The Advanced GIS course is a study of current theories and development in GIScience, with an emphasis on using many of the specialized tools in ArcGIS including Network, Spatial, 3D, and Geostatistical Analyst. While our students were gaining a significant amount of advanced skills in GIScience, they mostly understood the world through the ArcGIS lens. Because the job market is dominated by ESRI software, it was impractical in our undergraduate institution to ignore the importance of the ESRI software. However, to supplement the students' ArcGIS skills, we introduced modules using QGIS in order to replicate some of the laboratory exercises they had previously completed using the ESRI suite of products. The main goal was to show the students that open source options exist, and that these options were comparable to the commercial offerings. Because QGIS was yet another technology stack introduced within a semester-long course, we made our exercises more *cookie-cutter* in approach, just to give the students the opportunity experience the overall feel of QGIS, and experiment with the various plug-ins, comparable to the ArcGIS *Analyst* extensions.

The benefit of introducing FOSSG software into the Advanced GIS course not only provided the students with experience using another tool, but to also expanded their intellectual capacity to evaluate the benefit of using these tools within a GIS context. We wanted our students to be able to articulate the pros and cons of both commercial and open source GIS software. The results were remarkable. Not only did our students learn and understand FOSSG software, but during many job interviews, our undergraduates found themselves teaching their prospective employers about the availability of such software, and were also able to discuss the pros and cons of using the software within the employer's existing structure.

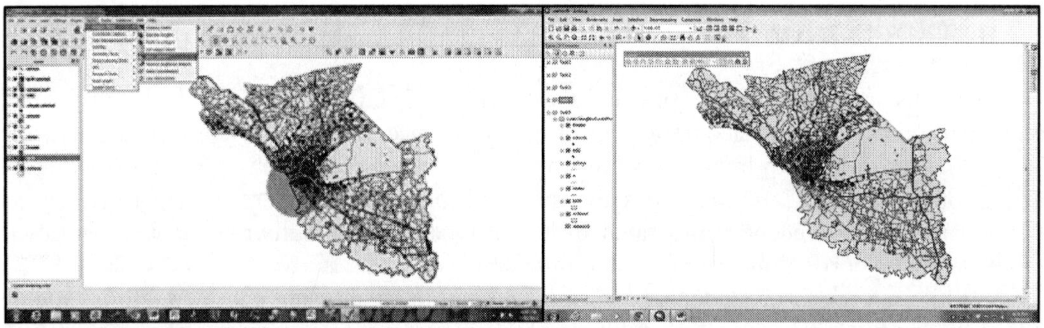

Figure 1 Richland Chemical Lab, performed by students using shown QGIS (left) and ArcGIS (right)

2.2 GIS Programming

Our GIS Programming course focuses on developing custom computer programs that address classical problems in geography and spatial analysis not ordinarily solved using out-of-the-box commercial GIS software. Here again, the dominant industry application is ArcGIS, so we focus 50% of the course on using Arcpy. The benefit is significant as few undergraduate geographers have experience programming in Arcpy. However, similar to our

Advanced GIS course, we recognized the need to expose our students to FOSSG software with respect to the programming environment. However, we did not want to simply replicate the Arcpy experience as a one-to-one match (like attempting to teach Qpy), but rather wanted to expose our students to technologies that the current ESRI offerings could not replicate.

Therefore, the remaining 50% of the course focused on the use of spatial SQL using PostGRES/PostGIS for solving spatial problems. For SQL, the focus was on traditional SQL statements, SQL data types, aggregate functions, and the spatial functions offered by PostGIS. Students worked through the book *How Do I Do That in PostGIS*, to learn how to perform the classic GIS tasks outlined in the book *How Do I Do That in ArcGIS/Manifold*, but within an SQL paradigm. Oftentimes, the laboratory exercises recreated the ArcGIS exercises in a SQL paradigm, allowing the students to see alternate, and oftentimes more efficient ways of completing a task.

Figure 2 Textbook used to teach students spatial SQL in PostGIS

Similar to the Advanced GIS course, the results were remarkable. In using SQL to solve complex GIS problems, our students discovered a confidence they hadn't experienced before. Also, during the independent final course project, most students opted to utilize SQL to complete the project as it was faster, more elegant in its approach, and easier to write.

2.3 Enterprise GIS

Our Enterprise GIS course is a special topics course offered every few years. The special topics course is usually only open to those students who have been introduced to SQL in the GIS Programming class. In this course, the students implement an enterprise GIS using PostGRES/GIS as the spatial data server, with ArcGIS, Manifold GIS, and Quantum GIS as the client applications. The students focus on database design and management of user roles, data views, triggers, and the development of end user applications. Typically, the design includes users (creatively named Mo, Larry, and Curly) with different editing capabilities. The context of the project was to create an Enterprise GIS in PostGRES for the Town of Denton, MD, as it was the same town the students created an ESRI file geodatabase for.Also, the students then perform various operations to understand the multi-user nature of spatial databases, such as multi-user editing, or perform large spatial selections using different underlying databases.

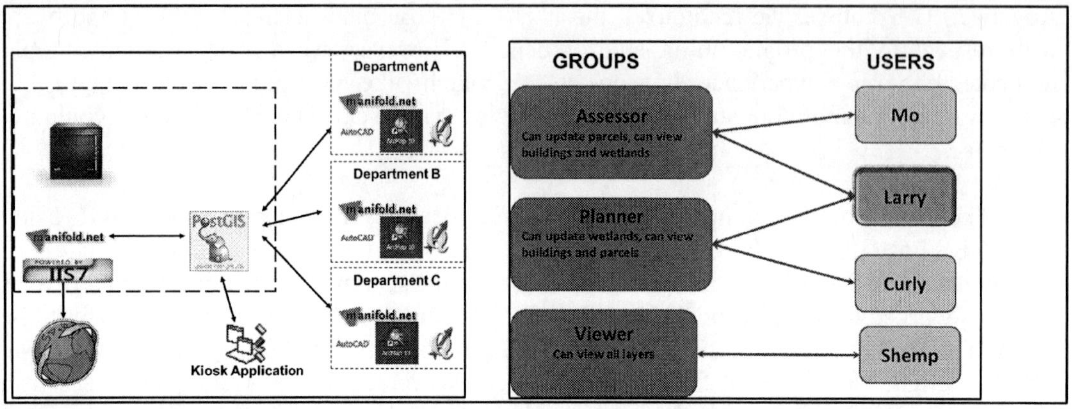

Figure 3 Enterprise GIS class designed and implemented a multi-user, multi-vendor client/server GIS

Again, the results of the course were remarkable, with our undergraduate students learning how to build an enterprise GIS with multiple users and multiple client applications. In addition, our students presented their work on implementing a multi-user, multi-vendor enterprise GIS at two regional GIS conferences. Finally, having previously built a File Geodatabase in the Advanced GIS class, students could articulate the strengths and weaknesses of both models.

2.4 Cartographic Visualization

The Cartographic Visualization class introduces students to the theory and application of cartographic principles and practices in cartographic design. Commercial products like Adobe Illustrator are used with ArcGIS for half of the course, as the students are already familiar with ArcGIS. However, the other half of the course introduces students to MapBox and TileMill so that students become familiar with open source cartography principles. Finally, student are required to make interactive and animated maps for final projects TileMill and MapBox.

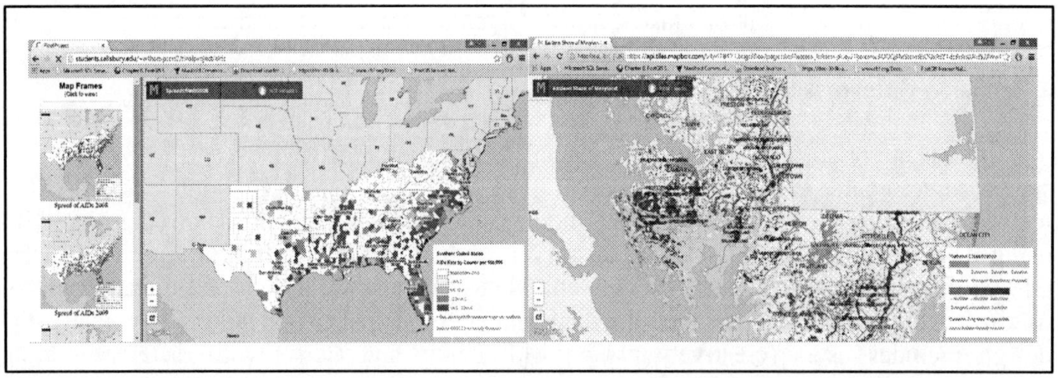

Figure 4 Examples of student final projects include interactive maps produced using MapBox and TileMill

3. RESEARCH

As an undergraduate institution, Salisbury University emphasizes research experiences for undergraduates. Research take the form of traditional sponsored research, and directed research projects by undergraduates with faculty mentors. The use of open source tools were used in sponsored research on parallel processing for the National Science Foundation (NSF) and food security mapping for the United States Department of Agriculture (USDA). Directed research was conducted as part of the Government and Non Profit Application Workshop (GNAppWorks) and use mapping APIs for Android devices.

3.1 The National Science Foundation

As part of a competitive grant, Salisbury University was awarded a 3 year NSF REU in parallel processing. The REU included 8 undergraduates each summer, working with 4 faculty mentors. Generally, the REU projects focused on classical parallel processing projects in the areas of cryptography, tomography, and graph theory for social networks. However, each summer, a parallel processing project in GIS was undertaken.

3.1.1 Parallel Processing with General Purpose Graphical Processing Units

An NSF REU gave undergraduates an opportunity to write raster based terrain algorithms utilizing GPUs on video cards. The students wrote the algorithms in C++, and utilized NVIDIA APIs. Throughout the process, the students not only improved the speed of the analysis, but also developed work-arounds for I/O bottlenecks to achieve greater speed improvements. The students developed algorithms for embarrassingly parallel applications including slope, aspect, and terrain ruggedness. For the kernel based terrain functions, students saw an improvement of 10x when using GPGU processing. However, when the students increased the kernel size (and thus the number of computations), the speed improvements approached 60x.

Figure 3 shows the general process used for performing the GPU analysis, along with a breakdown of times for input, output, and processing under a parallel and serial implementation. Most notable is the reduction in processing time of the GPU implementation compared to the serial implementation.

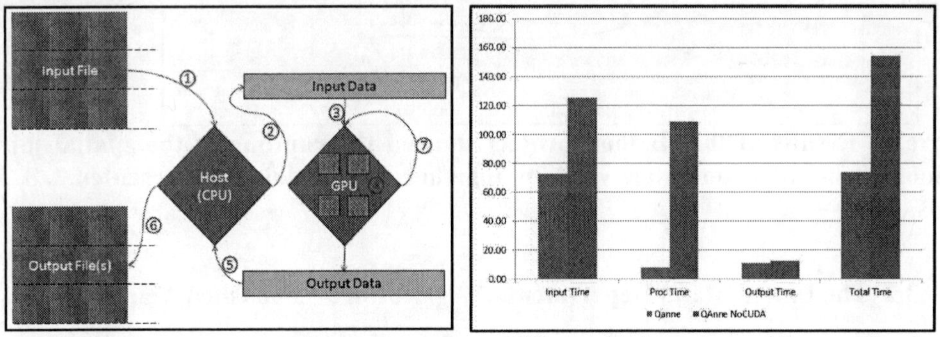

Figure 5 NSF research performed raster GIS analysis using video card showed dramatic speed improvements when utilizing GPUs on video cards

3.1.2 Parallel Processing with Multiple CPUs and Hadoop

In addition to GPU processing, students also experimented with distributed computing, utilizing the open source spatial Hadoop on very large spatial databases. The results demonstrated that the classic point-in-polygon test on tens of millions of points that required five days of processing with commercial GIS software could be completed in 200 seconds using Hadoop and special modifications to the underlying data structures. Within this project, students built their own cluster in the lab, in addition to renting time on an Amazon EC2 server. The ability to affordably rent 20 computers with SSDs on Amazon reduced the processing time to seconds, while our own cluster of 4 computers completed in 24 minutes.

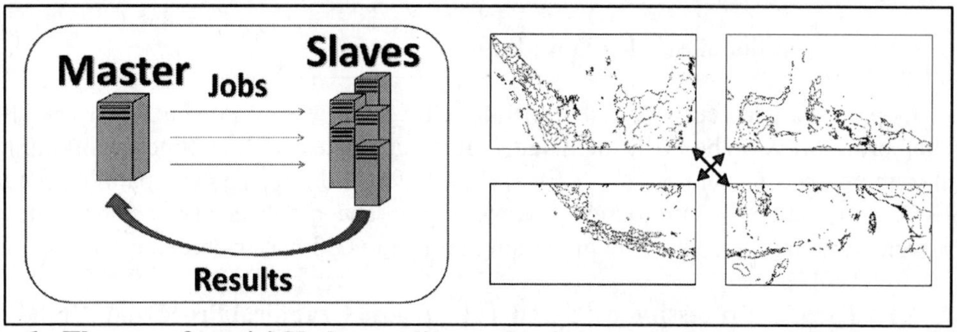

Figure 6. The use of spatial Hadoop splits up the data to work over multiple machines simultaneously to complete the task much faster than traditional GIS software on bigger data.

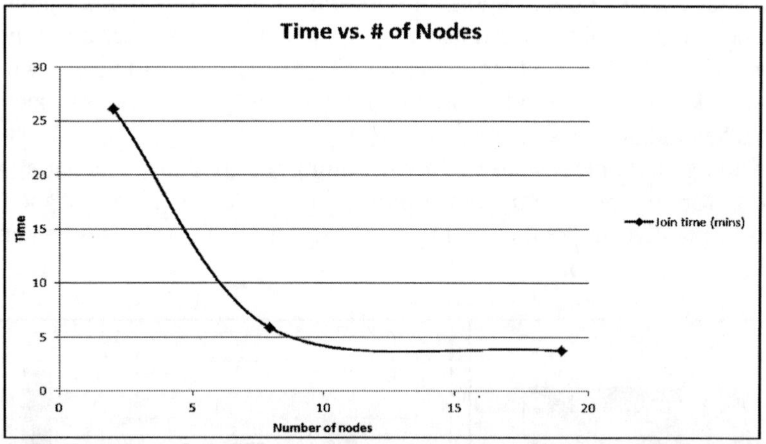

Figure 7 Results of the Hadoop project showed the runtime of the spatial join on a Hadoop cluster of n computers working together. The X values represented 2, 8, and 19 CPUs.

3.2 The United States Department of Agriculture – Foodshed Mapping

The Mapping Local Food Systems project, funded by the U.S. Department of Agriculture, examined New York's ability to supply its own food needs. The project was a joint effort between Salisbury University, Cornell University, and Tufts University. The Salisbury University portion was to create an Internet Mapping Service (IMS) to visualize the optimization algorithms developed as part of the research.

The IMS solution included the query and display of foodshed footprints for individual cities, and was implemented with GeoJSON as a datasource, and Leaflet.js as the client GUI.

Figure 8 The USDA Foodshed project utilized GeoJson data files, Leaflet, and OpenStreeetMap to display foodsheds for selected cities in New York State

3.1.3 GNAppWorks – The Sea Gull Century

In order to integrate our teaching and outreach programs, the Geography Department, the Henson School of Science, and the Eastern Shore Regional GIS Cooperative created the Government and Non-profit Application Workshop (GNAppWorks). GNAppWorks is a mobile application development initiative focused on providing applications to the government and nonprofit sector.Nonprofits with increasingly smaller budgets, citizens in lower socio-economic classes, and small government entities have needs for mobile applications, but lack the means to develop or fund development of high quality products. With the backing of Salisbury University, GNAppWorks provided computer science majors with an opportunity to work on real-world smartphone applications under the direction of a Faculty GIS mentor. Some of the clients have included The Nature Conservancy and the City of Havre de Grace.

As part of a research class, Computer science, Geography, and Business students worked collaboratively to develop Android (and iPhone) applications to guide cyclists participating in the Sea Gull Century bike ride.The application was used by thousands of riders, and showed specific rider locations along the three Lower Shore routes, displaying their speed, nearby vendors and rest stops. Geography undergraduates digitized the bike routes, the computer science students programmed the application, and the Business students worked with ride sponsors to participate in the marketing of the ride. They also worked with the University legal department to ensure that the application utilized the appropriate logos, colors, and descriptions. The students also gained an understanding about the collaboration required within a small start-up organization.

In addition, the research project, “Creating the Sea Gull Century Mobile Application – A Multi-Discipline Effort,” was presented at the University of Maryland Eastern Shore’s fifth annual Regional Research Symposium, and earned first place in the undergraduate oral presentation category.

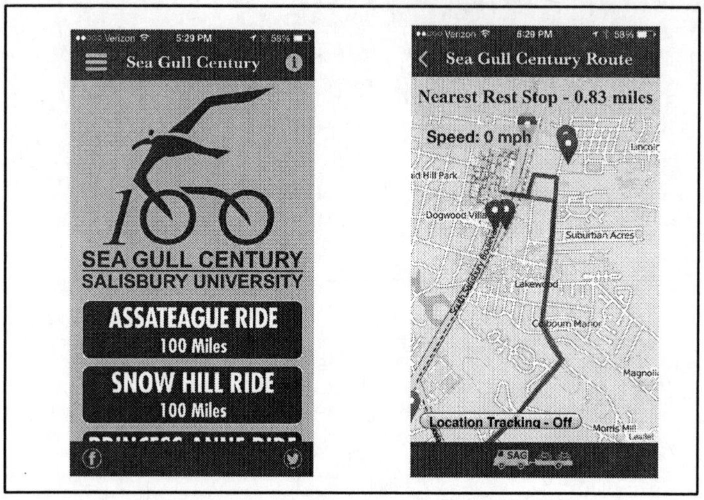

Figure 9 An Android application using Leaflet, GeoJSON, and OpenStreetMap was used to assist riders in the annual Sea Gull Century ride.

4. OUTREACH

4.1 The Eastern Shore Regional GIS Cooperative

The original goal of Salisbury University'sEastern Shore Regional GIS Cooperative (ESRGC)was to provide GIS and other high tech services to under-served markets on the Eastern Shore of Maryland. Over the years, the ESRGC has expanded to include larger work assignments, and now have a staff of 14 full time employees including geographers, computer scientists, and database analysts. As these work assignments grew into more IT related endeavors, our work transitioned from cartography and basic spatial analysis tasks to greater use of database technology, Internet mapping, and spatial analysis. As part of the educational experience, the ESRGC hires undergraduate Computer Science and Geography students to work collaboratively to develop IT based applications.

Early efforts at using commercial based GIS software for Internet delivery proved frustrating, as the commercial products inhibited the creativity the staff desired. So, while commercial GIS software did exist to support many of our projects, a decision was made to migrate almost all of the development work into a FOSS4G paradigm, as the open source solutions allowed us to provide greater flexibility and responsiveness to the application. In addition, the FOSS4G solutions allowed our clients to enter into Internet based IT solutions for a minimal cost.

Examples of some of the ESRGC activities with FOSS4G include our use of GeoDashboards for the State of Maryland, Oklahoma, and the Federal Government, the GNAPPWorks Initiative, and municipal applications for local and regional organizations.

4.1.1 GeoDashboards

Although each geodashboard we create are different, they follow the same general architecture:

1. Data is stored in PostGRES with PostGIS
2. The web server is node.js
3. Front end web pages are written in leaflet.js

4. Data is sent to the web pages either as GeoJSON directly from PostGIS, or as WMS services from TileMill.

The use of Leaflet allows us to create a very responsive GUI, and our connection with PostGRES/PostGIS allows us to issue virtually any kind of SQL or spatial SQL query.

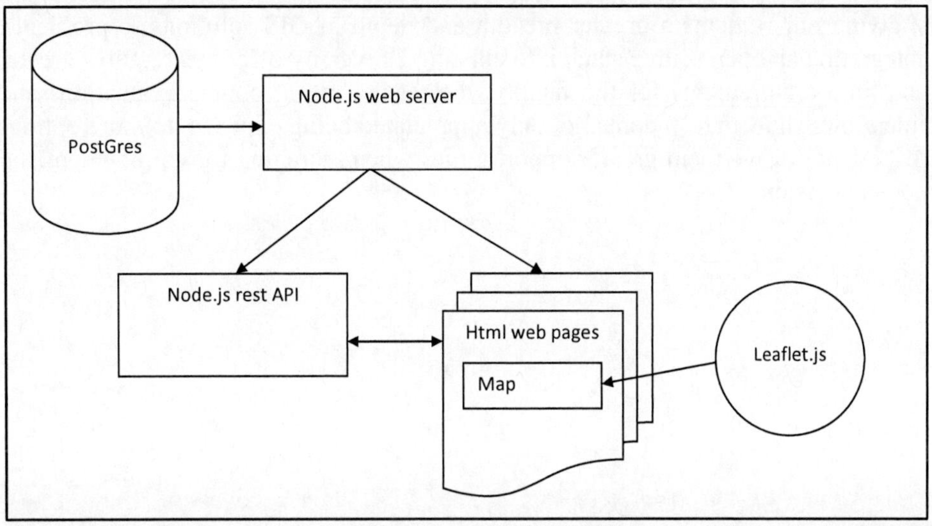

Figure 10 The general architecture of the ESRGC dashboards include the use of node.js as a web server, PostGRES as the back end database, and Leaflet.js for the client-side interface.

<u>Examples</u>

The ESRGC completed geodashboards for Maryland's BayStat program, the Maryland Department of Labor, Licensing, and Regulation (DLLR), the State of Oklahoma Licensing and Regulation, and the National Broadband Mapping Initiative, all using the architecture described in Figure 8.

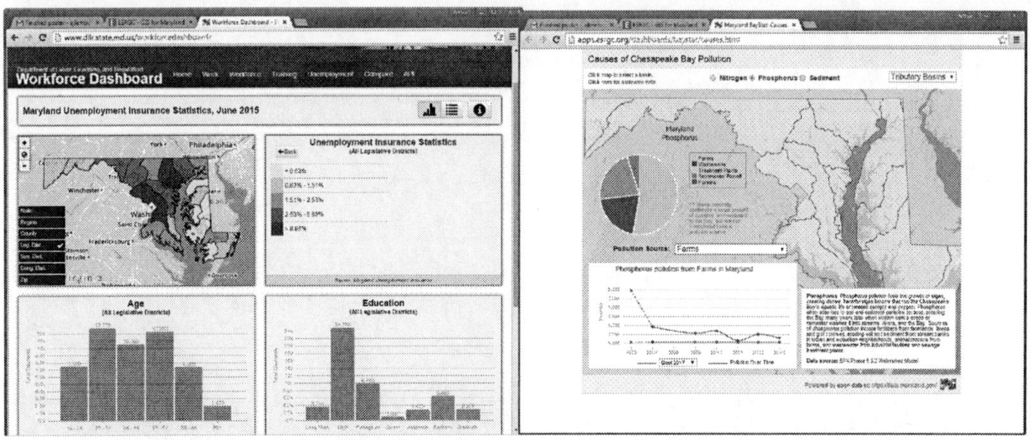

Figure 11 Example of Open Source Dashboards for the Maryland Department of Labor, Licensing, and Regulation, and the Maryland BayStat Program.

5. CONCLUSION

While most undergraduate GIScience programs focus on the use of a single commercial vendor, Salisbury University sought to be broader in our approach, introducing students to different commercial GIS software, in addition to including open source tools in our key GIS courses. In addition, faculty focused on research and outreach projects that included open source tools, giving our students a greater breadth and depth in GIS solutions to problems. Obviously, integrating an open source stack into existing University offerings requires greater dedication and time requirements for the faculty, the results in our course evaluations and Senior exit interviews show that students not only appreciated being exposed to more options, but that the experiences gave them greater opportunities when searching for employment and graduate school admission.

Free and Open Source Software for Geospatial (FOSS4G) Conference Proceedings

Volume 15 *Seoul, South Korea* Article 39

2015

OPENSENSEMAP - A Citizen Science Platform For Publishing And Exploring Sensor Data as Open Data

Matthias Pfeil
Institute for Geoinformatics, University of Münster

Thomas Bartoschek
University of Münster, Germany

Jan Alexander Wirwahn
Institute for Geoinformatics, University of Münster

AT4-01

OPENSENSEMAP - A CITIZEN SCIENCE PLATFORM FOR PUBLISHING AND EXPLORING SENSOR DATA AS OPEN DATA

Matthias Pfeil, Thomas Bartoschek and Jan Alexander Wirwahn

Institute for Geoinformatics, University of Münster
Heisenbergstr. 2, 48149 Münster, Germany
Email: matthias.pfeil | bartoschek | jan.wirwahn @uni-muenster.de

ABSTRACT

A plethora of citizen science sensor platforms for dierent use-cases already exist. They provide cheap, precongured, plug and playable hardware and software solutions. Using data from multiple platforms and resources can be a challenging task in respect of discovering, exploring, downloading and converting. This work aims to implement a one-stop-shop for sensor data that tries to tackle these problems. Therefore a basic data schema capable of metadata is established that allows publishing generic sensor platforms and sensor data. For exploration a web platform is implemented based on common web standards. This work describes the requirements, design and implementation of the OpenSenseMap. OpenSenseMap is following the clientserver model and its complete application stack is implemented in JavaScript. It describes the basic components and the underlying data model. All components are using open-source technology and are published under an open-source license too. The RESTful service is the heart of OpenSenseMap and implements the data publish and data retrieval functions. Developers may use it to build custom applications around OpenSenseMap or integrate OpenSenseMap into existing projects. The NodeJS application runs its own web server and implements the different HTTP request types. The OpenSenseMap concept is being evaluated with a technical and a user survey.

1. INTRODUCTION

Citizen science is often called "public participation in scientific research" (Hand, 2010) and describes the engagement by non-professional scientists in collecting and analysing data, decision making, developing technology and publication of these on a voluntary basis. The idea of involving citizen in scientific projects is not new. Two examples are the Christmas Bird Count (Silvertown, 2009) by the National Audubon Society in the USA, which started back in 1900 and is still ongoing, and the Galaxy Zoo project, which identifies and classifies galaxies on sky images taken at the Sloan Digital Sky Survey. An always discussed concern about citizen science is data quality (Wiggins et al., 2011). Due to limited knowledge of the volunteers and missing or questionable metadata, scientists often characterise data collected by citizens as valueless. In the beginning citizen science was connected with activities where humans were used as sensors (Goodchild, 2007). Nevertheless, the scientific community is becoming more and more interested in citizen science today and citizen science projects are part of complex research projects (Conrad & Hilchey, 2011); i.e. ambient environment monitoring. We think two reasons boosted this trend. First, citizens are getting more interested in their environment and its effects on daily life. Also many popular citizen science projects are related to ambient environment monitoring and topics around smart cities; i.e. the AirQualityEgg project (http://airqualityegg.com) and the SmartCitizen project (https:// smartcitizen.me). Second, the vast development of technology makes it easy and aordable to build the necessary tools for a citizen science project. Microcontroller platforms like Arduino or tessel.io and the quality and price of sensors make it nowadays easy to build sensor platforms, which are easy to use and affordable in price. By using open-source hardware and software, it is easier - especially for traditional non-technical research communities - to create custom sensor platforms. Furthermore, the web makes it easier to share and discuss data at global scale.

This work is closely related to SenseBox, which is a generic sensor platform for the Web Of Things (WoT) proposed by Bröring et al. (2012). The SenseBox is a tiny, lightweight and low-cost sensor platform being developed at the Institute for Geoinformatics, University of Münster. It is an ongoing citizen science project and enables citizen scientists to run private sensor stations for environmental observations like temperature and wind speed. Up to now, SenseBoxes are rather isolated: it is up to its owner to share the collected data with the world. Sharing the collected data for scientific purposes is a challenging task especially for non-professional scientists as the data needs to be structured and enriched with additional metadata. Structuring and annotating the collected data is not only a prerequisite to make the data usable for scientific research but also for making the data easily explorable for the involved citizen scientists. An active and involved community is one key for a successful citizen science project.

The main objective of this work is the design and implementation of a web application for sharing the data collected by sensor platforms like SenseBox; we call it OpenSenseMap. OpenSenseMap defines an extensible but yet tiny and lightweight encoding standard for spatio-temporal sensor data and metadata. With its RESTful API, OpenSenseMap fits the characteristics of WoT. These key features and its openness make it easy to adopt Open-SenseMap to many other sensor platforms and citizen science projects.

2. RELATED WORK

2.1 Internet of Things and Web of Things

The term Internet of Things (IoT) describes the mash-up of real-world objects with the virtualworld and was coined by David Brook (Kopetz, 2011 & Uckelmann et al. 2011]. Everyday life objects (things) like household devices are equipped with embedded computing systems. These embedded computing systems make things capable of sensing, processing and connectivity and are the building blocks of IoT (Kortuem et al., 2010) Most IoT services are custom implementations and require expertise for integrating and interoperating "which prevents users from creating small tactical, ad-hoc applications using sensor networks" (Guinard & Trifa, 2009).

To tackle this the next evolutionary step of IoT is theWeb of Things (WoT) (Gurnard et al., 2011). TheWoT paradigm proposes to use well-known web standards to realize IoT applications. Using web standards has the advantage that all things speak the same language and makes it easy to integrate them into the existing web. Each WoT enabled device is identifiable by its own unique Uniform Resource Identifier (URI).

A typical used web standard for communication is REST. REST stands for Representational State Transfer and is a common architectural pattern for building web applications (Fielding & Taylor, 2002). This approach uses the standardized Hypertext Transfer Protocol (HTTP) to transport data between a client and server. HTTP handles the communication by listening for requests and answering with responses. To identify the target of a request, REST uses the aforementioned URIs. The whole communication is stateless; that is, each request contains all necessary information to respond to a request. By using these standards all "things" speak the same language and make the custom IoT solutions redundant. FurthermoreWoT is more user-friendly and can be easily integrated into web projects.

2.2 Sensor networks

A sensor network (Faludi, 2010) typically contains two to many sensor nodes and uses a network to exchange data. Each sensor node is able of measuring distinct characteristics of its ambient environment and publishes its measurements inside the network. The amount of connected sensors is theoretically unlimited but limited in real-world applications by the technical specifications of the used hard- and software. According to Soharby et al. (2007) typical applications of sensor networks are data collection, monitoring or surveillance. Most sensor networks contain a logical entity called processing node or gateway (cf. Figure 1). The gateway is responsible to aggregate the measurements of the network's sensor nodes and provide them to interested clients. A client may be a real user or another software platform. In general a gateway is more powerful than the sensor nodes.

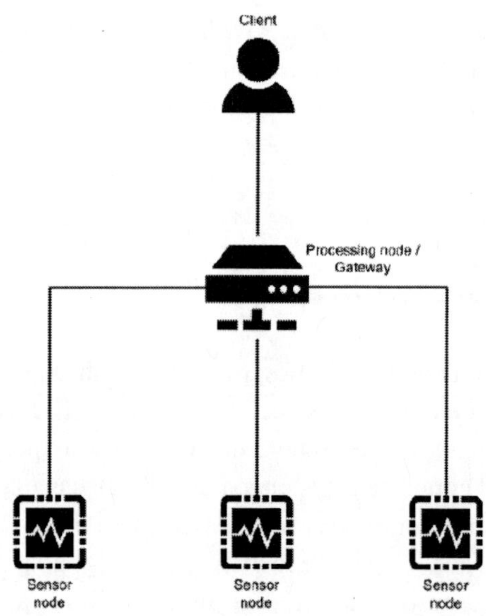

Figure 1. Sensor network with the central processing node / gateway and connected sensor platforms

Two approaches to realize a gateway are the Sensor Web Enablement (SWE) and Sensor Web portals; both will be described more detailed in the following sections.

2.3 Sensor Web Enablement

The Sensor Web Enablement (SWE) is a standardized framework developed by the Open Geospatial Consortium (OGC). SWE enables developers to make all types of sensors and their data discoverable, accessible and useable by standardized web technologies. The

SWE framework proposes the following characteristics for sensors (Bröring et al., 2011):

- sensors are annotated with a geo-location
- sensors are connected to the web
- sensors have adequate metadata
- sensors are readable remotely
- sensors may be controllable remotely

The SWE framework includes different standards to realize the proposed characteristics of sensors. The most relevant standards will be described in the following sections.

2.3.1 *Sensor Observation Service*

The Sensor Observation Service (SOS) is a standardized web service interface and represents the gateway of a SWE based sensor network. It provides services to store, request and filter sensor data. Furthermore, the SOS offers management capabilities for the sensor network itself and the connected sensors. This includes functionalities to register new sensors and to add new and delete existing observations.

The SOS builds upon further standards dened by the OGC and International Organization for Standardization (ISO) and provides a comprehensive variety of interoperable capabilities to interconnect in-situ or dynamic sensor platforms (Florence, 2012).

2.3.2 *Sensor Observation Service*

The Observation & Measurement (O&M) model is an essential component of the SOS. O&M is a standard to represent and exchange domain independent and spatio-temporal observations and measurements of a sensor. An observation is an act that takes place at a certain time and geo-location (spatio-temporal) and produces a measurement for the observed feature of interest.

2.3.3 *Sensor Model Language*

The SWE calls for sensors to have adequate metadata to understand how the sensor works and generates observations. To describe sensor metadata the OGC has dened the Sensor Model Language (SensorML). With SensorML it is possible to describe "geometric, dynamic, and observational characteristics of sensors and sensor systems" (OGC, 2007). Beside describing sensors, SensorML provides a way to apply additional sensor information and process descriptions. These information are a prerequisite to discover the sensor network and its sensors and the semantics of the observations. SensorML is encoded as XML.

2.4 Sensor Web Portals

Bröring et al. (2012) claim that so-called sensor web portals are emerging. These platforms act like the gateway of a sensor network and allow users to register sensors and publish the collected data. We currently see two kinds of sensor web portals: generic sensor web portals, which provide the architecture and infrastructure to connect any sensor to the web, and specialized sensor web portals, which focus on one specific use case. Usually they offer an API for registering sensors and publishing sensor data.

2.4.1 *OpenSensors.io*

OpenSensors.io is an IoT platform supported by the Open Data Institute (ODI)[8]. The ODI promotes the evolution of an open data culture. Open data empowers people to action on problems that concern themselves. Furthermore open data creates and spreads knowledge about local and global issues.

OpenSensors.io usesMQTT, which is a publish-subscribe based lightweight message protocol. Data is being published on OpenSensors.io in so-called topics. One can think of topics as the equivalent to Twitter hashtags. Devices publish data on OpenSensors.io by sending a MQTT message to a topic. Those interested may subscribe to a topic. Topics are the only (and rather limited) metadata supported by OpenSensors.io.

The user has full control over the gathered data and can restrict the data privacy. Publisher gain full public domain license over the published data and other users can make use of the data by attributing the work with the ODC Open Database License (ODbL) (http://opendatacommons.org/licenses/odbl/summary/)

OpenSensors.io is an example of a generic sensor web portal. It is a commercial sensor web portal but offers a free personal plan with a limited number of devices and storage.

2.4.2 *Xively*

Xively (http://xively.com) is also an example of a generic sensorweb portal. It is comparable to OpenSensors.io but offers extended functionality regarding connectivity and metadata handling. Additionally to MQTT, Xively supports REST and Sockets. Furthermore, it is possible to store and process metadata like geo-location on Xively. A popular citizen science project hosted on Xively is the AirQualityEgg[10] project.

Primarily Xively is a commercial sensor web portal but offers a free plan for personal use only with limited capabilities regarding exploring and sharing data.

2.4.2 *Smartcitizen.me*

Smartcitizen.me is the sensor web platform of the Smart Citizen project and focusses on urban ambient monitoring. The Smart Citizen project offers the so-called Smart Citizen Kit, which is an Arduino based sensor platform that can be purchased. It allows volunteers to capture data on temperature, humidity, light, sound, carbon monoxide and nitrogen dioxide.

The collected data may be explored by everyone but only owners of a Smart Citizen Kit can publish data on the Smartcitizien.me Sensor Web portal. Clearly, the SmartCitizen.me sensor web portal is an example of a specialized sensor web portal. Currently 1003 sensor platforms with 5015 sensors are online.

3. PRINCIPLES AND REQUIREMENTS

OpenSenseMap is meant to be a platform for professionals and non-professionals where it is possible to store, publish and explore sensor data. Therefore the requirements are split up into two sections. Section 3.1 focuses on the technical requirements for a gateway in a sensor network. Section 3.2 focuses on the non-professionals and what is needed to include citizens regarding design and interaction.

3.1 Technical Requirements

Steenkamp et al. (2009) state that a main task of a gateway in a sensor network is to connect contradictory networks. In this case that means to connect the sensor network and the client. Registered sensor platforms shall be able to publish their collected data on OpenSenseMap. Users shall be able to register new sensor platforms and to administer existing sensor platforms. Administration includes deleting sensor platforms and modifying the setup of a sensor platform. Table 1 lists all use cases for the user. This leads to the first requirement: users shall be able to register new sensor platforms and administer already created sensor platforms.

Table 1. List of interactions between user and OpenSenseMap

Use Case	Description
register new sensor platform	A new sensor platform is registered including metadata and connected sensors
unregister sensor platform	Remove a sensor platform from the sensor network
alter sensor platform properties	Alter the provided metadata for the sensor platform
add sensor	Add a new sensor to a sensor platform
remove sensor	Remove a sensor from a sensor platform
alter sensor properties	Alter the provided metadata for the sensor
get data	Retrieve data from sensor platform and sensors

Currently a lot of citizen science projects implement own infrastructures and are bound to a specific topic (Wiggins & Crouton, 2011). Approaches to build platforms to provide infrastructures for generic citizen science projects as described by Dickinson et al. (2012) exist but with a different focus. They focus on being a one-stop-shop for discovering citizen science projects and to get information about the project but not offering a standardized way to connect and publish sensor platforms and its data. Bröring et al. (2012) propose to separate the sensor platform logic from the gateway logic. By applying the Web of Things paradigm to the gateway all sensor platforms of the network get discoverable by other systems and users. Unfortunately, the unique addresses of the sensor platforms does not help to understand the underlying dataset, because important metadata information like the unit of measurement are not encoded in the address. Thus it is necessary to dene a data model that can handle the actual data and is capable of metadata. This puts two important requirements on the data model. It must be flexible to handle any kind of sensor data but also must be lightweight to allow low-cost and not well equipped microcontrollers to be part of the sensor network, leading to the requirement of a data model definition capable of storing actual data and metadata.

The gateway is the central entity of a sensor network. It should be connected to the internet all time and handles the requests of the sensor platforms (when publishing their collected data) and of the users (when exploring the collected data and/or performing administrative tasks). Therefore a reliable, robust and high-performance back-end is required. The back-end must handle bad requests without causing internal server errors and respond with appropriate error messages. Furthermore an authentication method is needed to ensure only registered sensor platforms are allowed to post data to the platform. Though the back-end shall be reliable, robust and performant.

All gathered data must be stored on the gateway. Therefore a reasonable database which is capable to handle geo-data and oer geospatial features for indexing and querying is needed. Additionally it must support a flexible schema. Considering the possible amount of gathered data, the size of the stored objects should not take up to much space. The database shall be capable of a flexible schema, small storage size and geo-data.

3.2 Citizen Science & Design

Running a successful citizen science project is not that simple. In the past, citizens were often allowed to contribute only to selected phases of a research project. Current evaluations of citizen science projects have shown that the stage of inquiry is one key for a successful citizen science project (Wiggins & Crowston, 2012). It should be the goal of every citizen science project to establish an active and involved community. In addition to that, Newman et al. (2012) recommend to build citizen science projects upon open-source solutions and encourage to pick-up the idea of open data. These recommendations lead to the requirement that the collected data shall be published as open data and the implementation shall use open-source technologies. The collected data shall be published as open data and open-source technologies should be used.

A typical characteristic of citizen science is often the heterogeneity of the participants. This is often an advantage but also a challenge. In this case, the Sensor Web portal will be used by users with different experience levels of web applications. Nielsen (1994) defines software usability as the sum of learnability, efficiency, memorability, errors, and satisfaction. This implies that software must be easy to learn, efficient to use, easy to remember, should not allow to make errors and should be fun to use. In case of OpenSenseMap, the user shall be guided through the registration of a new sensor platform and information on usage shall be provided throughout the Sensor Web portal. OpenSenseMap shall focus on usability by non-professionals.

To make the collected data available to a wide audience and encourage the participation of interested citizens, OpenSenseMap shall be useable free of charge. Free of charge means not only that no money is charged but also that OpenSenseMap is usable without prior registration.

4. OPENSENSEMAP - DESIGN AND IMPLEMENTATION

This chapter focuses on the design and implementation of OpenSenseMap. It describes the basic components and the underlying data model. OpenSenseMap is following the clientserver model (Berson, 1992) and its complete application stack is implemented in JavaScript (cf. Figure 2). All source code is published on GitHub. The server-side source code is available in the OpenSenseMap-API repository[12] and the client-side source code is available in the OpenSenseMap repository[13]. All components are using open-source technology and are published under an open-source license too. By this all interested citizens are invited to take the current state of the project and to contribute in either implementing new features, fixing bugs or suggesting and discussing future development of OpenSenseMap.

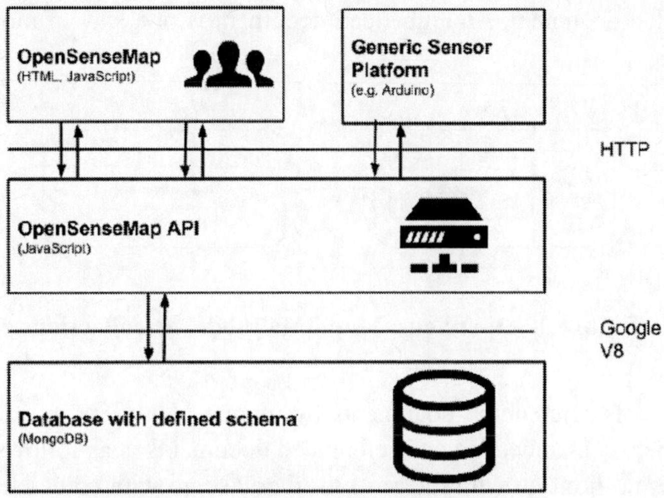

Figure 2. OpenSenseMap system architecture.

The performance of web browsers immensely increased over the last years. Today, it is possible to realize desktop-like web applications (Gutwin et al, 2011) due to highly efficient JavaScript engines and hardware acceleration. OpenSenseMap is completely written in HTML5 and JavaScript. This makes it runnable in any modern web browser on almost any device. As stated in chapter 3 not all provided functionality by the SWE framework are used and required by the OpenSenseMap. Therefore Bröring et al. (2012) advise that it may be easier to go with theWeb of Things approach. Additionally Demuth et al. (2013) stated that setting up OGC services and providing the right publishing methods are a challenging and cost-intensive task for a community. Therefore the application was developed from scratch and picks up the WoT paradigm in order to be applicable in citizen science projects. The development process of OpenSenseMap followed an agile development paradigm with early feedback from its potential users.

4.1 Database & Data model

As shown in Figure 3, OpenSenseMap uses MongoDB to store its data. MongoDB is a NoSQL Database. NoSQL stands for 'not only SQL databases' and are typically categorized into five different categories (Nayak et al., 2013) whereas MongoDB is a Document Database. Document Databases, store data in standard formats like JavaScript Object Notation (JSON) or XML. NoSQL Databases are well-suited for WoT applications: besides horizontal scaling, NoSQL databases support dynamic schemas. In contrast to traditional SQL databases, NoSQL databases require no structural definition of the data that is going to be stored. This makes NoSQL databases highly flexible because each data record in the database may be different in structure. This enables OpenSenseMap to keep pace with upcoming sensor platforms, which may include new sensors. OpenSenseMap's data model (cf. Figure 3) is derived from the O&M and SOS data model. It contains three collections and each collection is dened with a minimal schema; Table 2 summarizes the complete data model. Collections can be seen as tables in relational databases. Furthermore it is eye-catching in Figure 3, that the sensors collection is inside the sensor platform collection; this is called an embedded document. An embedded document is one way of modeling relations in NoSQL databases;

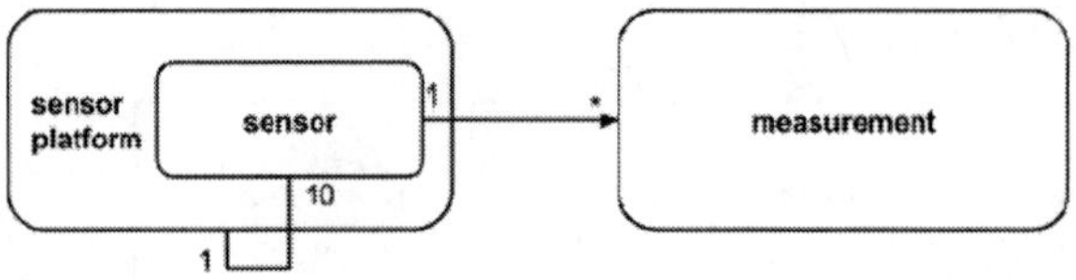

Figure 3. OpenSenseMap data model as collections.

an alternative is referencing documents by unique identifiers. A good rule (Iacobelli, 2014) to decide between embedded and referenced documents is as follows: if the amount of subdocuments has no limit go with references, otherwise go with embedded documents. The sensors of a sensor platform are modeled as an embedded document because the number of sensors is limited to ten for each sensor platform. The number of measurements is unlimited; therefore, the sensor's measurement collection is referenced. Furthermore, the entities sensor and sensor platform can be extended to include additional metadata. This could be detailed information such as the sampling rate of a sensor. In contrast, the entity measurement can not be extended.

Table 2. Complete OpenSenseMap schema for all collections

Collection	**Field**	**Required**
Sensor Platform	name	yes
Sensor Platform	location	yes
Sensor Platform	boxType	yes
Sensor Platform	exposure	yes
Sensor Platform	grouptag	no
Sensor Platform	sensors	yes
Sensor Platform	image	no
Sensor	title	yes
Sensor	unit	yes
Sensor	sensorType	no
Sensor	lastMeasurement	no

Measurement	value	yes
Measurement	sensor_id	yes

All data is stored as JSON in the database. JSON was selected because it was originally derived from JavaScript and is nowadays a lightweight data-exchange format. JSON can be used in JavaScript without implementing parsing and transformation operations. Furthermore, JSON is well-suited for use in environments with restricted resource and performance capabilities (Nurseitov, 2009).

4.2 Back-end / API

This section describes the back-end implementation of OpenSenseMap and its exposed API. OpenSenseMap makes use of NodeJS. NodeJS enables us to use JavaScript on the server-side and its event-driven and non-blocking I/O is well-suited for building real-time and data-intensive applications. Furthermore, NodeJS supports asynchronous I/O for sockets, files and HTTP communication; this makes a traditional web sever like Apache obsolete. NodeJS is actively developed and supported by its community and Joyent.

4.2.1 *RESTful Service*

The RESTful service is the heart of OpenSenseMap and implements the data publish and data retrieval functions. Developers may use it to build custom applications around OpenSenseMap or integrate OpenSenseMap into existing projects. The NodeJS application runs its own web server and implements the different HTTP request types. This makes the registered sensor platforms with their sensors and measurements accessible by unique URIs; Figure 4 shows the implemented URI schema.

http://opensensemap.org:8000/boxes/<platform-id>/sensors/<sensor-id>

Figure 1. URI schema to address single sensor platforms, its sensors and its measurements

The API currently implements the common HTTP-methods GET, POST, PUT, DELETE. These HTTP-methods realize the use-cases defined in section 3.1. If a resource hits the API and requests a resource, the request is processed at the NodeJS application and it responds with the data for the requested resource or an HTTP error code and message[20]. The GET-request is used to return different data from OpenSenseMap. Executed at root level, it returns a list of all registered sensor platforms. By adding the identifier of a selected sensor platform, it returns detailed information of the selected sensor platform and its latest measurements. To reduce the response size it is also possible to query individual sensors. The POST-request is used to register new sensor platforms and to publish new measurements. The PUT-request is used to update or modify an already registered sensor platform. The DELETE-request is used to delete a registered sensor platform and all of its published data. OpenSenseMap does not support a traditional user management but implements a simplified per sensor platform user management. To prevent abuse of the service, each modifying HTTP-method is authenticated using an unique api-key (Babal, 2015). The unique api-key identifies and authenticates the owner of a sensor platform.

4.2.2 *Data formats*

OpenSenseMap currently supports two different response formats: JSON and GeoJSON. By default, the response is formatted in JSON. By adding .geojson to a GET-request, the response is formatted in GeoJSON. GeoJSON is always a JSON object but encodes geographic data structures in a standardized way. Most Geographic Information Systems (GIS) are capable of processing GeoJSON. Both formats - JSON and GeoJSON - are an alternative to XML for data exchange on the web due to its compact representation (Crockford, 2006). That means it is more human-friendly in case of read- and writability and as well as for machines to parse and to generate. JSON is an open and language-independent data format.

Data is published on OpenSenseMap under the ODC Public Domain Dedication License (PDDL). The PDDL allows the reuse and modication of the original data without any restrictions.

4.3 Front-end

This section describes the graphical user interface (GUI) to interact with the back-end implementation, which was introduced in the previous section. The user-interface (cf. Figure 5) allows users to easily explore and contribute to OpenSenseMap. It fullfils two usecases: Exploring the already registered sensor platforms and their measurements and registering new sensor platforms. The functionality required for the two use-cases is implemented in separate modules, which will be described in the following two sections.

Figure 5. Explore view of the OpenSenseMap platform.

4.3.1 *Explore*

The explore view allows users to discover the registered sensor platforms and their published data; it is the default view when entering the OpenSenseMap website. It is a map based interface and supports basic spatial filtering using standard map interactions like panning and zooming. Additionally by clicking on the "magnifying glass" icon an address can be entered and searched for. Furthermore, users can browse the registered sensor platforms by searching for a sensor platform name or the measured phenomenons (cf. Figure 6a). While typing the filter term, the results are filtered and displayed as a list. To get detailed information about a sensor platform, a sub-view was implemented and can be accessed by clicking on a marker on the map.

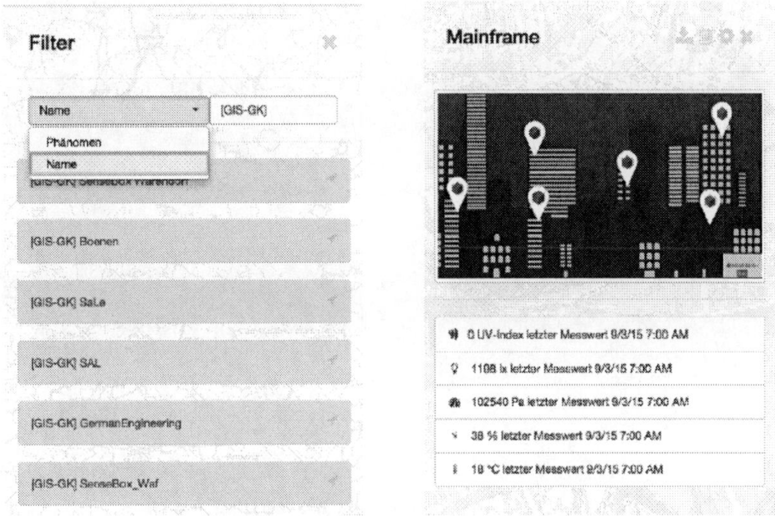

(a) Filter view with listed results.

(b) Detail view of a selected sensor platform.

Figure 6. Sub-views of explore view.

The detail-view (Figure 6b) shows all metadata, all sensors and their lastest measurements. The measurements are updated in real-time. Furthermore, the explore-view follows an URI pattern to make the boxes uniquely identifiable and open up the detail-view directly. According to the requirements of usability the front-end interface was build with a clean and easy to use design in mind. The main used framework is AngularJS22 to encapsulate the single modules and to provide a way to easily extend the modules. The map interface is provided by Leaet.js that oers an easy way to display JSON and GeoJSON data.

4.3.2 *Registration*

Users can add new sensor platforms using a guided step-by-step registration procedure (cf. Figure 7). The registration procedure is divided into five single steps and all required fields must be entered in order to proceed with the next step. According to Nielsen (1994) that users should not be able to make mistakes, this mechanism preserves the user against registering a non-functional sensor platform. As mentioned in section 3.2, no prior registration is needed to use the platform. The first step of the registration procedure requests the user to enter his name and email address. This information is only needed for registering a new sensor platform and to prevent people to upload spam and useless data. The email address is used to send the necessary api-key for authentication and a personalized sketch, described later in this section, to the owner of the sensor platform. In the following two steps of the registration procedure, the user has to specify the sensor platform by providing a name, exposure, geo-location and further optional metadata. The sensors of the sensor platform can be specified in two different ways:

- manual configuration
- plug & play solution

Figure 7. Step-by-step registration form.

If choosing manual configuration, the user has to specify for each connected sensor the measured phenomenon, the unit of measurement and the sensor type (cf. Figure 8). The number of connected sensors is limited to ten sensors per sensor platform. After registering a sensor platform using the manual configuration option, the user is able to upload data to OpenSenseMap using OpenSenseMap's RESTful API. Of course, the user needs programming knowledge to implement this functionality.

If choosing the plug & play configuration, the user can select from a list of supported sensor platforms. The plug & play configuration is based on pre-configured templates, which integrate the information made by the user during registration into an executable program for the selected sensor platform. An example for SenseBoxes, which are custom Arduino boards, shall illustrate this. Arduino requires a so-called Sketch, which is the program that runs on an Arduino board. It is responsible to interface with the connected sensors. The missing part is to push the collected data to OpenSenseMap using the exposed API. With knowledge of the underlying hardware setup and the user information provided during the registration procedure, a personalized sketch, which is capable of interfacing the connected sensors and pushing data to OpenSenseMap, can be generated automatically.

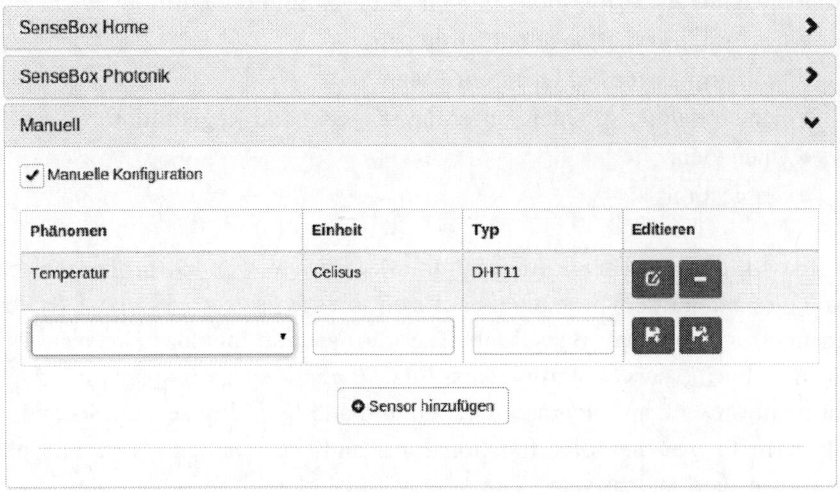

Figure 8. Registration of a generic sensor platform (manual conguration).

The described plug & play mechanism was implemented for the different SenseBoxes. In the last step where the sensor setup is specified, the user can select one of the four preconfigured SenseBoxes. By choosing one of these the registration procedure identifies the corresponding sketch template. The template is copied and personalized by inserting the IDs to reference the sensor platform and sensors. The proposed plug & play mechanism enables citizens without programming knowledge to participate in complex projects. What was exemplarily shown for the SenseBox hardware setup can be easily applied to different sensor platforms. To our knowledge, no Sensor Web portal offers this kind of configuration. The code generation for manual hardware setups is not possible right now and is a challenging task concerning the diversity of available microcontroller and sensors.

5. EVALUATION

This chapter evaluates the user interface design of OpenSenseMap. The user interface was evaluated by conducting a web-based survey. After a short description of principles and the survey design, the survey results will be analysed.

5.1 Principles and survey design

A web-based survey was conducted in order to evaluate the user interface design of OpenSenseMap. The survey follows the design principles for web-based surveys (Couper et al., 2001) and uses standardized survey components like scale, free-text and multiple choice inputs. Different statements on the user interface design of OpenSenseMap were questioned and the survey participants have to agree or disagree. The survey participants have to rate their agreement or disagreement on a six-level Likert scale (Wuensch, 2005) hence, no neutral rating is possible and a tendency must be stated by the survey participants. Furthermore, the survey participants have to perform two tasks and rate how easy it was to solve the task. The survey participants can give detailed feedback on each task by using the provided free-text input eld.

The survey contains 21 questions in total and is split into the following categories:

- General information about participant
- First impressions of OpenSenseMap
- Functionalities of OpenSenseMap (Explore and Registration)
- Open Data
- Suggestions

The first section on general information asks for age, gender, profession, expertise in web technologies and participation in other citizen science projects. The next section focusses on the first impression of OpenSenseMap. The third section includes two tasks, which have to be performed by the survey participants. First, the survey participants have to register a new sensor platform with an integrated temperature and humidity sensor. Second, the survey participants have to use the filter functionalities and the detailed view. The next section focusses on open data and privacy. The survey finishes with an open question on general suggestions and missing functionalities.

The user study was shared with all SenseBox customers without any knowledge about the person. The survey questionnaire and the results are attached in the Appendix.

5.2 Analysis

The web-based survey was conducted by 15 participants. The group of survey participants was made up of eleven males, three females and one unspecified gender participant. The average age of the survey participants was 34 years with a standard deviation s = 13;85. The age of the survey participants is in the range of 19 to 65 years. Eight survey participants are employed, five are students and two are pensioners. Seven survey participants say that they have no technical background. All survey participants claim to have experience with web technologies; 40% describe themselves as experts.

All survey participants rated OpenSenseMap's user interface design as well-arranged. Only the used icons were rated as being not intuitive. Three survey participants answered that "the red icons suggest that the corresponding sensor platform is offline or that an error occurred at the corresponding sensor platform". The legend icon, which shows up an explanation of the icons, in the left lower corner of the map is inconspicuous and should be placed more prominently on the site. Furthermore, the survey participants asked for an optional quick-tour when using OpenSenseMap for the first time.

All survey participants said that it was easy to register a new sensor platform. 67% of the survey participants fully agreed with the guided step-by-step registration procedure. The remaining 33% agreed with the guided step-by-step procedure but suggested to reduce the number of steps. Specifying the configuration of the new sensor platform was rated to be easy by 14 of the 15 survey participants. But one survey participant was not able to complete this task because of insufficient information; more detailed information should be added to the registration procedure. The perception of the explore view is similar. The survey participants mentioned that they had problems finding the filter function. After finding the filter function (magnifying glass icon), all survey participants were able to solve the task. The filter-as-you-type realization was perceived positively because of the instant feedback of the user interface.

Concluding with the questions regarding open data and privacy, 80% of the survey participants completely agree with the statement that they appreciate the free of charge model of OpenSenseMap. According to that 93% have no problem that all data is published as open data and is accessible and reusable by anyone. However, nearly 50% of the survey participants would favour the opportunity to mark data as private and not to publish it as open data.

6. CONCLUSION AND FUTURE WORK

This work presented OpenSenseMap - an open sensorweb portal for citizen science projects. OpenSenseMap is open in many ways. Open in that it builds upon an open-source software stack. Open in that the collected data is published as open data. Open in that its RESTful interface and lightweight data model make it possible to connect many different WoT-capable sensor platforms; we have already seen other sensor platforms23 connecting to OpenSenseMap. Open in that selecting a NoSQL document database and its flexible data model make it highly extensible.

The user-interface of OpenSenseMap was carefully designed in that it shall be easy and is fun to use. The survey confirmed that OpenSenseMap is easy to use but also showed what needs to be improved and gave ideas for future work. The users asked for an optional quick-tour when using OpenSenseMap for the first time, additional information when using the manual setup of a yet unknown sensor platform and more intuitive icons. Furthermore the survey revealed a contradictory meaning about open data: even if welcoming open data, users asked for some kind of privacy settings. A further analysis of this conict may be interesting and valuable.

Future work on OpenSenseMap should include the extension of the plug&play mechanism to further sensor platforms, data export from OpenSenseMap and exploration functionalities. We currently see two approaches to extend the plug&play mechanism: a configurator-based approach, which is capable of the most common microcontrollers and sensors, or a community-based approach, which allows users to create recipes for new hardware and provide them to the other users. Data export from OpenSenseMap should support geospatial formats like Shapefiles and common-use file formats like CSV or Excel. For exploration the user needs more functionalities like diagrams of historic data or advanced filter functionalities.

7. REFERENCES

Babal, H., 2015. Token-based authentication with angularjs & nodejs. http://code.tutsplus.com/tutorials/token-based-authenticationwith-angularjs-nodejs--cms-22543. last accessed 7.3.2015.

Berson, A., 1992. *Client/Server Architecture*. McGraw-Hill, Inc., New York, NY, USA, 1992.

Bröring, A., Echterhoff, J., Jirka, S., Simonis, I., Everding, T., Stasch, C., Liang, S., and Lemmens, R., 2011. New generation sensor web enablement. *Sensors* 11, 3 (2011), 2652–2699.

Bröring, A., Remke, A., and Lasnia, D., 2012. Sensebox–a generic sensor platform for the web of things. In *Mobile and Ubiquitous Systems: Computing, Networking, and Services*. Springer Berlin Heidelberg, 2012, pp. 186–196.

Conrad, C., and Hilchey, K., 2011. A review of citizen science and community-based environmental monitoring: issues and opportunities. *Environmental Monitoring and Assessment* 176, 1-4

(2011), 273–291.

Couper, M. P., Traugott, M. W., and Lamias, M. J., 2001. Web survey design and administration. *Public Opinion Quarterly* 65, 2 (2001), 230–253.

Crockford, D. 2006 The application/json media type for javascript object notation (json).

Demuth, D., Bröring, A., and Remke, A. A citizen science sensor platform as a live link from gis to the internet of things. Proceedings of AGILE 2013.

Dickinson, J. L., Shirk, J., Bonter, D., Bonney, R., Crain, R. L., Martin, J., Phillips, T., and Purcell, K., 2012. The current state of citizen science as a tool for ecological research and public engagement. *Frontiers in Ecology and the Environment* 10, 6 (2012), 291–297.

Faludi, R., 2010. *Building Wireless Sensor Networks: With ZigBee, XBee, Arduino, and Processing, 1st ed.* O'Reilly Media, Inc., 2010.

Fielding, R. T., and Taylor, R. N., 2012. Principled design of the modern web architecture. *ACM Transactions on Internet Technology* (TOIT) 2, 2 (2002), 115–150.

Florence, T., 2012. Sensor observation service. https://www.seegrid.csiro.au/wiki/Siss/SensorObservationService. last accessed 01.03.2015.

Goodchild, M. F., 2007. Citizens as sensors: the world of volunteered geography. *GeoJournal* 69, 4 (Aug. 2007), 211–221.

Guinard, D., and Trifa, V., 2009. Towards the web of things: Web mashups for embedded devices. InWorkshop on Mashups, Enterprise Mashups and Lightweight Composition on the Web (MEM 2009), in proceedings of WWW (International World Wide Web Conferences) (Madrid, Spain, Apr. 2009).

Guinard, D., Trifa, V., Mattern, F., and Wilde, E., 2011. From the internet of things to the web of things: Resource-oriented architecture and best practices. In *Architecting the Internet of Things*, D. Uckelmann, M. Harrison, and F. Michahelles, Eds. Springer Berlin Heidelberg, 2011, pp. 97–129.

Gutwin, C. A., Lippold, M., and Graham, T., 2011. Real-time groupware in the browser: testing the performance of web-based networking. In Proceedings of the ACM 2011 conference on Computer supported cooperative work (2011), ACM, pp. 167–176.

Hand, E., 2010. Citizen science: People power. *Nature* 466, 7307 (2010), 685–687.

Iacobelli, G., 2014. Mongodb schema design: Embedded vs references. https://coderwall.com/p/px3c7g/mongodb-schema-designembedded-vs-references. last accessed 16.02.2015.

Kopetz, H., 2011. Internet of things. In *Real-Time Systems*, Real-Time Systems Series. Springer US, 2011, pp. 307–323.

Kortuem, G., Kawsar, F., Fitton, D., and Sundramoorthy, V., 2010. Smart objects as building blocks for the internet of things. *Internet Computing,* IEEE 14, 1 (2010), 44–51.

Nayak, A., Poriya, A., and Poojary, D., 2013. Type of nosql databases and its comparison with relational databases. *International Journal of Applied Information Systems* 5, 4 (2013).

Newman, G., Wiggins, A., Crall, A., Graham, E., Newman, S., and Crowston, K., 2012. The future of citizen science: emerging technologies and shifting paradigms. *Frontiers in Ecology and the Environment* 10, 6 (2012), 298–304.

Nielsen, J., 1994. *Usability engineering*. Elsevier, 1994.

Nurseitov, N., Paulson, M., Reynolds, R., and Izurieta, C., 2009. Comparison of json and xml data interchange formats: A case study. Caine 2009 (2009), 157–162.

OGC, 2007. OPEN GEOSPATIAL CONSORTIUM. OpenGIS Sensor Model Language (SensorML) Implementation Specication. http://portal.opengeospatial.org/files/?artifact_id=21273. Version: 2007.

Silvertown, J., 2009. A new dawn for citizen science. *Trends in Ecology & Evolution* 24, 9 (2009),

467 – 471.

Sohraby, K., Minoli, D., and Znati, T., 2007. *Wireless Sensor Networks: Technology, Protocols, and Applications*. Wiley-Interscience, 2007.

Steenkamp, L., Kaplan, S., andWilkinson, R. Wireless sensor network gateway. In AFRICON, 2009. AFRICON '09. (Sept 2009), pp. 1–6.

Uckelmann, D., Harrison, M., and Michahelles, F., 2011. An architectural approach towards the future internet of things. In *Architecting the Internet of Things*, D. Uckelmann, M. Harrison, and F. Michahelles, Eds. Springer Berlin Heidelberg, 2011, pp. 1–24.

Wiggins, A., and Crowston, K., 2011. From conservation to crowdsourcing: A typology of citizen science. In System Sciences (HICSS), 2011 44th Hawaii International Conference on (Jan 2011), pp. 1–10.

Wiggins, A., and Crowston, K., 2012. Goals and tasks: Two typologies of citizen science projects. In System Science (HICSS), 2012 45th Hawaii International Conference on (2012), IEEE, pp. 3426–3435.

Wiggins, A., Newman, G., Stevenson, R., and Crowston, K., 2011. Mechanisms for data quality and validation in citizen science. In e-ScienceWorkshops (eScienceW), 2011 IEEE Seventh International Conference on (Dec 2011), pp. 14–19.

Wuensch, K. L., 2005. What is a likert scale? and how do you pronounce'likert?'. East Carolina University (2005).

Free and Open Source Software for Geospatial (FOSS4G) Conference Proceedings

Volume 15 *Seoul, South Korea* Article 40

2015

Evaluation Of An Open-Source Collaborative WebGIS Prototype In Risk Management With Students

Zar Chi Aye
Institute of Earth Sciences, Faculty of Geosciences and Environment, University of Lausanne

Marie Charrière
Water Resources Section, Faculty of Civil Engineering and Geosciences, Delft University of Technology

Roya Olyazadeh
Institute of Earth Sciences, Faculty of Geosciences and Environment, University of Lausanne

Marc-Henri Derron
Institute of Earth Sciences, Faculty of Geosciences and Environment, University of Lausanne

Michel Jaboyedoff
Institute of Earth Sciences, Faculty of Geosciences and Environment, University of Lausanne

Recommended Citation

Aye, Zar Chi; Charrière, Marie; Olyazadeh, Roya; Derron, Marc-Henri; and Jaboyedoff, Michel (2015) "Evaluation Of An Open-Source Collaborative WebGIS Prototype In Risk Management With Students," *Free and Open Source Software for Geospatial (FOSS4G) Conference Proceedings*: Vol. 15 , Article 40.
DOI: https://doi.org/10.7275/R5B27SH7
Available at: https://scholarworks.umass.edu/foss4g/vol15/iss1/40

AT6-03

EVALUATION OF AN OPEN-SOURCE COLLABORATIVE WEBGIS PROTOTYPE IN RISK MANAGEMENT WITH STUDENTS

Zar Chi Aye[1], Marie Charrière[2], Roya Olyazadeh[1], Marc-Henri Derron[1] and Michel Jaboyedoff[1]

[1]Institute of Earth Sciences, Faculty of Geosciences and Environment, University of Lausanne
Géopolis, UNIL-Mouline, 1015 Lausanne, Switzerland
Email: zarchi.aye@unil.ch; roya.olyazadeh@unil.ch; marc-henri.derron@unil.ch; michel.jaboyedoff@unil.ch
[2]Water Resources Section, Faculty of Civil Engineering and Geosciences, Delft University of Technology
Building 23, Stevinweg 1, 2628 CN Delft, The Netherlands
Email: m.k.m.charriere@tudelft.nl

ABSTRACT

Over the past decades, advancements in web services and web-based geospatial technologies have led to increasing delivery, access and analysis of rich spatial information over the web. With the use of open access data and open-source technology, it has become possible to make better, transparent and informed decisions for policy and decision makers. Under the framework of the European Marie Curie CHANGES project, a prototype web-based collaborative decision support platform was developed for the evaluation and selection of risk management strategies, mainly targeting flood and landslide hazards. The design of the conceptual framework was based on the initial feedback and observations obtained from field visits and stakeholder meetings of the case study areas of the project. A three-tier client-server architecture backed up by Boundless (OpenGeo) was applied with its client side development environment for rapid prototyping.

The developed prototype was tested with university students to obtain feedback on the conceptual and technical aspects of the platform as well as to analyze how the application of interactive tools in the exercise could assist students in studying and understanding risk management. During the exercise, different roles (authorities, technicians, community) were assigned to each group of students for identification and selection of risk mitigation measures in the study area: Cucco village located in MalborghettoValbruna commune of North-Eastern Italy. Data were collected by means of written feedback forms on specific aspects of the platform and the exercise. A subsequent analysis of the feedback reveals that students with previous experience in GIS (Geographical Information Systems) responded positively and showed interests in performing exercises with such kinds of interactive tools for learning, compared to the ones with fewer or no GIS experience. These results also show that the prototype is useful and supportive as a decision support tool in risk management while user-friendliness, interactivity and practical aspects of the platform could be further improved.

1. INTRODUCTION

There have been an increasing numberof disasters reported by natural hazards (EM-DAT, 2012). The frequency, magnitude and impactsof such hazards, especially in mountainous regions, are increased due to the changing patterns in climate and development (Sterlacchini et al., 2014).In the past years, structural control measures played an important role in the mitigation of disasters and protection of people. Few or no consideration was given to the long-term sustainability or social, cultural and environmental perspectives of the chosen risk reduction strategy(APFM, 2006).Only during the past decade, the need for an integrated risk management framework has been recognized, accounting for both temporary and permanent preventive measures in order to reduce the impact of natural hazards (Fuchs et al., 2012). Such kind ofintegrated approach calls for coordinated efforts in the selection of efficient and effective measuresin order to achieve the best combination of strategies and practicesregardless of the size of disasters (Hansson et al., 2008).It is important that diverse views and preferences of various stakeholders are considered in the decision-making process towards a common goal (Jankowski and Nyerges, 2001).

Natural hazards and associated risks are spatial in nature.Geographic information systems (GIS) arepowerful and valuable tools for spatial data analysis, manipulation and visualization in disaster risk management. Nowadays, with an increasing and emerging use of the web and geospatial technology, it has become possible not only to share, exchange and disseminate but also to analyze spatial information over the web.A web-GIS architecture thus allows users with different values from different organizations, located in different places at the same or different time, to seamlessly collaborate via a web environment (Dragićević and Balram, 2004). Moreover, the growing use of open access data and open-source technology makes it possible to enable a better, transparent and informed decision-making process for policy and decision makers.Many researches have initiated for decision support in the fields of planning, environmental and natural hazards management (Geertman and Stillwell, 2004;Salewicz and Nakayama, 2004; Sugumaran et al., 2004; Mysiak et al., 2005; Pasche et al., 2007; Zhang et al., 2011).Despite the variety of applications, there arefew systemswhichattempt to engage various stakeholders especially in planning of combined risk management strategies based on available risk information, through the integration of collaborative web-GIS based platformwith decision support tools.

Under the framework of CHANGES project, an online collaborative web-GIS platform was developed for risk management of hydro-meteorological hazards, in particular floods, debris flows and landslides. The aim of this decision support platform is to assist and integrate stakeholders' inputs into the formulation and selection of different risk management measures.To collect the preliminary feedback, the prototype platform was presented to the local and regional stakeholders of the case study sites(Romania, Italy and Poland) during the dissemination meetings of the project.As a further step, in this study, the prototype was tested with students from University of Lausanne (Switzerland) not only to obtain in-depths feedback on the different aspects of the platform (such as visualization, accessibility, usefulness, ease of use and so on) but also to analyze the potential of such interactive tools for students' learning process related to natural hazards and risk management.In thispaper, sections are organized as follows: section 2 introduces the collaborative web-GIS framework of the developed prototype.Section 3 presents the structure of the evaluation exercise carried out with students,along with the study area used for the exercise. The feedback results of the tested prototype arepresented and discussed in Section 4, and finally, we conclude the paper (section 5) with presented aspects and potential perspectives of the developed platform.

2. A COLLABORATIVE WEBGIS PROTOTYPE

The preliminary conceptual inputs of the platform were derived from initial field visits and stakeholder meetings at the three European case study sites of the project: Buzău County of Romania,the Friuli-Venezia-Giulia region of Italy and the MalopolskaVoivodeship of Poland. During the recent years, natural hazards such as floods, landslides and debris flows have occurredand affected mountainous communities in these study areas. Moreover, these areas suffered from the lack of funds, urging for a more efficient use of limited resources and better interactions between responsible authorities managing the risk (Prenger-Berninghoff et al., 2014).Although different platforms appeared to exist in the study areas, there is no collaborative platform in the selection of risk management alternatives, which brings together therelevant stakeholderswith different expertise in risk management (Aye et al., 2014). This collaborative framework, hence, specifically combines a web-GIS interface with a multi-criteria evaluation (MCE) tool to support stakeholders in decision-making process, allowing to identify and compare different alternatives in a participative manner, and this could ensure the proper and efficient use of limited funds and resources in risk management.

Figure 1 illustrates the main features of the platform, consisting of 1) Main navigation panel on the left with access to three main modules (i.e. *data management*, *risk management* and *user management*); 2) Map view panel in the center (with layer navigation and legend panel on the left) with tools for zooming, searching locations, styling, drawing and editing features, etc.; and 3) Data view panel in the south to show feature information of the respective (vector) map layers. The *data management* module includes the tools for uploading and visualizing of raster and vector maps as well as for creation of vulnerability curves. The main features of the *risk management* module are qualitative and quantitative risk analysis tools as well as tools for creation and selection of different risk reduction scenarios.The *user management* module is used for creating, assigning and managing user accounts and roles used in the collaborative process.

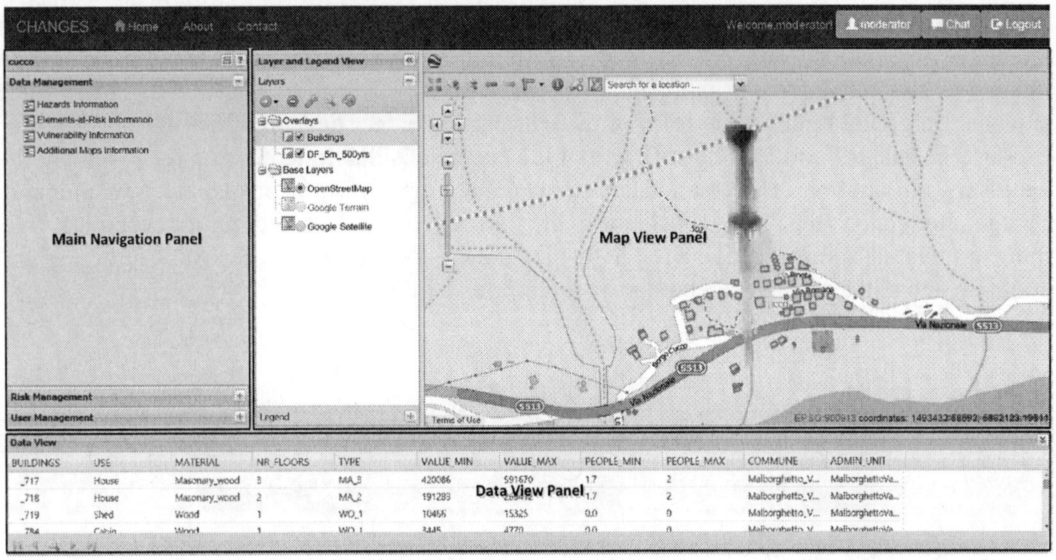

Figure 1. Main features of the prototype platform (for a user with admin rights).

The prototype is fully functional and developed based on theBoundless framework, i.e.an open-source geospatial software with modular components, and its client side software development kit (SDK). This was chosen due to a modular, customizable and extensible framework for the development of high-level web applications with built-in plugins and widgets for integration of existing map tools and functionality.A number of open source software solutions are employed: GeoServer for web map server;GeoWebCache for tile cache; PostGISdatabase for spatial data storage; GeoExt, ExtJSand OpenLayersfor user interface development and PHP server scripts.

This prototype was presented to the stakeholders from the three study areas and preliminary feedback was collected to obtain their opinions and suggestions. From the 49 collected forms, it was concluded that the platform is useful, innovative and supportive while user friendliness and practical aspects could be improved. Results obtained from this first round of feedback can be foundin the relatedpublications (Aye et al., 2014; Aye et al., 2015).

3. EVALUATION OF THE PROTOTYPE(WITH STUDENTS)

3.1 Structure of the evaluation exercise

To collect the feedback on conceptual and technical aspects of the collaborative platform, the prototype was further tested with university students. The evaluation of prototype with students is also believed to allow them in learning aboutrisk management with a "real world" problem of decision making. This kind of activity can be regarded as "active learning" in which students are involved "in doing things and thinking about the things they are doing" as defined in Bonwell and Eison (1991, p.2).Well-designed activities can contribute to the understanding of concepts to be learned (Wiggins and McTighe, 1998).

Thisevaluation exercise was carried out, during a morning session of a course on risk communication,with8 Master students(majoring in Geology, Risk analysis and monitoring,Environmental risks, Social environment) at the University of Lausanne. The exercise is composed of three main stages: 1) Identification of areas at risk; 2) Formulation of alternative scenarios; and 3) Selection of alternative scenarios. These steps followan integrated risk management approach with involvement of various stakeholder groups.The students (in groups) played the roles of different stakeholders depending on the stages of the exercise(i.e. stage 2 and 3). The structure of the exercise is illustrated in Figure 2.Some of the necessary information (e.g. creation of user accounts, uploading of maps, etc.) wasprepared by the moderator (teacher), considering the limited time allocated to the exercise. The moderator is an administrative user with the capacity to moderate the whole process, and could be one of the expert stakeholders in a real life setting.

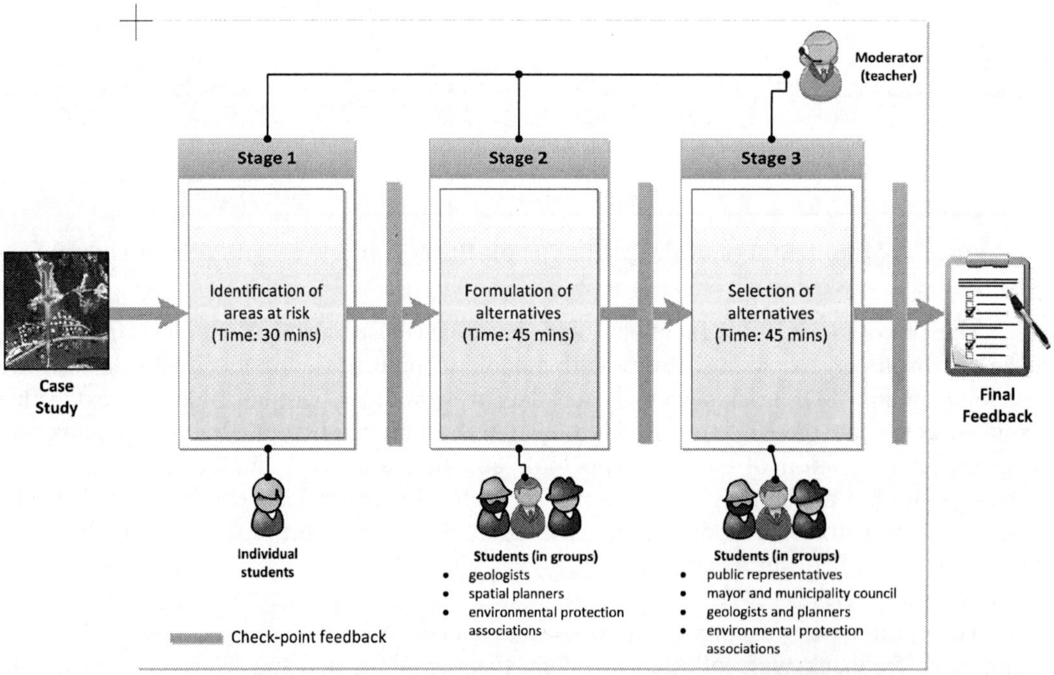

Figure 2. Structure of the evaluation exercise of the prototype.

As a guidance of the exercise, the following documents were handed out to the students at each stage of the exercise:

- *log-in* access information,
- a *scenario*sheet including the step-by-step instructions to follow by the students,
- a *role description* sheetincluding the description of the stakeholder roles for group works, and
- a*check-point feedback* form including the evaluation of certain aspects of the prototype, to be filled out by students at the end of each stage. It was composed of 1) an open question for the analysis of the presented problem; 2)five to ten rating questions (with a scale of 1 to 5) for specific evaluationson the interface and functionality; and 3) two open questions for improvements and suggestions on the presented stage of the prototype.

At the end of the exercise, the students were asked to fill two *final feedback* forms:*user evaluation* and *exercise feedback*. The *user evaluation feedback* evaluates the overall prototype such as innovativeness, interactivity, usefulness, user-friendliness, satisfaction and support as a decision support tool, rather than detailed aspects like in thecheck-point feedback. The *exercise feedback* evaluates the exercise itself in order to have the understanding and opinions of students on the presented aspects such as usefulness forlearning and understanding,helpfulness in understanding of how real world situation works, stimulation of interests in risk management topic andin doing further exercises which involve interactive tools.

3.2 Study area of the exercise

The Cucco village, Malborghetto-Valbruna municipality located in North-Eastern Italy, was used as the study area for the exercise. This area has been affected by debris flows in August 2003 and some houses were damaged. After this event, new mitigation measures were placed by the Civil Protection. This debris flow event is estimated to have a return period of 500 years according to the rainfall data analysis and potential future scenario was modelledto identify remaining risk and assess the effects of existing mitigation measures in the area (Hussin et al., 2014).

3.3 Three stages of the exercise

In the *first* stage (see Figure 2), individual students were asked to identify the areas at risk in Cucco village.For this purpose, debris flow hazard and building footprints mapswere uploaded beforehand into the platform by the moderator. The students conducted their analysis by simply overlaying these two layers and visualizing the areas being touched by debris flow in the web-GIS interface of the platform.

After identifying the areas at risk, the next step is to determine the possible measures to protect those areas at risk. Therefore, in the *second* stage, students worked in three groups assigned with a stakeholder role that are representative of a real-life situation: geologists, spatial planners and environmental protection associations. The task of each group is to design its own alternative scenario, which is a combination of possible risk reduction measures (both structural and/or non-structural measures).Potential structural measures include creation of new mitigation measures, structural adjustments of the existing measures or houses. Non-structural measures concernednon-physical actions such as relocation of houses, natural regeneration in the area or establishment of an early warning system. Each group of students proposed its own alternative scenario by creating (sketching) measures in the platform.In the check-point feedback form of this stage, the students were asked to explain why their scenarios should be considered as the most appropriate compared to other groups, along with other specific assessments of the functionality provided.

The alternative scenarios proposed by different groupswere evaluated and ranked in the *third* stage of the exercise in order to select one single alternative scenario.The decision-making process that is needed to achieve this selection benefits from using MCE methods. These methods consider the inclusion of additional important criteria such as social and environmental impacts than the traditional cost-benefit analysis (Munda, 2004) in evaluating the performance of alternatives. In the prototype platform, Compromise Programming (CP) method (Zeleny, 1973; Simonovic, 2010) is used to calculate the ranking of alternatives. This method identifies alternatives which are the closest to the ideal solution by means of distance values. The ideal solution is the one for which performance values of all considered criteria are maximized.The alternative with the minimum distance value to the ideal situation is considered as the "best compromise solution".For the simplification, within this exercise, criteria are pre-defined by the moderator (in real life, an expert) to evaluate the performance of the alternativesderived from the second stage. The alternative scenarios and their correspondingperformance values against criteria are evaluated by the moderatorin advance due to the time constraints of the exercise.Four groups of students are re-assigned in this stage to work in groups: public representatives, mayor and municipality council, geologists and planners, and environmental protection associations.The task of each group is to rank the alternatives by assigning weights to the defined criteria. In other words, depending on the role of each group, the students are asked to classify the importance of the criteria (with a scale of 1: the least important to 5: the most important criteria). Within the platform, each group can assign weights and visualize their ranking outcomes of alternatives in comparison with the ones of the other groups. A negotiation process(using the chat function) is started with the other groups to try to achieve a final ranking of the alternatives on which every group agree. In the follow-up check-point feedback form, students commented on the results of their given weights and ranking outcomes as well as on the certain aspects of the interface such as visualization of charts for criteria weights and alternative rankings.

4. RESULTS AND DISCUSSION

4.1 Check-point feedback

The check-point feedback evaluates certain parts of the prototype in three sections as explained in Section 3. To demonstrate the*open analysis question*, the results obtained from the second stage of the exercise are presented. As an example, the measures designed by the geologist group is illustrated in Figure 3. This group of students proposed the combination of measures which included the improvement of the retention basin, the barrier nets and the protection forests in the area. In their opinions, structural measures are effective and durable despite bearing the high cost in the implementation of such measures. Similarly, the planner group also proposed structural measures such as the structural adjustments ofhouses and implementation of individual measures such as small walls and metal plates for the protection of houses. However, as opposed to other groups, the group of environmental protection associations proposed non-structural measures such as awareness raising, early warning system and relocation of houses which are believed to be better than structural measures as the later might give theillusionto the people that such measures fully protect them).These feedback results show that students performed wellin role-playing and proposed differentmitigationmeasuresaccording to their assignedroles of stakeholders. Importantly, this participatory exercise with engagement of expert stakeholders also demonstrated why anefficient, combinedand coordinated risk management strategy is important inrisk management.

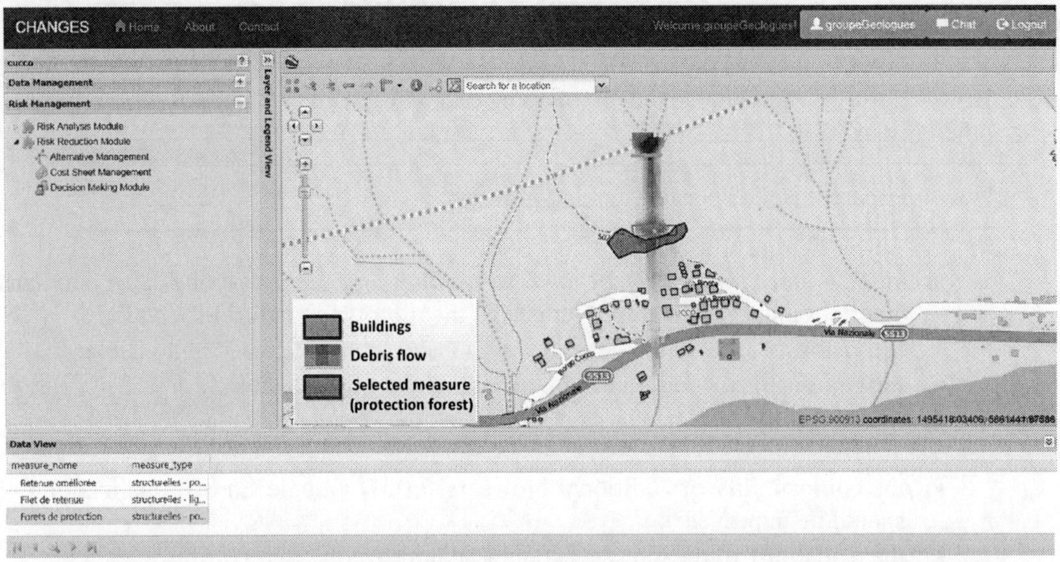

Figure 3. An alternative scenario as designed by a group of students (geologists).

The *rating questions*were related to various aspects of the interfaces, tools and functionality such as user-friendliness, satisfaction, usefulness, supporting ability, relevance and understanding of contents and so on. A total of 21 rating questions were collected for three stages of the exercise. Table 1 presents some extracted check-point feedback with respective average scores given by the participating students. The rating score ranges from a scale of 1 (Not at all) to 5 (Absolutely). As can be seen in the table, the user friendliness of sketching interface could be improved (average score = 3.5). This is maybe due to the confusing layer style option. This can be improved by restricting the sketching tool to allow only either point, line or polygon geometry within the same layer. However, students mentioned thatsketching tools (i.e. "create" and "editing" feature) are useful for designing measures (average score = 4). The chart options were also found helpful to visualize criteria weights and compare alternatives with others (average scores> 4) in the third selection stage of the exercise. Overall, the transparency of decision-making process achieved an average of score of 3.9, in which students with experiences in GIS (75%) scored 4.7 and the rest (25%) scored only 1.5. This explained why high transparency score is not achieved as expected.

Table 1. Selected check-point feedback(with average scores) given by students.

Selected questions	Average scores
How easy was it to find the maps you needed to visualize?	3.9
How easy was it to sketch measures in the map interface?	3.5
How useful is the "create" and "editing" feature tools?	4

How helpful would it be if a toolbox of mitigation measures was available?	4.5
How understandable is the weighting scale?	4.3
How helpful is the pie-chart visualization (criteria weights)?	4.4
How helpful is the comparison of ranking outcomes with other groups?	4.3
How transparent is the decision-making process?	3.9

Regarding the *open improvement and suggestion questions*, feedback of students provided an important input for the improvement of the prototype platform such as:

- the visibility of layer and legend view tabshould be expanded and visible;
- the geographical coordinates on the map should be available;
- the visibility of the hazard zone in the image should be made more understandable;
- the compatibility of additional browsers for 3D Google Earth visualizationtool should be improved;
- the readability of the interface should be improved;
- the visibility and friendliness of tools for the creation of alternative scenarios should be enhanced;
- the weighting scale of criteria should be indicated;
- the (stacked) bar chart visualization for ranking outcomes should be clearer;
- the explanation of the terminology usage in the interface should be provided andof the chat option should be made better accessible.

4.2 Final feedback: prototype

In the first section of the *evaluation feedback* form, students were asked to explain, in a few words, their understanding of the tested prototype. Students mentioned that it is a good decision tool for formulation and selection of scenarios with all the concerned actors in the presented risk zone. Moreover, it was stated that the tool is not only useful to communicate hazards and related impacts but also to enhance the collaboration betweenthe different experts in risk management. It was also mentioned that the tool allows to include different privileged criteria for the parties in decision making for the selectionof alternatives.According to feedback responses, the purpose of the platformwas well-understood and thus one of the important evaluation aspects of the exercise and platform was fulfilled.

The average scores for overall aspects of the prototype are shown in Figure 4.As can be seen in the figure, the students found the platform useful (average score = 4.5), and supportive as a decision support tool (average score = 4). Meanwhile, user friendliness of the interface and the usefulness of themain left navigation panel could especially be improved (average scores of 3.1 for both). The feedback results also show that the prototype platform is successful in performing its intended task (average score = 3.9).The overall user satisfaction achieved an average score of 3.5 (i.e. more than enough in the scale of 1-5), which is acceptable considering theunavailability of tutorial documentation and training sessionsfor students before the exercise.

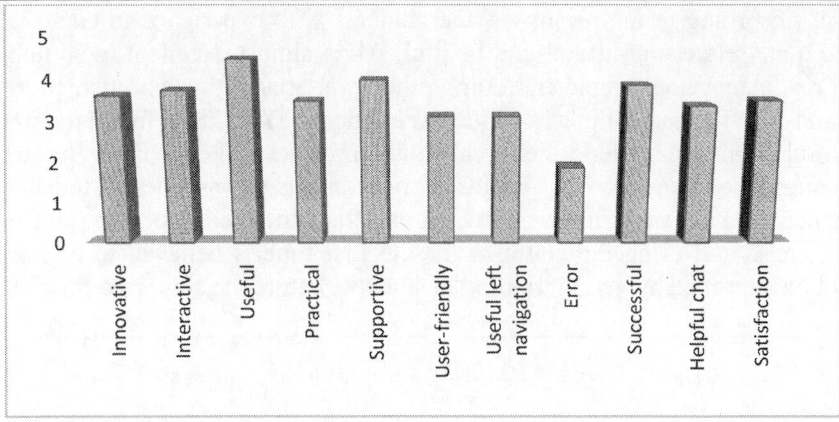

Q1. Is the prototype innovative?

Q2. Is the prototype interactive?

Q3. Is the prototype useful?

Q4. Is the prototype practical?

Q5. Is the prototype supportive as a decision support tool?

Q6. Is the prototype easy to use?

Q7. How useful is the left navigation panel to find information needed?

Q8. How often the prototype have errors and need to refresh?

Q9. How successful is the prototype in performing its intended task?

Q10. How helpful is the chat functionality between users?

Q11. Are you satisfied with the prototype, in overall?

Figure 4. Feedback on the platform (overall) based on the rating scale of 1 to 5.

For the overall aspects to improve, students stated that the tool needs to be more interactive, accessible, user-friendly and intuitive. During the short discussion, they also mentioned that the availability of a step-by-step video explanation, manual or training documentation of the tools and modules would be helpful in using the platform.Moreover, some mentionedthat the applicability of the platform in real world by the authority could be limited as risk management is quite "tricky", and we agree that potential ways of how to encourage and engage stakeholders in the process should be explored. However, it should also be noted that the purpose of a decision support platform is to assist the stakeholders in making better-informed decisions by providing necessary information and tools. Therefore, the legal responsibility of stakeholders and the primary decision makerremains depending on the institutional context of the study area where this platform is applied.

4.3 Final feedback: exercise

The first section of the *exercise feedback* formasked students of what they learned from this exercise. Note that all participating students had previous experience in role-playing. Students mentioned that it is a multi-disciplinary tool for decision making and the exercise was very useful as they were able to see the effects of the same problem from different points of views.They have also learned how an optimal decision can be reached considering different aspects at the same time in comparison with other stakeholders.

Figure 5 shows the average ratings of the five questions asked in the second section of the exercise. According to the responses, the students with experience in GIS (75%) found the exercise quite interesting, useful and helpful; while almost excellent in stimulating their interests in risk management topic and doing other exercises with such interactive tools. On the other hand, the students with few or little experience in GIS (25%) found that the exercise is quite helpful in understanding of how real situation works while results were quite low for the other questioned aspects.This result is not surprising as these students did not haveexperience working with similar software, and thus, the feeling of being uncomfortable doing the exercise and using the platform for the first time is believed to be normal. This aspect could be improved by giving training to students before the actual exercise.

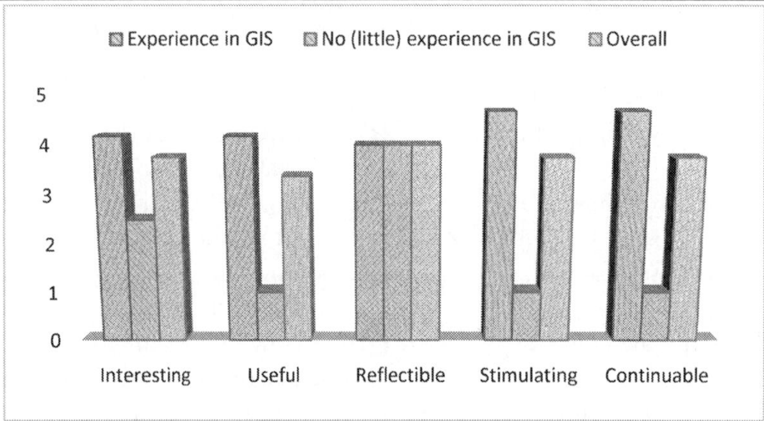

Q1. Is this exercise interesting?
Q2. Is this exercise useful for your learning and understanding?
Q3. Is this exercise helpful in understanding of how real situation works?
Q4. Does this exercise simulate your interests in risk management topic more?
Q5. Would you like to do other exercises with interactive tools?

Figure 5. Feedback on the exercise based on the rating scale of 1 to 5.

Regarding aspects of the exercise to be improved, students commented that more time for the exercise is needed in order to present and discuss the results with others (to reach a best consensus solution at the end). It has also been mentioned that the exercise gave them a good ideaof the difficulties that can be facedin risk management based on participative decision making. Students with no or few experience in GIS stated that use of the platform could be more simple and adaptable for those who never worked in a mapping environmentbefore (or alternatively, training can be given to those). Nevertheless, less-experienced students found the exercise interesting to understand scenarios of risk management because this topic is not addressed in their major, i.e. social environment.

For the allocated time frame of the exercise, at least one half-day should be allocated so that sufficient time is given not only for the good explanation of the theoretical framework of the platform but also for the follow-ups discussion with students at every stage of the exercise. This could especially enhance the usefulness aspect of the exercise in learning and understanding of the presented topic. During the exercise, it was observed that students were less confused andadapted to the exerciseas the time went on. The students especially enjoyed the third stage of the exercise in assigning preference weights on criteria according to their playing roles and in comparing the outcomes with other groups for the selection of alternative scenarios.This is, maybe, due to the simplicity and less complicated steps needed to perform in this stage of the platform. Or maybe because this is the most interesting part of the exercise asinteractions take place between groups (unlike in other stages: individually or within groups). Discussion within some groups also got little heatedas they debateoverwhich measures to propose in the second stage of the exercise. However, all groups managed to finish the tasks within the specific time frame. Interestingly, one student, in particular, expressed that "collaboration is hard" when being asked to explain what they learn from this exercise. In addition, some students raised questions and showed their interests in the approach used for the decision-making process.

5. CONCLUSION

In this paper, we presented how the evaluation of a collaborative decision support platform was carried out with Master students from University of Lausanne majoring in environmental topics. Thisevaluated prototype was developed based on the open-source framework, and combined web-GIS interface with a MCE tool for decision support in selection of risk management strategies for natural hazards like floods and landslides. The role of students in this exercise was to evaluate the prototype as well as to learn the process of risk management through the evaluation process, which allows them to analyze the presented problem, propose and select a solution by working together with other students. Students in groups (towards a common goal)brought their own experiences and background knowledge as they come from different specialized majors. As discussed before, conflicting interests and values between different groups were observed, for example, structural measures were more favored by geologist group while nature group favored non-structural ones. To achieve the most appropriate solution, all potential alternatives should be considered and compared against each other in terms of economic, social and environmental criteria. This allowed to reflect the real inter-disciplinary situation in which the involvement of various experts, decision makers and the community is crucial to achieve a sustainable and combined risk management strategy,especially for the case study areas where limited funds are available and weak link of interaction activities exist between risk management stakeholders.

In overall, the analysis of feedback results especially shows that the prototype is quite supportive and useful as a decision support instrument with good performance in carrying out its intended task. However, aspects such as user-friendliness, interactivity and practical aspects of the platform could be further improved.As students suggested, provision of manual documentation and or video demonstration would be helpful in using the platform. Importantly, students benefit learningfromthe prototype evaluation. During the exercise feedback,75% (i.e. 6 out of 8 participated students) responded positively and showed great interests in active learning with such interactive tools, compared to the rest which had no or few GIS experience. However, this can be improved by giving training to those who are not familiar with GIS applications, if such innovative hands-on exercises were to be developedforrelevant courses at the university. Nevertheless, all students agreed that this exercise reflected the real situation and helped them in improved understanding of the decision-making process in risk management. This feedback provided an important inputnot only for further improvements on presented aspects of the prototypebut alsofor potential application of the platform for activelearning with the students. Some of the improvements are already considered in the next version of the prototype and it is also planned to adapt the platform for environmental risk related exercises with Bachelor students at the university.

6. ACKNOWLEDGEMENTS

We would like to thank all the participated students of the risk communication course(spring semester, 2015) at University of Lausanne for giving their valuable feedback and suggestions on the prototype. We also thank the CHANGES team for sharing of their research results in the case study area of Italy. The authors also acknowledge the funding provided by European Commission for FP7 Marie Curie ITN CHANGES project (Grant No. 263953), which made possible to carry out this research for prototype development of a decision support system in risk management.

7. REFERENCES

APFM, 2006. Social Aspects and Stakeholder Involvement in Integrated Flood Management. WMO: Gevena, Switzerland.

Aye, Z.C., Sprague, T., Prenger-Berninghoff, K., Cortes, V.J., Jaboyedoff, M., Derron, M-H., 2014. Collaborative decision making in selection of risk management strategies through an interactive web-GIS based platform. In: *Proceedings of the International Conference on Analysis and Management of Changing Risks for Natural Hazards*, Padua, Italy.

Aye, Z.C., Jaboyedoff, M., Derron, M.-H., and van Westen, C.J., 2015.Prototype of a Web-based Participative Decision Support Platform in Natural Hazards and Risk Management. *ISPRS Int. J. Geo-Inf.*, Vol 4, 1201-1224.

Bonwell, C. C., and Eison, J., 1991. Active Learning: Creating Excitementin the Classroom. ASHE ERIC Higher Education Report No. 1,George Washington University, Washington DC, USA.

Dragićević, S., andBalram, S., 2004. A Web GIS collaborative framework to structure and manage distributed planning processes. *Journal of Geographical Systems,* Vol 6(2), 133-153.

EM-DAT, 2012.The OFDA/CRED international disaster database.Available at www.emdat.be. Last access: 22.07.2015.

Fuchs, S., Keiler, M., Sokratov, S., and Shnyparkov, A., 2012. Spatiotemporaldynamics: the need for an innovative approach in mountainhazard risk management.*Nat. Hazards*, 68, 1217–1241,doi:10.1007/s11069-012-0508-7.

Geertman, S. and Stillwell, J., 2004. Planning support systems: an inventory of current practice.*Comput. Environ. Urban Syst.*,28(4), 291–310, doi:10.1016/S0198-9715(03)00024-3.

Hansson, K., Danielson, M., andEkenberg, L., 2008. A framework for evaluation of flood management strategies. *Journal of Environmental Management*, 86(3), 465-480.

Hussin, H.Y., Ciurean, R., Frigerio, S., Marcato, G., Calligaris, C., Reichenbach, P., Van Westen, C.J., and Glade, T., 2014. Assessing the effect of mitigation measures on landslide hazard using 2D numerical runout modelling. In: *Landslide Science for a Safer Geoenvironment*, K.Sassa, P. Canuti, Y. Yin, Eds., Springer International Publishing, 679-684.

Jankowski, P., andNyerges, T., 2001. GIS-Supported collaborative decision making: results of anexperiment.*Annals of the Association of American Geographers*, 91 (1), 48–70.

Munda, G., 2004. Social multi-criteria evaluation (SMCE): methodological foundations and operational consequences. *European Journal of Operational Research* 2004, 158, 662 – 677.

Mysiak, J., Giupponi, C., and Rosato, P., 2005. Towards the development of a decision support system for water resource management. *Environ. Model. Softw.*, 20, 203–214, doi:10.1016/j.envsoft.2003.12.019.

Pasche, E., Küpferle, Ch., and Manojlovic, N., 2007. Capacity Building of Spatial Planners for Flood Risk Management in Urban Environment through Decision Support Systems and Interactive Learning. International Symposium on New Directions in Urban Water Management, UNESCO Paris.

Prenger-Berninghoff, K., Cortes, V. J., Sprague, T., Aye, Z. C., Greiving, S., Głowacki, W., and Sterlacchini, S., 2014. The connection between long-term and short-term risk management strategies for flood and landslide hazards: examples from land-use planning and emergency management in four European case studies.*Nat. Hazards Earth Syst. Sci.*, 14, 3261-3278, doi:10.5194/nhess-14-3261-2014.

Salewicz, K.A., and Nakayama, M., 2004. Development of a web-based decision support system (DSS) for managing large international rivers. *Glob. Environ. Change,* 14, 25–37. doi:10.1016/j.gloenvcha.2003.11.007.

Simonovic, S.P., 2010.*Systems Approach to Management of Disasters: Methods and Applications.* Wiley & Sons Inc., New York.

Sterlacchini, S., Akbas, S. O., Blahut, J., Mavrouli, O. C., Garcia, C., Luna, B. Q., and Corominas, J., 2014. Methods for the characterization of the vulnerability of elements-at-risk. In: *Mountain Risks: From Prediction to Management and Governance*, Springer Netherlands, 233-273.

Sugumaran, R., Meyer, J. C., and Davis, J., 2004. A web-based environmental decision support system (WEDSS) for environmental planning and watershed management. *Journal of Geographical Systems*, 6(3), 307-322.

Wiggins, G. and Mc Tighe, I., 1998. *Understanding by Design.* MerrillEducation/ASCD, College Textbook Series, ASCD, Alexandria,Virginia, USA.

Zeleny, M., 1973. Compromise programming. In: *Multiple Criteria Decision Making*, J.L. Cochrane, M. Zeleny, Eds., University of South Carolina Press, Columbia, South Carolina.

Zhang, Y., Sugumaran, R., McBroom, M., DeGroote, J., Kauten, R. L., and Barten, P. K., 2011. Web-Based Spatial Decision Support System andWatershed Management with a Case Study. *International Journal of Geosciences*, 2(3), 195.

Free and Open Source Software for Geospatial (FOSS4G) Conference Proceedings

Volume 15 *Seoul, South Korea* Article 41

2015

Developing a land use database of the Kanto Region, Japan in the 1880's

Nobusuke Iwasaki
National Agriculture and Food Research Institute, Institute

David Spraguea
National Agriculture and Food Research Institute, Japan

Naoko Fujita
Faculty of Design, Kyushu University

Ikuhiro Teramoto
Western Region Agricultural Research Center, National Agriculture and Food Research Organization

Hiroshi Yamaguchi
Art Research Center, Ritsumeikan University

AT5-02

Developing a land use database of the Kanto Region, Japan in the 1880's

Nobusuke Iwasaki[1], David S. Sprague[1], Naoko Fujita[2], Ikuhiro Teramoto[3] and Hiroshi Yamaguchi[4]

[1] Ecosystem Infomatics Division, National Institute for Agro-Environmental Sciences
3-1-3 Kannondai, Tsukuba 305-0031, Japan
Email: niwasaki@affrc.go.jp, sprague@affrc.go.jp
[2] Faculty of Design, Kyushu University
4-9-1 Shiobaru, Minami-ku, Fukuoka 815-8540, Japan
Email: fujita@design.kyushu-u.ac.jp
[3] Western Region Agricultural Research Center, National Agriculture and Food Research Organization
6-12-1 Nishifukatsu-cho, Fukuyama 721-8514, Japan
Email: yellow@affrc.go.jp
[4] Art Research Center, Ritsumeikan University
56-1 Toujiinn-kitamachi, Kita-ku, Kyoto 603-8577, Japan
Email: yamahirog@gmail.com

ABSTRACT

Historical land use records are valuable information for biodiversity protection, disaster management, rural area planning and many other uses. The Rapid Survey Maps (RSM) that were surveyed in the 1880's (early Meiji Era), are the first modern cartographical map series of Japan and important sources of information on traditional land use in early modern Japan. We had been analyzing these maps based on polygon data and raster based Web-GIS System to disseminate the Rapid Survey Maps using FOSS4G, but, these are difficult to apply for quantitative analyses of land use change. Thus, we developed a grid based land use database using QGIS and PostGIS, and published the database using GitHub.

First, we developed a land use data input system consisting of a client and server. The client was developed using QGIS API and the server was a PostGIS database. Point data as a 100 m grid was stored in the PostGIS server and land use category underneath each point was input using the QGIS application. About 1,400 thousand records (70%) have already been inputted. Error of grid based land use data is less than 1% compared with vector based land use data.

We analyzed land use change from the 1880's to 1975's. The most significant difference between the 1880's and 1970's is the area of urban land use and "rough land" such as grassland and bush. Urban area increased remarkably and grassland area almost disappeared. That does not mean grassland changed to urban area. Most grassland changed to agricultural land uses and forest, and urban area was formerly mainly agricultural land use and forest.

Some inputted data have been copied to GeoJSON and uploaded to GitHub as open data (Creative Commons BY 2.1 Japan). A tentative data browsing site was constructed with Leaflet. In this site, it is possible to compare point type land use data in the 1880's and present topographic map/RSM raster data. We hope that this database contributes to not only academic research, but also business, government, and public interest.

1. INTRODUCTION

Historical land use records are valuable information for biodiversity protection, disaster management, rural area planning and many other uses. For these purposes, accurate survey maps are neccessary. In the case of Japan, because the national isolation policy continued until the end of the Edo era, there was not accurate map until the 1880's. At this time, the Japanese government didn't have national mapping agency. Thus, the regional divisions of the Japanese army surveyed their own territory. The method of survey was called the Rapid Survey Maps Methods (*Jinsoku-sokuzu-hou*) and the maps which were surveyed by these methods, were called the Rapid Survey Maps (*Jinsoku-sokuzu*). In the Kanto region, around Tokyo, the Rapid Survey Maps (RSM) that were surveyed in early Meiji Era, are the first modern cartographic map series of Japan. The survey extent of the RMS is show show in Figure 1. The most of the Kanto plain and the Boso peninsula were included survey extent.

Figure 1. Coverage of Rapid Survey Maps. **Figure 2. Image of Rapid Survey Map.**

Figure 2. shows an image of a topical RSM. The land use categories were delineated by color with Chinese characters marking land use. In some RSM, there was a scenery picture outside of map extent. The pictures are also important sources for reconstructing the land use and landscape in early modern Japan.

The RMS is important sources of information on traditional land use in early modern Japan. There are many studies about the RSM and land use in early Meiji era. Shirai (2002) analyzed land use in the Shimousa plateau. Ichikawa (2006) described transition of the satoyama landscape in the west part of Tokyo. Sprague and Iwasaki (2009) were conducted time serise analysis of land use change in southern Ibaraki prefecture. Koyanagi *et al.* (2012) evaluated impacts of past landscape on present distribution of species in the Inashiki plateau. Iwasaki and Sprague (2005) examined the relationship between land use change and habitat of Japanese macque in the Boso peninsula. But it is difficult to view land use for entire Kanto plain. Therefore, we developed Web-GIS system (Historical Agro-Environmet Browsing System, HABS) for disseminating RSM data using FOSS4G (Iwasaki *et al*, 2009). After the release of HABS, many people used RSM data not only for academic purpose, such as biodiversity conservation and disaster management, but also for hobby use, for example, exploring old pilgrimage path, abandoned roads and castle ruins of old castles. However, it is still difficult to apply quantitative analyses of land use change.

Thus, a purpose of this paper is developing a quantifiable land use database and publish the database as open data.

2. LAND USE INPUT SYSTEM AND ACCURACY ASSESSMENT

2.1 Development of land use data input system

First, we examined a data model of land use data. In a previous study, we developed polygon-based land use data (Sprague *et al.*, 2007, Sprague and Iwasaki, 2009). But, constructing polygon data is required intensive work. In the case of the RSM, digitizing required a month per map. A few decades may be necessary for inputting the polygon data of all RSM, because total number of RMS is about 900 maps. In this research, we input land use data as 100m grid point data.

We developed a land use data input system consisting of a client and server. The client was developed using QGIS API and the server was a PostGIS database. Point data as a 100 m grid was stored in the PostGIS server and land use category underneath each point was input using the QGIS application. This application makes inputting land use data easier and faster than polygon based methods.

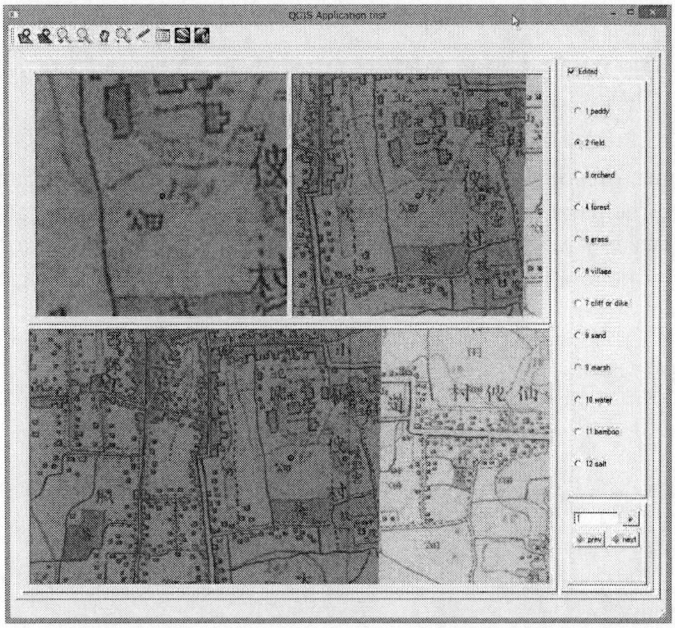

Figure 3. Land use input application developed by QGIS API.

2.2 Accuracy assessment of point based land use data.

The point sampling method did not illustrate an actual land use in RSM. Thus, an accuracy assessment was conducted based on the comparison between polygon data and point data. In the assessment, we compared the land use category ratio of point and polygon data in the Ushiku area, southern part of Ibaraki Prefecture. The details of the area was described in Sprague and Iwasaki. (2009).

Table 1. Land use ratio of point data and polygon data

	Area (%)	
Land use	**Point based data**	**Polygon based data**
Paddy	14.17	13.56
Upland field	18.39	19.11
Tea, orchard	0.14	0.00
Wood	40.11	40.24
Grass, bush	18.64	18.46
Village, road	2.80	2.59
Marsh	0.77	1.02
Water	4.99	5.00

Table 1 shows the results of comparison. Relatively large error was observed in linear land use features, such as paddy field. But error of grid based land use data is less than 1% compared with vector based land use data and overall accuracy was enough to evaluate the long term land use change.

3. PROGRESS OF DATABASE DEVELOPMENT AND APPLICATION FOR LAND USE CHANGE EVALUATION

The total number of points in the RSM database is about 2 million and it is too much to operate at once. Thus, the point data were divided based on Japanese topography map grids (about 9 km north-south, 14 km east-west) and we inputted land use data by the map grid. Figure 4 shows the current progress of data input. About 90% of map grids, which showed gray color in Figure 4, has already been inputted.

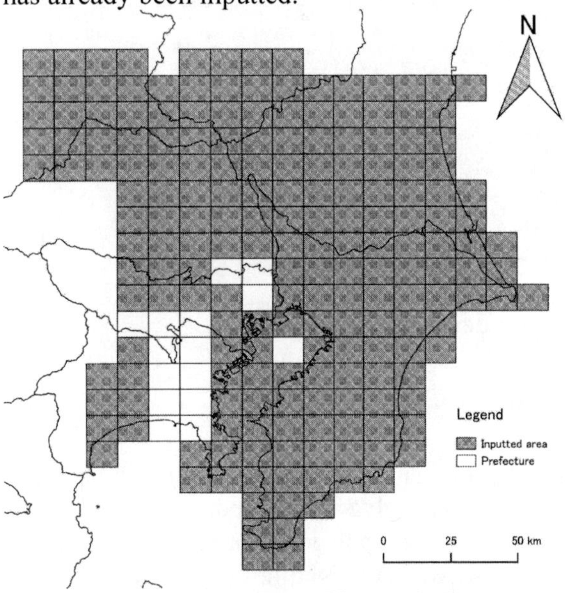

Figure 4. Progress of land use data input

We analyzed land use change from the 1880′s to 1975's using the east part of inputted data. The land use data in the 1975's was obtained from the "National Land Numerical Information Land utilization segmented mesh Data". The segmented mesh is a grid type data and size of grid is about 100 by 100 m.

The land use in the 1880's are thought to represent a traditional rural land use pattern in Japan. And the land use in the 1975's might be under the strong influence of high economic growth in Japan. Therefor, land use change from the 1880's to 1975's might be reflect Japanese rapid modernization and economical growth.

The table 2 shows land use in the 1880's and 1975's. The most significant difference between the 1880's and 1975's is the area of urban land use and "rough land" such as grassland and bush. Urban area increased remarkably from 7.3% to 13.4%. On the other hands, grassland area decreased significantly from 11.1% to 1.1%. The area of increased urban land use and decreased grassland were similar value. The area of forest and upland fields don't show large change. But that does not mean grassland changed into urban land use and forest and upland field were stable.

Table 2. Land use in 1880's and 1970's

	Area (%)	
Land use	**1880's**	**1975's**

Paddy	19.4	23.7
Upland field, orchard	13.5	17.4
Wood	31.7	30.4
Grass, bush	11.1	1.1
Village, urban	7.3	13.4
Water, mash	17.0	14.6

Table 3 is land use change matrix between two times and Figure 5 is land use maps in the 1880's and 1975's. About 60% of grassland in 1880's changed to forest and 22% became to upland fields. In 1880's grassland mainly occupied hill slope and plateau. Grassland on the hill slope and plateau was converted to forest and upland fields, respectably. Regarding urban land use, 16.6% of upland fields and 12.6% of forest in 1880's changed to urban land use in 1975's. We can concluded that there were complex land use change pattern in the Kanto region.

Table 3. Land use change matrix from 1880's to 1970's

	Land use in 1970's						
Land use in 1880's	**Paddy**	**Upland field**	**Forest**	**Grass, bush**	**Urban area**	**Water, mash**	**Total**
Paddy	71.9	5.9	10.1	0.7	8.9	2.5	100.0
Upland field, orchard	22.4	41.5	15.8	0.7	16.6	2.6	100.0
Forest	8.3	21.3	55.9	1.4	12.6	0.6	100.0
Grass, bush	5.5	21.9	61.3	2.1	8.4	0.8	100.0
Village, road	21.7	15.9	17.2	0.6	39.7	4.9	100.0
Water, mash	11.0	0.9	2.3	0.8	8.8	76.2	100.0

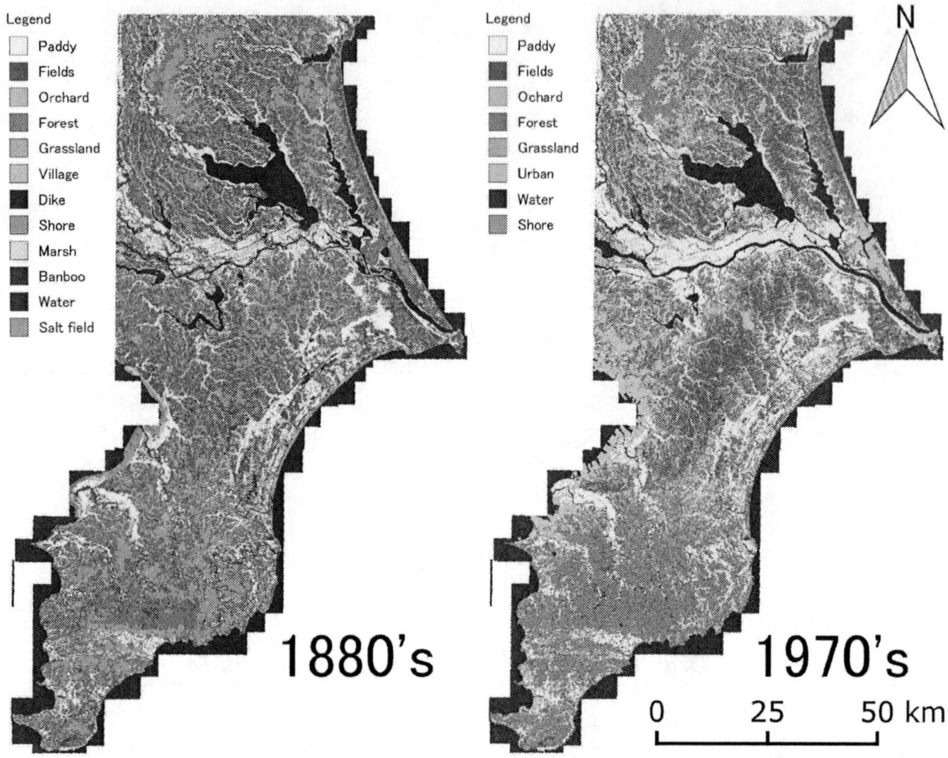

Figure 5. Land use mapns in 1880's and 1970's

4. Dissemination data as Open Data

To examine an effective methods for disseminating data, a subset of the constructed database had been published. For the mobile device, point data of the Boso Peninsula were converted to 100m grid polygon data as mbtiles format. RSM raster data were also provided as mbtiles.

For the Web and GIS analysis, we had been converted database to GeoJSON and uploaded to GitHub. The license of the uploaded data is Creative Commons BY 2.1, Japan. Currently, 27 map grids were uploaded. GeoJSON data can be displayed in GitHub page by default. But, all of point data were displayed same legend. Therefor, a tentative data browsing site was constructed with Leaflet (Figure 6). In this site, it is possible to choice present topographic map/RSM raster data as a base layers and display the point type land use data in the 1880's. By overlaying the present and passed information, it makes easier to recognize local land use change.

In GitHub, a user can check change history of data and also report error of data. We would like to make a communication with user using these function and develop a participatory accuracy assessment system, in the future works.

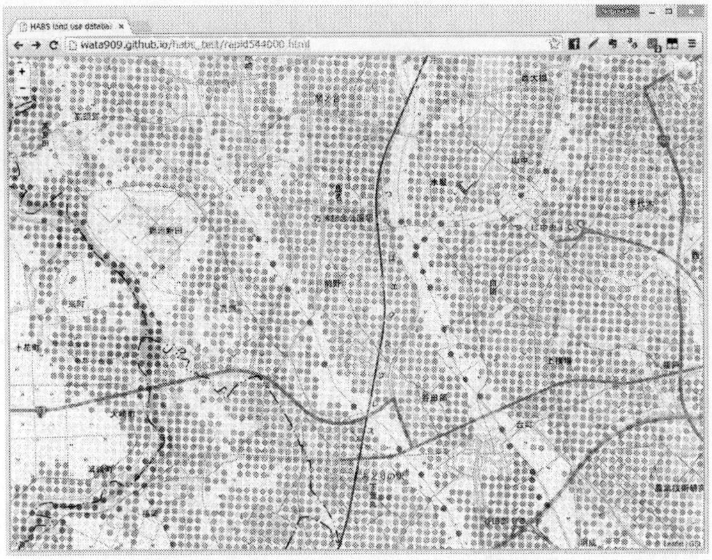

Figure 6. Overlay point based land use data and present topography map

5. Conclusion

In this paper, we described about development of the land use database derived from RMS of the Kanto region, Japan in the 1880's. We developed land use database as 100m grid point and about 90% of database has been inputted. The accuracy of point data are reliable for examining land use change and we conducted land use change analysis of east part of the Kanto region. Then, a subset of the constructed database had been published for examining an effective methods for disseminating the database.

In previous study, we published RMS raster data as Web-GIS and also provided tile map data. These data promoted utilization of RMS not only for academic use, but also for an individual interest. We hope that this database also contributes for disseminatin of RSM utilization not only for academic research, but also for business, government, and public interest.

6. REFERENCES

Shirai, Y., 2002. Firewood and charcoal production, and land use change in the western Shimosa upland during the second decade of the Meiji period 1878-1887: an analysis of the *Jinsokusokuzu* (Concise Chart) and the *Teisatsuroku* (Scouting Report). *The Historical Geography* 44, 1-21 (in Japanese).

Ichikawa, K., Okubo, N., Okubo, S. and Takeuchi, K., 2006. Transition of the satoyama landscape in the urban fringe of the Tokyo metropolitan area from 1880 to 2001. *Landscape and Urban Planning* 78, 398-410.

Sprague, D. S., Iwasaki, N. and Takahashi, S., 2007. Measuring rice paddy persistence spanning a century with Japan's oldest topographic maps: georeferencing the Rapid Survey Maps for GIS analysis. *International Journal of Geographical Information Science* 21, 83-95.

Sprague, D. S. and Iwasaki, N., 2009. Time series GIS analysis of rural land use change in southern Ibaraki Prefecture with historical maps starting with Rapid Survey Maps. Journal of The Japanese Institute of Landscape Architecture 72 623-626 (in Japanese).

Koyanagi, T., Kusumoto, Y., Yamamoto, S., Okubo, S., Iwasaki, N. and Takeuchi, K., 2012. Grassland plant functional groups exhibit distinct time-lags in response to historical landscape change. *Plant Ecology* 213, 327-338.

Iwasaki, N. and Sprague, D. S., 2005. A study of the land use history in the Japanese macaque habitat expansion area of the Boso Peninsula, Japan. *Journal of Rural Planing* 24, S1-S6 (in Japanese).

Iwasaki, N., Sprague, D. S., Koyanagi, T., Furuhashi, T. and Yamamoto, S., 2009. Development of the Historical Agro-Environment Browsing System constructed by FOSS4G. *Theory and Applications of GIS* 17, 83-92 (in Japanese).

Free and Open Source Software for Geospatial (FOSS4G) Conference Proceedings

Volume 15 *Seoul, South Korea* Article 42

2015

Landmark Based Path Planning And Linear Path Generation For Mobile Map Applications

Thejaka Mahaulpatha
Faculty of Computing, Sri Lanka Institute of Information Technology

Pasindu Chandrasekara
Faculty of Computing, Sri Lanka Institute of Information Technology

Dananjaya Thathsara
Faculty of Computing, Sri Lanka Institute of Information Technology

Irendra Koswatta
Faculty of Computing, Sri Lanka Institute of Information Technology

Nimalika Fernando
OSGeo Lab of SLIIT

Recommended Citation

Mahaulpatha, Thejaka; Chandrasekara, Pasindu; Thathsara, Dananjaya; Koswatta, Irendra; and Fernando, Nimalika (2015) "Landmark Based Path Planning And Linear Path Generation For Mobile Map Applications," *Free and Open Source Software for Geospatial (FOSS4G) Conference Proceedings*: Vol. 15 , Article 42.
DOI: https://doi.org/10.7275/R5KK990T
Available at: https://scholarworks.umass.edu/foss4g/vol15/iss1/42

AT4-03

LANDMARK BASED PATH PLANNING AND LINEAR PATH GENERATION FOR MOBILE MAP APPLICATIONS

Thejaka Mahaulpatha, Pasindu Chandrasekara[1] , Dananjaya Thathsara ,Irendra Koswatta and Nimalika Fernando[2]
Faculty of Computing, Sri Lanka Institute of Information Technology, Malabe, Sri Lanka
[1]mail2psc@gmail.com, [2]nimalikaf@gmail.com

ABSTRACT

Landmarks are yet to be integrated with mainstream mobile phone based navigation aids. In geographical regions where land marks are commonly used by the community for navigation support, the lack of them in electronic navigation aids make them less useful for such communities. In this study a land marks based navigation model is derived considering the value of them for local community in Sri Lanka. The landmarks can be prominent or not, make sense only during certain time of the day or been important differently for people with different age groups. We assume that the attributes of landmarks can be used to give a strength value for them for navigation. In this study three parameters, the visibility of them at different time of the day, the horizontal spread of the landmark and the height of them are considered as attributes which gives strength to a landmark.

First, to give more importance to landmarks, we have developed an algorithm where not only the distance of a route but the strength of landmarks is also considered when selecting the best route to navigate. The Dijikstra algorithm is used as the base which output possible shortest paths considering only the distance. This algorithm was enhanced to output the optimum paths considering both the distance and the strength of landmarks along it. If the route is having more strength related to landmarks, it is prioritized. The route's strength is defined based on number of landmarks visible along it and the strength of them. In order to calculate the number of landmarks along a route a landmark buffer is used. The day/night visibility and the height/spread are used to calculate the strength of the landmarks along the route.

Second, to utilize the mobile screen in more effective manner which has the size limitation, map generalization approach is used when showing the navigation path. We have identified that after placing landmarks on a mobile screen, the map become too congested and it becomes difficult to read the navigation path. This is more prominent when the path is having many turns. The selected path is reduced to a linear map which shows the path reducing curves while emphasising the turns by markers. The linear map, which is a schematic representation of the actual map, reduces the path between junctions to straight lines. The junctions and turns are also incorporated to the resulting map. The landmarks around significant turns are also provided and guidance is provided based on landmarks. Douglas-Peucker algorithm is used to derive the linear path.

A prototype implementation is done using mobile web approach to reduce the platform dependency. In the simple mobile web application developed, jQuery mobile, and php are used for the user interface development and server side implementations respectively. PostgreSQL with postGIS capabilities and pgRouting is used as a spatial database. Web services and smart queries are used to implement the basic functionalities communicating with the spatial database and the front end. The application is still being verified and tested in Sri Lanka at the moment.

1. INTRODUCTION

Use of landmarks in way finding is considered as old as human civilization. Path integration and land mark based navigation are two common approaches used by humans in way finding and sometimes these two approaches are used combined (Fallah, 2013). In landmark based way finding, the land marks are used as reference points and recalling the land marks can assist in identifying the current position and the orientation of a traveller (Loomis, 1999 as cited by Fallah, 2013).

Therefore, landmarks, if incorporated properly can enhance support provided by navigation support systems. The landmarks are having high social and cultural sensitivity and therefore considering the local perspective is important. OSM data comes with few parameters relate to landmarks (OSM Landmark tags, 2015) .This paper attempts to adopt a landmark based navigation model for OSM based path planning. As the attributes already available are not sufficient to provide a weighted model, more tags are introduced to capture more detailed information of landmarks using OSM. The optimum path for navigation is calculated based on both distance and the landmarks accordingly.

When displaying the optimum path on a mobile screen, it is identified small screen can be congested with too many road details and landmarks on top of them. Therefore, the map simplification method is proposed to make use of the mobile screen effectively. The resulting linear map is supported with landmarks and real time guidance.

This paper presents a brief introduction about the proposed navigation model and the design and implementation of the linear path generation approach adopted. This is part of an on-going research study on identifying landmark models suitable for Sri Lankan local contexts and the other sections of the study will be reported separately.

2. BACKGROUND

2.1 Landmark classifications and navigation models

Landmarks can be identified as everything that sticks out from the background can serve as a landmark (Presson & Montello, 1988). Classification of landmarks is important to represent them in a meaningful manner for a navigation model. How landmarks get highlighted from the rest is considered via salience of them. The salient features of a landmark include three important attributes, visual, semantic, and structural attributes (Sorrows & Hirtle 1999). The most discussed visual features of landmarks are height, proportion of height and width, façade area, façade structure and style of roof (Eva NUhn *et al.*2012). According to (Fang, Li *et al.* 2011), there are several attempts to integrate landmarks for route instructions (Caduff & *Timpf,* 2005; Richter, 2007; Duckham et *al.*, 2010). Though there have been plenty of work done under landmark salience, considering the local context of a country or a region for the significance of landmarks is less known in this domain.

2.2 Map simplification for mobile maps

Generalisation is the decreasing of the level of detail on a map so that it remains uncluttered when its scale is reduced. Several simplifying routines use complex geometric criteria (angle measurements and distance) to select critical or meaningful spots. Line simplifying has been deeply studied and it is highlighted as the most commonly found simplification method (Mariane et al., 2008). Douglas-Peucker algorithm is one of the most commonly used algorithms for line simplification (Saalfeld 1999).

3. METHODOLOGY

3.1 Proposed Landmark based model

The nodes connecting road sections can be considered as a graph and then graph based algorithms can be used to calculate the best path to travel between starting node S and ending node E by minimising the cost of travel. The cost of traveling is commonly reduced to distance and the Dijkstra (Daniel Kastl, Frederic Junod 2014) and A* algorithms can be adopted straightforward manner in such a case. If n nodes exists between nodes *S* and total cost for travel T is defined as $T = \sum_{i=1}^{n} c_i$, where *c* is the cost of traversing between two consecutive nodes.

The proposed model considers the two factors, the distance to travel (*d*) and the strength of landmarks (*w*) both and it is assumed that

$$c = {}^{d}/_{w} \tag{1}$$

The strength of land marks are defined in terms of the land mark count and the prominence. It is assumed that a path with a landmark along it is more suitable than a path without a landmark alone it. A weight between 1 and 6 is used for the count of the land mark and if no land mark is available per 1km road section, value 1 is assigned for that section. The prominence is measured considering the spread, height, social significance and cultural significance. Accordingly, the landmark model is defined as

$$Ss = \frac{\sum(\text{landmark count/length}) + \text{salience rankings}}{\text{no. of attributes}} \tag{2}$$

Ss in equation (2) is the score for significance in landmarks along the path. Here the w in equation (1) which is the strength is defined using the value obtained in Ss in equation (2) and a defined weighted model which differs according to the value of Ss in equation (2). Additionally, the day/ night visibility also is considered. As there is little reported work on combining landmarks and distance, the equation (1) and (2) are used as a starting point and it is expected to be refined with evaluations. A field survey and interviews are being used to identify a suitable weight of the count of the landmarks. In this stage of the study, emphasize is given to the implementation of the model using OSM data and the linear path generation. The validity checking and the refinement of the model will be reported in a separate article later.

3.2 Simplified path generation

In simplification of the path, main goal is to generalize the path so that the only the most required information for travel would be visualized considering the limited mobile screen size. The bend removal and turn simplification are main approaches used. The main target was to reduce the bends between the source and the destination, while maintaining straight lines between the junctions. Junctions are given priority as they would lead to turns and therefore become decision points. Douglas Peucker algorithm is used for bend simplification while maintaining junctions. Junctions are made with any interconnection between road segments and some intersections can be between minor roads and the traversing road which would be not important when traveling on major roads. Therefore an approach is requited to differentiate the junctions which made with major roads and that of minor roads.

In Sri Lanka, there are 6 different classes of roads, namely A, B, C, D and E. Class A and B are national level roads, C & D are provincial level roads and E is local government authority level road (SLRRP 2012). Recently, another category of roads, Expressways (highways) are introduced and E class is now using for expressways. In OSM, the road categories of Sri Lanka are identified as six classes and tags are defined for them (OSM 2015). Accordingly, the junctions which made by connecting E, A or B roads (major roads) and junctions made with other categories of roads (minor roads) are differentiated. If the selected path is not going to use minor roads, the junctions made with minor roads are not be shown by default yet it is possible to user to switch on this layer if they wish. The resulting map would have three layers, the selected path as interconnected straight lines, the major junctions and all junctions which lead to turns when following the selected path as points and the minor junctions and junctions which do not lead to turns as points. Figure 1 illustrates a selected path and a simplified with turns as per the methodology adopted.

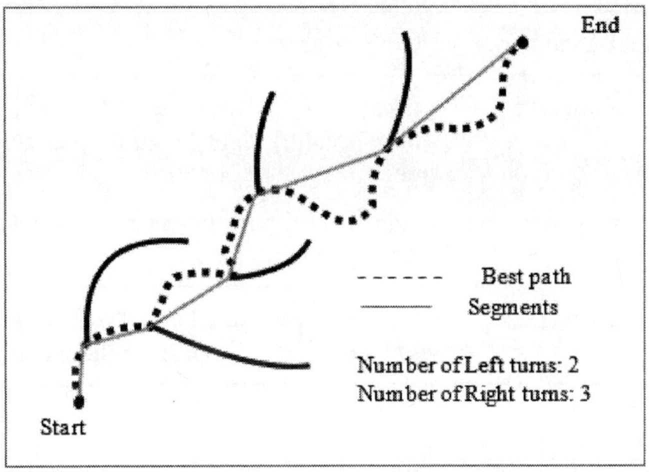

Figure 1. Selected path and simplified path with turns

3.3 Navigation guidance

Navigation guidance is done to support ,(a) establishing the orientation when following a path without any turn nearby and, (b) easy and clear identification of correct turns when closing to a junction to take a quick decision on the turns. Landmarks and road interconnection data are used to provide above guidance.

When the path is selected, the land marks along the path and around the junctions are extracted before simplifying it using buffer functions. The road interconnections are also extracted. When following a path without a requirement to take turns, the information about land marks are given. Side of the road they can be seen and the identification details are provided as information. Time of travelling is compared with day/night visibility tag and more information is derived accordingly.

When reaching each junction, based on the pre-calculated simplified path, junction's road interconnection data and the real time location information of the traveller, the decision on which turn to take is calculated. The azimuth of the traveller in real time and known azimuth of each interconnecting road at a junction is used to identify whether the left, right turn or no-turn (straight forward) is to take at a junction. Same comparison method is used when at a junction, more than one road is towards one direction and one of them is the road to take. The landmark information around a turn is used to enhance the instructions about correct turns in more practical manner. Set of word tokens are used to link turn information with land mark information. For an example, sample navigation instruction without and with landmark information would be as shown in Table 1. Token words are indicated between < > and navigation words are shown within [ ].

Table 1. Comparison of sample navigation instructions

Type	Sample Navigation Instruction
Without Landmarks	<take> [left, {first, second..} left, right, {first, second} right] <turn>
Without Landmarks	<Go Straight>
With Landmarks	<take> [left, {first, second..} left, right, {first, second} right] < turn> <near> [Bo tree] [Milk bar] <will be seen> [right side] <after> < taking turn>
With Landmarks	<Go Straight> [Bare land] <will be seen> [right side] <after> <passing the junction>

3.4 Map display

The main intention of simplifying the map is make it more suitable for mobile screen. The simplified path is shown on the mobile screen with orientation adjustment, which shows the path always towards the travellers heading direction along the mobile screen. The initial zoom level of the map is defined based on the selected path so that the whole path can be seen on the screen. However, as this may lead to a very small scale map when path is connecting large distance cities, the navigation mode change the map display to a large scale such as 1: 5000, so that 3.5 km length can be shown on a screen with 7cm height.

The layout is designed in such a way that the visibility of different layers such as path without simplification, path with simplification, major junctions, all junctions, landmarks alone path, and landmark near junctions can be controlled by the user. The orientation of the map is designed to self-rotate to adjust for taking turns. The mobile screen, therefore would show straight-line with heading towards the navigation direction, always.

4. IMPLEMENTATION

4.1 Technology Selection

The implementation of the proposed navigation model is totally done via free and open source technologies. To start come up with a basic navigation application we needed a data set, a spatial database, a map server and a map viewer. The OSM data set of Sri Lanka is used as the basic data set since it is open source and we could manipulate them as intended. As the user contribution grows, the data set would become rich and updated so that the application also would become up to date. In the initial stage, a copy of the OSM data stored locally is used.

PostgreSQL database with PostGIS extension is used as the database to sore OSM data and landmark data both. The PostGIS capabilities are used extensively for implementation of spatial manipulations. OSM2po tool was used to load the OSM data set to the database since it creates a basic network topology which enables routing on a road network. In order to get the routing capabilities, pgRouting extension was used which consist of a variety of routing algorithms that can be used for specific problems.

GeoServer is selected as the map server for render the maps based on data available on local database and real time OSM data. GeoSerever is an open source server for sharing geospatial data and is designed for interoperability and it can publish data from any major spatial data source using open standards. OpenLayers3 was used as the map viewer of the application. OpenLayers3 is the latest descendent of the OpenLayers family which is an open source, high-performance, feature-packed JavaScript library for mapping needs. HTML5 with JQuery Mobile is used to implement the front end of the application along with OpenLayers3. The use of JQuery Mobile paves the way to the mobile web approach which makes the application platform and device independent. The JQuery Mobile platform presents a rich set of UI elements which can be used alongside with HTML5 to give the attraction that the users always seek. PHP was used as the server side language to handle operations related to application logic, which are not dependent on spatial data or map visualization.

QGIS was the helper application used to view, edit and test spatial data and operations during implementation. QGIS is an open source GIS application which enables us to create, edit, visualize and analyse spatial data. Using the helper application saved a lot of time during implementation since it has a lot of capabilities to interact with spatial data without linking to the real application logic.

4.2 Implementation of the landmark model

The landmark model is implemented based on the OSM tag scheme. New tags are defined to record the landmark details as attributes. These include distribution of places of worship in to religions, a new categorization of building tags to match local context and a few more tags to represent locally specific content as well. (e.g.: place-of-worship: Buddhism, place-of-worship: Hinduism, place-of-worship: Islamic, place-of-worship: Catholic). All these tags comprehend to the osm tagging system (k: v).

4.3 Sample Results

Figure 2 shows the comparison of number of point geometries in sample selected path with and without simplification. Figure 3 shows a sample output of a path between A and B with distance minimisation and with the landmark model. Figure 4 shows a sample output of a path with proposed simplification and with major junctions over layered with landmarks.

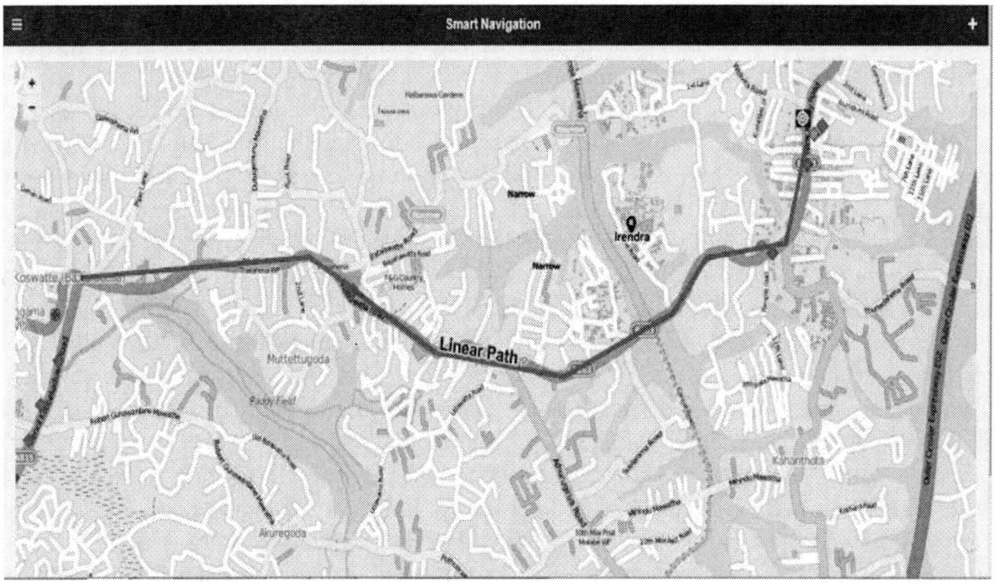

Figure 2: Comparison of point geometries after simplification

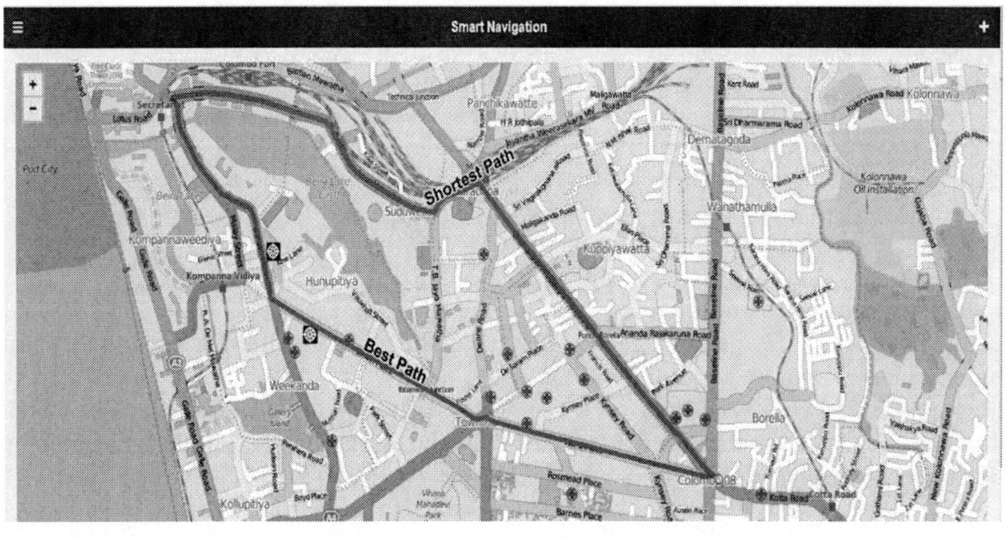

Figure 3: Comparison between best path and shortest path

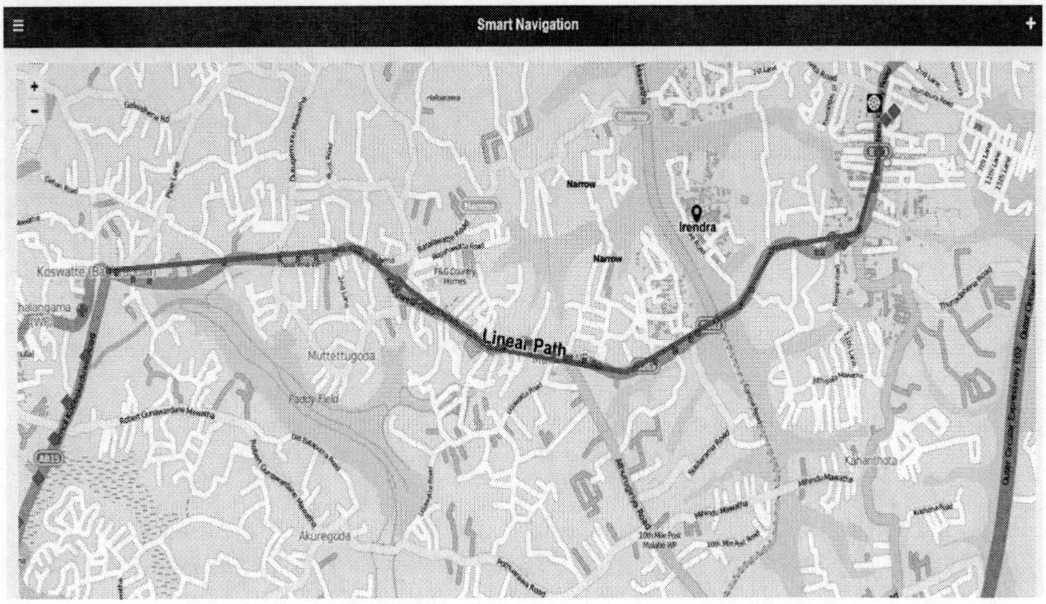

Figure 3: Linear map output with simplification and other supporting layers

5. CONCLUSION

This paper presents about implementation of a landmark based navigation model based on OSM data. New tags are introduced to represent attributes of local landmark information to the data model. The model implemented using open source technologies to generate navigation path considering landmarks also. The selected path is then simplified to generate a linear map which can be use more effectively on mobile screens. The technical implementation is tested and verified with OSM data and mobile map concepts.

The model and the linear path need to go through functionality and user evaluation. The comparison of the proposed path with distance based paths and the user-friendliness of the linear map needs to be evaluated.

6. ACKNOWLEDGEMENTS

This paper is based on a study supported by a SLIIT Research Grant received on 2015.

7. REFERENCES

Caduff, D. and Timpf, S., 2005. *The landmark spider: representing landmark knowledge forwayfinding tasks. In: T. Barkowsky, et al., eds. Reasoning with mental and external diagrams: computational modeling and spatial assistance. Stanford, CA: AAAI Press, 30–35.*

Duckham, M., Winter, S., and Robinson, M., 2010. *Including landmarks in routing instructions.Journal of Location Based Services, 4 (1), 28–52.*

Eva NUhn,Wolf Reinhardt,Benjamin Haske.,2012. *Generation of landmarks from 3D city models and OSM data*

Fallah, N., I. Apostolopoulos, K. Bekris and E. Folmer (2013). *"Indoor Human Navigation Systems: A Survey." Interacting with Computers 25(1): 21-33.*

Mariane Alves Dal Santo, Francisco Henrique de Oliveira, Guilherme Clasen Wosny.,2008 *Algorithms for Automated Line Generalization in GIS [online] http://events.esri.com/uc/2008/proceedingsCD/papers/papers/pap_1801.pdf*

Loomis, J.M., Klatzky, R.L., Golledge, R.G. and Philbeck, J.W. 1999. *Human Navigation by Path Integration. In Golledge, R.G. (ed.), Wayfinding: Cognitive Mapping and other Spatial Processes, pp.125–151. Johns Hopkins University Press, Baltimore, MD.*

OSM Land mark tags., 2015 [online] http://wiki.openstreetmap.org/wiki/OpenSeaMap/Landmarks

OSM. (2015, 25 June 2015). *"Sri Lanka Tagging Guidelines." [online] Retrieved August 2015, from http://wiki.openstreetmap.org/wiki/Sri_Lanka_Tagging_Guidelines.*

Saalfeld, A. 1999. *"Topologically Consistent Line Simplification with the Douglas-Peucker Algorithm." Cartography and Geographic Information Science 26(1): 7-18.*

SLRRP 2012. *Sri Lanka Road Rehabilitation Project, Ex-post Evaluation Report, The Export-Import Bank of Korea.,*

Daniel Kastl, Frederic Junod *pgRouting Workshop Manual 2014*

Free and Open Source Software for Geospatial (FOSS4G) Conference Proceedings

Volume 15 *Seoul, South Korea* Article 44

2015

An Open-Source WebGIS Platform for Rapid Disaster Impact Assessment

Roya Olyazadeh
Institute of Earth Sciences, University of Lausanne

Zar Chi Aye
Institute of Earth Sciences, University of Lausanne

Michel Jaboyedoff
Institute of Earth Sciences, University of Lausanne

Marc-Henri Derron
Institute of Earth Sciences, University of Lausanne

Recommended Citation

Olyazadeh, Roya; Aye, Zar Chi; Jaboyedoff, Michel; and Derron, Marc-Henri (2015) "An Open-Source WebGIS Platform for Rapid Disaster Impact Assessment," *Free and Open Source Software for Geospatial (FOSS4G) Conference Proceedings*: Vol. 15 , Article 44.
DOI: https://doi.org/10.7275/R5V1230N
Available at: https://scholarworks.umass.edu/foss4g/vol15/iss1/44

AT6-04

An Open-Source WebGIS Platform for Rapid Disaster Impact Assessment

Roya Olyazadeh, Zar Chi Aye, Michel Jaboyedoff and Marc-Henri Derron

Institute of Earth Sciences, University of Lausanne
Géopolis, UNIL-Mouline, 1015 Lausanne, Switzerland
Email: roya.olyazadeh@unil.ch; zarchi.aye@unil.ch; marc-henri.derron@unil.ch; michel.jaboyedoff@unil.ch

ABSTRACT

Natural disaster impacts have increased worldwide in the past decades. Earthquake is one of the disasters that have been studied for real-time analysis and crisis management. Disaster-related losses have been examined by the damage extent of the houses, infrastructures, fatalities and injuries converted to financial losses. WebGIS technologies provide a wide range of solutions to map those damages, analyse data and publish the results. Open-Source tools and data have been widely used today because they stay free and facilitate access to data especially significant in developing countries. This research presents a WebGIS prototype using Open-Source Geo-Spatial technologies such as PostGIS, Geoserver, Geoexplorer and OpenStreetMap (OSM) to evaluate the rapid impact of naturally produced disasters for the total damages. For this purpose, expert knowledge, such as earthquake intensities and vulnerability inputs are imported into the system and the loss of the damage is rapidly estimated. This work is part of a project for catastrophe modeling based on Open-Source data and software. We hope that applying Open-Source knowledge and solutions will decrease the time and efforts needed for rapid disaster and catastrophe management.

1. INTRODUCTION

Generally, disaster crisis happened with a complete mixture of human actions and natural hazards that directly result in a vital change during a short period of time such as death, disease, displacement including damage to infrastructure and economic loss (Wisner, et al. 2003). During the past decades, hazard events, namely as earthquakes, droughts, floods, storms and fires have produced significant loss of people, properties and environmental damage. By understanding the past hazards and anticipating the future events, the risk disasters can be minimized. As a result, disaster assessment should be a repetitive and remaining process (International Federation of Red Cross 2000) including underlying causes, dynamic pressures and unsafe conditions; this refers to a relationship between disaster, hazard and vulnerability (Blaikie, et al. 1994). Different types of assessment are based on different type of disasters and available resources. The initial assessment can be carried out quickly, and when more information are available, this can be improved (International Federation of Red Cross 2000).

During the past years, more than 1,100 dense earthquakes have occurred, causing more than 1,500,000 casualties and collapsing buildings of more than 90% (Lantada, Pujades and d Barbat 2009). Recently, the available fundamental information immediately after an earthquake is its magnitude, depth and epicenter provided by U.S. Geological Survey (USGS) data (http://earthquake.usgs.gov). However, the damage patterns are not an easy process and it requires more detailed information on-site (James and Pascale 2012). Besides, hazard map production has long process and include lots of efforts, and therefore, they cannot be available quickly and freely. Besides building and vulnerability information need a huge database. As a regard, open data, such as OpenStreetMap (OSM) and USGS data, including shake maps can be integrated directly in the Web-GIS application and will decrease the time and efforts needed for the analysis. This data will be used to estimate damage and loss of the event.

The WebGIS platforms, spatial data infrastructures, Geo visualization tools and GUI (Graphical User Interface) in the field of risk management have been applied in numerous related works (see (James and Pascale 2012), (van Westen, et al. 2014), (Aye, et al. 2015), (FEMA 2008) and (Open Quake 2015) among others). The web-GIS systems can support disaster assessment of an earthquake immediately and facilitate the analysis. In addition, different GIS prototype systems ((James and Pascale 2012) and commercial technologies ((FEMA 2008), (Esri White Paper 2008), (Open Quake 2015) and (InaSAFE 2015) have been proposed. Despite the variety of systems, there is no system which challenges the application of open data like OSM in planning of rapid disaster assessment, combined with loss and risk estimation based on available risk information, through the application of a web-GIS platform.

In this work, web-GIS technologies play a fundamental role both in rapid disaster assessment and loss estimation mainly for earthquake. This prototype application calculates loss of damage by importing data from OSM and adding other information such as earthquake intensity, vulnerability and the value of the buildings. The application is implemented based on the open source framework, namely OpenGeo (Boundless).

Section 2 of this paper begins with the methodology for rapid disaster assessment and conceptual framework of the system. In section 3, the background architecture and implementation are proposed, and section 4 is devoted to describing the data that has been used to test the application and discuss an initial result of the development. In the last part, the conclusion of this study and future works for the catastrophe management and modeling platform are reported.

2 The Conceptual Framework

The central goal of this study is to develop an integrated system for catastrophe management in case of an earthquake, focusing on rapid disaster impact assessment. The entire system plans to improve rapid assessment when there is a lack of information and data. The functions of the system are related to four main phases:

1. Hazard
2. Elements-at-risk
3. Vulnerability
4. Loss.

Intensity can be defined as a major disturbance created by a disaster. Dealing with earthquake, general information does not give an indication about the frequency. Generally, the hazard is stated in terms of the incidence rates of intensity values (Cardona, Ordaz and Marula 2008). Additionally, vulnerability is the characteristics in terms of the ability to resist and improve the effect of a hazard (Blaikie, et al. 1994). Vulnerability functions are extremely hazard related. For example, some buildings can be very vulnerable to earthquake and less to other hazards like floods (Cardona, Ordaz and Marula 2008). Likewise, the loss computation for events like earthquake is problematic, because of the lacking vulnerability information of buildings (objects at risk) or hazard intensity. Due to the uncertainties of this process, loss can be known as a probabilistic distribution in the shaking area. This methodology contains modules that permit building information such as area and prices to be added for object at risk (i), to estimate damage and loss, primarily using the shaking intensity ($I(x_i)$) at the location of the object i within the database. Using a probabilistic approach the loss of the ith object for one simulation is defined by:

$$Loss_i = P(0/1)_i \times fV_i(I(x_i, RND)) \times W_i \quad (1)$$

P(0/1): Probability of an object i to be affected (yes or no) depending on the knowledge
$fV_i()$: Vulnerability function of the considered object depending on the intensity I
$I(x_i, RND)$: Intensity function depending on the location x_i and a random value RND [0-1]
W_i: Value of the object at risk (mainly buildings)

The total loss:

$$TL = \sum Loss_i \quad (2)$$

Figure 1 (The loss assessment flowchart) allows the end users to visualize and understand an integrated rapid disaster assessment framework.

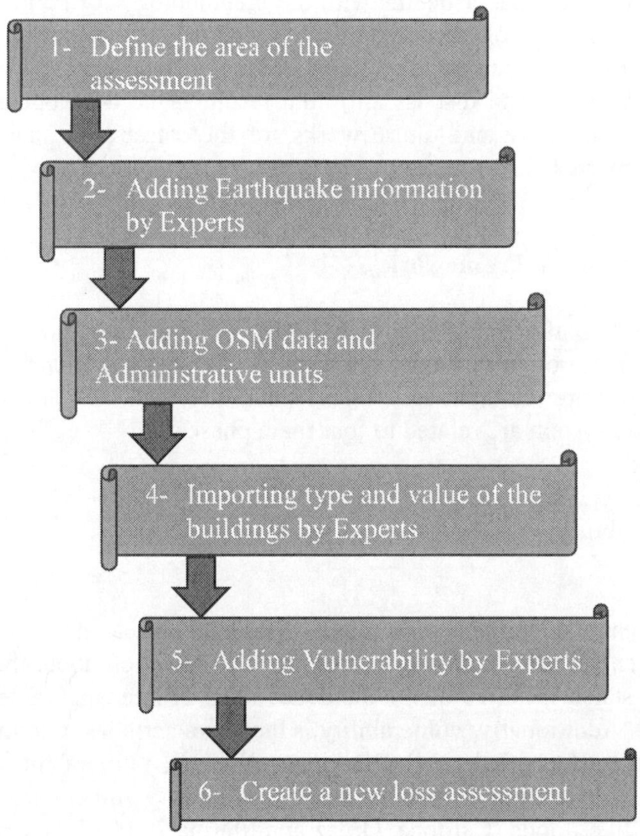

Figure 1: The loss assessment process

3. Implementation

The main features of the platform (Figure 2), containing: 1. Top panel that focuses on user management and documentation. The *user management* module is used for creating, assigning and managing user accounts and permissions. 2. Main panel that is divided to Map view, Layer view, Legend and Data view in a table format. The map view panel is located in the center with tools for zooming, searching locations, styling, drawing and editing features, etc.; and Data view panel is located in the south to visualize feature information about the particular (vector) layer on the map. The main features of the *loss calculation* module are added to the top bar of Map view for adding a new shake map layer as well as for calculating loss.

The GIS system is designed to process multiple events with different magnitudes for different epicenters. Considering the spatial data stored in the Geodatabase component, the system has been structured to obtain earthquake information such as magnitude, epicenter, intensity, vulnerability tables and OSM data. In this way, it is possible to have a preliminary assessment of the damage. For this reason, the vulnerability has been imported in terms of probabilistic percentage or damage function for different buildings.

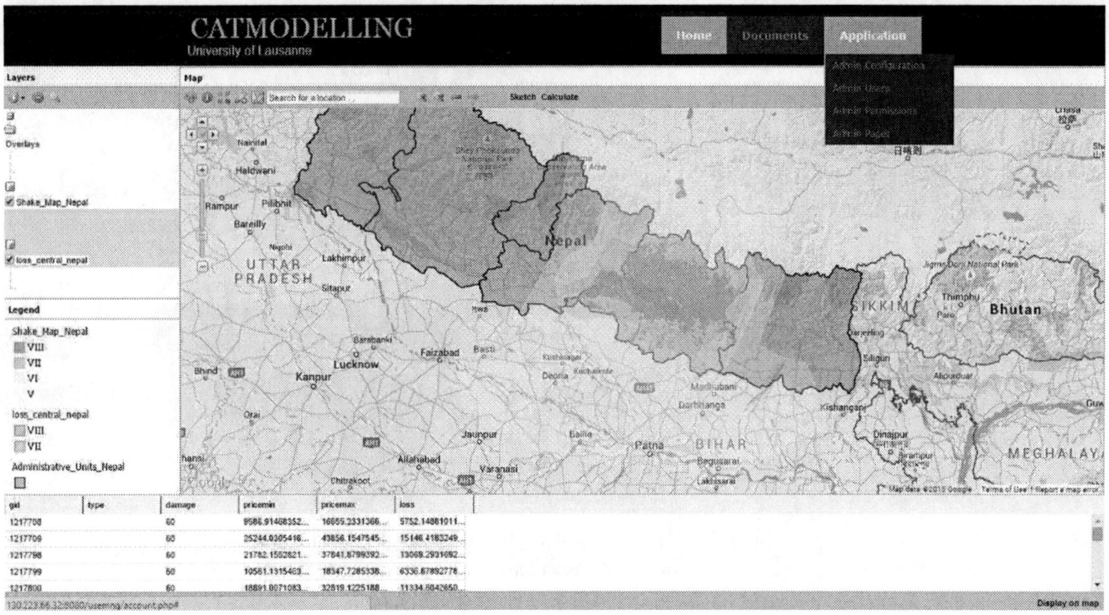

Figure 2: The main view of the platform with admin login

3.1 Architecture

The GeoSpatial analysis and visualization plays a fundamental role in disaster and post-event management. For this purpose, the GeoSpatial technologies have been combined into the architecture as a Geo-Visualization interface. The information and maps are stored and managed within a GeoSpatial database, therefore, it is possible to visualize and request the data by a map viewer. The designed global architect (Figure 3) is implemented by using free open source software (FOSS). The platform consists of the following components:

1. Geo-database (PostgreSQL and PostGIS);
2. Geographic user interface, including maps (Extjs, OpenLayers and GXP)
3. Web-GIS server (Geoserver)
4. User management with SQL (UserCake: http://usercake.com/)
5. Data analysis and process in PHP

The user sends a request login to the Apache web server using the user interface of the web browser. The request passes through the PHP to SQL server and login is succeeding. Afterwards, the requests to the Geoserver are being handled (e.g. Layer query, overlay of different layers, etc.). The data are being analyzed and processed on the web server and sending back to the format into HTML pages. Lastly, the results are displayed in the forms of maps and tables by using OpenLayers.

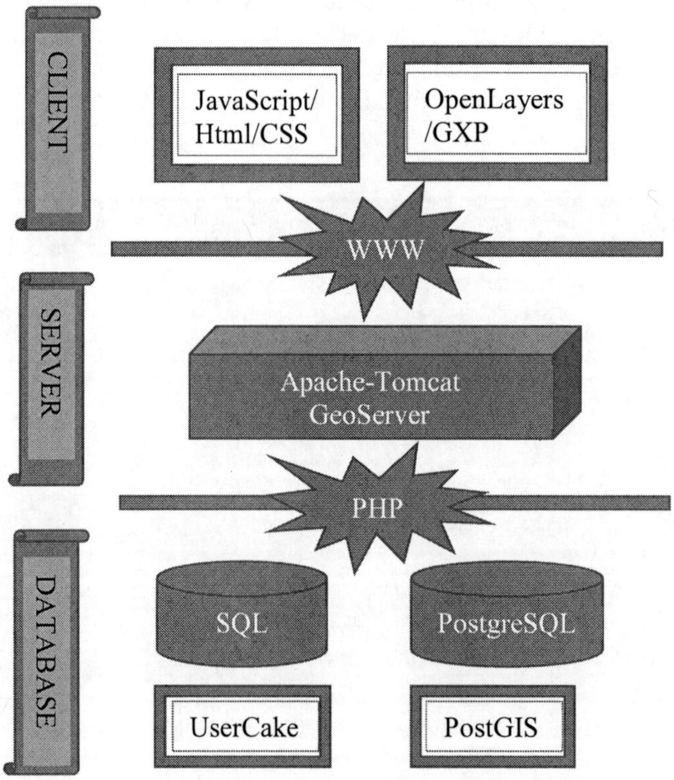

Figure 3: Architecture of the WebGIS platform

3.2 Procedures

The disaster assessment method includes four main phases (Table 1). Once the user enters into the system, the first phase is to fill the information about earthquake, importantly epicenter, depth and Magnitude. Then, the expert will update the intensity of each earthquake by uploading a SHP file or drawing on the map canvas. Following, the OSM data will be uploaded to the system as a SHP file layer. This data can be generated in QGIS and imported as spatial database.

Phase 1: Earthquake data		
Input Data	**Tools**	**Output**
Upload or Draw 1. Shake Map 2. Earthquake information		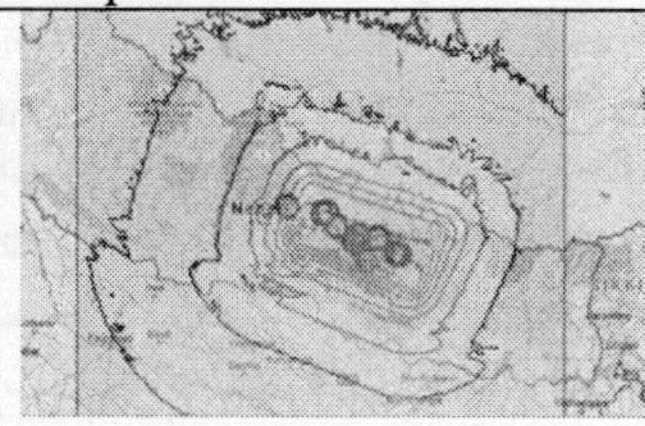
Phase 2: Building Data		
Input Data	**Tools**	**Output**
Upload 1. OSM building 2. Administrative Units	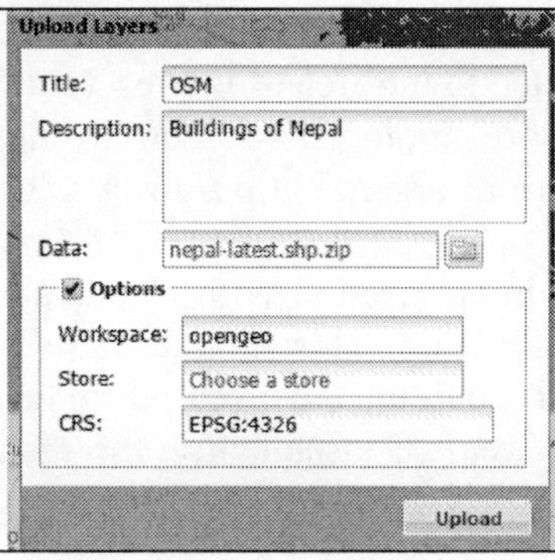	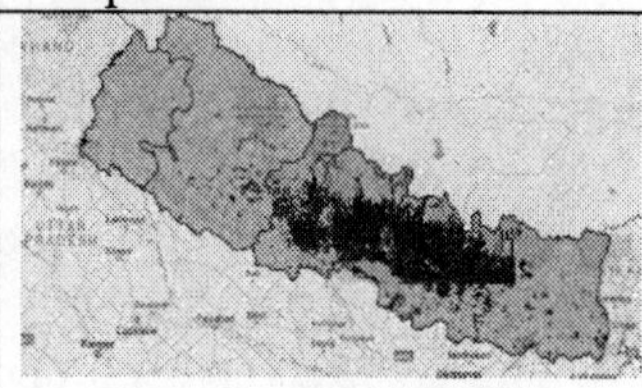
Phase 3: Vulnerability		
Input Data	**Tools**	**Output**
CSV file 1. Value of building 2. Damage probability in the affected area	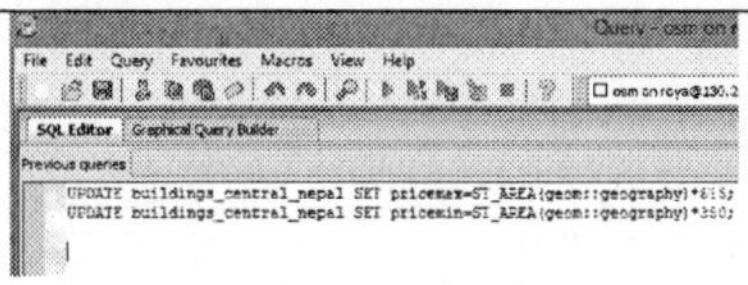 PostGIS	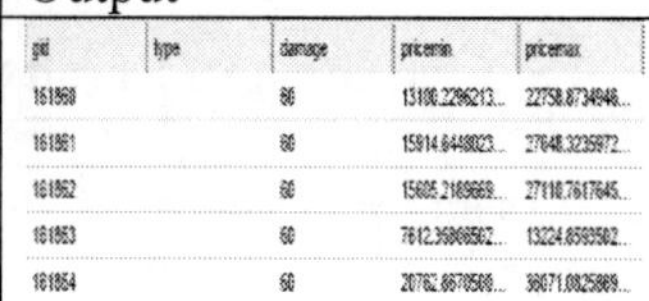
Phase 4: Loss estimation		
Input Data	**Tools**	**Output**
1. Earthquake 2. Buildings 3. Type of damage	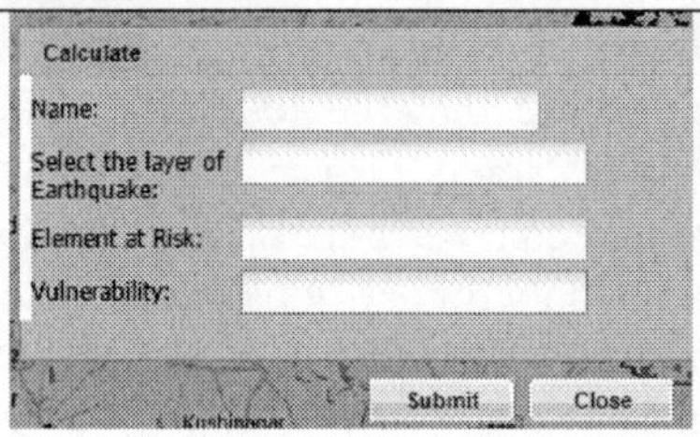	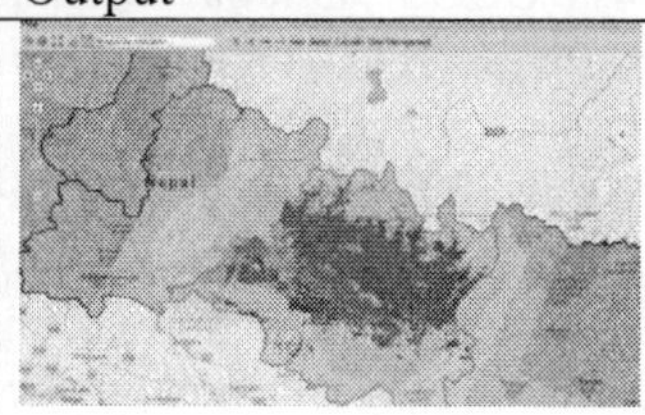

Table 1: The main steps of the loss calculation in the platform

Before moving into the calculation phase, vulnerability information is required to be linked to the building data. This data derive from a table in CSV file mentioning the type of the building and damage probability or defining the probability of the damage area. The value of each building is being calculated and added to the system for the final phase (Area multiply to price per square meter). Consequently, the user can select earthquake layer, building layer and vulnerability type after naming the calculation process. As a result, the loss is computed for each building and in overall. Besides, the computed loss can be shown as a map layer in the system. These procedures are the basic steps in the existing system, however, it is planned to enhance more detailed data and information on how the system can simulate and estimate intensity and vulnerability information for an improved analysis.

3.3 Data Model and GeoServer

Geodatabase is designed to integrate and incorporate Geo-spatial data delivered as an input to the system, including the data linked to Earthquake (e.g., magnitude, epicenter and intensity) and specific data connected to the area of interest (Building information, OSM data, vulnerability tables). The FOSS technologies chosen to develop this component were PostgreSQL/PostGIS (www.postgresql.org and http://postgis.net/).

The GeoServer component, in connection with Geodatabase (PostGIS), is provided to process spatial analysis and visualization. This module delivers a complete and up-to-date description of the different layers like earthquake layer, building layer and, as a result, the maps of expected financial loss in that area. Consequently, the results are stored and visualized through GeoServer WMS and WFS features (http://geoserver.org/)

4. Study Data and Results

As an example, the 2015 Nepal earthquake, which is the largest occurred during the last 50 years in this country, is applied to this study (Earthquake Track 2015). In April 2015, a massive earthquake of 7.8 M happened in Nepal as a result of faulting between the India plate and Eurasia plate. At least 8,702 people were killed, 22,493 injured (UNHCR 2015) 500,717 buildings were destroyed and 269,190 damaged in this earthquake and the M 7.3 aftershock on May (USGS National Earthquake Information Center 2015). Despite Nepal is one of the countries in the world with the lowest economy, the main source of economy is agriculture and tourism. However, reports demonstrate that the loss affected by recent earthquakes could considerably set back the economy of Nepal (Grossi 2015).
Figure 4 displays the location of the recent earthquake and the aftershocks in Nepal (Robertson and Koontz 2015). The shake map of data for earthquake of 7.8 M was imported into the system by drawing the area in a polygon format. This data can also be downloaded from USGS website as different formats such as vector or raster (USGS 2015).

Figure 5 indicates the shaking intensity affected by the Earthquake of 7.8 M (Grossi 2015).

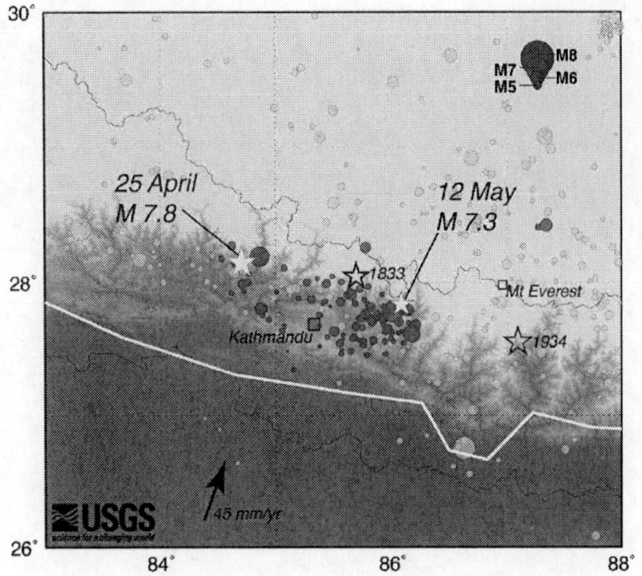

Figure 4 . Magnitude 7.8 Earthquake in Nepal & Aftershocks (from (Robertson and Koontz 2015))

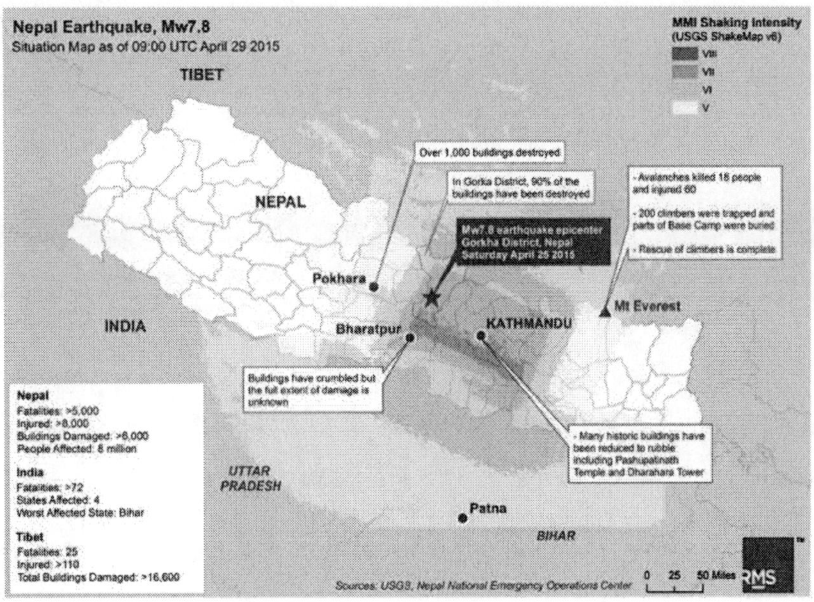

Figure 5: Shaking intensity map of Earthquake 7.8 M (from (Grossi 2015))

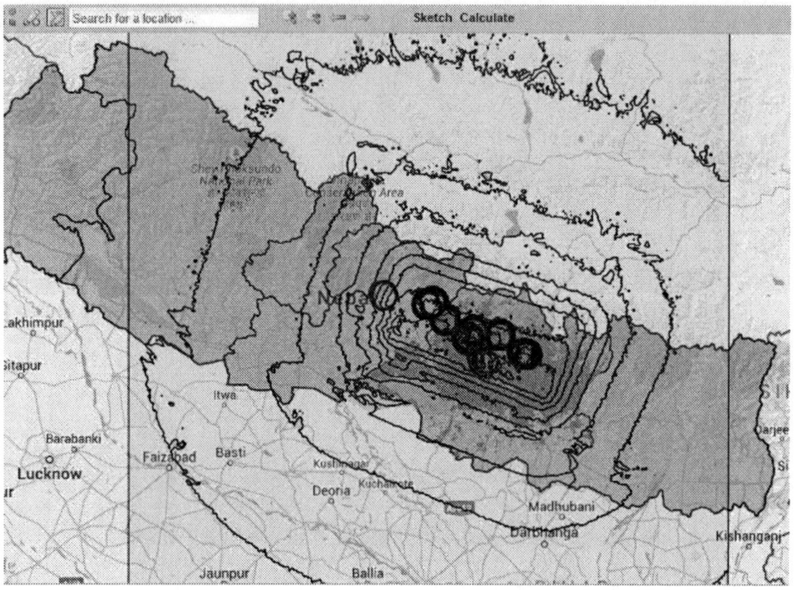

Figure 6: Shake map of Earthquake 7.8 M imported to the system

Figure 7 shows the shake intensity map has been sketched in the system and the building layer of the country has been uploaded as a SHP file to the system (GeoFabrik 2015). The building information is a huge file, hence, we decided to limit the study on a specific administrative unit of Nepal that is affected mostly by the earthquake called as Central. The administrative units' map of Nepal was downloaded as a vector layer (Administrative units: GADM database http://www.gadm.org/).

Figure 7 demonstrates the buildings that are overlaid with the central administrative unit.

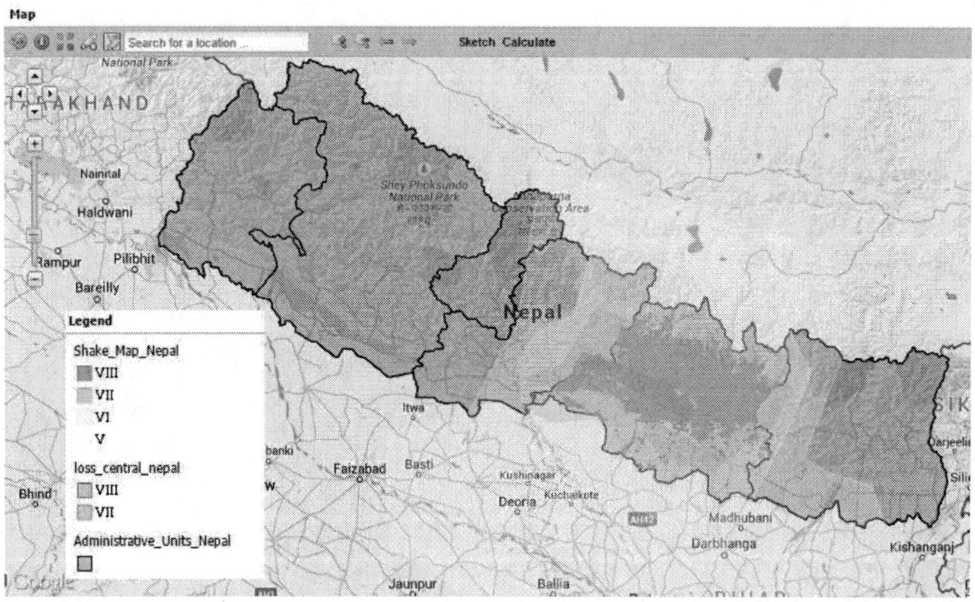

Figure 7: Overlay of buildings with central administrative unit in the map view of the system

After adding all the layer maps, there is a requirement for the update of the building information. The price per square meter in Nepal is estimated between 350 to 615 US dollars (Numbeo 2015). By having the area of each building in the database, the price of each house can be estimated roughly. Finally, the loss of each building and overall loss for the whole region are calculated. Table 2 indicates the total number of houses that are exposed in the event as well as the overall loss. This result is the preliminary result of the system based on open data and expert knowledge. The authors, therefore do not recommend others to use this result. According to the government, the earthquake destroyed 160,786 houses and more than 3 million houses were damaged (UN 2015). The result of this study shows that 3.1 million houses were exposed in the earthquake and 124,000 of the houses were exposed to the higher risk or in the absence of other information, the maximum loss shows 124 000 buildings were destroyed.

Intensity Zones	Number of Houses Exposed in the area	Loss Price: Minimum cost per square meter 350 $	Loss Price: Maximum cost per square meter 615 $
VIII	124 000	2.1	3.6
VII	957 000	12.2	21.3
V and VI	2 000 000	13.3	23.3
Overall	3 100 000	27.6 Billions	48.2 Billions

Table 2: Overall Loss for different intensity layers of the shake map

5. Conclusion

Even though different tools have evolved in this field, their practice is restricted due to the complex design of the systems, flexibility and usability. There are many challenges in developing a WebGIS system. The most important ones are the lack of information and data. In order to utilize GIS and other Geo-Spatial technologies, a variety of spatial data is required. The elements-at-risk information is critical for disaster impact assessment and in this study, OSM data are used. Though this data is not a complete set, it provides a basic and more information can be added based on expert and local knowledge. Besides the estimation of loss by using open data (e.g., OSM, USGS, Numbeo and GADM database) do not involve extensive collection and can be performed fast with a modest budget. The more accurate loss estimation requires an extensive inventory at additional cost to the end user and can be employed in future works. To be mentioned, stories of the house, lifelines like water supply and transportations were not considered in this study and will be applied in further stages of this system. This paper presented the initial implementation and the background framework of a web-based GIS system for rapid impact assessment and demonstrated a preliminary result of the case study area in Central Nepal for the recent earthquake on April 2015. The full functional system is still under development in order to fulfill the lack of information, the user requirements and skilled knowledge of the earthquake experts.

References

Aye, Zar Chi, Michel Jaboyedoff, M. H Derron, C. J van Westen , H. Y Hussin, R. L Ciurean, S Frigerio, and A Pasuto. 2015. "An interactive web-GIS tool for risk analysis: a case study in the Fella River Basin, Italy." *Nat. Hazards Earth Syst. Sci. Discuss* 4007-4057.

Blaikie, P, T Cannon, I Davis, and B Wisner. 1994. *At Risk: Natural Hazards, People's Vulnerability, and Disasters.* London: Routledge.

Cardona, Omar D, Mario G Ordaz, and Mabel C Marula. 2008. "Estimation of Probabilistic Seismic Losses and the Public Economic Resilience—An Approach for a Macroeconomic Impact Evaluation." *Journal of Earthquake Engineering* 60-71.

Earthquake Track. 2015. *Biggest Earthquakes Near Nepal.* Accessed July 2015. http://earthquaketrack.com/p/nepal/biggest.

Esri White Paper. 2008. "Geographic Information Systems Providing the Platform for Comprehensive Emergency Management."

FEMA. 2008. *HAZUS-MH Estimated Annualized Earthquake Losses for the United States .*

GeoFabrik. 2015. *OpenStreetMap Data Extracts.* http://download.geofabrik.de/.

Grossi, Patricia. 2015. "The Nepal Earthquake: What We Know So Far." *The RMS blog.* April 29. Accessed July 15, 2015. http://www.rms.com/blog/2015/04/29/the-nepal-earthquake-what-we-know-so-far/.

InaSAFE. 2015. . http://inasafe.org/en/.

International Federation of Red Cross. 2000. *Disaster Emergency Needs Assessment.* Disaster Preparedness Training Programme.

James, Valentina, and Carmine Pascale. 2012. "Emergency, Open source GIS Tools to Map Earthquake Damage Scenarios and to Support." *The Fourth International Conference on Advanced Geographic Information Systems, Applications, and Services.* GEOProcessing. 152-157.

Lantada, N, L. G Pujades, and A. H d Barbat . 2009. "Vulnerability index and capacity spectrum based methods for urban seismic risk evaluation." *Nat Hazards* 501-524.

Numbeo. 2015. *Cost of Living in Nepal.* July. http://www.numbeo.com/cost-of-living/country_result.jsp?country=Nepal.

Open Quake. 2015. *GLOBAL EARTHQUAKE MODEL.* January 21. Accessed July 15, 2015. http://www.globalquakemodel.org/openquake/about/platform/.

Robertson, Jessica, and Heidi Koontz. 2015. *Magnitude 7.8 Earthquake in Nepal & Aftershocks (USGS).* May. Accessed July 2015. http://www.usgs.gov/blogs/features/usgs_top_story/magnitude-7-8-earthquake-in-nepal/.

UN. 2015. *Nepal earthquake: More than 3,00,000 houses damaged, says UN report.* May. http://www.firstpost.com/world/nepal-earthquake-more-than-300000-houses-damaged-says-un-report-2224632.html.

UNHCR. 2015. *Nepal: 2015 Earthquakes.* June. Accessed July , 2015. http://data.unhcr.org/nepal/.

USGS. 2015. *Earthquake Hazards Program: Shake map.* Accessed July 2015. http://earthquake.usgs.gov/earthquakes/shakemap/global/shake/20002926/.

USGS National Earthquake Information Center. 2015. *USGS.* April. Accessed July 2015. http://earthquake.usgs.gov/earthquakes/eventpage/us20002926#general_summary.

van Westen, C.J, W.H Bakker, V Andrejchenko, K Zhang, J Berlin, I Cristal, and R Olyazadeh. 2014. "User Guide: RiskChanges: a Spatial Decision Support System for analysing changing hydro-meteorological risk." Accessed 2015. http://www.changes-itn.eu/RiskChanges/UserGuide/tabid/159/Default.aspx.

Wisner, Ben, Piers Blaikie, Terry Cannon, and Ian Davis. 2003. *At Risk: natural hazards, people's vulnerability and disasters (Second Edition).* UNDP.

Free and Open Source Software for Geospatial (FOSS4G) Conference Proceedings

Volume 15 *Seoul, South Korea* Article 43

2015

An Open Source Web Service For Registering And Managing Environmental Samples

Anusuriya Devaraju
CSIRO Mineral Resources

Jens Klump
CSIRO Mineral Resources

Pavel Golodoniuc
CSIRO Mineral Resources Flagship

AT5-01

AN OPEN SOURCE WEB SERVICE FOR REGISTERING AND MANAGING ENVIRONMENTAL SAMPLES

Anusuriya Devaraju, Jens Klump and Pavel Golodoniuc

CSIRO Mineral Resources Flagship,
PO Box 1130, Bentley,Western Australia, 6102.
Email: {anusuriya.devaraju, jens.klump, pavel.golodoniuc}@csiro.au

ABSTRACT

Records of environmental samples, such as minerals, soil, rocks, water, air and plants, are distributed across legacy databases, spreadsheets or other proprietary data systems. Sharing and integration of the sample records across the Web requires globally unique identifiers. These identifiers are essential in order to locate samples unambiguously and to manage their associated metadata and data systematically. The International Geo Sample Number (IGSN) is a persistent, globally unique label for identifying environmental samples. IGSN can be resolved to a digital representation of the sample through the Handle system. IGSN names are registered by end-users through allocating agents, which are the institutions acting on behalf of the IGSN registration agency. As an IGSN allocating agent, our goal is to implement a web service based on existing open source tools to streamline the processes of registering IGSNs and for managing and disseminating sample metadata. In this paper, we present our ongoing work onthe design and development of the web service, and its data schema and database model for capturing key aspects of environmental samples. We show how existing controlled vocabularies can be incorporated into the service development to support the metadata registration of different types of samples. The proposed sample registration and curating approach has been trialed in the context of the Capricorn Distal Footprints project on a range of different sample types, varying from water to hard rock samples. The initial results demonstrate the effectiveness of the service while maintaining the flexibility to adapt to various media types, which is critical in the context of a multi-disciplinary project.

1. INTRODUCTION AND MOTIVATION

In the earth science disciplines, physical samples are essential for understanding the complexity of our environment and its resources. For example, direct sampling techniques such as drilling and core sampling are employed to identify mineral and energy resources,insects and plants specimens are collected tounderstand their complex molecular interactions, and scientific ocean drilling programs provide insights into the seafloor and sub-seafloor microbial communities.While huge efforts have been put into collecting physical samples, these usually lie isolated; they may be kept by sample curators, laboratories, state agencies, or museums, and their metadata may existon proprietary data systems or on researchers' personal hard drives.Environmental samples are also diverse, as their collection, curation, and analysis vary from one domain to another. Each curator may follow their own way of documenting the samples. The descriptions may be incomplete, and no common service is available to discover the samples and their descriptions. It is also possible that the sample naming conventions used are only unique locally. According to Ramdeen (2015), "this can lead to complicated hybrid collections [...] that, owing to their complicated curation schemes and lack of standardization may become lost". Inaccessible, irrecoverablesamples are unusable and have little value until they are released to scientists and public access(Geological Society of America (GSA), 2012)(Ramdeen, 2015). When sharingmetadata of samples in a globally distributed environment, the unique identification of samples becomes essential. Persistent identifiers such as Digital Object Identifiers (DOI)[1] have proven successful in enabling users to access a digital resource via a persistent link on the Web. In a similar manner, assigning globally unique identifiers to physical samples will facilitate access to manysamples that are scattered across various repositories. Theseidentifiers can be used to locate samples unambiguously and to gather their associated metadata and data systematically. As a result, this new way of publishing samples and their metadata will create opportunities to re-use existing samples in terms of"new societal issues, environmental concerns, scientific interpretations, and analytical techniques"(Geological Society of America (GSA), 2012).

The International Geo Sample Number (IGSN) is a persistent and unique alphanumeric codefor identifying environmental samples and specimens. A sample's IGSN number can be resolved to a digital representation of the sample through the Handle[2] system. The IGSNsof physical samples are registered by end-users (e.g., individual researchers, data centers and projects)through allocating agents. Allocating agents arethe institutions acting on behalf of the main registration agency (IGSN e.V.)[3]. For further details on IGSN, see Section 2.

The IGSN is based on precursor work at Lamont-Doherty Earth Observatory (LDEO), which was developed as the System for Earth Sample Registration (SESAR)[4]. SESAR was developed with the requirements of individual investigators in geochemical research in mind. There are shortcomings in its technical implementation. Our work is expansion of precursor work in SESAR to suit Australian geo community needs. The Internet of Samples in the Earth Sciences (iSamples)[5] is an EarthCube Research Coordination Network programthat aims toconnect sample collections across the Earth sciences through cyberinfrastructures. One of the key questions addressed by the program is identifying metadata profiles and tools required to facilitate samples discovery and interoperability across domains. Our work complements this issue by developing a metadata schema,which can be used to represent various physical samples, and by implementing relevant tools, e.g., an allocating service and a metadata store.

This paper presents our ongoing work on developing an allocating agent web service and a metadata schema to streamline the processes of registering IGSNs and managing and disseminating sample metadata in CSIRO. The service supports a Representational State Transfer (REST) API for namespace governance, registration of IGSN and associated metadata. The schema defines descriptive metadata that are required for IGSN registration by the allocating agent. We will use the registration of water samples collected as part of the Capricorn Distal Footprints project[6]to demonstrate the solutions developed.The paper is organized as follows: Section 2 provides an overview of IGSN e.V., including organization and governance, syntax, and metadata. Section 3 presents a description of the service and the schema developed. This is followed by their implementation and results in Section 4. Finally, Section 5 concludes the paper with some directions for future work.

[1]http://www.doi.org/
[2]http://www.handle.net/
[3]http://www.igsn.org/
[4]http://www.geosamples.org/mysesar
[5]http://earthcube.org/group/isamples
[6]http://www.sief.org.au/FundingActivities/RP/Distal.html

2. IGSN OVERVIEW

Figure 1 shows the entities that form the IGSN architecture. The IGSN Registration Agency (IGSN e.V.) is theimplementing organization that operates the main registry and resolver services. At this level, the registry only imposes registration metadata (e.g.,the IGSN number and its landing page, the registrant, and the registration date). The Handle system is similarto Domain Name System (DNS), which resolves a URL to an Internet Protocol (IP).An allocating agent is the member institution that is authorized by the IGSN e.V. to register the IGSN in the permitted namespace. Examples of allocating agents are the Interdisciplinary Earth Data Alliance (IEDA), the Commonwealth Scientific and Industrial Research Organisation (CSIRO), and the German Research Centre for Geosciences (GFZ).[7]The difference between the top-level service and the allocating service is that the former only requires basic IGSN *registration metadata*[8], whereas the latter captures samples' *descriptive metadata*.Clients (e.g., individual data centers or sample curators) register samples together with metadatathrough an allocating agent. To enable this, an allocating agentshould providea service that registers identifiers from the top-level registry, keeps a repository of samples registered through the agent, and maintains a metadata portal to make samples searchable and accessible.This paper presents the development of an allocating service and itsdescriptive metadata schema.The metadata portal is not covered in this paper, but it is subject to future investigation. A client also hosts the landing pages of the registered samples. These pages may include more detailed metadata of samples (e.g., a description of the variables, the sample data provider,and links to actual datasets and related documentations).

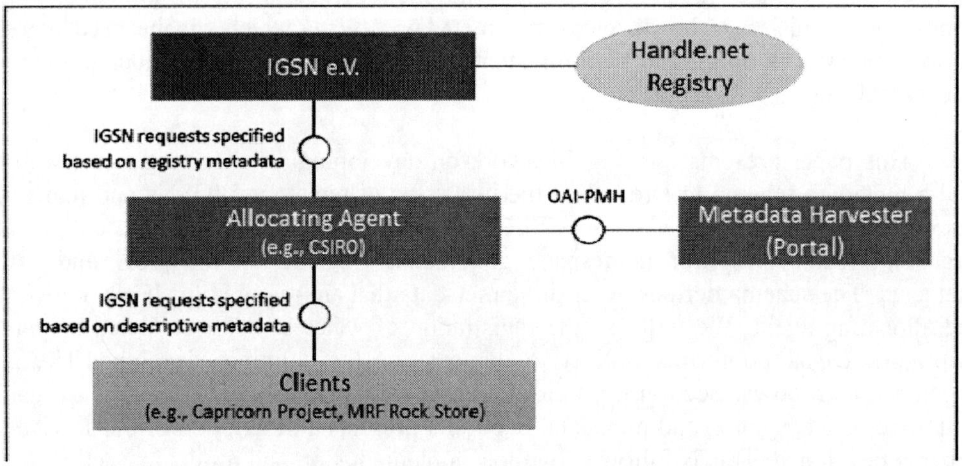

Figure1. Hierarchical architecture of the IGSN registration.

An IGSN must be unique and is case insensitive. It consists of *<namespace><code>* (see Table 1).The allocating agent should be certain that the element *<code>* is unique within its namespace. The recommended format of the IGSN is a nine-character string that comprises a namespace identifier and a sample number. However, it is also stated that the communities may adopt their own IGSN formats if necessary(IGSN e.V., 2015). In our service, we do not restrict the total length of an IGSN, as we see the possibilities of projects

[7]Existing allocating agents : http://dokuwiki.gfz-potsdam.de/datawiki/doku.php?id=igsn:allocating_agents
[8] For further details on the IGSN registry metadata schema, see http://trac.gfz-potsdam.de/igsn/wiki

registering samples with longer numbers.

Table 1. IGSN syntax with examples.

Element	Restriction	Descriptions and Examples
<IGSN>	*<namespace><code>*	An IGSN takes the form of an alphanumeric character divided into two elements: *<namespace><code>*. An example of IGSN is CSCAP00001.
<namespace>	UPPER (A-Z)	A namespace refers to the prefix of an allocating agent (e.g., CS stands for CSIRO).
<code>	Any combination of (A-Z), (a-z), (0-9), hyphen-minus (-), and dot (.)	A code is a combination of the sub-namespace and the sample number specified by the client. For example, 'CAP' stands for Capricorn Distal Footprints project,and '00001' is the local sample number assigned by the client.

3. DEVELOPING AN ALLOCATING SERVICE FOR CSIRO

This section describes the descriptive metadata model and summarizes operations supported by the allocating service.

3.1 Descriptive Metadata

One important concern for building the allocating service is thedescriptive metadata model (schema) and its database model. The schemadefines the essential characteristics of samples. The allocating service leverages the schema to validate and handle sample information sent by a client. The contributions made in terms of the schema development are summarized as follows:

a. Since the potential projects that will use the allocating service involvevarious types of samples, it is important to ensure that the schema can be adapted to respond to the samples' diversity and requirements.Therefore, we have used the records of different types of existing samples and inputs from the domain experts (e.g., sample collectors and curators) as a basis to identify the core elements representing a sample. Figure 2 illustrates a partial view of the descriptive metadata schema developed.Metadata of multiple samples can be defined based on the descriptive schema. The elements in the schema are generic enough to be reused for various samples. They can be grouped into identification, collection, curation, and related resources (for examples, see Table 2).
b. Some of the elements in the metadata schema are specified as mandatory (e.g., *sampleNumber, sampleName, isPublic, landingPage, sampleType* and *sampleCuration*).A client must supply these elements during the sample registration process. They are alsousefulin finding and retrieving physical samples via a metadata harvester (see Figure 1). The curation information is also required, as it indicates the storage and access to the physical samples.
c. A predefined list of values for the element*sampleTypes* and the attribute *featureType*has been specified, based on the ODM2 controlled vocabularies[9], to ensure the consistency and correctness of theirvalues. Here, the principle of Linked Data has been applied; as such,the predefined terms are identified by corresponding URIs. Therefore, more meaningful information about the terms can be determined.
d. We have re-used elements (e.g.,*sampleNumber* and *relatedResourceIdentifier*) and data

[9]http://vocabulary.odm2.org/

types (e.g., *dateType* and *relationType*) from the IGSN registry schema in our descriptive schema. We have extended*dateType*to describe both time instant and interval.

e. The schema also provides options to representdifferent types of location. For example, the location where a sample is collected can be specified as absolute location, bounding box, locality and relative location.

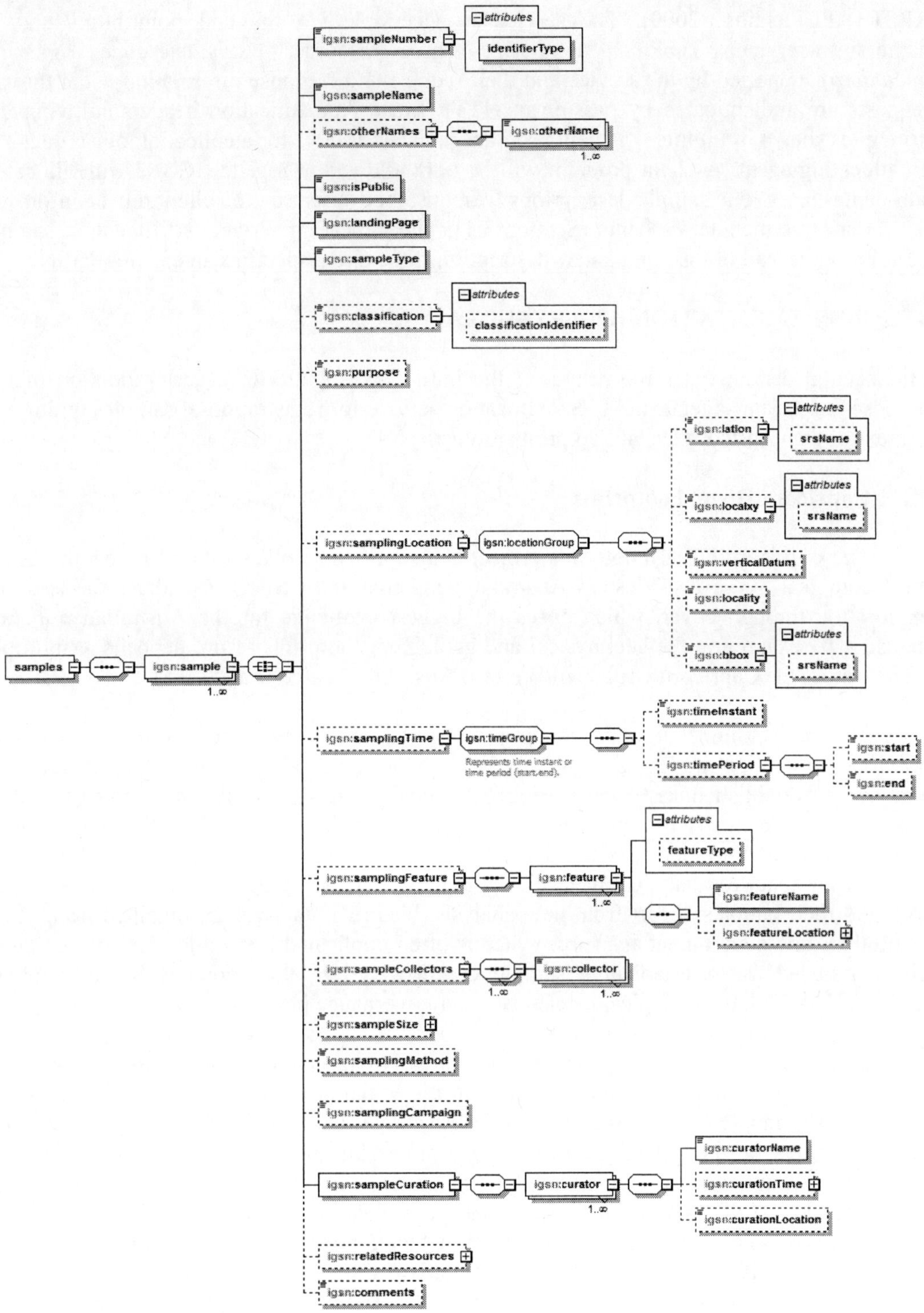

Figure 2. A partial view of the descriptive metadata schema.

3.2 Allocating Service

The allocating service enables a client to register sub-namespaces and multiple samples and retrieve the metadata of a particular sample programmatically. The service implementsa

REST API(Fielding, 2000). The service is accessible via the end point:http://{server-name}/igsn-service/. Table 3 summarizes three resources (*/subnamepsace, /igsn,*and */metadata*) supported by the service and their request and response information. Note that all requests are authenticated by passing in HTTP Basic Authentication headers.Follwing the principles shown in Figure 3, a client program registers IGSNs for a collection of samples via an allocating-agent. A client program will be only allowed to register IGSNs with allocated sub-namespaces. The sample descriptions (response body) sent by the client must conform to the descriptive metadata schema (Section 3.1). The allocating service will update a sample description instead of inserting a new description if the client specifies an existing IGSN.

4. IMPLEMENTATION AND RESULTS

This section demonstrates the results of the initial investigation and early adoption of the proposed metadata schema and IGSN allocation service for registration of samples within the context of the Capricorn Distal Footprints project.

4.1 Capricorn Distal Footprints

The Capricorn Distal Footprints project is a significant collaboration between industry, the Geological Survey of Western Australia, academia and CSIRO to address the issue of exploration through cover, which poses the biggest challenge for the Australian minerals industry, by examining the geophysical and geological footprints of ore deposits at multiple scales across the Capricorn Orogen in Western Australia (Pearce et al., 2015).

The project aims at understanding the metallogenic evolution of the orogen and provides exploration models to aid future discoveries. Regional geophysical interpretation is analyzed to establish links to the hydrogeochemical, regolith, and resistate mineral studies around known deposits to gain insight into the key indicators of mineralisation, not just within a few hundred meters of the deposits but much further afield. The regional study of such a scale requires analytical data of a large number of samples that include water, plant, soil, and rock samples. Apart from new analysis, historical datasets are obtained from other institutions and government agencies, which is often duplicated or hard to identify. All these difficulties led us to establish a sample curation system that would provide a robust framework of both technological mechanism and governance policies.

IGSN system offers a viable and broadly accepted solution for geo-sample identification in various science disciplines. In the initial stage of the project, the proposed approach was adopted for identification of water samples using IGSN. The proposed solution not only offers the persistent identifiers for samples but also offers a robust yet flexible mechanism to associate rich self-descriptive metadata with the sample. The Capricorn Distal Footprints project offers an ideal environment for application of IGSN system. The diversity of sample types requires adoption of multiple metadata profiles to provide ability to capture sample type specific intrinsic properties. The IGSN registration service is currently designed to handle generic metadata schema and the problem of sample type specific properties is yet to be solved.

Table 2. Examples of elements of the descriptive metadata. The rightwards arrow (→) indicates a child element.

	Examples (elements/attributes)	Obligation	Max Occurrence	Data Type
Identification	*sampleNumber*	M	1	CharacterString[a]
	landingPage	M	1	URI
	isPublic	M	1	Boolean
	sampleType	M	1	CV[b]
Collection	samplingFeature → feature : *featureType*(attribute)	O	1	CV[c]
	samplingMethod	O	1	CharacterString
	sampleCollectors → *collector*	O	N	CharacterString
Curation	sampleCuration →curator→ *curatorName*	M	1	CharacterString
Other related information	relatedResources →*relatedResourceIdentifier*	O	N	CharacterString
	comments	O	1	CharacterString

[a]The following XSD restriction applies to a sample number: <xs:pattern value="([A-Z]{2}[A-Za-z0-9\-.]*)"/>
[b] Controlled vocabulary of sample types: http://vocabulary.odm2.org/medium/
[c] Controlled vocabulary of feature types: http://vocabulary.odm2.org/samplingfeaturetype/

Table 3. A summary of the allocating service API.

Request URI	Descriptions	Method	Request Headers	Request Body	Response Headers	Response Body
/igsn-service/**subnamespace**/	This request will register a sub-namespace if the specified sub-namespace does not exist.	POST	Content-Type: text/plain	subnamespace={subnamespace}	-	HTTP status code (201 or non-201 with a short explanation of the status code)
/igsn-service/**subnamespace**/all	This will retrieve all registered sub-namespaces.	GET	Accept: application/json	-	Content-Type: application/json	HTTP status code (200 with JSON representing registered namespaces or non-200 status with explanation)
/igsn-service/**igsn/**	This request will register IGSNs for samples. It will update the sample description instead of inserting a new record if the IGSN exists.	POST	Content-Type: application/xml	UTF-8 encoded descriptive metadata	Content-Type: application/json	HTTP status code (201 with a list of successful and unsuccessful samples or non-201 with a short explanation of the status code)
/igsn-service/**metadata**/{igsn}	This request returns the descriptive metadata associated with a given {igsn}.	GET	Accept: application/xml	-	Content-Type: application/xml	HTTP status code (200 with XML representing a dataset or non-200 status code with explanation)

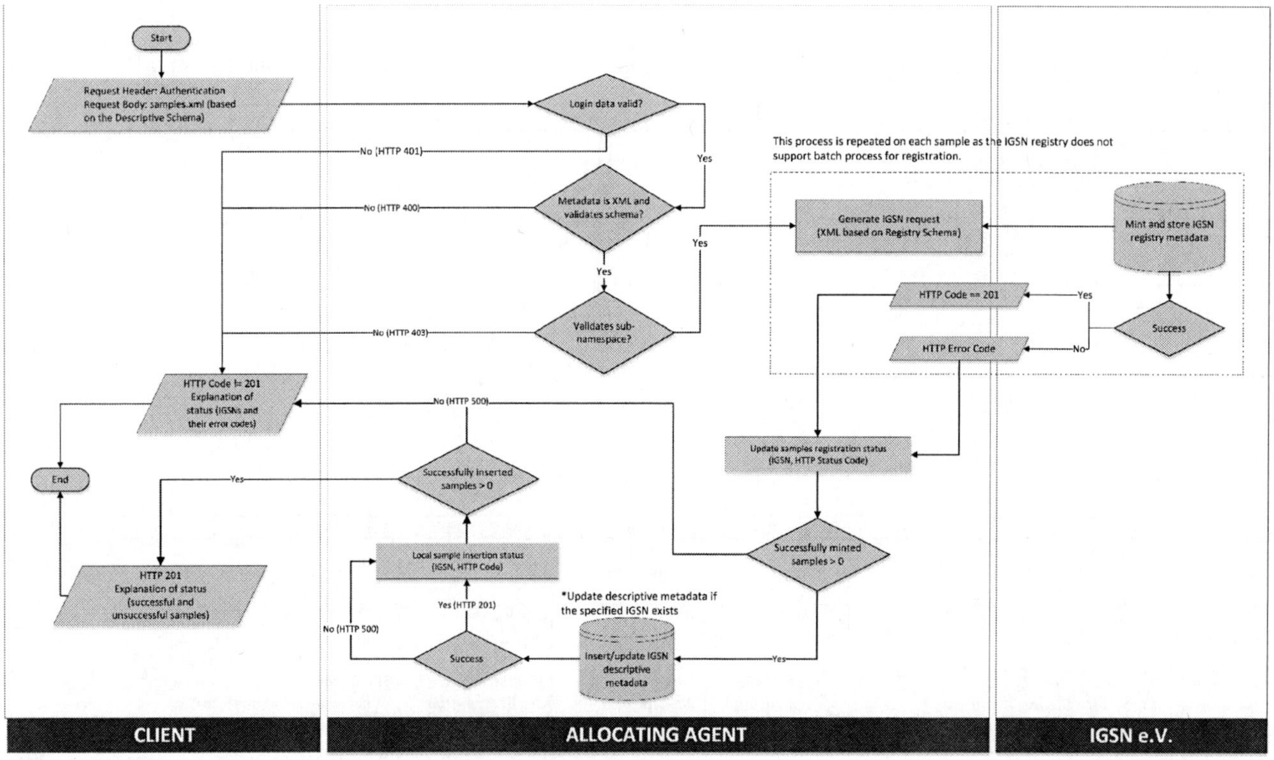

Figure 3. IGSN registration by a client program.

4.2 System Architecture and Results

Figure 4 illustrates the system architecture of the service developed. A relational database model was developed usinga MySQL database to store descriptions of samples and data centers. The allocating service is hostedon the Tomcat application server. All registration requests from the allocating service to the IGSN registry are sent via HTTPS protocol. The project curation system makes requests based on the API on Table 3. A client can also use the service in test mode by settingan optional query parameter, e.g., */igsn?testMode=true*. This means that the service will not register IGSNs from the top-level registry nor modify the descriptive metadata store. Figure 5 shows an example of descriptive metadata sent by the client program to register IGSNs for water samples from the Capricorn Distal Footprints project.TheIGSN registry shows that these samples have been successfully registered.An example of the sample IGSN is "10273/TEST/CSCAP0003", and its actionable link is: http://hdl.handle.net/10273/TEST/CSCAP0003.Note that the samples have been registered with the special test prefix "10273/TEST" via the allocating service. This prefix is reserved by the IGSN registry for testing purposes. We will register actionable IGSNs once the allocating service is publiclyavailable and the Capricorn Footprints project sample curation system is completed.

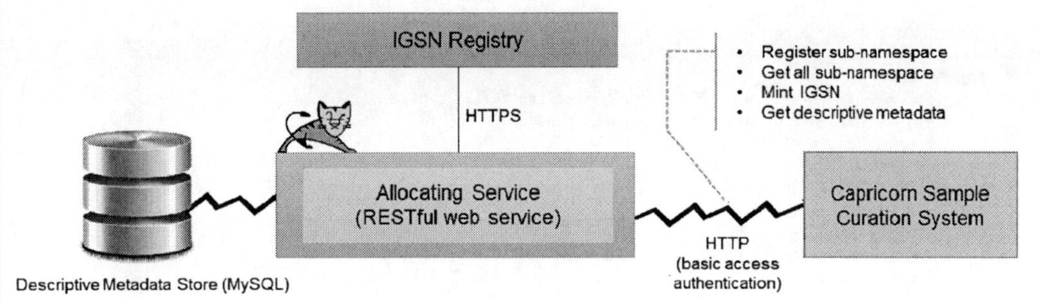

Figure 4. System architecture.

5. CONCLUSIONS

This paper describes the development of a web service for the registration of geological samples by CSIRO at an organizational scale. The solutions developed areuseful to connecting physical samples across the Earth sciences to the Web in a systematic manner. The proposed service and the schema were trialed within the scope of the Capricorn Distal Footprints project.Samples collected during the field sampling campaigns were registered through the CSIRO IGSN allocating service and therefore uniquely identified with global IGSN numbers. The descriptive schema has been developedbased on existing samples and insights from the domain experts. Although the results shown are based on water samples, the schema is extensible and can be applied to various sample types, which is essential in the context of a multi-disciplinary project.The descriptive schema is available through the GitHub repository[10]. While the basic requirements of the service have been achieved, there arestill areas for further development before making the service publically available:

a. The service will be tested with data from other sample stores in the Mineral Resources Flagship (MRF) of CSIRO, e.g. the sample repository of the Australian Resources

[10]https://github.com/kitchenprinzessin3880/csiro-igsn-schema.git

Research Centre (ARRC) and the National Collection for Mineral Reflectance Spectra. The IGSN has also been incorporated in laboratory information management systems at ARRC.

b. The allocating service will be extended to support updates and deletions of sample registrations. Following the IGSN top registry, the delete operation should only marka registered sample as "inactive". A mechanism to activate the sample again will be included. A retrieval of IGSNs and their associated metadata based on a user defined polygon will also be supported.

c. The mapping between the descriptive metadata elements and the concepts specified in existing data standards (e.g., the Observations and Measurements—Sampling Features, ISO 19115 and GeoSciML) should also be documented to ensure the correct application of the schema across different science domains.

d. Currently, the descriptive schema incorporates the ODM2 controlled vocabularies to represent sample and feature types. The schema should be extended to vocabularies describing units, spatial reference system, vertical datum, and classification.

e. We willdevelop a web portal which harvests metadata from repositories of different allocating agents in Australia via the Open Archives Initiative Protocol for Metadata Harvesting (OAI-PMH).

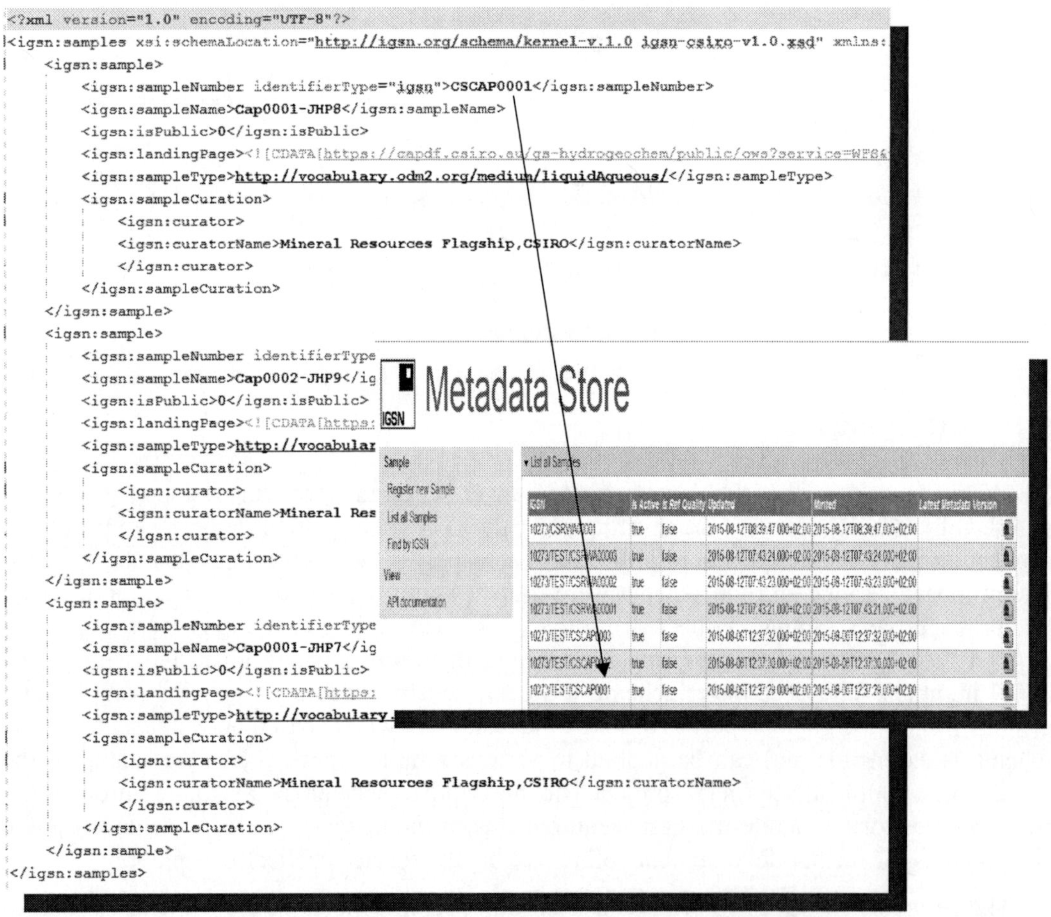

Figure 5. Sample registration request and results.

6. REFERENCES

Fielding, R. T., 2000. Chapter 5: Representational State Transfer (REST), *Architectural Styles and the Design of Network-based Software Architectures*. University of California, Irvine.

Geological Society of America (GSA), 2012. Position Statement - Geoscience Data Preservation. The Geological Society of America. Retrieved from http://www.geosociety.org/positions/position9.htm

Pearce, M. A., Hough, R., Ley, Y., Spinks, S. C., Thorne, R., White, A. J. R., Golodoniuc, P., Gray, D., Munday, T., 2015. The Capricorn Distal Footprints Project: An Orogen-scale approach to mineral systems. Proceedings of the GAC-MAC-CGU-AGU Meeting.

Ramdeen, S., 2015. Preservation challenges for geological data at state geological surveys. *GeoResJ*, 6, 213 – 220. http://doi.org/http://dx.doi.org/10.1016/j.grj.2015.04.002

Free and Open Source Software for Geospatial (FOSS4G) Conference Proceedings

Volume 15 *Seoul, South Korea* Article 45

2015

Advances in Civic Co-management within the Geospatial Ecosystem Applied to Disaster Risk Management

Tomas Holderness
University of Wollongong

Etienne Turpin
SMART Infrastructure Facility, University of Wollongong

Rohan Wickramasuriya
SMART Infrastructure Facility, University of Wollongong

Matthew Berryman
SMART Infrastructure Facility, University of Wollongong

AT6-01

Advances in Civic Co-management within the Geospatial Ecosystem Applied to Disaster Risk Management

Tomas Holderness, Etienne Turpin, Rohan Wickramasuriya, and Matthew Berryman

SMART Infrastructure Facility, University of Wollongong
Northfields Avenue, Wollongong, New South Wales 2522, Australia
Email: tomas@uow.edu.au

ABSTRACT

The use of mobile devices for identifying risk and coordinating disaster response is well accepted and has been proven as a critical element in disaster risk management. As new tools, applications, and software are adopted by municipal governments and NGOs for the identification and management of urban risk, the need for greater integration of the various data they collect becomes acute. While the challenge of integrated data management is substantial, it is aided by the fact that many new tools have been developed to include an Application Programming Interface (API), which allows the machine-to-machine (i.e. automated) sharing of open data. While some proprietary platforms for the management of urban data are currently available, they are extremely costly and very limited in terms of data inputs; to date there are no open source geospatial software tools for the integrated management of various API sources to evaluate hazards for disaster response.

A key to improving disaster risk management as an element of risk identification is the development of an integrated open source Decision-Support Risk Evaluation Matrix that enables: 1) automated integration of multiple geospatial and non-geosapatial API sources into a low cost, user-oriented dashboard; 2) backend database and software design for the Risk Evaluation Matrix that enables data sources to be parameterized and interrogated; 3) the development of an output API stream that allows additional secondary applications to optimize their evaluations and analyses through open access to critical risk information. To address these challenges this paper presents an open source Risk Evaluation Matrix, currently in development, which aims to provide situational oversight of flood hazards from multiple data-sources, including social media, in the city of Jakarta, Indonesia.

1. INTRODUCTION

The use of mobile devices for identifying risk and coordinating disaster response is well accepted (Meier, 2015), and has been proven as a critical element in government disaster risk management. In environments where limited formal geographical data available, government agencies and organizations tasked with responding to natural disasters are increasingly relying on crowd-sourced data from mobile devices to inform decision makers in real-time (Holderness, 2014; Meier, 2015). However, as new tools, applications and software are adopted by municipal governments and non-governmental organizations for the identification and management of urban risk the need for greater integration of the various data they aid in collecting becomes acute. While the challenge of integrated data management is substantial, it is aided by the fact that many new tools have been developed to include an Application Programming Interface (API), which allows the machine-to-machine (i.e. automated) sharing of open data. Such architectures therefore afford the opportunity of creating an integrated information system, which automates requests for real-time information from multiple sources simultaneously, and integrates them to provide comprehensive decision support during disasters.

Jakarta and its surrounding conurbation (Jabodetabek) has the highest rate of urbanisation in the world and comprises the second-largest contiguous settlement on Earth. With a greater metropolitan area hosting 13 rivers, 1100 kilometres of canals, over 28 million residents, and 40% of the city below sea level, the annual monsoon season affects all residents (Jakarta city: 14 million, Jabodetabek: 28 million) and frequently necessitates evacuations of thousands of people (Peters, et al., 2015). For example, the 2007 monsoon season saw flooding cause the death of 80 residents, displace more than 340,000 people and resulted in economic loss in the capital of $453,000,000 (Ward et al., 2011). However, because flooding in Jakarta is mostly a function of infrastructure fragility at a given point in time, prediction and modelling of future flood situations is near-impossible (Turpin et al., 2013). The problem is further compounded by an extreme lack of information available on the hydrological infrastructure, the urban fabric, and population distribution required to undertake risk-based hazard exposure planning. The severity of the information shortage in Jakarta is so critical that the World Bank classifies Jakarta as “Data Starved” (Baker, 2012).

In contrast, Jakarta has one of the highest rates of mobile phone penetration worldwide, contributing to a significant social media footprint. The city alone accounts for 2% of the world's total Twitter activity (Semiocast, 2012). As such the potential for crowd-sourced situational information, from citizen reporters, is significant and proven in the context of improving flood response in Jakarta (Holderness & Turpin, 2015). The combination of unique geography and prevalence of formal and informal data-sources mean that Jakarta is a key case study for the development of improved risk management through new tools and open source software. Open source software software solutions are critical in Jakarta because cost-prohibitive private products are unrealistic under current budget constrains. Risk information and coordination through open data protocols is also critical to support decision-making about disaster response, emergency planning, and community resilience. Furthermore, rich suites of open and accessible geospatial risk data generate activity in NGOs and the private sector, empowering longer term planning tools such as InaSAFE and economic calculators such as JakSAFE. In contrast to systems using formal hazard data, the PetaJakarta.org project successfully demonstrated the utility of social media as a method for crowd-sourcing confirmed flood reports from the public, for rapid situation assessment by the Jakarta Emergency Management Agency (BPBD DKI) (Holderness & Turpin, 2015). Furthermore, PetaJakarta.org enabled two-way communication between Jakarta's citizens and the government, complimenting formal information dissemination process carried out by BPBD. In this context, improving the veracity and integration of informal data, such as those captured by PetaJakarta.org, is paramount to maximizing the potential of information from new media in an operational manner.

This paper presents a prototype open-source Risk Evaluation Matrix, currently in development, which aims to provide a real-time decision support system for flood hazard data in Jakarta. Given the prevalence and importance of geospatial data in disaster risk management, the system is structured using a geographical information systems paradigm where locational information is at the heart of the data structure, and the primary output of information is in the form of a real-time hazard map. This paper details how the proposed GIS is designed to integrate existing formal and informal data sources used for flood hazard management in Jakarta. Following this the paper sets out the technical architecture of the proposed system and a description of how it will be used by the emergency management agency for decision support in response to flooding in Jakarta.

2. JAKARTA'S DISASTER RISK MANAGEMENT ECOSYSTEM

Table 1 shows a summary of data sources available to BPBD DKI Jakarta used for decision support during flood events. Currently, all these information systems operate independently, with human operators corroborating data from the different systems manually. The two principle systems used by BPBD DKI are: (1) the Disaster Information Management System which provides a non-spatial record of reports of flooding, and (2) Flood Gauges which show the movement of water throughout the city's hydrological network. Data from these two systems is analyzed in the BPBD control room, and further verified against secondary sources of information such as CCTV imagery. Analysts at BPBD DKI use this information to manually assess the hazard across the city, which in turn is used to action response depending on the severity of the situation.

Despite significant recent advances in the use of GIS, and sharing of open data within the Jakarta government (Holderness & Turpin 2015), because there is no integrated approach to information systems within the control room, the process of compiling a spatial record of flood affected areas is carried out manually for each of Jakarta's 2,700 sub districts over a six hour period. The spatial records are created using the open source QGIS software to generate a series of static maps; whilst these are useful records of the flood situation at a given point in time, due to the time lag in information processing the outputs are often not used in decision making in the control room, which are frequently required to happen in near real-time. This solution is particularly unsustainable as the Jakarta government strives to improve the spatial resolution of hazard reporting down to the ward level – a ten-fold increase in data volume. The problem will be further compounded as a new generation of data sources such as social media crowd-sourcing come online and feed into the disaster risk management decision making process (Jain et al., 2015; Meier, 2015).

Table 1. Hazard Data Sources Available to BPBD DKI Jakarta.

Data Source	Description	Contents	API
DIMS	Disaster Information Management System	Confirmed reports of flooding (e.g. "911" calls).	Yes
Flood Gauges	Flood Gauge Alerts	Empirically derived risk metric based on upstream river heights	Yes
KoBo Toolbox	Reports from Emergency Service Personal	Ground-truthed hazards and needs assessments	Yes
PetaJakarta.org	Crowd-sourced reports of flooding	Citizen confirmed reports submitted by social media	Yes
OpenStreetMap	Crowd-sourced maps	Urban infrastructure (e.g. roads)	Yes
CCTV	Live imagery of major road infrastructure	Real-time video feeds	No

The PetaJakarta.org project was a proof-of-concept pilot study undertaken in collaboration with BPBD DKI Jakarta and Twitter Inc., to examine the potential and utility of crowd-sourcing flood reports from citizens via social media (Holderness & Turpin, 2015). In particular, the system demonstrated that using a GeoSocial Intelligence approach, that is, requesting confirmed reports directly from users via established communication networks, provided BPBD DKI with the most up to date information on locations of flooding within the city

(Holderness & Turpin, 2015). However, in the critical review of the system the validation and consolidation of the high number of reports from PetaJakarta.org (over 1,000 during five flood events), against existing formal sources of data such as DIMS, continued to be a labor intensive process (Holderness & Turpin, 2015). In short, while the PetaJakarta.org pilot study proved the utility and veracity of crowd-sourcing hazard information from social media, it essentially created a new stream of data, independent from other data streams (see Table 1) which had to be manually processed by BPBD.

In addition to PetaJakarta.org, the Kobo Toolbox is listed in Table 1 as it will be used by the Jakarta government for flood reporting during the 2015-2016 monsoon season for the first time. The Toolbox will be used by BPBD staff to collect pre-flood preparedness assessments, hazard and exposure surveys during flooding, and damage and loss assessments after flood waters have receded. The Kobo platform features an API for data access (Kobo Toolbox, 2015) making it interoperable with an integrated information system. In a related development a new API for the Disaster Information Management System (DIMS) has also been created, to help facilitate more integrated analysis of available data within BPBD. This initiative is also supported by the creation of a unique primary key index for each of Jakarta's 2,700 districts, meaning that non-spatial sources of information (e.g. from the DIMS API) can be joined to spatial records for cartographic visualization and analysis.

Therefore, to build on the success of new data inputs and technologies such as crowd-sourced reports from social media via PetaJakarta.org, and the Kobo Toolbox data, the process of integrating and producing hazard record information presents a significant opportunity for improvement. In this context, leveraging the established use of free and open source geospatial software and open data at BPBD will help maximize transparency and utility of the system for decision support during flooding. Furthermore, as an integrated information management system the Risk Evaluation Matrix will have its own API to facilitate improved sharing of hazard information in real-time, both to the public (e.g. via the BPBD website) and to other partners within the DRM ecosystem in Jakarta (Holderness & Turpin, 2015).

3. GEOSPATIAL ARCHITECTIRES FOR INTEGRATED RISK EVALUATION

3.1 Proposed System

To reap the benefits of accessible, low cost mobile devices and their various applications, emergency management agencies like BPBD DKI Jakarta require an open source Risk Evaluation Matrix for the management of their API-derived data. *CogniCity* is a GeoSocial Intelligence Framework currently used to deliver PetaJakarta.org – a decision support system developed by the SMART Infrastructure Facility with BPBD DKI Jakarta based on citizen reporting of flood events via social media. This project will extend CogniCity to gather data from additional APIs so as to deliver field reports from other sources in a real-time manner to aid BPBD's decision making during flood events. The system will leverage CogniCity's geospatial capabilities to present reports via the existing map interface used by BPBD DKI Jakarta. In this way, CogniCity would enable critical data integration, by extending the information ecosystem to fuse data gathered from citizens via social media and formal reporting via API-derived sources into a single Risk Evaluation Matrix. The resulting decisions will then be recorded by the system for dissemination to secondary users via a robust API, further helping to foster the disaster risk management ecosystem in the region.

The aim of the system is to improve BPBD DKI's ability to identify and confirm the need of government response of flooding across the city of Jakarta. The system will support the human-based decision making process by effectively parallelizing the analysis of multiple data streams amongst BPBD DKI staff; instead of six staff working to corroborate the flood situation from the six data feeds listing in Table 1, one district at a time, six staff could use the Risk Evaluation Matrix to evaluate all the available information for six districts concurrently. This process will be supported by a cartographic data interface, which will enable BPBD DKI verification of reports, to qualify areas of the city flooded and requiring assistance. The resulting verified flood hazard geospatial layer will be shared for communication with the public and other government agencies via a real-time map on the BPBD DKI website.

Figure 1 is a schematic representation of the CogniCity Risk Evaluation Matrix, showing input from the four API-enabled data sources listed in Table 1. Once these data are received by the system they are displayed on an interactive map, which plots hazards from each of the four data sources against their location within the city. Hazard reports are organized by city district, and presented in an accompanying matrix table which lists the details of hazard reports from each data source for each district. The matrix view and map are linked interactively, such that interrogation of hazard reports in the matrix will be reflected in the map, for example through a drill-down operation, commonly used in decision support (Wickramasuriya et al., 2013). The matrix facilitates decision support by allowing a BPBD DKI analyst to use the existing data sources available to gain an on-the-ground situational insight, based on the combination of hazard reports displayed on the map. The analyst is then able to evaluate all data sources, except for CCTV imagery, for each area, in an integrated manner. BPBD DKI can then toggle whether an area under examination is currently exposed to the hazard (i.e. flooding), generating a corresponding record in the system. Records for all areas are then automatically combined into a single spatial dataset at regular, frequent intervals, representing the current flood extents in the city. New, incoming reports from any of the four streams which have not been checked are subsequently highlighted, alerting analysts to the presence of new data within the matrix. Thus, the system helps support the identification of hazards in the city by complimenting the existing socio-technological structure of information management system at BPBD DKI, where critical decision making on hazard exposure is a human-led activity (Holderness & Turpin, 2015). The resulting data are subsequently shared through the open data API in real-time, and are consumable by other information systems such as the InAWARE and InaSAFE situational overview and real-time risk response and planning systems, used within the Indonesian government.

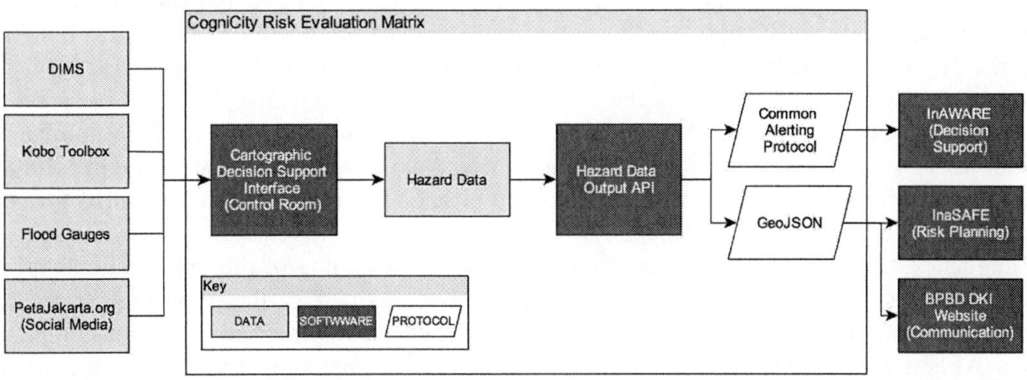

Figure 1. Schematic Showing Open Source CogniCity Risk Evaluation Matrix

3.2 System Architecture

The system is modeled on a web-based geographical information system architecture, comprised of four components. (1) A reports module to gather data from the input API streams. (2) A spatial database to cache incoming hazard reports, and store incidents of flooding as verified by the user interface. (3) A web-server to serve both the data for the risk matrix, and the API of resulting hazard layer. (4) A web-based user interface consisting of the Risk Evaluation Matrix for use by BPBD DKI. The system will be based on the aforementioned CogniCity framework, which provides a scaleable client-server architecture for handling real-time geospatial reports of flooding, powering the crowd-sourced flood map at PetaJakarta.org. For the Risk Evaluation Matrix, CogniCity will be extended to support the additional three API data sources. The reports module, created using the NodeJS platform will periodically poll each of the APIs for data updates, and send this data to the PostgreSQL/PostGIS spatial database for caching. Once data is received by the database, the server module, also written in NodeJS, pushes this data to the client interface within the BPBD DKI control room (Figure 1). The use of NodeJS and the open source PostgreSQL/PostGIS is a proven technology stack in this context, effective at handling high-throughput of information and capable of scaling to meet demand of concurrent users. For example the PetaJakarta.org CogniCity instance deployed during the 2014-2015 was shown to be capable of handling 250 tweets per second, and more than 10,000 concurrent website users (Holderness & Turpin, 2015).

The client interface will also extend the technology developed for PetaJakarta.org, using the LeafetJS library and HTML5 technology to provide a scaleable "slippy" map of the city of Jakarta, as well as embedded data table forming the Risk Evaluation Matrix. Once an area is marked as flooded in the interface, this information is pushed back to the database, whereupon it is available via the system API in the Common Alerting Protocol or GeoJSON formats for consumption, primarily by the InAWARE and InaSAFE software and BPBD website. The Risk Evaluation Matrix is open source software, and while the output hazard data will be freely accessible under an open license, due to sensitivity of incoming reports and potential malicious use, access to the operational system itself will be restricted to the BPBD control room. It is envisaged, however that by making the software open source and available online other organizations, such as other regional governments in Indonesia, could implement the system as part of their own disaster risk management systems.

4. CONCLUSIONS

This paper presents an open source system, currently in development, which has the potential to significantly improve information management and resulting response to flooding in the city of Jakarta, Indonesia. This is an important evolution of the disaster risk management ecosystem in the region as the number of new and innovative hazard data, such as crowd-sourced information, comes online. This approach is applicable across a number of Southeast Asian coastal megacities, many of which will bear the brunt of future climate change and resulting extreme weather events. Further, it is important to recognize that the system has been designed in conjunction with, and based on the requirements of, the Jakarta Emergency Management Agency, using a detailed understanding of the organization's information flows and institutional ethnography carried out as part of the preceding PetaJakarta.org project. However, through the use of open source software and standards-compliant data APIs it is envisaged that the system will be readably transferable to other cities, hazards, and languages in the region.

8. REFERENCES

Baker, J. L., 2012. (editor), *Climate change, disaster risk, and the urban poor: cities building resilience for a changing world.* The World Bank, Washington, D.C.

Holderness, T., 2014. Geosocial Intelligence. *Technology and Society Magazine*, IEEE, 33, 17-18.

Holderness, T., Turpin, E., 2015. *White Paper – PetaJakarta.org: Assessing the Role of Social Media for Civic Co-Management During Monsoon Flooding in Jakarta, Indonesia.* University of Wollongong, Wollongong.

Jain, A., Adebayo, J., de Leon, E., Li, W.; Kagal, L., Meier, P., and Castillo, C., 2015. Mobile Application Development for Crisis Data. *Procedia Engineering*, 107, 255-262.

Kobo Toolbox, 2015. *Kobo Toolbox | Data Collection Tools for Challenging Enironments.* Available online at http://www.kobotoolbox.org/. Accessed 24th July 2015.

Meier, P., 2015. *Digital Humanitarians: How Big Data Is Changing the Face of Humanitarian Response.* CRC Press, Boca Raton.

Peters, G., Butsch, C., Krachten, F., Kraas, F., Namperumal, S. & Marfai, M. A., 2015. Analyzing Risk and Disaster in Megaurban Systems-Experiences from Mumbai and Jakarta. *Planet@ Risk*, 3.

Semiocast, 2012. *Twitter reaches half a billion accounts.* Available online at http://semiocast.com/en/publications/2012_07_30_Twitter_reaches_half_a_billion_accounts_140m_in_the_US. Accessed 20 April 2015.

Turpin, E., Bobbette, A., Miller, M., 2013. (editors) *Jakarta: Architecture + Adaptation. Jakarta.* Universitas Indonesia Press, Jakarta.

Ward, P. J., Marfai, M. A., Yulianto, F., Hizbaron, D. R., Aerts, J. C. J. H., 2011. Coastal inundation and damage exposure estimation: a case study for Jakarta. Natural Hazards 56(3) 899- 916.

Wickramasuriya, R., Ma, J., Berryman, M., Perez, P., 2013. Using geospatial business intelligence to support regional infrastructure governance. *Knowledge-Based Systems*, 53, 80-89.

Free and Open Source Software for Geospatial (FOSS4G) Conference Proceedings

Volume 15 *Seoul, South Korea* Article 46

2015

Flood Hazard Mapping of Lower Indus Basin Using Multi-Criteria Analysis

Saba Zehra
University of Karachi, Pakistan

Sheeba Afsar
University of Karachi, Pakistan

AT6-02

Flood Hazard Mapping of Lower Indus Basin Using Multi-Criteria Analysis

Saba Zehra*, Sheeba Afsar

University of Karachi, Pakistan

Corresponding Author : Saba Zehra (ssabazehra@gmail.com)

Abstract

The Super Flood, which hit Sindh in 2010, has turned out to be a wakeup call and has underlined the overwhelming challenge of natural calamities, as 2010 flood and the preceding flood in 2011 caused a huge loss to life, property and land use. These floods resulted in disruption of power, telecommunication, and water utilities in many districts of Pakistan, including 22 districts of Sindh (Memon, 2012). These floods call for risk assessment and hazard mapping of Lower Indus Basin flowing in the Sindh Province as such areas were also inundated in 2010 flood, which were not flooded in the past in this manner. This primary focus of this paper is the use of Multi-criteria Evaluation (MCE) methods in integration with the Geographical Information System (GIS) for the analysis of areas prone to flood (Papaioannou et. al. 2015). This paper makes use of an open source geographic information system SAGA GIS (System for Automated Geoscienific Analyses) for carrying out MCE (Pickup, 2008). This research demonstrated how SAGA GIS tools can be used to produce map of flood vulnerable areas using MCE techniques. Slope, Aspect, Curvature, Soil, and Distance from Drainage, Land use, Precipitation, Flow Direction, and Flow Accumulation are taken as the causative factors for flooding in Lower Indus Basin. Analytical Hierarchy Process-AHP was used for the calculation of weights of all these factors. Finally, a flood hazard Map of Lower Indus Basin was generated which delineates the flood prone areas in the Sindh province along Indus River Basin that could be inundated by potential flooding in future. It is aimed that flood hazard mapping and risk assessment using open source geographic information system can serve as a handy tool for the development of land-use strategies so as to decrease the impact from flooding.

Academic Discipline and Sub-Disciplines : Flood modeling; Open source

Keywords : Geospatial Modeling; Flood hazard mapping

Free and Open Source Software for Geospatial (FOSS4G) Conference Proceedings

Volume 15 *Seoul, South Korea* Article 47

2015

Dynamic Styling For Thematic Mapping

Simon Moncrieff
Department of Spatial Sciences, Curtin University

Elizabeth-Kate Gulland
Cooperative Research Centre for Spatial Information (CRCSI)

AT2-04

DYNAMIC STYLING FOR THEMATIC MAPPING

Simon Moncrieff[1] and Elizabeth-Kate Gulland[2]

Department of Spatial Sciences, Curtin University
Kent Street, Bentley WA 6102, Australia
and Cooperative Research Centre for Spatial Information (CRCSI)
[1]Email: s.moncrieff@curtin.edu.au
[2]Email: e.gulland@curtin.edu.au

ABSTRACT

Current web standards have facilitated the online production and publication of thematic maps as a useful aid to interpretation of spatial data and decision making. Patterns within the raw data can be highlighted with careful styling choices, which can be defined for online maps using tools such as Styled Layer Descriptor (SLD) XML schema. Dynamic generation of maps and map styles extends their use beyond static publication and into exploration of data which may require multiple styles and visualisations for the same set of data.

This paper explores the application of thematic styling options to online data, including mapping services such as Open Geospatial Consortium (OGC)-compliant Web Mapping and Web Feature Services. In order to be relevant for both user-specified and automated styling, a prototype online service was developed to explore the generation of styling schema when given data records plus the required output data type and styling parameters. Style choices were applied on-the-fly and to inform the styling characteristics of non-spatial visualisations.

A stand-alone web service to produce styling definitions requires a mechanism, such as a RESTful interface, to specify its own capabilities, accept style parameters, and produce schema. The experiments in this paper are an investigation into the requirements and possibilities for such a system. Styles were applied using point and polygon feature data as well as spatially-contextual records (for example, data that includes postal codes or suburb names but no geographical feature definitions). Functionality was demonstrated by accessing it from an online geovisualisation and analysis system. This exploration was carried out as a proof of concept for generation of a map styling web service that could be used to implement automated or manual design choices.

1 INTRODUCTION

In information visualisation, graphical representations are used to explore data and look for meaningful patterns (Mazza, 2009). Thematic maps such as choropleths are a popular visualisation of spatial data (Rautenbach *et al.*, 2012; Yau, 2011). Whilst choropleths, which represent an attribute value spatially by shading regions to show the value's relative intensity, are often published in (potentially static) reports to support decision-making, the interactivity enabled by a slippy map client presents an opportunity to extend thematic mapping to a more dynamic representation of spatial information. In both static and dynamic visualisations, choices made about how to present information visually - the styling - influence the effectiveness of the graphic for communicating information.

This paper explores dynamic styling of data to produce thematic maps as images. Styling parameters are processed on-the-fly, a process that is suitable for online spatial visualisation and analysis systems, as seen for example in Moncrieff *et al.* (2014) and Moore *et al.* (2013), as well as online mapping applications. Different visualisation styles are tested over data records in order to explore possible capabilities to build into an independent dynamic styling agent.

Thematic styling affects interpretation by specifying how to display spatial data, including features such as colour, border style and rules for different map scales. Standard formats such as the Open Geospatial Consortium (OGC) Styled Layer Descriptor (SLD)

(OGC, 2007), part of the OGC Web Mapping Service (WMS) standard, can specify symbols according to attribute values to produce thematic maps from different data sources. A variety of open source desktop applications and servers, including QGIS, GeoServer[1] and MapServer, implement SLD or an alternative styling schema format. Mapping services that comply with standards such as WMS or OGC Web Feature Services (WFS), whilst affording communication across processes and machines, have a limited ability to automatically adapt styling to best represent their contents. For example, GeoServer combines an SLD file with a WMS layer's data to produce and present a map image using pre-set styles, or it can provide WFS layer data to be displayed in flexibly by processing styling instructions on a client machine. Some orchestrated services use a script or service to produce more dynamic style definitions but these are specialised instructions designed for specific circumstances (Rautenbach *et al.*, 2012). Difficulties faced in the development of a more general dynamic styler include the wide variety of choices and the at times complex relationship between style and data. In this paper, choices that would be useful for a styling service - including possibilities for combining dynamic and server-side styling - are investigated from the perspective of an online spatial visual analytics engine.

In the proposed method, valid style instructions can be produced for a variety of spatial data and visualisations. A styling request must specify the data to be mapped as lists or vectors of text-based records and can also specify styling options including colour schema and the method for classifying data features. The fundamental styling was then extended to explore additional methods for styling and presenting information and data visually. This caters for experimentation with different styles over a set of data in order to develop a more informative map. It also allows styling decisions to be made separately from the specific data records or processes they are being applied to. Styling information is used to produce visualisation images in a similar manner to a WMS layer, but with more flexibility for styling options and no need for manual specification of complex styling schema syntax. Where a style parameter is not specified, default values can be used depending upon other input parameters. For example, a diverging colour scheme such as red-to-blue could be used as a default for a data style to highlight values above and below a pivot value (such as mean temperature).

In this paper, section 2 discusses map styling choices and previous research into styling in automatic mapping systems. In section 3, the design of the new styling service is described along with a case study implementation. Results from testing the case study are discussed in section 4, and final conclusions are outlined in section 5.

2 BACKGROUND

The way in which data is presented has a profound effect on the way it is interpreted. Although this paper focusses on styling possibilities for digital thematic maps, display options are equally important in printed maps and non-spatial data visualisations.

Many elements are needed to create a meaningful map. In a thematic map, design decisions about the type of map, such as choropleth or point symbols; classification method and number of classes; and colour scheme influence perceptions about its message and purpose. Mapping applications make it easier to create maps but do not offer guidance on choices such as colour scheme, which can result in confusing maps (Harrower and Brewer, 2003). Design issues such as these affect automated map production as much as manually created maps. Possibilities for an automated system include predefining styles for each (or

[1] GeoServer http://geoserver.org/about/

all) datasets on offer or applying manual or calculated style options to the map data as required.

2.1 Styles in Mapping Systems

Thematic styling definitions can be used directly to transform data to meet specific interpretation needs. Take, for example, *rendering transformation* functions in GeoServer. A function can be embedded in a SLD file so that raw spatial data is transformed and then styled to give a meaningful, visual result. An example is gs:Contour, which can restyle a raster image as a contour map with specified contour line values.

Thematic styling files have been applied to multi-part systems that coordinate several spatial services. One example is a multipublishing environment (Kettunen *et al.*, 2012), which uses style rules to cater for different types of display devices, such as touch screen, mobile phone or print map. However, as with many online systems, the SLD files used are predefined and static.

Another application of thematic styling is on mobile mapping devices. In Weiser and Zipf (2007), a tool was developed to produce SLD for an OGC Web Map Service (WMS) compliant output. Styling information was based on their original maps, produced on proprietary desktop software (Esri ArcMap). A user can change styling options via the tool and the resulting SLD information is saved as an external file. This file could then be accessed by a WMS client or used in a local application on a mobile device.

The importance of easily-interpretable visuals is paramount in spatial decision support systems (SDSS). People making high-level decisions about issues, such as service accessibility, need access to images that correctly highlight the current state of a region. Visualisation decisions such as using a network distance rather than a straight-line distance can make an enormous difference to the practicality of results. Burdziej (2012) incorporated dynamic SLD production into an SDSS for accessibility information, producing an SLD on demand by a PHP script. This enhanced the system by creating maps suited to the users and avoided the necessity of pre-storing an unreasonably large number of SLD files to allow for the wide range of options.

Iosifescu-Enescu *et al.* (2010) developed a 'Map and Diagram Service' that extended WMS and SLD standards and incorporated style information to produce a map image on-the-fly via a 'getMap' operation. It also defined a 'getStyle' operation to retrieve styling information about a layer in a map visualisation. They found that styling formats in existing OGC standards were suited for topographic maps, but less appropriate for thematic maps. They added a new 'Diagram Symbolizer' to the Symbol Encoding part of the SLD standard to allow for symbols such as pie charts and classification schemes like 'quantile'. The service was tested within an SDSS based on sensor readings. The same team also showed that styling information to be sent to a mapping service could be updated by users via graphic user interfaces (Iosifescu *et al.*, 2013).

3 APPROACH

A wide variety of styling options was trialled to test the visualisation possibilities for a dynamic styling service. In particular, options for attributes at point locations or within regions, including classification schemes and buffer overlays, were examined.

A number of factors need to be considered in order to generate style definitions in a flexible manner. A styling web service needs to respond to a number of different input types for spatial and other data content. It also must be able to adapt to different ways of interpreting the input data vectors, including a variety of spatial analysis algorithms. Finally, it must be able to return output and styles in a specified format, such as SLD, WFS and/or WMS. For usability of the experimental styling service, it is also important that the features it supports - including available styling methods and output formats - are discoverable. This could be achieved through parameters to a RESTful interface, such as a GetCapabilities request, as defined in the OGC Web Processing Service (WPS) standard for returning the service's capabilities.

3.1 Architecture

A proof of concept web service was developed using the Django[2] web framework, which is suited to generating the output of such a web service (see Figure 1). The web service employs established map classification techniques, implemented through APIs provided by the PySAL[3] spatial analysis python library. It applies thematic colour schemes identified by ColorBrewer[4] and derived using Matplotlib[5] colour schemes to encapsulate a higher number of colour gradients within each colour space. As a proof of concept, the style descriptors were encapsulated within Mapnik[6], although the principles in a dynamic styler design can be applied equally to alternative style schema formats such as SLD.

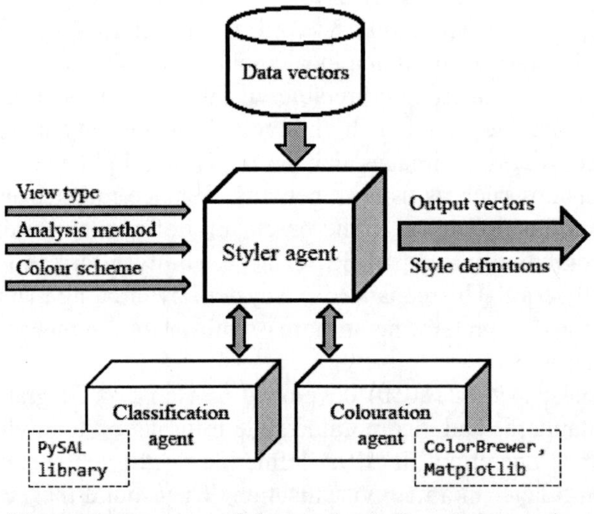

Figure 1: Process of dynamic styling web service

3.2 System inputs

System inputs include the data to style, either using a text based format or a link to the data to be styled (for example, a WFS dataset), along with any required styling options. A final parameter for the service is the output format required, for example an image or a WFS incorporating a colour attribute. In the case of an image output, functionality enabling a style descriptor input was also included. Separate agents were designed to interpret map classification analysis algorithms and colour schemes to be called upon from the styler.

---

[2] Django https://www.djangoproject.com/
[3] PySAL Python library https://pysal.readthedocs.org/en/latest/
[4] ColorBrewer http://colorbrewer2.org/
[5] Matplotlib http://matplotlib.org/
[6] Mapnik http://mapnik.org/

3.2.1 Data inputs

The styler accepts geometry vectors of point or polygon features with attributes, or a list of records, each containing an attribute list of key-value pairs. While styling parameters can be specified through a web service REST API, incorporating meta-data directly into the input data was also explored.

3.2.2 Style parameters

Up to three styling parameters can be specified to determine the generation of the thematic map: 1) analysis method; 2) view type; and 3) colour scheme. Default values are assigned for style choices, depending upon the type of input vectors.

The *analysis method* available depends upon the input data vector. First, the attribute(s) to map must be specified. In implementing the polygon region mapping algorithm, only univariate data is considered with respect to styling. However, point data can be visualised in multiple dimensions, as discussed in section 3.3. Other options include size of feature buffers (if any) and the method for classifying features. Once data classes have been assigned, the colour scheme chosen is also important for conveying meaning.

The *view type* relates to symbolisation changes as the map window changes, such as zooming in or out. The options are global, local, and static. Global applies a style across the entire dataset; if the map view is focussed on a small subregion, it will still use the global values to define the range of the colour bins used. In a local view, symbols will be adjusted to the range of values in currently visible features within a map extent, or within a given bounding box search area. Static mode is an extension of local; a local view will be applied to an initial map extent and this styling can then be maintained during any subsequent changes to the extent, for example if zoom or pan actions are used within a slippy map. This method is appropriate for analysis of regions such as a metropolitan area in which a large number of geographic regions may be defined, for example city blocks. See Figures 2 and 3 for a comparison between local and global styling methods.

Colour scheme can be selected from available schemes as described in Section 3.1.

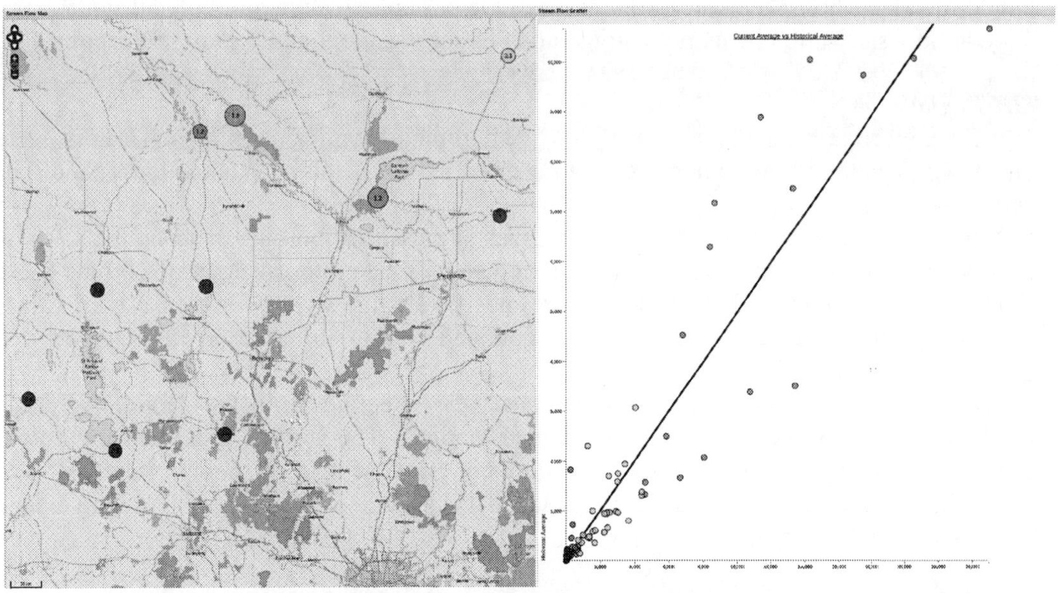

Figure 2: Legend Extent: Map classification by global extent (entire dataset).

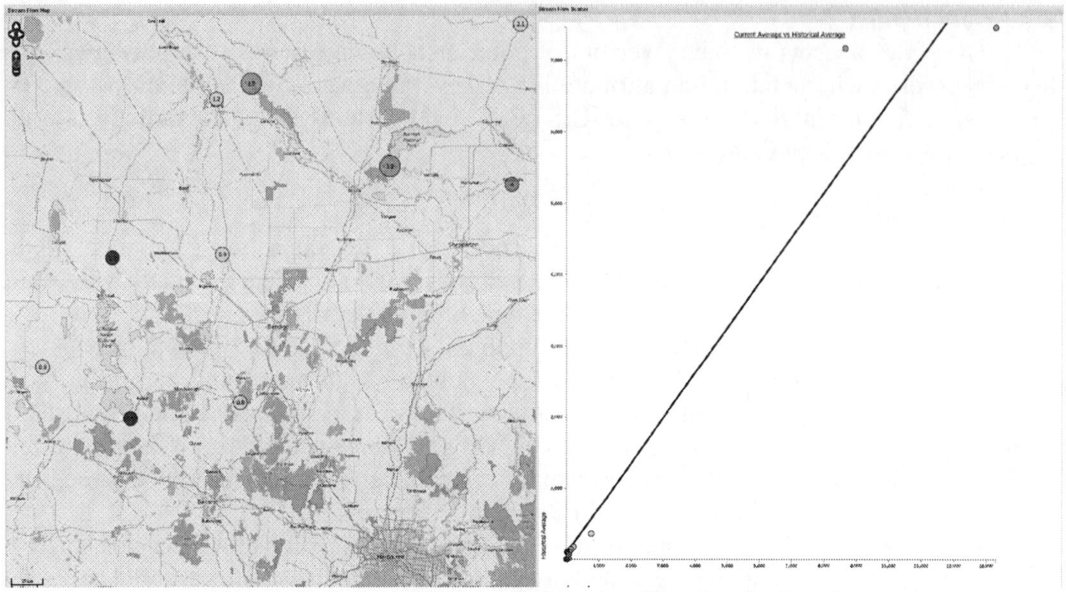

Figure 3: Legend Extent: Map classification by local extent (dataset represented in map extent).

3.3 Thematic Mapping

To visually represent thematic data, two types of information were considered: spatial data, and spatially contextual data. The latter is data that has a spatial attribute imposed on the data, such as population counts by postcode. In this example, a postcode can be linked to a spatial region for calculation and display of population density. The styler can produce styles for region data, point data and non-spatial data. It is possible to show more than a single dimension in a thematic map; for regions, represented by polygons, this can be achieved using a bivariate choropleth and for point data, it is possible to show multiple attributes for a single point visualisation by manipulating variables such as the x and y radii of the point.

3.3.1 Regions

To convert a data input vector into a choropleth map, a number of factors were considered including: how should the data be classified for display? What colours should be used?

Inputs to the data styler include the number of classification bins to separate data into, and the map classification scheme technique to use. For example, an input of 10 bins and a quantile scheme would produce deciles. More specifically, the styler produces a set of values representing the decile class for each polygon and styling information defining how to display it as a choropleth map.

A number of PySAL map classifications were integrated into the service including Jenks natural breaks, quantiles, and standard deviations. The service uses thematic colour schemes identified by ColorBrewer, and derived using Matplotlib colour schemes. The PySAL library includes a wide range of spatial analysis methods which can be used to extend the capabilities of the styler in the future.

3.3.2 Points

Three methods of styling point data were examined: a thematic visual representation, a contextual representation, and a relative size point representation. A number of parameters can be used to alter the presentation of point data. For example, the attribute(s) used to determine the radius (both in the x and y directions) can be specified along with the presence, or absence, of a label and the content of the label, and the attribute used to determine the fill colour of the point. For the first two methods the point size is determined using the minimum and maximum values of the attribute feature space to scale the radius of each point from a minimum set radius to a maximum radius. The visual representation emphasises the attribute over which the thematic styling was determine, by default, the radius is determined using the same attribute (x=y). The contextual representation uses multi-variate data to incorporate further context into the thematic map, in this case, a label and the radius (x, y) are included, and can correspond to different variables. The relative size point representation incorporates a radius determined in metres, providing a spatial context.

Once the colour space and filter bins have been determined, layer styling options that can be applied to the thematic visualisation include the colour of the boundary of the polygons, and the layer alpha. To define styles for point data, PySAL, ColorBrewer and Matplotlib are used to select appropriate colour schemes in a similar manner to that described for polygon data. Circle sizes of individual point symbols can also be scaled according to the range (minimum to maximum) of the attribute of interest in the input dataset.

Multivariate mapping extends thematic mapping techniques to two or more attributes and also has a range of options. For instance, using scaled ellipses as described earlier to compare two attributes. Style information on a set of points in Mapnik can be displayed by altering colour, radius x, radius y and contextual information such as text. The styler can reinforce a message by associating the same attribute with symbol colour and size, or assist in comparisons by using different attributes for each display variable. Adding complexity to the symbology increases the amount of information that can be portrayed, but at a potential increased cognitive cost to the user who needs to interpret it (Mazza, 2009).

3.4 Extended Styling

To increase the flexibility of the styling service, a number of different methods were explored to extend the base styling, detailed in Section 3.2.2. While these methods will be detailed in Section 4, two such extensions are discussed in detail in this section.

3.4.1 Meta-layers

Meta layers are layers of ancillary data that can be viewed concurrently to the thematic map to provide context. Consequently, maps that incorporate meta-layers can also accept more than one set of data as input. For an example of a meta-layer, consider public transportation coverage data (how many households are within walking distance of each bus stop) that may use a point vector to represent bus stops and their coverage values. But a meta-layer used to calculate this result - in this case house locations - can also be included to add meaning to the final result. The styler can combine these two sets of data inputs into a single map output.

3.4.2 Beyond spatial

Thematic styles can also be applied to non-spatial visualisations. For example, a plot can be styled so that points on the plot are colour-matched to regions shown on a map, with the region name or postcode acting as a common property. Further, cartographic styling techniques can be used to thematically present non-spatial data. Colour schemes generated in this manner, for both spatial and non-spatial data, can be extended to be applied to other visualisations, such as scatter plots. The method can also be applied to visualisations of multivariate data, to highlight a relationship between the styling attribute and the remaining attributes; examples of such visualisations include a scatter plot matrix and parallel coordinate plots. This method is implemented by appending a colour attribute to a given feature space, for example incorporating a colour attribute into a JSON array, with D3[7].

4 EXPERIMENTAL RESULTS

Styling options were tested on spatial data from a number of sources, to examine the extent of styling available through the service. The results in this section represent a subset of the styling methods that were implemented and investigated, and are presented to demonstrate the differences in interpretability between applying different styling methods to different data layers. All base layer map images are sourced from Open Street Maps[8].

4.1 Analysis Method

As well as the view types, as shown in Figures 2 and 3, different map classification algorithms were also explored, in particular, the efficacy of different approaches with respect to the underlying data being visualised and the subsequent impact on the interpretation of the data. Figures 4 and 5 show the same data, with the same number of bins (five), but styled differently; equal intervals was applied in Figure 4, while Figure 5 was generated using quantiles. Both maps show the smoothed rate of blood borne diseases derived from synthetic data designed to realistically simulate hospitalisations for Western Australia, with greyed out areas representing no results for the area. These figures demonstrate the importance of the map classification to the interpretation of the data; in this case, an outlier skews the map classification in the case of Figure 4, limiting a visual interpretation of the underlying disease rate.

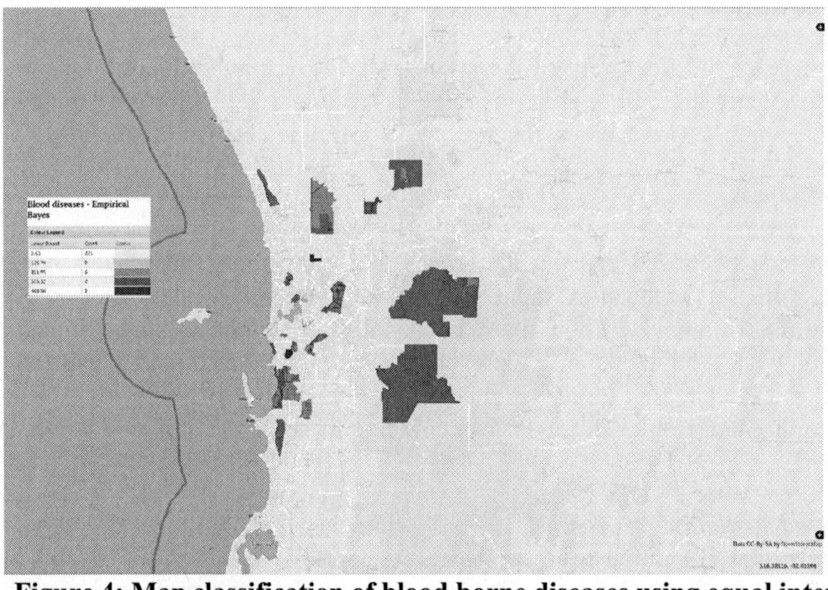

Figure 4: Map classification of blood borne diseases using equal interval, with 5 intervals.

[7] Data driven documents http://d3js.org
[8] OpenStreetMap http://www.openstreetmap.org/copyright

Figure 5: Map classification of blood borne diseases using quintiles; that is, quantiles with 5 intervals.

However, there are situations where equal intervals map classification is suitable. Figure 6 displays a probability calculated for each geographic feature displayed, using equal intervals with 10 intervals has the effect of partitioning the probability space represented into deciles. Additionally, the legend represents a coarse histogram of the distribution of the probabilities.

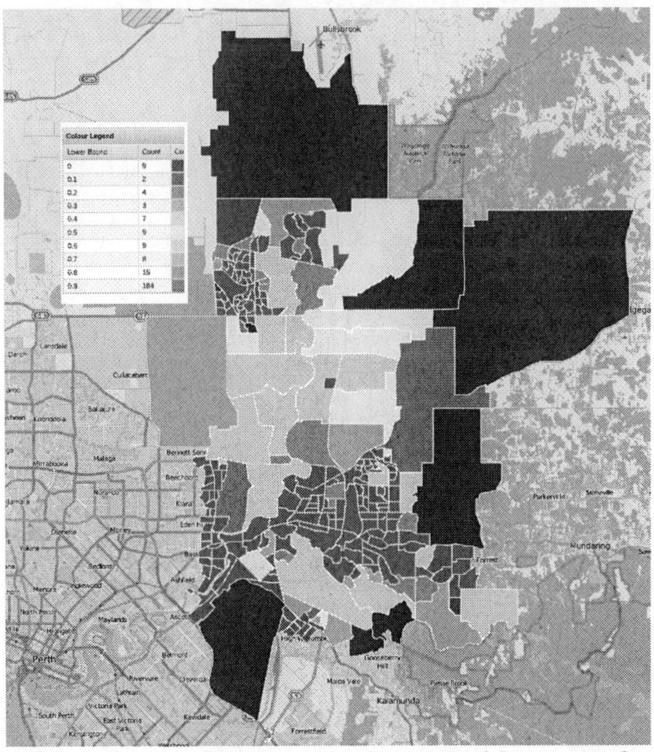

Figure 6: Map classification of a data set with a range from 0 to 1.

4.2 Colour scheme

Figures 7 and 8 represent the application of two different colour schemes to the same set of data. Figure 7 uses a red, yellow, green colour scheme, delimited using natural breaks. In contrast, Figure 8 uses a diverging scale around a pivot point for the same data set. In this case, 0.75 was selected as a significant value and therefore applied as the pivot point. Consequently the thematic map highlights the values above and below 0.75. This example shows why the method used to present the data should be tied to the purpose of the resulting visualisation.

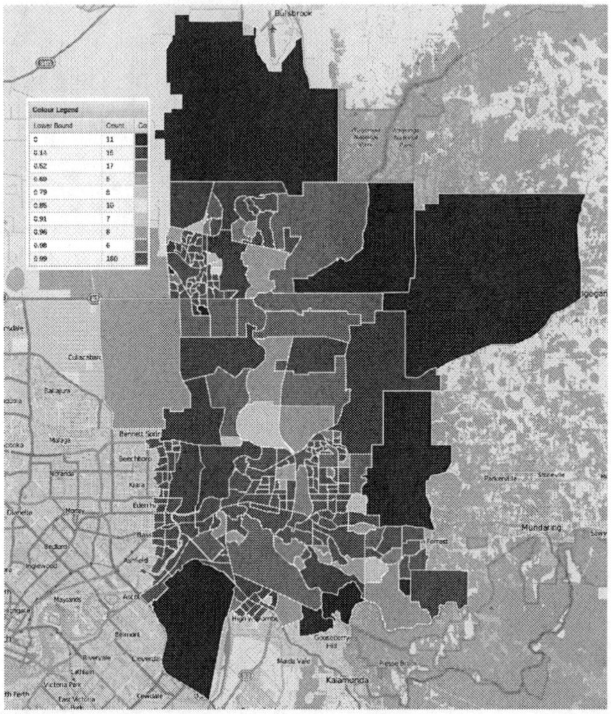

Figure 7: Red-green colour scheme over natural breaks.

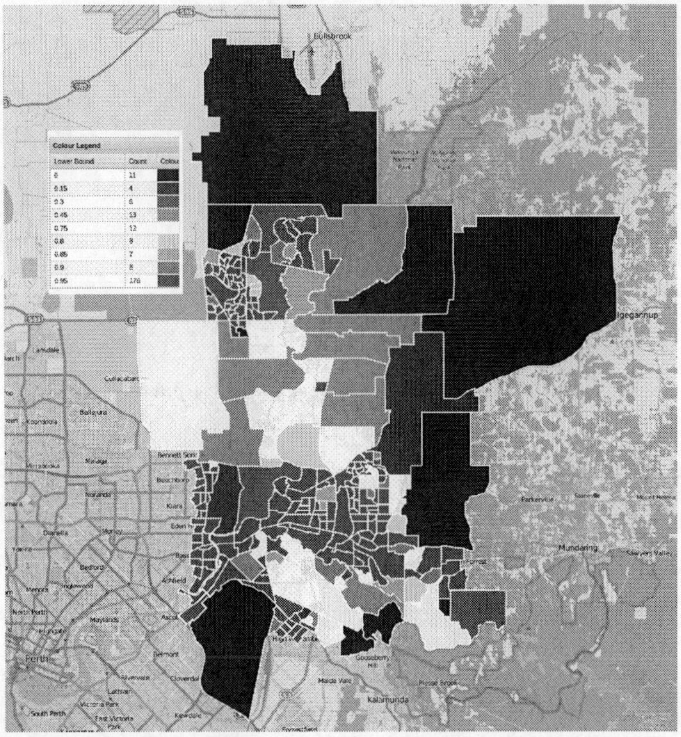

Figure 8: Diverging colour scheme around a pivot point value of 0.75.

4.3 Point data

Figures 9 and 10 show examples of thematic visual styling and contextual representation respectively for data set comprising multivariate point data, in this case the long and short term average of time series data, and the proportion of the two. Quintiles map classification was used in both cases, along with a red, yellow, green colour scheme. Figure 9 shows solely the proportion of the short term to long term averages. Figure 10 shows the same information, with the proportion being included as the label, and the radius determined by the short term average. While a static radius was used for both figures, the minimum radius can be tied to zoom level to dynamically adjust radii of points as a user zooms in.

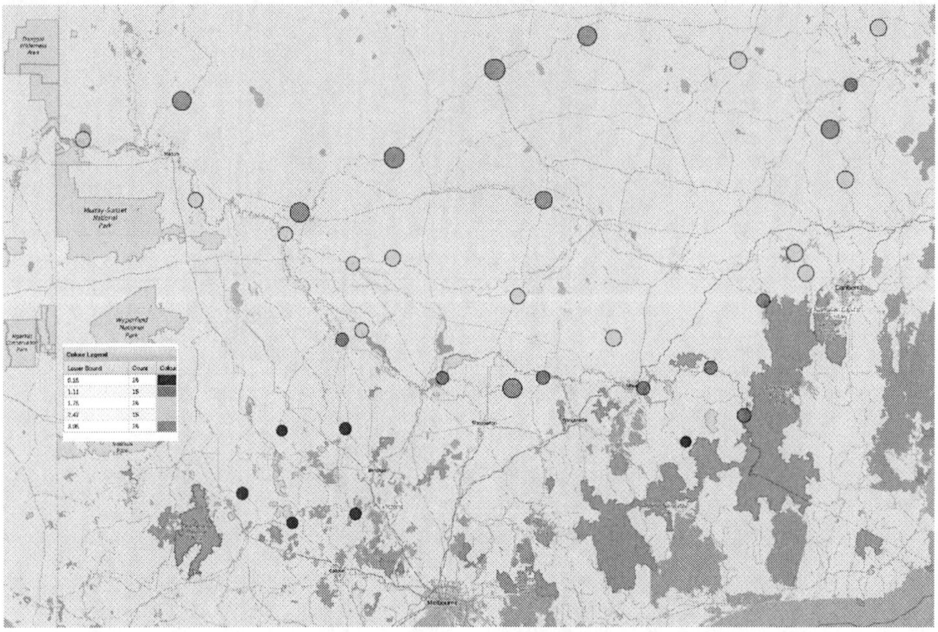

Figure 9: Thematic visual styling: proportion of short to long term averages.

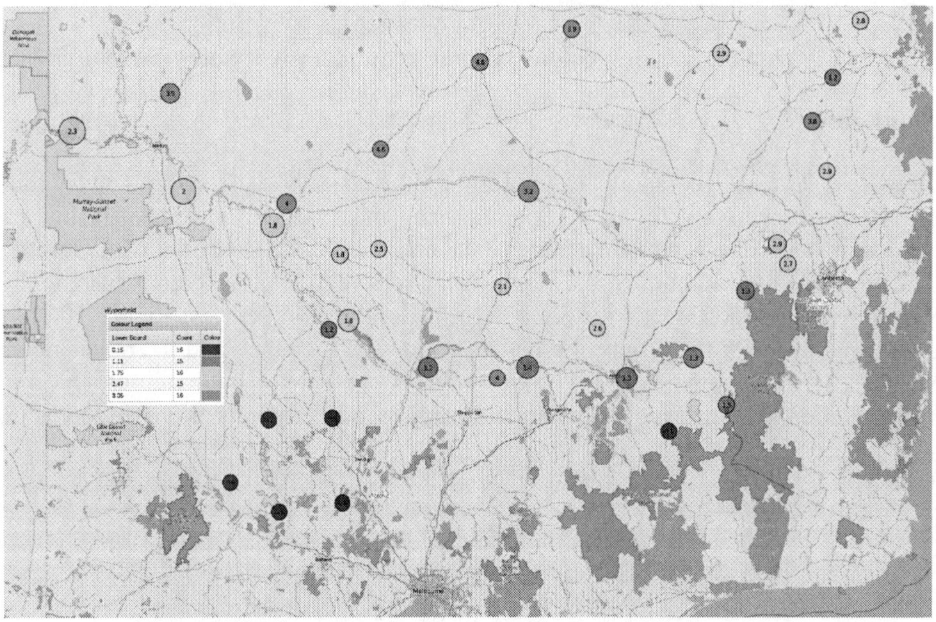

Figure 10: Contextual representation: labelled proportion of short to long term averages.

The styling of points was extended to incorporate a relative size on the map for a point by using a buffer size to determine the radius. For example, Figure 11 shows the results of displaying a set of points, corresponding to a public service, with a 500m buffer as the point radius that can be used to approximate coverage for each service. Due to the large overlap of points visualised in this manner, a high transparency was applied to ensure overlapping points remained visible.

Figure 11: Point data plotted using a 500m radius.

4.4 Non-spatial visualisations

Figure 12 displays an example of applying cartographic styling techniques to non-spatial visualisations. This methods inserts the colour scheme into a JSON based spatial format, which can then be interpreted by JavaScript visualisation libraries such as D3 to produce a visualisation combining the colour scheme and a visualisation technique with the aim of embedding the map classification colour scheme within the visualisation.

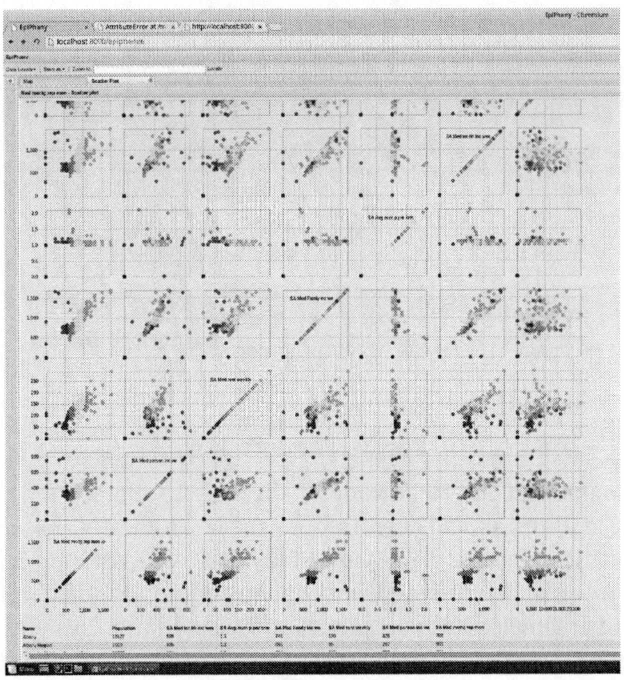

Figure 12: Scatter plot matrix generated using cartographic styling for the plot colours.

4.5 Meta-layer

This method refers to incorporating an ancillary layer within a data layer, with both layers being viewed simultaneously so that the meta-layer can provide context. In general, the meta-layer should correspond to underlying data used in the derivation of the data layer. Consequently, when shown with minimal styling, the spatial relationship between the initial data and the subsequently derived data can be seen. Figure 13 shows a grid data visualisation with a bespoke defined style applied that defines a class for each grid; in this case, the classes were derived from the trend associated with each grid. For this dataset, the trend was derived by determining a combined trend associated with points contained within each grid. Consequently, this data is shown as a meta-layer to provide context.

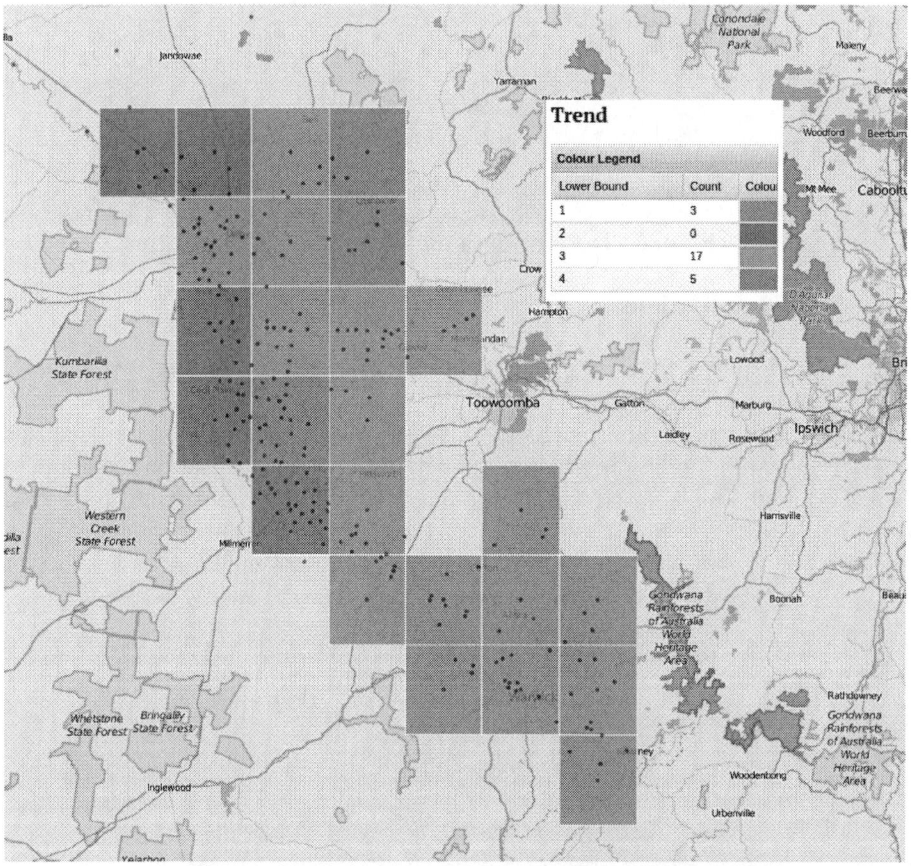

Figure 13: Point data plotted using a 500m radius.

5 CONCLUSIONS

No single visualisation can be considered the "most suitable" for all thematic mapping problems. Mapping choices should therefore be made on-the-fly, based on the data itself, rather than predefined and static. This case study has demonstrated how the use of open source tools can be coordinated to produce web map styling information on-the-fly. This will enhance the scope of automated spatial analysis and visualisation systems as demonstrated on the case study using Mapnik. Currently, only point and polygon geometry features can be interpreted by the dynamic styler. However, methods to extend dynamic styling to line visualisations, such as for road or waterway data, will be explored in future work.

6 ACKNOWLEDGEMENTS

The Cooperative Research Centre for Spatial Information has supported this work, whose activities were funded by the Australian Commonwealth Cooperative Research Centres Programme.

7. REFERENCES

Burdziej, J., 2012. A web-based spatial decision support system for accessibility analysis - concepts and methods. *Applied Geomatics*, 4(2):75-84.

Harrower, M. and Brewer, C. A., 2003. Colorbrewer.org: An online tool for selecting colour schemes for maps. *The Cartographic Journal*, 40(1):27-37.

Iosifescu, I., Iosifescu, C. M., Panchaud, N., Eichenberger, R., Sieber, R., and Hurni, L., 2013. Advances in web service-driven cartography. *International Cartographic Conference*, Dresden Germany.

Iosifescu-Enescu, I., Hugentobler, M., and Hurni, L., 2010. Web cartography with open standards - a solution to cartographic challenges of environmental management. *Environmental Modelling and Software*, 25(9):988-999.

Kettunen, P., Sarjakoski, L. T., Ylirisku, S., and Sarjakoski, T., 2012. Web map design for a multipublishing environment based on open APIs. In Peterson, M. P., editor, *Online Maps with APIs and Web Services*, chapter 12:177-193. Springer, Berlin Heidelberg.

Mazza, R., 2009. *Introduction to Information Visualization*. Springer, London.

Moncrieff, S., West, G. A. W., Cosford, J., Mullan, N., and Jardine, A., 2014. An open source, server-side framework for analytical web mapping and its application to health. *International Journal of Digital Earth*, 7(4).

Moore, A., de Oliveira, M., Caminha, C., Furtado, V., Basso, V., and Ayres, L., 2013. Applying geovisual analytics to volunteered crime data. In Moore, A. and Drecki, I., editors, *Lecture Notes in Geoinformation and Cartography: Geospatial Visualisation*, chapter 9,185-209. Springer-Verlag, Berlin.

Moreland, K., 2009. Diverging color maps for scientific visualization. In *5th International Symposium on Advances in Visual Computing (ISVC `09)*, II:92-103, Las Vegas, USA.

OGC, 2007. *OpenGIS Styled Layer Descriptor Profile of the Web Map Service Implementation Specification*. Technical report, Open Geospatial Consortium.

Rautenbach, V., Coetzee, S., Strzelecki, M., and Iwaniak, A., 2012. Results of an evaluation of the orchestration capabilities of the ZOO project and the 52° North framework for an intelligent geoportal. *ISPRS Annals of Photogrammetry*, Remote Sensing and Spatial Information Sciences, I-4:163-168.

Weiser, A. and Zipf, A., 2007. A visual editor for OGC SLD files for automating the configuration of WMS and mobile map applications. In Gartner, G., Cartwright, W., and Peterson, M. P., editors, *Location Based Services and TeleCartography*, 265-278. Springer, Berlin Heidelberg.

Yau, N., 2011. *Visualize This*. Wiley Publishing, Inc, Indianapolis.

Free and Open Source Software for Geospatial (FOSS4G) Conference Proceedings

Volume 15 *Seoul, South Korea* Article 48

2015

A Comparison of Image Aligning and Correcting Software with an Unmanned Aerial System

Kangsan Lee
Department of Geography, Kyung Hee University

Jinwoo Park
Department of Geography, Kyung Hee University

Jinmu Choi
Department of Geography, Kyung Hee University

Recommended Citation

Lee, Kangsan; Park, Jinwoo; and Choi, Jinmu (2015) "A Comparison of Image Aligning and Correcting Software with an Unmanned Aerial System," *Free and Open Source Software for Geospatial (FOSS4G) Conference Proceedings*: Vol. 15 , Article 48.
DOI: https://doi.org/10.7275/R5C24TNM
Available at: https://scholarworks.umass.edu/foss4g/vol15/iss1/48

AT3-05

A Comparisonof Image Aligning and Correcting Software with an Unmanned Aerial System

Kangsan Lee[1], Jinwoo Park[2] and Jinmu Choi[3]

[1]Department of Geography, Kyung Hee University
26 Kyungheedae-ro, Dongdaemun-gu, Seoul, Rep. of Korea
Email: gi.star@khu.ac.kr
[2]Department of Geography, Kyung Hee University
26 Kyungheedae-ro, Dongdaemun-gu, Seoul, Rep. of Korea
Email: zawoon96@khu.ac.kr
[3]Correspondence, Department of Geography, Kyung Hee University
26 Kyungheedae-ro, Dongdaemun-gu, Seoul, Rep. of Korea
Email: cjm89@khu.ac.kr

ABSTRACT

In the past few decades, many kinds of UAS for image acquisition have been developed. But software for image aligning and correcting is mostlyhigh-cost commercial. This problem caused the cost-problem in using UAS. Nowadays, a variety of software, not only commercial but also open-source, provides powerful image processing tool. There are a number of software to support image processing. In this study,five popular programs are testedfor comparison. The goal of this study is to compare popular open-source software based on the ease use, overall accuracy and processing time for chunk of images from UAS.

1. INTRODUCTION

Traditionally, as a data collection platform, aircraft and satellite have a major role in remote sensing. However, these remote sensing platforms have structural problems: time and cost. More specifically, researchers have to spend a lot of money to get high-resolution images. Furthermore, because of some constraints, such as weather condition, sometimes these platforms or images from the platforms may be useless(Jensen, 2009). In these circumstances, the new remote sensing platform has been rising; UAS, Drone.

The word, UAS is an abbreviation of Unmanned AircraftSystem(Marshall *et al.*, 2015). Many researchers have been focusing on 'Unmanned'(Martinez Rubio *et al.*, 2005; Vierling *et al.*, 2006; Zhang and Kovacs, 2012). Especially, it is necessary in a certain situations where people cannot access. For example, researchers used UAV as a landslide investigator in the Southern French Alps (Niethammer *et al.*, 2012).

After gathering images from UAS, the researchers should stitching images for further research. This is because gathered images from UAS is not a whole scene of the entire research area; UAS is moving around over the ground while the shutter of the camera worked. Therefore, the single images should be matched among each other to composite large single image dataset.

There are a variety of software which can handle this work, many researchersuse these image processing programs in various areas: Engineering(LeBlanc *et al.*, 2013), Medicine(Wójcicka and Wróbel, 2012), and Geomorphology (Niethammer *et al.*, 2012).However, we don't know much about which software we choose to process our UAS images. Thus, this study aims to compare several image stitching software using sample images gathered by UAS. It is expected that the result of this comparison provides a solution which software is better for UAS-gathered image stitching.

2. DATA ACQUISITION

2.1 Image gathering

Before image alignment and stitching, the images were collected by UAS. In this study, one of the famous open-source UAS, 3DR Robotics IRIS+, and small digital camera, Gopro, were used. IRIS+ is composed of four propellers and Pixhawk controller. The benefits of the Pixhawk system include integrated multithreading, a Unix-like programming environment, completely new autopilot functions such as sophisticated scripting of missions and flight behavior(Meier *et al.*, 2012). With these hardware environments, 'Mission Planner' has a significant role in data acquisition. In fact, there are so many limitations in manual control.The UAS should be controlled by the user-defined flight route to get accurate images in the research area. However, in manual control mode, it is impossible to maintain fixed altitude, azimuth and speed. Therefore, this study uses the Mission Planner to control autonomous flight. These hardwares and software are shown below (Figure 1).

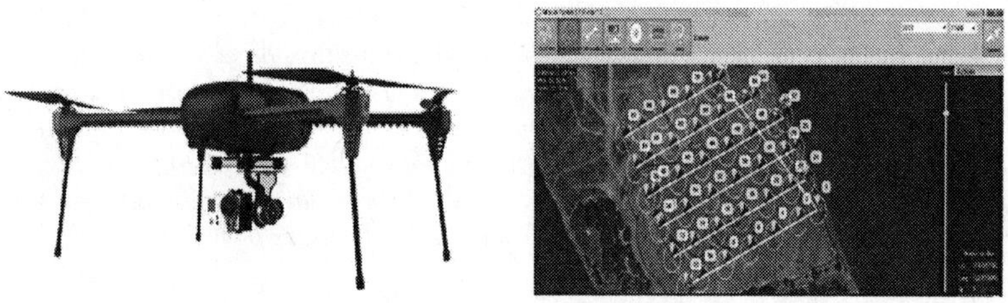

Figure 1. IRIS+ and GoPro (Right), Mission Planner for autonomous flight(Left)

Using the hardware and software system, approximately 200 images were collected in the sampling area. Regarding to the law of Korean Aviation, controlling UAS in urban or public area is illegal. Accordingly, the sample images were taken in the rural area, Baramarae beach, located in the west coast of South Korea.

2.2 Correcting lens distortion

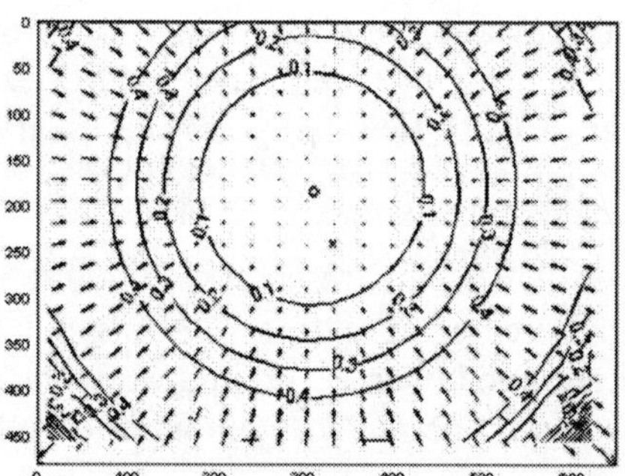

Figure 2. Radial distortion (California Institute of Technology, 2011)

GoPro, small digital camera has a lens distortion(Tauro *et al.*, 2014). As shown in Figure 2, lens distortion should be removed before image alignment. Although many image processing programs provide fixing tool for lens distortion, the Adobe Lightroom program is selected to correct lens distortion. This is because some program shows slow processing speed and do not remove the distortion efficiently. Photoshop Lightroom, shows high-speed performance with pre-defined lens information: focal length and distortion factor. As a result, all sample images were transformed into adjusted images with the same quality automatically (Figure 3).

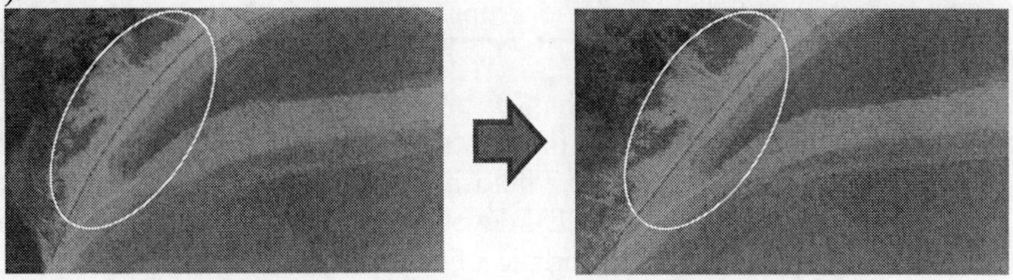

Figure 3. Lens distortion correction (before – later)

3. IMAGE ALIGHNMENT

The main purpose of this study is comparing several image alignment software with sample images. Specifically, many commercial and open-source image alignment program have been developed for image alignment. But each program has their own characteristic; for example, some of them is only suitable for making panoramic photos, image stitching and image mosaicking. So, many researchershaveencountered that 'Which software should I select to do image alignment for the UAS-gathered images?'

One of the solution tothis question is comparing several software at the same hardware environment. The workstation environment is stated in Table 1.

Table 1. Specification of the testing environment

Component	Specification
CPU	Intel Xeon CPU E3-1231 3.40GHz
RAM	16GB(DDR3, Dual Channel)
GPU	GeForce GTX 750Ti
Cuda Core	640
Storage	SSD, 256GB

3.1 Image alignment software

Among the variety of image alignment software, five major software are tested for comparison.Specific information about these programs are stated in Table 2. The first one isAgisoft Photoscan, one of the famous commercial image stitching program. It is mainly used for three dimensional image making based on a conjugate-point calculation between adjacent images. After building point clouds and image pyramids, anorthophotograph can be generated.

Table 2. Software information

Name	Developer	License	Notes
Photoscan	Agisoft	Proprietary	Using photometry method
Photoshop	Adobe	Proprietary	User-friendly interface
Image Composite Editor	Microsoft	Proprietary (Free for non-commercial use)	Variety of pre-defined projection

Hugin	Pablo d'Angelo et al.	GPL	Based on panorama tools
Pix4Dmapper	Pix4D SA	Proprietary	Optimized for UAS

Second, Adobe Photoshop also has a photomerge function in automatic image processing tab. This is not a professional image stitching program, but many people use it to enhance image quality. In this perspective, Photoshop may be a useful software for public.

Third, Image Composite Editor (ICE), developed by Microsoft, is not an open-source program, but free software. The characteristic of this program is that it has plenty of pre-defined projection. Also, it provides lots of image format. So, users can choose what they want to make just using a user-friendly GUI system.

Fourth, Hugin is the most famous open-source image stitching program. Both the amateur photographers and professional researchers used to utilize this software. The biggest advantage of Hugin is that this is a free open-source program; users can use it free-of-charge.

Lastly, Pix4D image processing software is a mapping and modeling solution to convert thousands of images into a georeferenced 2D mosaic and 3D models. In these days, several UAS companies have a collaboration with the company of Pix4D software. It shows that this is a powerful image stitching software with a UAS system. But this commercial program has a cost-problem.

3.2 Comparing image stitching software

With these five image stitching programs, several comparisons have been done; image stitching time, image volume, and image quality.

Two hundredsample images were gathered usinga UAS. To compare image stitching processing-time for each software, a comparative experiment was designed; the amount of input data had been changed from 50 to 200 increased by fifty. It is reasonable that the bigger the volume of input data increased, the longer the processing time required. However, this experiment is useful to figure out individual software's performance (stitching time) and accuracy (quality of result). Figure 4 describes a concept of the experiment.

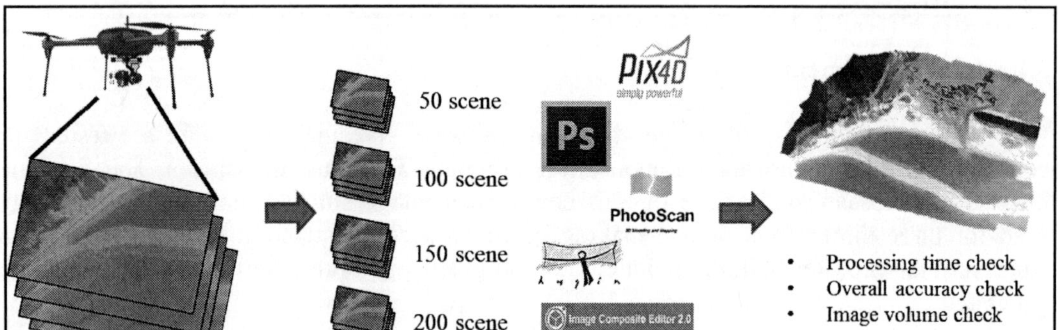

Figure 4. Concept of software comparing experiment

3.3 Result

Unfortunately, Hugin among tested five programs doesn't show powerful functions when it comes to stitch images. This is because its original purpose is not a photogrammetry work, but making a panoramic image. More specifically, Hugin does not estimate conjugate points between adjacent images that have coarse overlapping terrain features, then it cannot calculate a relative orientation with a chunk of shoreline images. Moreover, Photoshop cannot success to stitch 200 images because of hardware limitation: not enough memory and CPU performance.

The resulting graphs (Figure 5) show that how processing time and image volumes had been changed depending on the number of input images and processing programs. In brief summary, processing time and image volumes are gradually increased in all softwareaccording to the number of images, but not proportionalto the number of input images. But the rate of increase of time and volume shows differences among the program. MS ICE shows the fastest performance and has the least image volume. On the contrary, Pix4D is getting much slower and the results of Photoscan have a huge image volume.

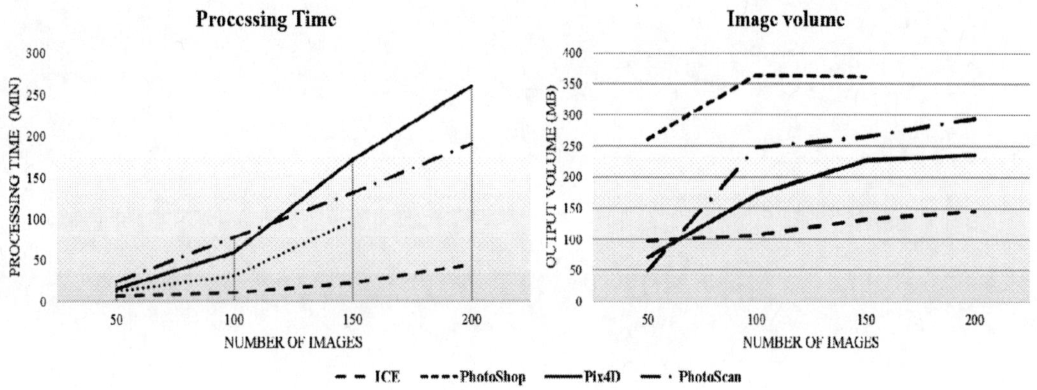

Figure 5. Processing time and image volume

Additionally, the third comparing point is checking a visual accuracy of the result.Following images (Figure 6) show how different the visual accuracy of the result image is.

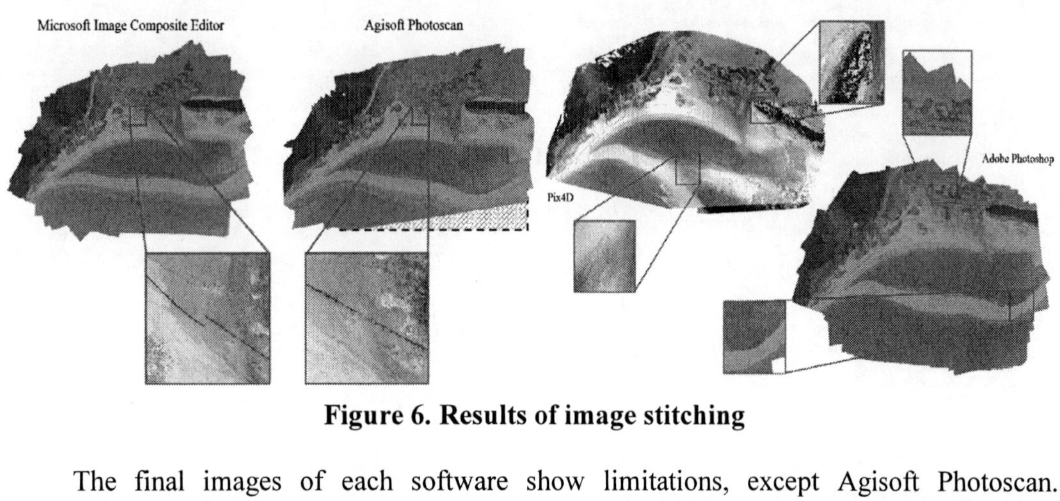

Figure 6. Results of image stitching

The final images of each software show limitations, except Agisoft Photoscan. Microsoft ICE and Adobe Photoshop don't make a smooth surface of the images. There are plenty of distorted connections among the adjacent images. Also, in Photoshop, the color correction does not match well with the other images. This problem can be solved using a vignetting effect, but it takes too much time to calculate. The result image of Pix4D is shoddy butit may be improved using many stereographic UAS images to cover the entire research area. Photoscan, however, shows the best result: greater visual accuracy and good color correction. Althoughthe processing time and the image volume of itare higher than those of MS ICE, they are lower than those of Pix4D.

4. CONCLUSION

In these days, researchers can easily find alternative open-source softwarethat can save their costrequired for high-cost commercial software. But it is true that some of the open-source software cannot show as good performance as commercial one. Thus, some of the commercial software is suitable for complex and elaborate research work. Therefore, it is valuable to test programs for a specific processing.

One of the famous open-source image stitching software, Hugin, does not show its powerful performance in an academic experiment. Although assuming so many reasons, one of the main reasons is its limitation in calculation; this software is not suitable for complex image stitching.On the other hand, many open-source programs are available especially for image stitching. Visual SFM, for example, became the best 3D image reconstruction open source software. Many researchers use this program rather than buying an expensive commercial program.

The sales market of UAS, called Drone, have been increased geometrically in many countries with cheap price. Many people use this small plane and a camera to capture their neighbor or journey. Therefore, the needs of image stitching will be increased in the near future. For these reasons, the characteristics of open-source programsshould be compared fro public to choose the best program for their specific work.

5. REFERENCE

California Institute of Technology., 2011-last update. Available: http://me2.do/5aNxf3Nc

Jensen, J. R., 2009. *Remote Sensing of the Environment: An Earth Resource Perspective 2/e*. Pearson Education India.

LeBlanc, B., Niezrecki, C.,Avitabile, P.,Chen, J., and Sherwood, J., 2013, Damage detection and full surface characterization of a wind turbine blade using three-dimensional digital image correlation. *Structural Health Monitoring*, 12(5-6), 430-439.

Marshall, D. M., Barnhart, R. K., Shappee, E., and Most, M. T., 2015, *Introduction to unmanned aircraft systems*, CRC Press, New York.

Martinez, R. J., Lahoz, J., Aguilera, D., and Codes, J., 2005, IMAP3D: Low cost photogrammetry for cultural heritage.*Proceedings of the CIPA 2005 XX International Symposium,* Torino, Italy.

Meier, L., Tanskanen, P., Heng, L., Lee, G. H., Fraundorfer, F., and Pollefeys, M., 2012, PIXHAWK: A micro aerial vehicle design for autonomous flight using onboard computer vision.*Autonomous Robots*, 33(1-2), 21-39.

Niethammer, U., James, M. R.,Rothmund, S.,Travelletti, J., and Joswig, M., 2012, UAV-based remote sensing of the Super-Sauze landslide: Evaluation and results, Engineering Geology, 128, 2-11.

Tauro, F., Porfiri, M., and Grimaldi, S., 2014, Orienting the camera and firing lasers to enhance large scale particle image velocimetry for streamflow monitoring, Water Resource Research, 50(9), 7470-7483.

Vierling, L. A., M. Fersdahl, X. Chen, Z. Li, and P. Zimmerman, 2006, The Short Wave Aerostat-Mounted Imager (SWAMI): A novel platform for acquiring remotely sensed data from a tethered balloon, Remote Sens Environ, 103(3), 255-264.

Wójcicka, A., and Wróbel, Z., 2012, The panoramic visualization of metallic materials in macro-and microstructure of surface analysis using microsoft image composite editor (ICE), *Information Technologies in Biomedicine*, Springer Berlin Heidelberg, 358-368.

Zhang, C. and Kovacs, J. M., 2012, The application of small unmanned aerial systems for precision agriculture: a review, *Precision agriculture*, 13(6), 693-712.

Free and Open Source Software for Geospatial (FOSS4G) Conference Proceedings

Volume 15 *Seoul, South Korea* Article 49

2015

Generating Geospatial Footprints For Geoparsed Text From Crowdsourced Platial Data

Ahmad O. Aburizaiza
Department of Geography & Geoinformation Science, George Mason University

Matthew T. Rice
Department of Geography & Geoinformation Science, George Mason University

Michael F. Goodchild
Department of Geography, University of California, Santa Barbara

Recommended Citation
Aburizaiza, Ahmad O.; Rice, Matthew T.; and Goodchild, Michael F. (2015) "Generating Geospatial Footprints For Geoparsed Text From Crowdsourced Platial Data," *Free and Open Source Software for Geospatial (FOSS4G) Conference Proceedings*: Vol. 15 , Article 49.
DOI: https://doi.org/10.7275/R5GT5KC1
Available at: https://scholarworks.umass.edu/foss4g/vol15/iss1/49

AT1-02

GENERATING GEOSPATIAL FOOTPRINTS FOR GEOPARSED TEXT FROM CROWDSOURCED PLATIAL DATA

Ahmad O. Aburizaiza[1,2], Matthew T. Rice[1], and Michael F. Goodchild[3]

[1]Department of Geography & Geoinformation Science, George Mason University
Suite 2400, Exploratory Hall, Fairfax Campus, 4400 University Dr. Fairfax, VA 22030, USA
Email: aaburiza@gmu.edu
[2]Department of Geomatics, King Abdul Aziz University
P.O. Box 80210, Faculty of Environmental Designs, Jeddah 21589, Saudi Arabia
[3]Department of Geography, University of California, Santa Barbara
1832 Ellison Hall, UC Santa Barbara, Santa Barbara, CA 93106, USA

ABSTRACT

The research paper reports on the generation of geospatial footprints from geoparsed text associated with geocrowdsourced platial data collected and stored in the George Mason University Geocrowdsourcing Testbed (GMU-GcT). The GMU-GcT facilitates study of social dynamics, quality assessment, data contribution patterns, and position validation for geocrowdsourced geo data, with a primary purpose of mapping transient obstacles and navigation hazards in a dynamic urban environment. This paper reports on the automated generation of spatial footprints using open-source software, and discusses the role of automated spatial footprints in quality assessment for automated position validation. A detailed, local gazetteer is used to store placenames and placename variants including abbreviated, slang, former, and jargon-based instances. Obstacle reports containing location descriptions are geoparsed and processed with the help of the GMU-GcT gazetteer to generate geospatial footprints, which are used in a quality assessment process to validate the position of obstacle reports. Continuing research with the GMU-GcT has produced fifteen characteristic footprints types, which are generated and grouped into simple, complex, and ambiguous categories. The open-source tools used for generating these footprints are MapBox, MapBox.js, TURF.js, jQuery, and Bootstrap.

1. INTRODUCTION

George Mason University, is the largest public university in the Commonwealth of Virginia, hosts 33,000 students and several thousands faculty and staff, in a dynamic urban environment outside Washington, D.C. The campus, adjacent to the City of Fairfax, is the site of near-constant construction and expansion. This changing urban environment presents difficulty for students, faculty, and staff who are visually- or mobility-impaired and depend on familiar navigation pathways to get to and from work or home. Construction barricades, sidewalk obstructions, and detours are commonplace. The temporary, unplanned nature of these disruptions make them nearly impossible to capture and map using traditional GIS workflows, and a crowdsourced approach is one of the only practical ways to provide the information in a timely manner. Geocrowdsourcing and volunteered geographic information (VGI), introduced by Goodchild (2007), and reviewed by Elwood (2008), Haklay (2010) and others, represents an opportunity to extend traditional mapping through open-source software and novel geocrowdsourcing workflows. The GMU Geocrowdsourcing Testbed (GMU-GcT), developed by Rice *et al.* (2012, 2013, 2014), introduces a comprehensive geocrowdsourcing approach for collecting and quality-assessing transient obstacle data. Related work by Laakso

(2013), and Karimi *et al.* (2014) provides a useful look at how the accessibility domain benefits from novel open-source mapping applications and data modeling. The GMU-GcT is based on the early work in geoparsing, gazetteers, and geocrowdsourcing (Aburizaiza 2011, Rice at al. 2011, Rice 2012a) and extended to include quality assessment, routing, and visualization (Rice *et al.* 2013a, 2013b, 2014).

Contributors to the GMU-GcT include students, faculty, staff, and members of the public, who submit obstacle reports containing the location, the basic characteristics, and images of the obstacles (Figure 1). This information is processed in a preliminary quality assessment procedure and displayed on a map as a provisional obstacle report. Student moderators field check the reports and provide a comprehensive quality assessment through ground truth. The valid reports are maintained in the system and displayed to the public as confirmed reports.

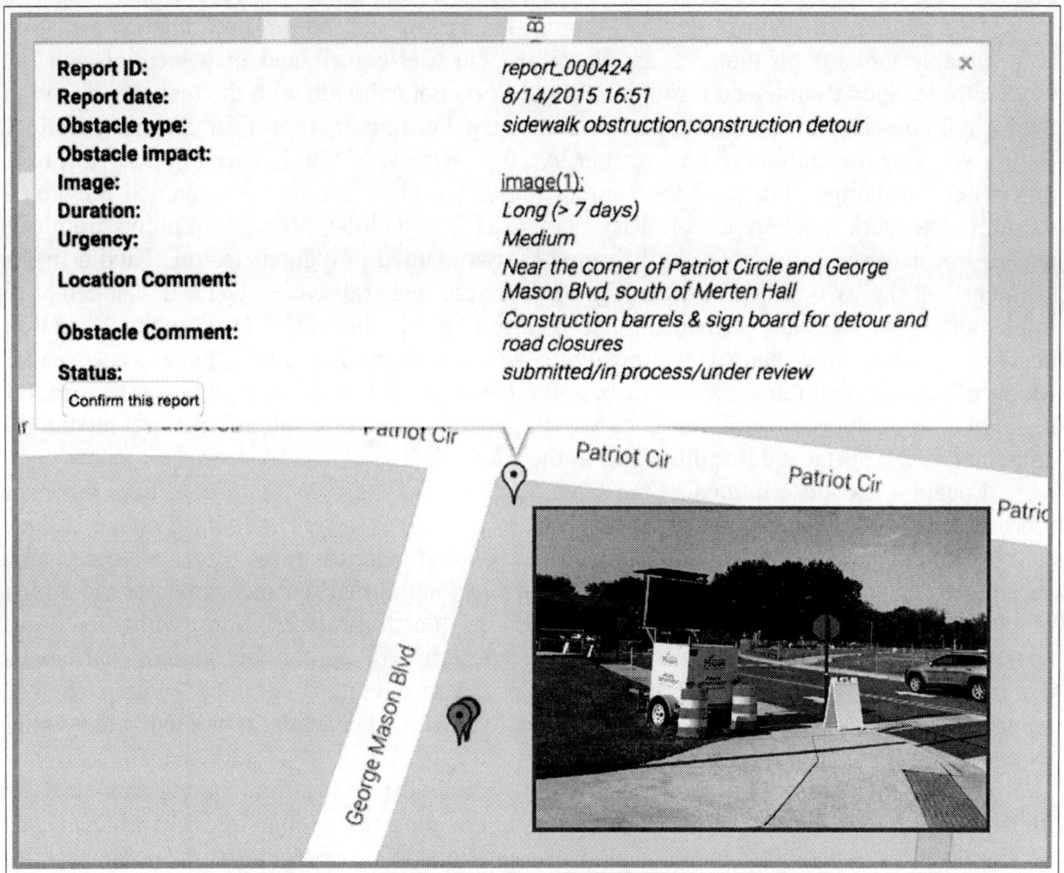

Figure 1: A GMU-GcT obstacle report with image

2. POSITION VALIDATION AND LOCATION DESCRIPTION TEXT IN THE GMU-GcT

Qin *et al.* (2015) present the quality assessment procedures in the GMU-GcT, including assessments of location, time, and attribute, the three primary facets of the atomic view of geographic information noted by Longley *et al.* (2011). Modeled after the comprehensive quality assessment of geocrowdsourced data by Girres *et al.* (2010) and Haklay (2010), Qin *et al.* develop assessments of position and categorical attribute agreement, and use this quality assessment procedure to create a composite score for geocrowdsourced data.

Rice *et al.* (2015) discuss the details of the positional validation procedures in the GMU-GcT, by introducing a concept of multi-position validation in the GMU-GcT. Reports contributed by the public are checked for position through three comparisons. First, they are compared to moderated ground truth, where the mapped position of a report contributed by the end-user is compared with the location established through a moderated field check. Second, the images contributed by the end-user are processed to extract embedded geotags and orientation data, which can be combined with geotags and orientation data from multiple reports to create an image geotag-based footprint. Finally, the location description provided by the contributed is processed as a geospatial footprint using extracted placenames, prepositions, distances, directions, landmarks, addresses, and other feature names.

This geospatial footprint developed from geoparsed placenames and other information is a valuable tool for position validation. When the user-contributed map location and the moderator location established through field check do not coincide with the features named in the location description, the inconsistency can be used to flag the report for closer inspection. Earlier work by the authors (Aburizaiza *et al.* 2011, Rice *et al.* 2011, Rice *et al.* 2012a) used this general technique, but used the simplest cases based on proximity to named polygonal features. The work was extended in Rice *et al.* (2015) to include intersection points from two named linear features, a convex hull formed by two named polygonal features, and a linear segment cut by two named linear features. These general cases were developed and implemented so that any crowdsourced contribution to the GMU-GcT with a location description containing named features with this configuration could have a geospatial footprint developed automatically. The ongoing recent work in this paper extends this general technique to include several more cases of general named or platial feature layout, as contained in the obstacle description text in the GMU-GcT.

Location text descriptions in the GMU-GcT contain references to one, two, three, or more places, and these place references can be in various forms including abbreviations, slang, former (old) names, and colloquial variants of standard names. Many of these name variants are contained in the GMU-GcT gazetteer. In additional to placenames or variations of placenames, the GMU-GcT frequently contains locations descriptions with directions, distances, and spatial prepositions. Moreover, a description can contain features that have a distinct group identity and associated name, or a named feature contained within another named feature, or an unnamed places, e.g. walkways, with a geospatial relationship to another named feature.

The combinations of features in the GMU-GcT's location description text is understandably complex. This paper focuses on extending the geoparsing capabilities based on simple and complex placename or platial orientations to generate more cases of geospatial footprints, and instantiating a reference library of the footprints for future research. Several cases are identified and more are being discovered. In the methodology section below, the cases are explained in three categories: simple, complex, and ambiguous. The GMU-GcT gazetteer used in this work is actively updated to reflect naming changes and instances of abbreviation, slang, and colloquial place references. The individual entries for the gazetteer are stored as a JSON array value stored in the GeoJSON properties.

3. METHODOLOGY

A web application was developed to run and test the geoparsing algorithms for the distinctive cases. The application was designed as a mobile web application to permit mobile and tablet users to utilize it through mobile browsers. HTML5, jQuery, and Twitter's Bootstrap were used to build the interface and adjust the site components according to the screen size.

MapBox is a well-known and powerful web mapping technology with a fast map tiling capability. MapBox has a JavaScript Library called MapBox.js which permits programmers to build web mapping applications for geospatial data visualization. It also allows programmers to customize its map object and navigational tools.

MapBox.js uses the TURF.js JavaScript Library, which is a geospatial analysis library that can run on client side, server side, or both. It uses the GeoJSON format as the input and the output. It is a very rich library and runs efficiently since geospatial analysis can run on the client side without connecting to server.

MapBox.js with TURF.js were used to build and visualize the geospatial footprints after extracting placenames, prepositions, and bearing words using jQuery and AJAX. Currently, campus data are stored in a GeoJSON file residing in the server. The plan in the future is to store the campus data in a MongoDB database since MongoDB stores data in JSON format rather than relational database table format.

The geospatial footprints developed are categorized into three categories: simple, complex, and ambiguous. Each category is explained in details. Some cases are similar in concept but are different in their algorithm structure. Such cases are explained in sync while their differences in the algorithm are covered after.

3.1 Simple geospatial footprints

The first simple case (Figure 2) is only one point type or one polygonal place mentioned in a text message with no preposition or bearing. Some landmarks are digitized as either point or polygon features. Other polygonal places are buildings and group of buildings. The point or the polygon is buffered and then both the place itself and its buffer are plotted on the map. The only difference between point and polygon cases is the zoom level set after the geoparsing process is finished.

The system also creates a Bootstrap modal object, similar to a JavaScript's alert, informing the user that a name of one landmark, or building, etc.. was found in the message. Also, the modal informs the user that there were no preposition or bearing words in the message. In other words, the algorithm's detailed steps are explained to the user. This explanatory Bootstrap modal is given in each footprint case. It also notifies the user if no text was entered, or if placenames are not found in the message. Figure 2 displays an example of the first simple case with its explanatory modal.

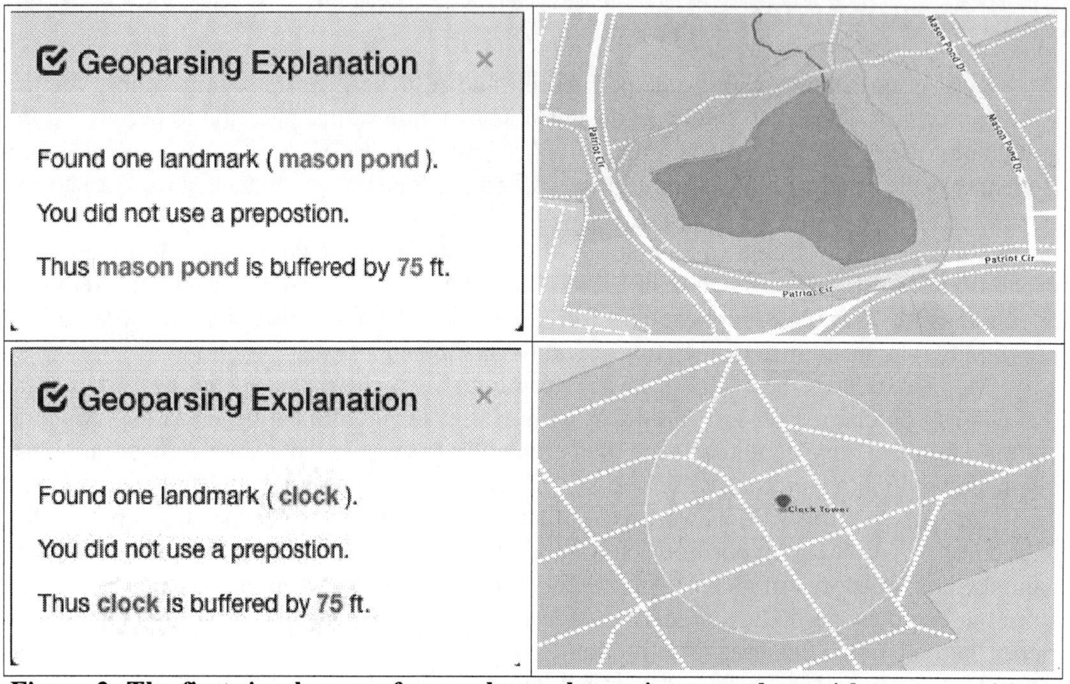

Figure 2: The first simple case of one polygonal or point type place with no prepositions nor bearing words

The next simple case (Figure 3) is finding only two names of intersected linear places. The algorithm searches the message for terms such as "intersection" or "corner of" as well. Even if two linear places are specified without intersectional terms, the algorithm would find the intersection of the two linear places. The explanation modal will clarify if intersection terms are found or not, in addition to the two linear places names. The intersection point is buffered and then the two linear places, the intersection point, and the intersection's buffer are all plotted on the map. Rarely, two linear places have two intersections instead of one. The algorithm is also taking care of this and the web application will zoom to the center point between the two intersections with an appropriate zoom level. Figure 3 illustrates two examples of this algorithm of one intersection and two intersections.

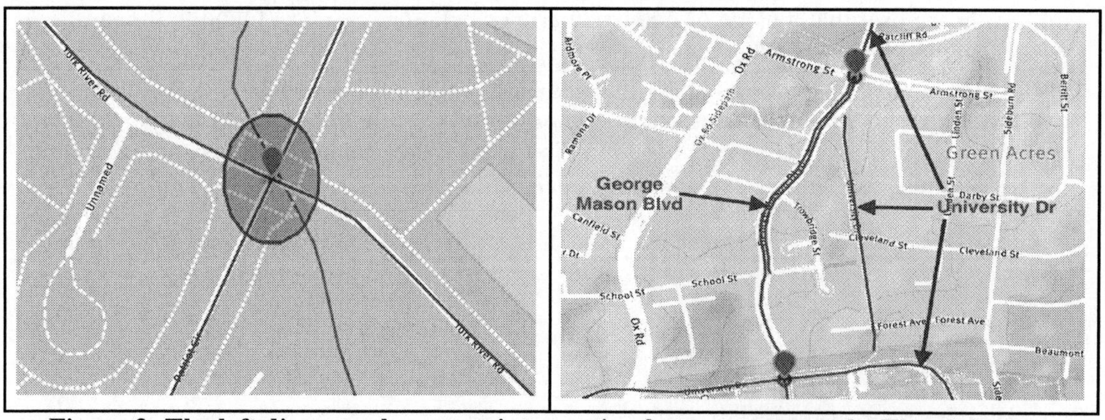

Figure 3: The left diagram shows one intersection between two roads. The right one demonstrates two intersections because one of the roads has three separate segments

3.3 Complex geospatial footprints

The first complex case (Figure 4) is extracting either one point type place and one polygonal place, or two polygons. A polygon can be either a placename of a single feature or a group of placenames. The vertices of the polygonal places are extracted to an array, using a Turf function named "explode". In the case of a point and a polygon, the point is added to the exploded array of points afterward. A convex hull is created based on the array points and then buffered negatively. The reason of buffering negatively is to avoid areas that are not exactly in between the two places. In the case of one point and one polygon, The buffer is clipped by the polygon since the location would be between the polygon and the point and not inside the polygonal feature itself. As for the case of two polygons, The buffer is clipped twice with the two polygons. After clipping, the result is a multipolygon GeoJSON object. Only one polygon is extracted from the multipolygon which intersects with a linestring connecting the polygon's centroid and the point feature or the two polygons' centroids in case of two polygons. This polygon is referred to as the in-between polygon. It covers the possible location indicated by the user. Figure 4 explains the details of selecting the in-between polygon. The explanatory modal, similarly to all cases, describes the details of the algorithms.

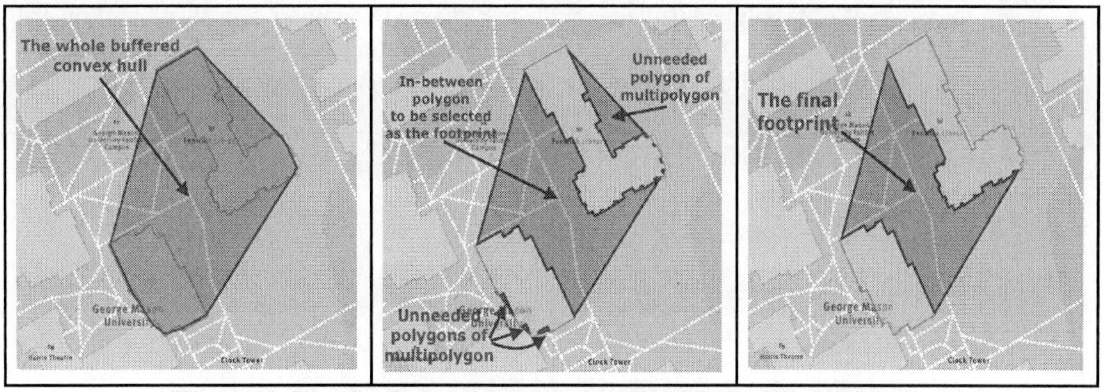

Figure 4: The final transitions in the algorithm of highlighting a location in-between two polygonal places

There are unnamed places such as walkways used to describe a location in relation to named places (Figure 5). The algorithm to select the unnamed places begins similarly to the previous case. Volunteers reported several comments about obstacles on walkways between two polygonal places. The points of the two places are exported to an array. Then a convex hull is created and negatively buffered. The buffer is also clipped by the two polygonal places. The result is a multipolygon and the in between polygon is selected using the linestring connecting the two places' centroids. The unnamed walkways are stored in a separate GeoJSON file on the server. Using jQuery and AJAX, the code iterates the walkways file and uses turf geospatial functions to collect the walkways within the in-between polygon. Figure 5 describes an example of selecting walkways between two places with its explanatory modal. One quick note is that the official building names are David King Hall and Robinson Hall A. The gazetteer has other possible names such as slang placenames, jargon-based placenames, and colloquial or information names. Rob A is a jargon-based name used by students in GMU. King Hall is a common name not the official name.

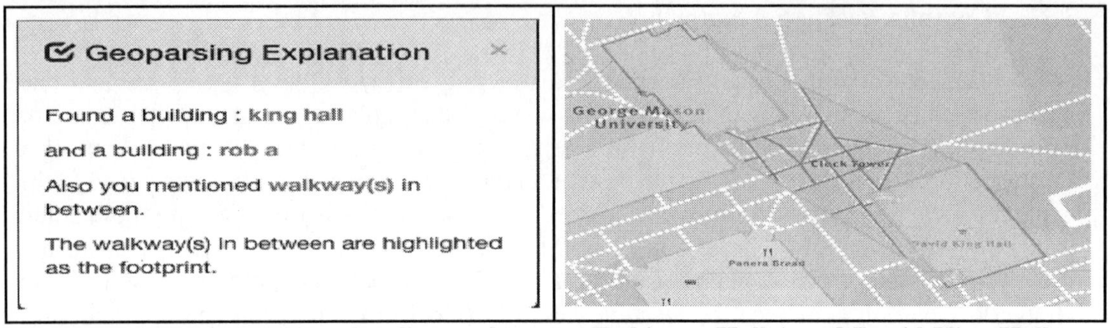

Figure 5: The footprint of walkways between Robinson Hall A and David King Hall

The fourth complex case (Figure 6) is explaining a location along a linear place with proximity to a polygonal place. The algorithm starts with finding the nearest point on the linear places to the centroid of the nearby polygon. Currently the algorithm will select the previous two segments and the next two segments starting from the nearest point selected before. After that, the points of both the polygonal place and the extracted linear segments are extracted to generate the convex hull. The convex hull is clipped by the polygonal place. The clipped convex hull, the polygonal place, and the extracted segments are all plotted on the map. Figure 6 demonstrates the result of the following volunteer's comment: "Directly in front of GMU commerce building, university Dr, chipped pavement along the sidewalk".

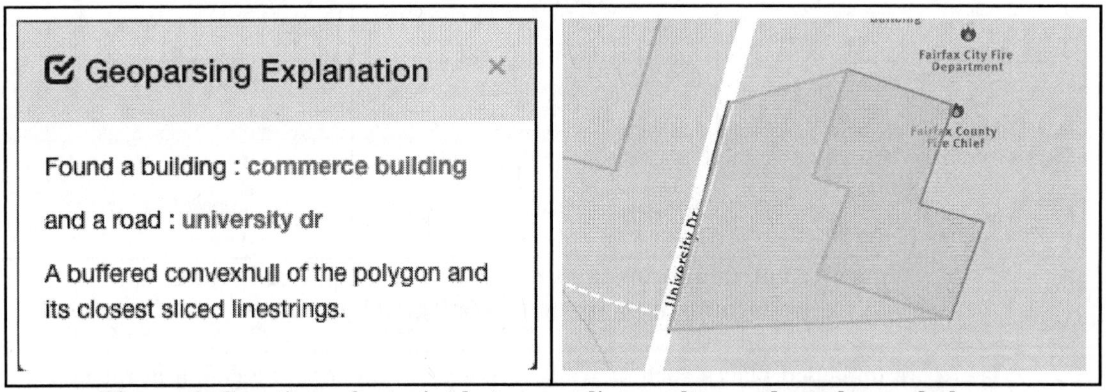

Figure 6: Creating a footprint between a linear place and a polygonal place

Creating a footprint of the case of a bearing word such as north, northwest, west in relation to a polygonal place is challenging (Figure 7). The algorithm begins by creating a rectangular envelope around the polygon. Then the width and length of the envelope are computed to calculate the diagonal of the envelope. Again this is all done through TURF.js geospatial functions. A point is created in the direction of the bearing word with half the diagonal as the distance. The new created point is then buffered, but the buffer could intersect with the polygonal place. If so, the algorithm will clip the polygonal place's part out to give the final geospatial footprint. Figure 7 demonstrates the algorithm's results for three directions: north, northwest, and west, for comparison. The polygonal place part is clipped out in all three cases. A fourth example is also illustrated showing the use of bearing word in relation to a point type place.

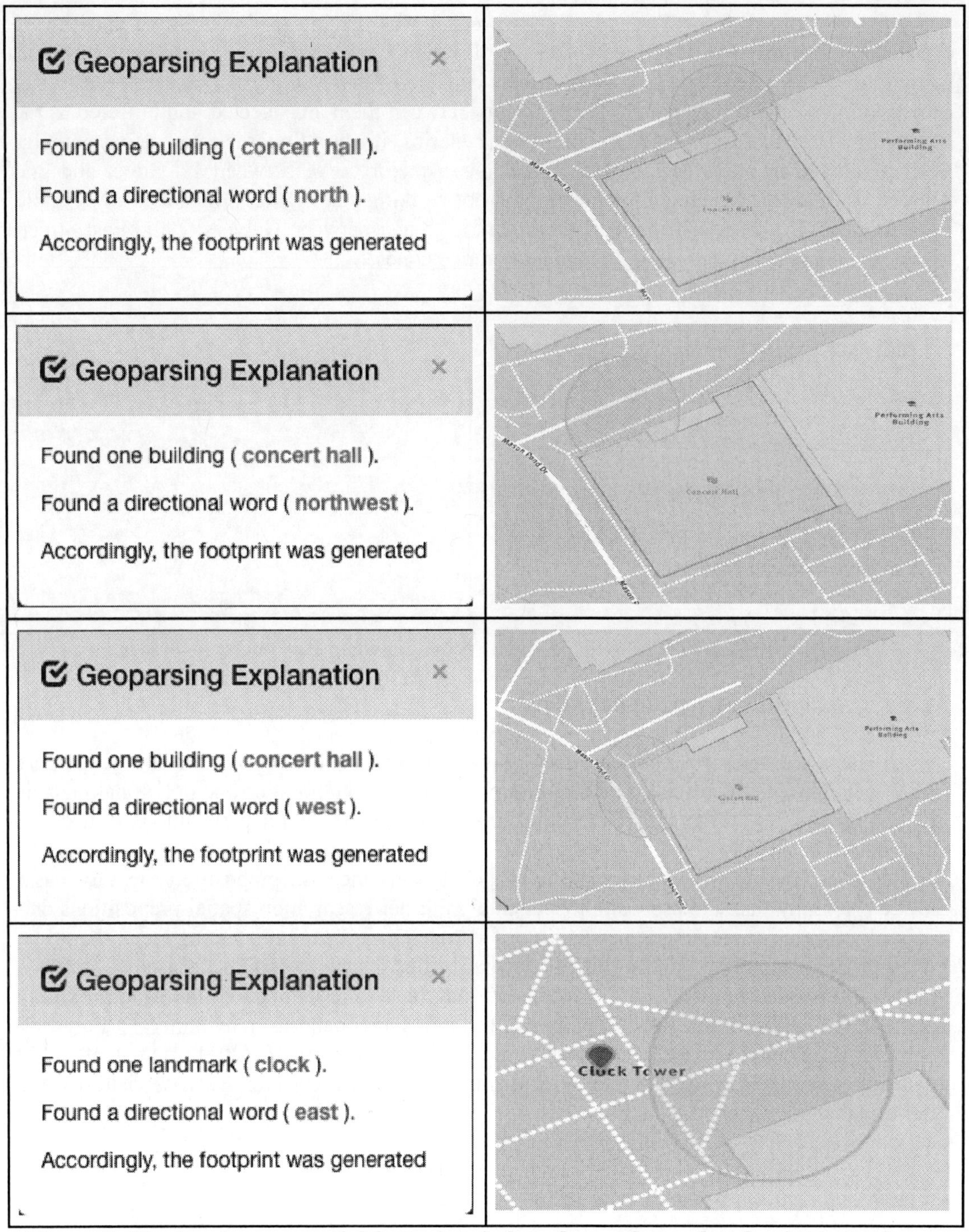

Figure 7: The bearing geospatial footprints in based on polygonal and point type places

The final complex case (Figure 8) is describing a location along a linear place between two intersections with two other linear places. A common example message would be "An X obstacle on Main Street between 1st Street and 2nd Street". The algorithm first determines which linear place should be highlighted. According the previous example, a volunteer could say instead "Between 1st Street and 2nd Street, I was driving on Main Street and saw X obstacle". The order of the linear places is different between the two messages. Different examples were tested and the algorithm was capable of determining the correct order. After the two intersections are created, the segments between them are selected and buffered as the footprint. The algorithm is still missing two scenarios, if Main Street is intersecting with 1st Street more than once, and if Main Street has a major curve between 1st Street and 2nd Street. Both cases are being implemented but not working yet. Figure 8 illustrates an example of this case. The original volunteer's comment is "Sager between University and East Street, fractured concrete covered by plywood and orange cones…".

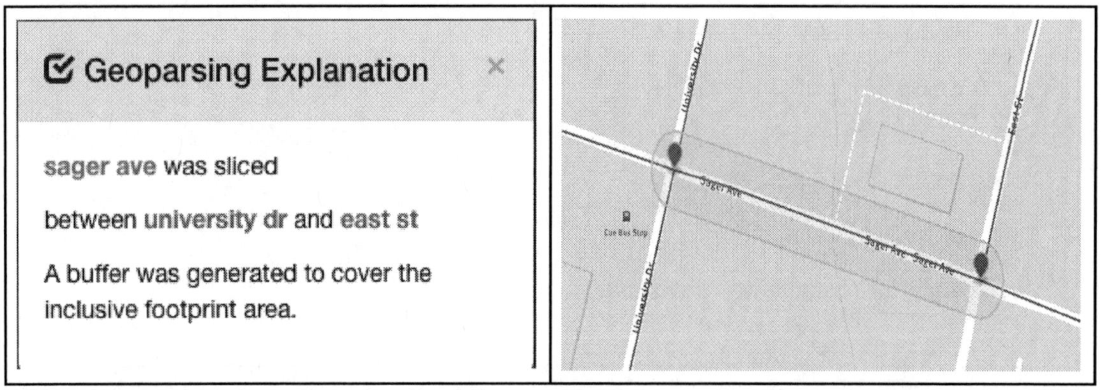

Figure 8: A footprint of a sliced linear place between two linear intersections

3.4 Ambiguous geospatial footprints

There are many examples on the GMU-GcT where the location text and geoparsing code yield ambiguous results. Finding a preposition with one point or one polygonal place is an ambiguous case (Figure 9). Currently the algorithm searches for prepositions in text and then accesses a JSON object that stores buffer distances based on the preposition found. The prepositions are categorized in proximity ranges. For instance, the preposition "next to" has a lower proximity than the preposition "close to". Further research on spatial prepositions and natural language processing is needed in order to determine what processing steps should be undertaken in this case, and specifically, what buffer distances should be used to buffer a place. A surveying of GMU-GcT contributors may be able to help determine what proximity is intended with certain spatial prepositions and some patterns may emerge. Figure 9 emphasizes examples based on different prepositions. Rice *et al.* (2011) note that buffer distances associated with spatial prepositions may be related to factors such as visibility and lighting that are not captured in the GMU-GcT.

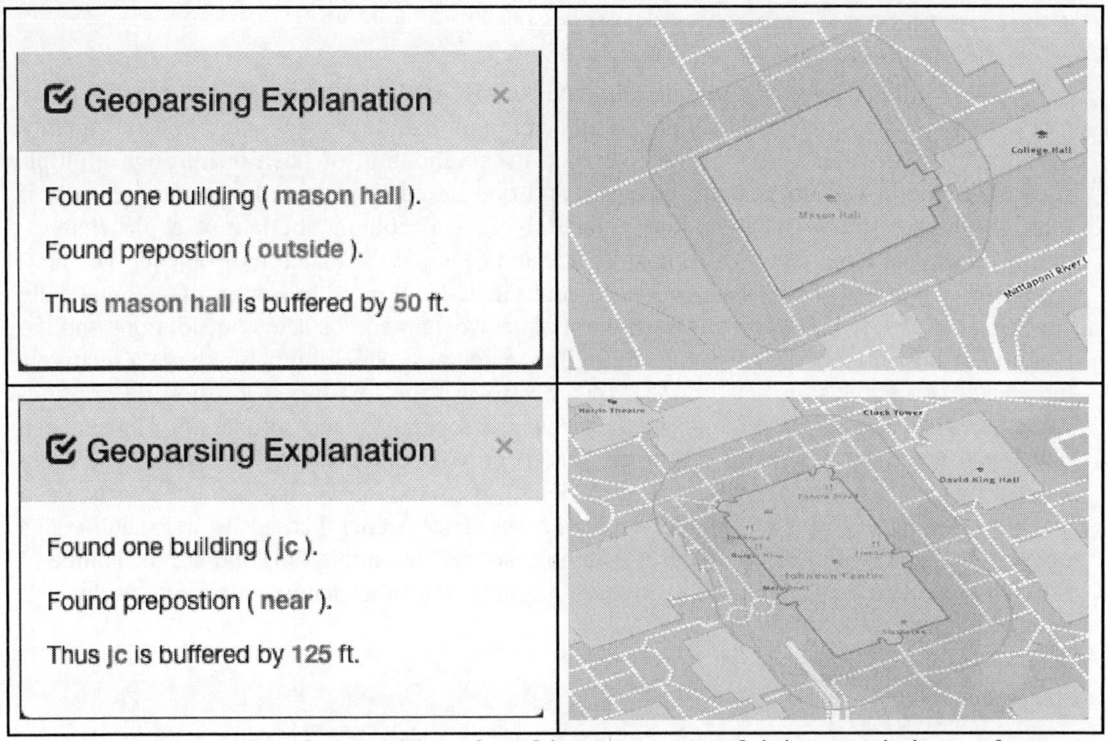

Figure 9: Examples of prepositions found in messages explaining proximity to places

The other ambiguous case is explaining the location by only one linear place(Figure 10). For instance, a volunteer reporting an obstacle on Main St. Main St in Fairfax VA is roughly 3 miles long. The location of the obstacle cannot be determined unless the user gives another placename to clarify the location. The code will inform the user about this ambiguity in a warning modal without generating a geospatial footprint. Figure 10 shows an example of such warning modal. This warning modal may be unnecessary or may be suppressed if additional positional information (such as an image geotag or user-asserted map position) is present. In these instances, the ambiguous geospatial footprint provides only general confirmation that the obstacle is close to or associated in some way with the linear place.

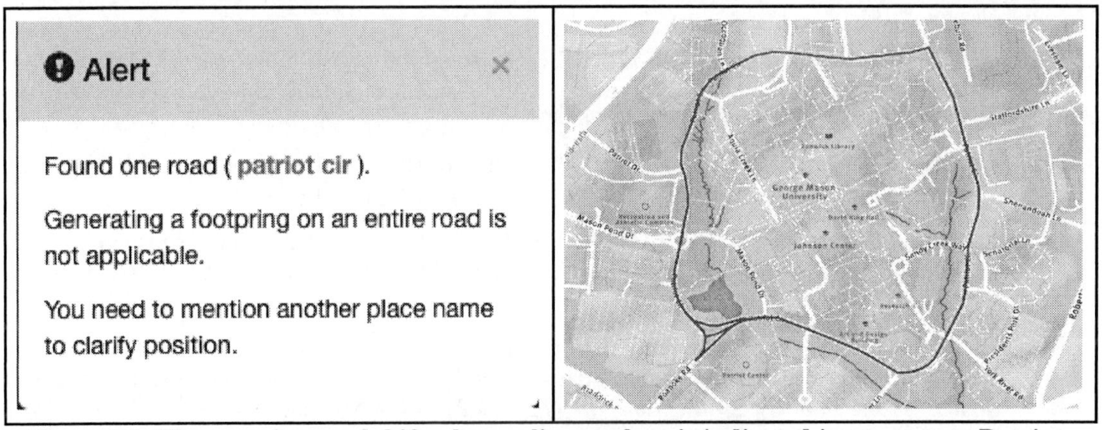

Figure 10: The warning modal if only one linear place is indicated in a message. Patriot Circle is highlighted on the right image

4. DISCUSSION AND CONCLUSIONS

The GMU Geocrowdsourcing Testbed (GMU-GcT) was developed to provide a mechanism to map transient navigation obstacles in real time. A comprehensive quality assessment system has been developed which uses validation of position through multiple sources. A useful way to validate position, and one based on natural human expression, is look at the descriptive location text entered by a contributor. Humans use placenames, prepositions, and directional words to organize and describe the location of objects. The text-based descriptions of obstacle locations in the GMU-GcT provide a way of automatically checking the asserted, map-based positioning of crowdsourced obstacle contributions and the position of obstacles as determined through image geotags, which by nature have a relatively high level of imprecision (Rice 2015). In order to use the text-based location descriptions contributed to the GMU-GcT, gazetteer-based geoparsing and processing have been developed to create geospatial footprints for geocrowdsourced obstacle reports, which are used to validate position. Hundreds of obstacle reports to the GMU-GcT have been processed and analyzed to develop the presented general cases. These general cases allow for automated position validation in a variety of cases. Some cases remain ambiguous and are the source of present and future work, which will involve linguists and other language experts to develop additional general cases.

5. REFERENCES

Aburizaiza, A.O., and M.T. Rice, 2011. "VGI and Geotechnology for Supporting Blind and Vision-Impaired People using a Localized Gazetteer." *FOSS4G: Free and Open Source Software for Geospatial, 2011.* Denver, Colorado.

Elwood, S. (2008). Volunteered geographic information: key questions, concepts and methods to guide emerging research and practice. *GeoJournal*, 72(3-4), 133–135. http://doi.org/10.1007/s10708-008-9187-z

Goodchild, M. F. (2007). Citizens as sensors: The world of volunteered geography. *GeoJournal*, 69(4), 211–221.

Haklay, M. (2010). How good is volunteered geographical information? A comparative study of OpenStreetMap and Ordnance Survey datasets. *Environment and Planning. B, Planning & Design*, 37(4), 682.

Karimi, H. A., Zhang, L., & Benner, J. G. (2014). Personalized accessibility map (PAM): a novel assisted wayfinding approach for people with disabilities. *Annals of GIS*, 20(2), 99–108. http://doi.org/10.1080/19475683.2014.904438

Longley, P. A., Goodchild, M. F., Maguire, D. J., & Rhind, D. W. (2011). *Geographic Information Systems and Science* (3rd edition). Hoboken, New Jersey: John WIley & Sons.

Qin, H., Rice, R., Fuhrmann, S., Rice, M., Curtin, K., & Ong, E. (2015). Geocrowdsourcing and accessibility for dynamic environments. *GeoJournal*, 1–18. http://doi.org/10.1007/s10708-015-9659-x

Rice, M. T., Aburizaiza, A. O., Jacobson, R. D., Shore, B. M., & Paez, F. I. (2012a). Supporting Accessibility for Blind and Vision-impaired People With a Localized Gazetteer and Open Source Geotechnology. *Transactions in GIS*, 16(2), 177–190. http://doi.org/10.1111/j.1467-9671.2012.01318.x

Rice, M. T., Curtin, K. M., Paez, F. I., Seitz, C. R., & Qin, H. (2013a). Crowdsourcing to Support Navigation for the Disabled: A Report on the Motivations, Design, Creation and Assessment of a Testbed Environment for Accessibility (US Army Corps of Engineers, Engineer Research and Development Center, US Army Topographic Engineering Center Technical Report, Data Level Enterprise Tools Workgroup No. BAA: #AA10-4733, Contract: # W9132V-11-P-0011) (pp. 1–62). Fairfax, VA: George Mason University. Retrieved from http://oai.dtic.mil/oai/oai?verb=getRecord&metadataPrefix=html&identifier=ADA588474

Rice, M. T., Hammill, W. C., Aburizaiza, A. O., Schwarz, S., & Jacobson, R. D. (2011). Integrating User-contributed Geospatial Data with assistive Geotechnology Using a localized Gazetteer. In A. Ruas (Ed.), *Advances in Cartography and GIScience*. Volume 1 (pp. 279–291). Springer Berlin Heidelberg. Retrieved from http://dx.doi.org/10.1007/978-3-642-19143-5_16

Rice, M. T., Jacobson, R. D., Caldwell, D. R., McDermott, S. D., Paez, F. I., Aburizaiza, A. O., … Qin, H. (2013b). Crowdsourcing techniques for augmenting traditional accessibility maps with transitory obstacle information. *Cartography and Geographic Information Science*, 40(3), 210–219. http://doi.org/10.1080/15230406.2013.799737

Rice, M. T., Paez, F. I., Mulhollen, A. P., Shore, B. M., & Caldwell, D. R. (2012b). Crowdsourced Geospatial Data: A report on the emerging phenomena of crowdsourced and user-generated geospatial data (Annual No. AA10-4733). Fairfax, VA: George Mason University. Retrieved from http://www.dtic.mil/dtic/tr/fulltext/u2/a576607.pdf

Rice, M. T., Paez, F. I., Rice, R. M., Ong, E. W., Qin, H., Seitz, C. R., … Medina, R. M. (2014). Quality Assessment and Accessibility Applications of Crowdsourced Geospatial Data: A report on the development and extension of the George Mason University Geocrowdsourcing Testbed (Annual No. BAA: #AA10-4733, Contract: # W9132V-11-P-0011) (p. 91). Fairfax, VA: George Mason University.

Rice, Rebecca M. (2015).Validating VGI Data Quality in Local Crowdsourced Accessibility Mapping Applications: A George Mason University Case Study (Master of Science Thesis, July 2015). George Mason University, Fairfax, VA.

Rice, R.M., A.O. Aburizaiza, M.T. Rice, and H. Qin. (2015) "Position Validation in Crowdsourced Accessibility Mapping", *Cartograpica* (in press).

Free and Open Source Software for Geospatial (FOSS4G) Conference Proceedings

Volume 15 *Seoul, South Korea* Article 50

2015

Performance Analysis of MongoDB Vs. PostGIS/ PostGreSQL Databases For Line Intersection and Point Containment Spatial Queries

sarthak agarwal
International Institute of Information Technology Hyderabad Gachibowli, Hyderabad, India

KS Rajan
International Institute of Information Technology Hyderabad Gachibowli, Hyderabad, India

AT1-03

Performance Analysis of MongoDB Vs. PostGIS/PostGreSQL Databases For Line Intersection and Point Containment Spatial Queries

Sarthak Agarwal[1] and KS Rajan[2]

[1]Lab for Spatial Informatics, International Institute of Information Technology
Gachibowli, Hyderabad 500032, India
Email: sarthak.a@research.iiit.ac.in
[2]Lab for Spatial Informatics, International Institute of Information Technology
Gachibowli, Hyderabad 500032, India
Email: rajan@iiit.ac.in

ABSTRACT

Relational databases have been around for a long time and Spatial databases have exploited this feature for close to two decades. The recent past has seen the development of NoSQL non-relational databases, which are now being adopted for spatial object storage and handling too. And this is gaining ground in the context of increased shift towards GeoSpatial Web Services on both the Web and mobile platforms especially in the user-centric services, where there is a need to improve the query response time. While SQL databases face scalability and agility challenges and fail to take the advantage of the cheap memory and processing power available these days, NoSQL databases can handle the rise in the data storage and frequency at which it is accessed and processed - which are essential features needed in geospatial scenarios, which do not deal with a fixed schema(geometry) and fixed data size.

This paper attempts to evaluate the performance of an existing NoSQL database 'MongoDB' with its inbuilt spatial functions with that of a SQL database with spatial extension 'PostGIS' for two primitive spatial problems - LineIntersection and Point Containment problem, across a range of datasets, with varying features counts. For LineIntersection function, the dataset consisted of two independent layers of horizontal lines and vertical lines with incremental lengths and their size varied from ten lines to million lines in each layer and another dataset with two layers, one of random lines of variable size and shape and another layer of a single line which is intersecting many lines of layer1. For Point Containment problem, the dataset consists of two layers, one of polygons in space of different shape and size and another layer of random points in the space, some inside the polygons and some outside.

All the data in the analysis was processed In-memory and no secondary memory was used. Initial results suggest that MongoDB performs better by an average factor of 25x for Line Intersection Problem and 10x for Point Containment Problem which increases exponentially as the data size increases in both indexed and non-indexed operations. Given these results NoSQL databases may be better suited for simultaneous multiple-user query systems including Web-GIS and mobile-GIS. Further studies are required to understand the full potential of NoSQL databases across various geometries and spatial query types.

1. INTRODUCTION

Traditionally Databases were designed to structure and organize any form of data. But as the database size increased and to optimize databases for Geo Spatial domain we use Spatial Databases. Satellite images are one prominent example of spatial databases. To extract spatial information from a satellite image it has to be processed with respect to a spatial frame of reference. But satellites are not the only type of spatial databases. Maps are also stored in spatial database.

Like other database systems Spatial databases have also relied on the Relational databases to handle and manage spatial objects and their associated attribute information. They have been of great value in cases where we have a defined structure of our schema, including geometrical characteristics of the spatial objects. While these have been able to store and manage very large datasets, their inherent need to pre-define its structure, database schema, makes it difficult to remodel or change it as per the evolving need, especially in the context of user-driven data collation and collaboration. In addition, in many spatial applications we do not always have a fixed schema, there can be many geometries with different shape and the requirements evolve as the data size increases depending on the case scenario where relational Databases can limit the potential use or design of the solution.

While NoSQL is an umbrella term for a loosely defined class of non-relational datastores and best described as 'Not only SQL'. Examples are Google's BigTable and Amazon's Dynamo. The recent past has seen the development of NoSQL non-relational databases, which are now being adopted for spatial object storage and handling too. And this is gaining ground in the context of increased shift towards GeoSpatial Web Services on both the Web and mobile platforms especially in the user-centric services, where there is a need to improve the query response time, these databases can handle the rise in the data storage and frequency at which it is accessed and processed - which are essential features needed in geospatial scenarios, which do not deal with a fixed schema(geometry) and fixed data size.

Relational databases are not designed to function with distributed systems; combining multiple tables across different computers is difficult and gives mostly a decrease in performance while NoSQL are distributed databases which can be spread over multiple servers and queried simultaneously. Relational databases are good for structured data. For example, sales figures fit well in organized tables. Unstructured data, such as points and lines over a network are not very suitable for a RDBMS, while NoSQL database systems are schema less database systems where we can append any number of geometry columns in a table and store multiple geometries within a same column in a table which is very essential property with respect to spatial context.

NoSQL data stores may provide advantages over relational databases, but generally lack the robustness of relational databases for those advantages. The aim of this paper is to discover some advantages of a selected NoSQL data store as compared to a traditional relational database when storing and querying spatial vector data.

This paper attempt to evaluate the performance of an existing NoSQL database and SQL database with respect to LineIntersection problem and Point Containment problem across a range of datasets. We here are interested in observing which data model performs better with these problem in terms of time complexity without accounting for indexing of the data.

LineIntersection query is the most popular spatial query in which we report the Intersection between two spatial geometries. Geometries can be anything from a point to a polygon. Line Intersection helps in routing as well as mapping various geometries. Suppose we have to make a route from point A to point B ,then from every we point we will find out the lines intersecting it to create a graph using this technique and apply our routing algorithm on that graph.

Like Intersection, Point containment problem is also a very important and popular spatial query problem which reports whether a given geometry lies inside another geometry or not. This problem helps in modeling and analyzing spatial data over a area. For instance, we want to report how the number of houses have changed over a time in a city by analyzing the spatial data of that city. We can count the number of points(which represents each house in this case) in the polygon(city) for all the years using point within a polygon problem.

2. REVIEW OF SPATIAL DATABASES

Similar to traditional databases, Spatial databases were also written initially in RDBMS but as technology evolved they are being now adapted to NoSQL .

2.1 RDBMS(PostGIS / PostgreSQL)

PostgreSQL is a powerful, object-relational database management system (ORDBMS). It is released under a BSD-style license and is thus free and open source software. PostGIS adds support for geographic objects to the PostgreSQL object-relational database. The interface language of the PostgreSQL database is the standard SQL, which allows for inserts, updates and queries of data stored in relational tables. Data operations for PostgreSQL are created in SQL.

For several decades, the most common data storage model for geographic data and associated attribute data has been the relational data model.

The functions in PostGIS can be put into following 5 categories.

- Management: Functions that manage information about spatial tables and PostGIS administration.
- Conversion: Functions that convert between geometries and external data formats.
- Retrieval: Functions that retrieve properties and measurements of a geometry.
- Comparison: Functions that compare two geometries with respect to their spatial relation.
- Generation: Functions that generate new geometries from others

Management depends on other features of a database. Conversion and Retrieval are relatively simple and they only rely on the data format. An example could be to answer "what's the area of my current state?" Comparison and Generation are about geometry relationships, which is the hardest part.

2.2 B. NoSQL (MongoDB)

MongoDB is a document oriented datastore. It is written in C++ and its main focus lies on high performance and retaining some friendly properties of SQL.

Currently, MongoDB uses GeoJSON objects to store spatial geometries. GeoJSON is an open-source specification for the JSON- formatting of shapes in a coordinate space. The GeoJSON spec is used in the geospatial community and there is growing library support in most popular languages.

Each GeoJSON document (or sub document) is generally composed of two fields:

1. Type – the shape being represented, which informs a GeoJSON reader how to interpret the “coordinates” field.

2. Coordinates – an array of points, the specific arrangement of which is determined by “type” field.

Overview of MongoDB's current implementation

- There are multiple geospatial indexes per collection in MongoDB- 2d, 2dsphere and geoHaystack
- Data can be imported from CSV files by converting it into GeoJSON objects.
- However MongoDB does not support R-trees.

Currently in MongoDB, containment and intersection tests are supported for both flat and spherical geometries. Sorting by distance is supported too. This is because these operations are the most popular operational queries for a modern application. They meet most requirements of typical location based applications.

3. COMPARATIVE ANALYSIS OF SPATIAL FUNCTIONS

Each of the implementation of spatial databases has a different approach for handling geometries.

3.1 Geometry / Flat Shape

PostGIS supports various operations for geometry , while MongoDB only supports storing points and containment / intersection queries with box/circle/polygon. 2d index is designed to speed up this case. Although it is possible to leverage open source libraries like GEOS, MongoDb don't plan to expand the features for flat geometry, because most LBS applications work with GPS data, which nicely fits with spherical model.

3.1 Geography / Spherical Shape

MongoDB supports storing and querying spherical shapes in GeoJSON. 2dsphere index is for this case. PostGIS only supports containment / intersection natively, which are supported by MongoDB as well. Distance between two objects is also supported in PostGIS.

4. DATASETS

For this analysis the dataset was custom generated.

4.1 Line Intersection

The Line Intersection problem works for any geometry in the space and report whether or not that geometry is being intersected by some other geometry. This problem is very important in real time scenarios for example - we want to know on which side of the road are we and also to avoid accidents we need intersection of various geometries. So we generated several synthetic datasets keeping in mind worst case scenarios.

The dataset consisted of two independent layers of horizontal lines and vertical lines with incremental lengths and their size varied from ten lines to ten million lines in each layer. Each line intersects at least one line in the other layer with some lines intersecting multiple lines. The above dataset emulates most of the real time cases with worst case scenarios. All the data in the analysis was processed using In-memory and no secondary memory was used.

4.2 Point Containment Problem

Like the previous problem Point Containment problem also works for any geometry and reports whether the given geometry is completely inside another geometry or not. This is a very important and popular problem with respect to spatial databases. It is useful in the domain of map generations, modeling, analyzing spatial data over a area for example we want to report how the number of houses have changed over a time in a city by analyzing the spatial data of that city. We can count the number of points(which represents each house in this case) in the polygon(city) for all the years using point within a polygon problem.

The dataset consisted of two independent layers, one of horizontal lines of incremental length and the size of the layer varied from 10 lines to 10 million lines. Another layer consisted of square boxes of different perimeter scattered both sequentially and randomly over the space with few lines completely inside the box, few intersecting the box and few completely outside the box. All the data in the analysis was processed using In-memory and no secondary memory was used.

5. PERFORMANCE

Tests were run on both of the database systems with same datasets one with Index and other without index. The time recorded are in seconds for both indexed and non-indexed analysis which are mentioned in table 1 below.

5.1 Line Intersection

Table 1. Time vs Intersection Table

No. of Intersections	MongoDB		PostGIS	
	Non-Index	*Index*	*Non-Index*	*Index*
21	1	1	1.546	9.721
1875	18	20	88.695	48.461
195691	185	190	13364.596	1963.123
< (1000*1000)	4093	3140	>1500000	172048

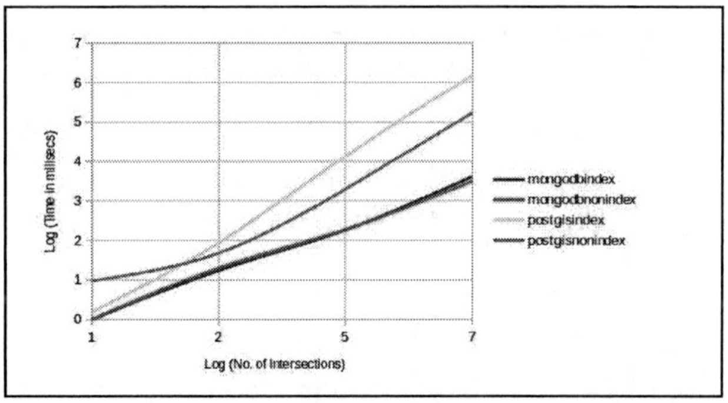

Figure 1. Graph of Intersections vs time (for all implementations)

5.2 Point Containment Problem

Table 1. Points within a polygon vs Time for all implementation.

No of Points Within	Mongo Non-Index	Mongo Index	PostGIS Non-Index	PostGIS Index
5002	6.75	7.75	52.95	54.97
5471	12	12	85.422	77.508
17004	20.8	24	142.53	158.76
45999	27.66	27.33	160.02	168.23
46008	50	51	675.33	631.556
122006	51.667	53.33	1339.23	1110.42
122016	96	99	2783.66	2609.47
204024	99.667	106	3991.564	3121.11223
298.23	125	112	-	-

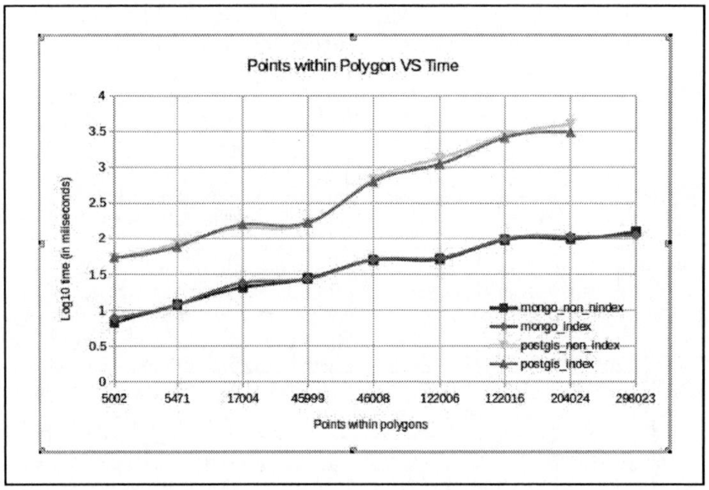

Figure 1. Graph of Point within Vs Time

6. RESULTS

Firstly the performance was analyzed without indexing any geometry and time was observed which suggests that MongoDB performs better as the datasize increases whereas PostGIS fails at very large datasets and time complexity increases exponentially. But after indexing the geometries performance of both the database engine improved by a substantial factor.

If we observe the results above we will notice that indexed datasets performs better than non indexed dataset and PostGIS time increases exponentially as size of dataset increases whereas MongoDB still performs within some bounds.

7. DISCUSSIONS AND CONCLUSIONS

The above analysis was necessary to evaluate the performance of NoSQL with respect to spatial databases and to tell whether NoSQL performs better than SQL in our current scenario. These results suggest that MongoDB performs better by an average factor of 25x which increases exponentially as the data size increases in both indexed and non-indexed operations for both LineIntersection and Point Containment problem. Given these results NoSQL databases may be better stated for simultaneous multiple-user query systems including Web-GIS and mobile-GIS.

This implies that non-relational databases are more suited to the multi-user query systems and has the potential to be implemented in servers with limited computational power. Further studies are required to identify its appropriateness and incorporate a range of spatial algorithms within non-relational databases. In future we are planning on expanding our study to other spatial query functions as well as spatial algorithms such as shortest problem to evaluate the performance of NoSQL on such platforms.

8. ACKNOWLEDGMENT

We would like to acknowledge the responses and interactions with MongoDB developers through emails which helped during this research project.

9. REFERENCES

References in the text should follow the standard format (Brown, 1997). As Brown (1997) suggested, they should be in the format shown immediately below. References are in 11 pt type.

Journal titles should be in full. Journal and book titles should be italicised. Multiple authors should all be listed.

Baas, B. Quak, W. Van Oosterom, P. De By, *NoSQL spatial: Neo4j versus PostGIS*

Cláudio de Souza Baptista, Maxwell Guimarães de Oliveira ,Tiago Eduardo da Silva. *Using OGC Services to Interoperate Spatial Data Stored in SQL and NoSQL Databases*

Dr. Johannes Scholz, TU Vienna, Gusshausstr. *Coping with Dynamic, Unstructured Data Sets – NoSQL: a buzzword or a savior?*

Zhifeng Xiao, Yimin Liu. *Remote sensing image database based on NOSQL database.*

http://nosql.mypopescu.com/post/300199706/geo-nosql-couchdb-mongodb-tokyo-cabinet

AT2-01

Geographical Weighted Regression Model for Improved Near-shore Water Depth Estimation from Multispectral Imagery

Poliyapram Vinayaraj[1] , Venkatesh Raghavan[1] and Shinji Masumoto[2]
[1]Media Center, Osaka City University
3-3-138 Sugimoto, Sumiyoshi-ku, Osaka 558-8585, Japan
[2] Graduate School of Science, Osaka City University, Japan
Email: vinay223333@gmail.com

ABSTRACT

There is often a need for making a high-resolution or a complete bathymetric map based on sparse point measurements of water depth. The common practice of previous studies has been to calibrate a single global depth regression model for an entire image. The performance of conventional global models is limited when the bottom type and water quality vary spatially within the scene. For a more accurate and robust water-depth mapping, this study proposes a regression model for a geographical region or local area rather than using a global regression model. Here , two different methods were discussed according to the availability of reference depth data to ensure the accuracy of the Geographical Weighted Regression (GWR) model. The entire data analysis workflow was carried out using GRASS GIS Version 7.0.1. GWR is seen to be effective in addressing the problem introduced by heterogeneity of the bottom type and provide better bathymetric estimates in near coastal waters. The study was carried out at Puerto Rico, Northeastern Caribbean Sea. High spatial resolution (RapidEye) and Medium spatial resolution (Landsat 8) images were used to estimate the depth. in order to calibrate and evaluate estimated depth, collected high accuracy LiDAR depth data (4m resolution) provided by NOAA. Comparison of results indicates that the GWR model improves the depth estimation significantly. The maximum depth calculated by this methid is 20 meter. In order to estimate the accuracy of the results in terms of quantitative analysis have been carried out by R,R^2 and RMSE. The results reveal the significant improvements by GWR model than Global model from both Landsat 8 [GWR model (R=0.97,R^2=0.93 ,RMSE=1.41m), Global model (R=0.88,R^2=0.78 ,RMSE=2.63m)]and RapidEye [GWR model (R=0.97,R^2=0.93 ,RMSE=1.34m), Global model (R=0.88,R^2=0.78 ,RMSE=2.60)] data.

1. INTRODUCTION

Accurate and high-resolution bathymetric data are a necessity for a wide range of coastal oceanographic research, but become particularly critical in predictive modeling approaches for the proactive management of impacts on near-shore. Mapping shallow water depth data from ships by Sonar is a quite an expensive task. Many shallow water areas are not accessible by hydrographic ships due to rocks, coral reefs or simply due to inaccessibility by boats due to shallow depth of the water. Airborne Light Detection and Ranging (LiDAR) can provide complete and accurate bathymetric measurements in shallow areas, but availability of this technology is currently limited and also involves significant cost. Thus, alternative method to estimate high resolution depth data need to be devised and validate. Until now, several algorithms have been proposed to estimate bathymetry from single band [5] and multispectral bands (Lyzenga, 2006, Stoffle, 1991 and Stumpf, 2004). The multispectral algorithms were commonly practiced to estimate good accuracy depth estimates by addressing the problem introduced by heterogeneous bottom type. Therefore this study was carried out to investigate depth estimation from multispectral imagery by conventional global regression and Geographically Weighted Regression (GWR) model.

2. MATERIALS AND METHODOLOGY

The study was carried out at Puerto Rico, northeastern Caribbean Sea. Two different satellite data were collected in order to test the algorithm with high and moderate spatial resolution data (RapidEye and Landsat 8). RapidEye data has 12-bit radiometric resolution and 5 meter spatial resolution. Even though the Landsat 8 data also has 12-bit radiometric resolution, it provides 30m spatial resolution. In order to calibrate and evaluate estimated depth, collected high accuracy LiDAR depth data (4m resolution) provided by NOAA (https://catalog.data.gov/dataset/). Five spectral bands collected from Landsat data are used for depth estimation (visible band (0.43-0.67μm)and Near-infra red band (0.85-0.88μm)) and Short wave infrared band is effectively used for correction as shown in equation 1. In the case of RapidEye data four visible bands (0.44-0.73μm) were used for estimation and Near infrared band (0.76-0.85μm) used for correction as shown in equation 1.

2.1 Pre-proccesing

The radiance observed from space using a visible light sensor over optically shallow water primarily consists of four components: atmospheric scattering component (path radiance), surface reflection component, in-water volume scattering component, and bottom reflection component. Primarily, a correction method is applied remove atmospheric scattering component, surface reflection component and in-water volume scattering component to retrieve only the bottom reflection component which is linearly related to depth Kanno, 2012). The correction algorithm (shown below) is a modification to the Kanno *et al.*, 2012.

$$X(\lambda)i = log\ [(L\ \lambda\ i - \alpha 0 - \alpha 1\ L\lambda IR)/L\lambda\ i] \tag{1}$$

Where, X(λ) i is the log transformed corrected band i, L λi is the radiance of band i, α0, α1 are the coefficients, LλIR is the radiance of infrared band. More detailed description about the algorithm is given in (Vinayaraj *et al.*, 2014). Here, LλIR was the infrared band used to correct the atmosphere, water surface and water column reflectance. The wavelengths used for correction of Landsat 8 and RapidEye images are 1.57-1.65 μm and 0.76 – 0.85 μm respectively. The previous study (Vinayaraj *et al.*, 2014) reveals that short wave infrared band is better to correct the radiance than near infrared band. Since the RapidEye image does not have a short-wave infrared band Near infrared band (0.76–0.85 μm) has been used for correction process.

2.2 Depth estimation

The two different method applied in this study are the conventional global regression model and GWR model.

2.2.1 *Global regression model*

The conventional regression models are a common practice by several authors (Kanno, 2012 and Lyaenga, 2006) in order to address the heterogeneity by bottom type and water quality. It is a simple multiple linear regression where, multispectral bands were taken as independent variables. The equation is shown below,

$$h=\beta 0+\beta 1X(\lambda)1+\beta 2X(\lambda)2+\ldots\ldots+\beta nX(\lambda)m \tag{2}$$

Where, h is the estimated depth, β0 and βn are the intercept and slope derived from the multiple regression. X(λ)n is the log transformed bands (derivation is shown in equation (1)). The conventional global regression model uses a single set of coefficients (β0, βn) for the whole image.

2.2.1 *GWR model*

To address the inadequacy of the global bathymetric inversion model, we apply the GWR model that was originally was proposed by (Brunsdon, 1996) for improved depth retrieval in complex and heterogeneous near coastal environments. The spatial non-stationarity in the image scene was addressed by this method. Similar to the kernel regression method, the GWR model uses a window to define the local neighborhood. Kernel size was decided according to the density of the reference depth and pixel size of the satellite imagery used. The equation is denoted as follows,

$$hj=\beta 0j+\beta 1jX(\lambda)1j+\beta 2jX(\lambda)2j+\ldots\ldots+\beta njX(\lambda)mj \qquad (3)$$

Where, j is the respective pixel, therefore coefficients for each pixel (j) was derived by assigning a fixed bandwidth. The bandwidth represents the size of the kernel used for regression. The pixels closer to the kernel center were assigned higher weights. The script called "r.gwr" available in GRASS GIS Version 7.0.0. was used to calculate GWR model.

3. DATA ANALYSIS

Data analysis in this research is carried out by considering the reference depth at different level in such a way that the different cases of common availability of reference depth could be addressed. Meaning that commonly available reference depth data are of two types, one is high-resolution echo sounder or airborne LiDARs available for a small region and the second one is low resolution data extracted from bathymtric charts. The reference depth used in this study for depth estimation analysis was derived from LiDAR survey. Study consider these scenarios of reference depth availability and evaluate the performance of both GWR model and Global model in order to estimate the near-shore depth from Landsat 8 and RapidEye images. Scenario 1 is to estimate depth from randomly distributed reference depth points, scenario 2 is to estimate the depth from 300 meter interval reference depth points, scenario 3 is to estimate depth from 600 meter interval reference depth points and scenario 4 is to estimate (extrapolate) depth from reference depth point available at small area to a complete near-shore (<=20m depth) region covering by the image scene. This scenario 4 is basically a new method proposed by this study to estimate the bottom class based coefficients and derive depth where no reference depth available. By using maximum likelihood analysis bottom classification of the study area is carried out and six classes were derived. Then, coefficients for each class is determined by averaging the coefficients derived from GWR model from each classes. Further, these class based coefficients are used extrapolate the depth from small region to all study area where no reference depth available. Here onwards, scenario numbers mentioned above are used to describe results in each case.

4. RESULT AND DISCUSSION

Previous studies have been observed (Vinayaraj *et al.*, 2014) that the global model is not efficient to address the non-stationarity in the near-shore due to the difference bottom type and water quality. Also observed that global model using single coefficients derived from each band cannot be sufficient to address the difference in the reflectance of each band. Hence, this study utilize the efficacy of GWR model in addressing the non-stationarity (Brunsdon, 1996) to near-shore water depth estimation algorithm. Here, four different scenarios were tested to evaluate performance of the GWR model and compare the derived results with different scenarios were carried out. The minimum and maximum of the reference depth value used for calibration of algorithms was set to 1m and 20m respectively. The minimum set as 1m in order to avoid error would have occurred due to tidal variation between data sets processed. The evaluation tests of the estimated results are carried out by using randomly selected data which were not included in the calibration or estimation process.

Scenario 1: Randomly selected 60,000 and 10,000 reference depth points were selected and used for calibration and coefficient determination from RapidEye and Landsat 8 data respectively. Scenario 1 was carried out to examine the efficacy of GWR model in estimating depth in near-shore waters.

Table 1. Results of evaluation of depth estimates derived from Landsat 8 and RapidEye data at Scenario 1

Satellite data	Global model			GWR model		
	R	R^2	RMSE (m)	R	R^2	RMSE (m)
Landsat 8	0.88	0.78	2.63	0.97	0.93	1.41
RapidEye	0.88	0.78	2.48	0.97	0.93	1.34

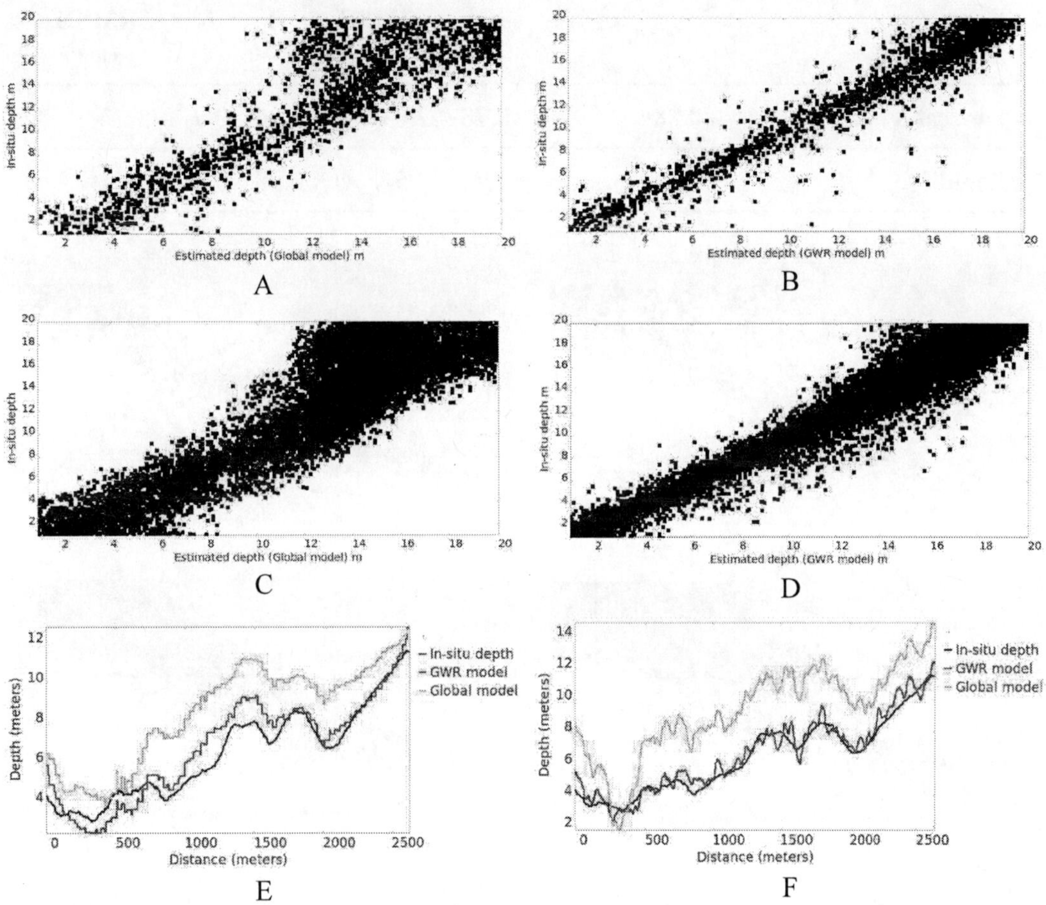

Figure 1. Evaluation of estimated depth derived from Landsat 8 and RapidEye data with reference depth for Scenario 1. Scatter plot of Landsat 8 (Global (A), GWR(B)), scatter plot of RapidEye (Global (C), GWR(D)) and profile of drawn on Landsat 8 (E) and RapidEye (F).

The results shown in Table 1 derived from both Landsat and RapidEye demonstrate the efficacy of GWR model. The comparative study has shown that GWR model improves the accuracy of the estimation significantly than the conventional global model. A bi-variate scatter plot shown in Figure1 and demonstrate the strong linear relationship with reference depth in GWR than and global model. Profile of the depth estimate also suggest that GWR profile is closer to the reference depth than global model.

Scenario 2: reference depth points were collected at an interval of 300 meter between the neighboring pixel for both Landsat 8 and RapidEye. 1750 were used to estimate the depth by both GWR model and global model and evaluated. The result of evaluation tests shown in Table 2 suggests that the GWR model significantly improve the accuracy even though less number of reference depth available. In Figure 2, bi-variate scatter plot of GWR is showing strong linear relationship with reference depth than global model. Figure 2 shows the profile of the depth estimates, demonstrate that GWR model more synchronize with reference depth than global model.

Table 2. Results of evaluation of depth estimates derived from Landsat 8 and RapidEye data at Scenario 2

Satellite data	Global model			GWR model		
	R	R^2	RMSE (m)	R	R^2	RMSE (m)
Landsat 8	0.88	0.78	2.63	0.96	0.91	1.72
RapidEye	0.89	0.78	2.51	0.95	0.91	1.61

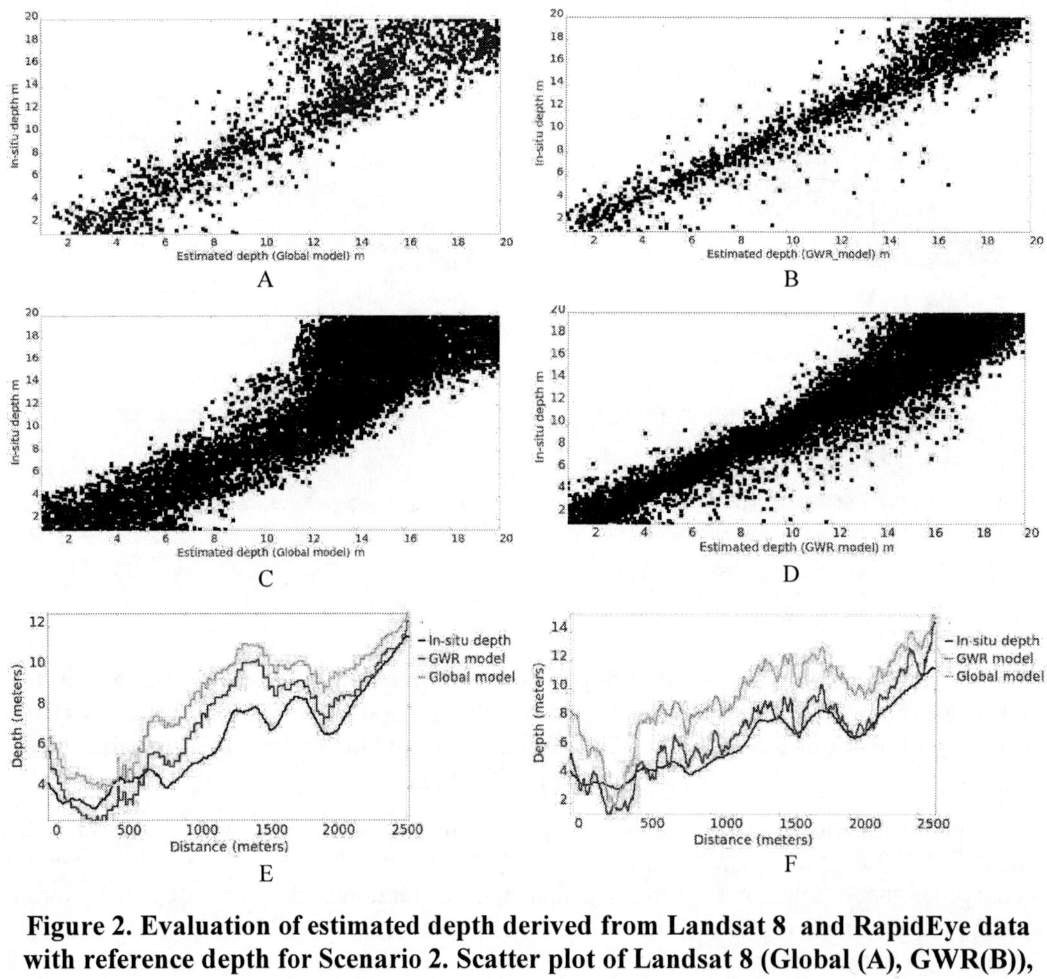

Figure 2. Evaluation of estimated depth derived from Landsat 8 and RapidEye data with reference depth for Scenario 2. Scatter plot of Landsat 8 (Global (A), GWR(B)),

scatter plot of RapidEye (Global (C), GWR(D)) and profile of drawn on Landsat 8 (E) and RapidEye (F).

Scenario 3: reference depth points were collected at an interval of 600 meter between the neighboring pixel for both Landsat 8 and RapidEye. Only 450 reference depth points were sufficient to cover all the study area. Even though the counts of reference depth points very less and interval between the neighboring pixels are far, the evaluation of the results showing that the GWR model is providing better estimates (Table 3). Bi-variate scatter plot in Figure 3 also show that GWR model estimates has strong linear relationship with reference depth. Observation carried out by a profile in Figure 3 (E, F) also indicate the improvement in accuracy of GWR model.

Table 3. Results of evaluation of depth estimates derived from Landsat 8 and RapidEye data at Scenario 3

Satellite data	Global model			GWR model		
	R	R^2	RMSE (m)	R	R^2	RMSE (m)
Landsat 8	0.88	0.78	2.62	0.94	0.88	1.91
RapidEye	0.88	0.78	2.48	0.93	0.87	1.87

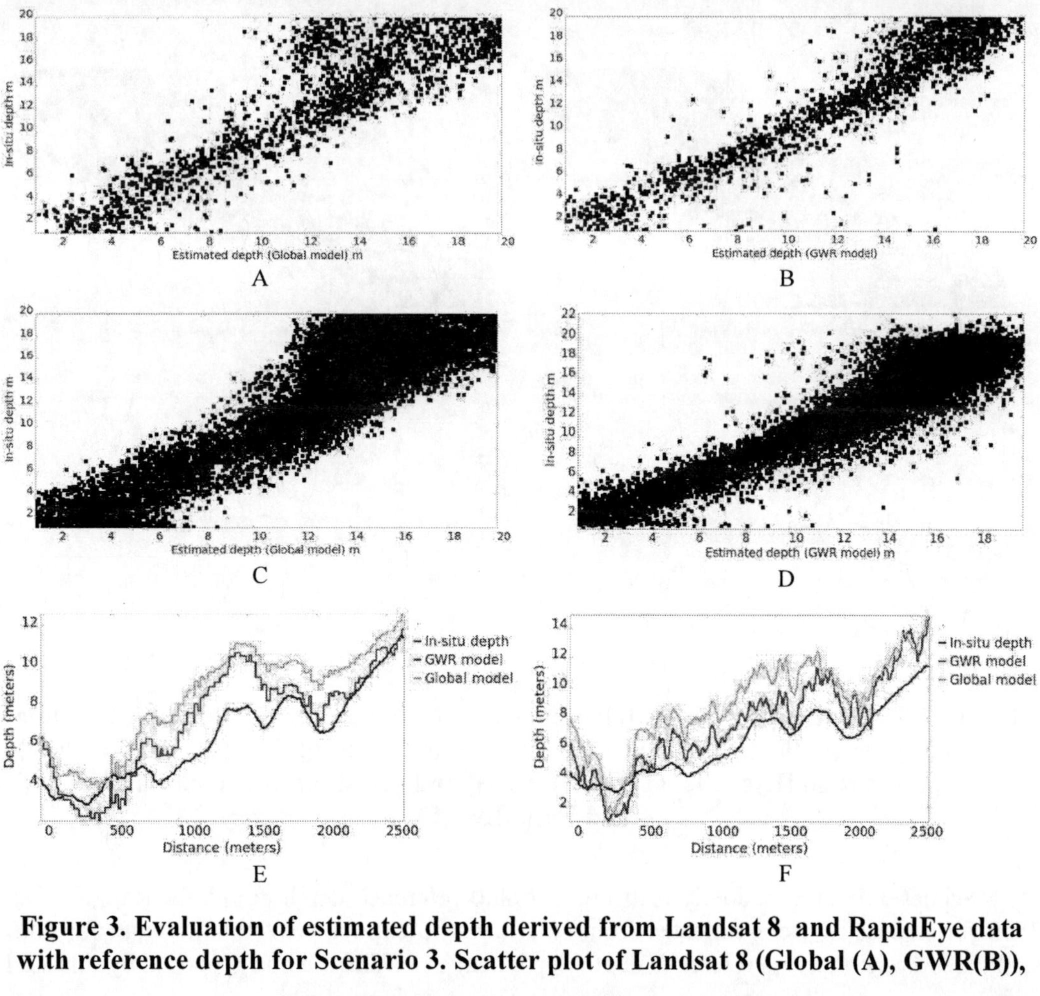

Figure 3. Evaluation of estimated depth derived from Landsat 8 and RapidEye data with reference depth for Scenario 3. Scatter plot of Landsat 8 (Global (A), GWR(B)),

scatter plot of RapidEye (Global (C), GWR(D)) and profile of drawn on Landsat 8 (E) and RapidEye (F).

Table 4. Results of evaluation of depth estimates derived from Landsat 8 and RapidEye data at Scenario 4

Satellite data	Global model			GWR model		
	R	R^2	RMSE (m)	R	R^2	RMSE (m)
Landsat 8	0.83	0.68	3.49	0.87	0.75	2.86
RapidEye	0.88	0.78	2.60	0.90	0.81	2.26

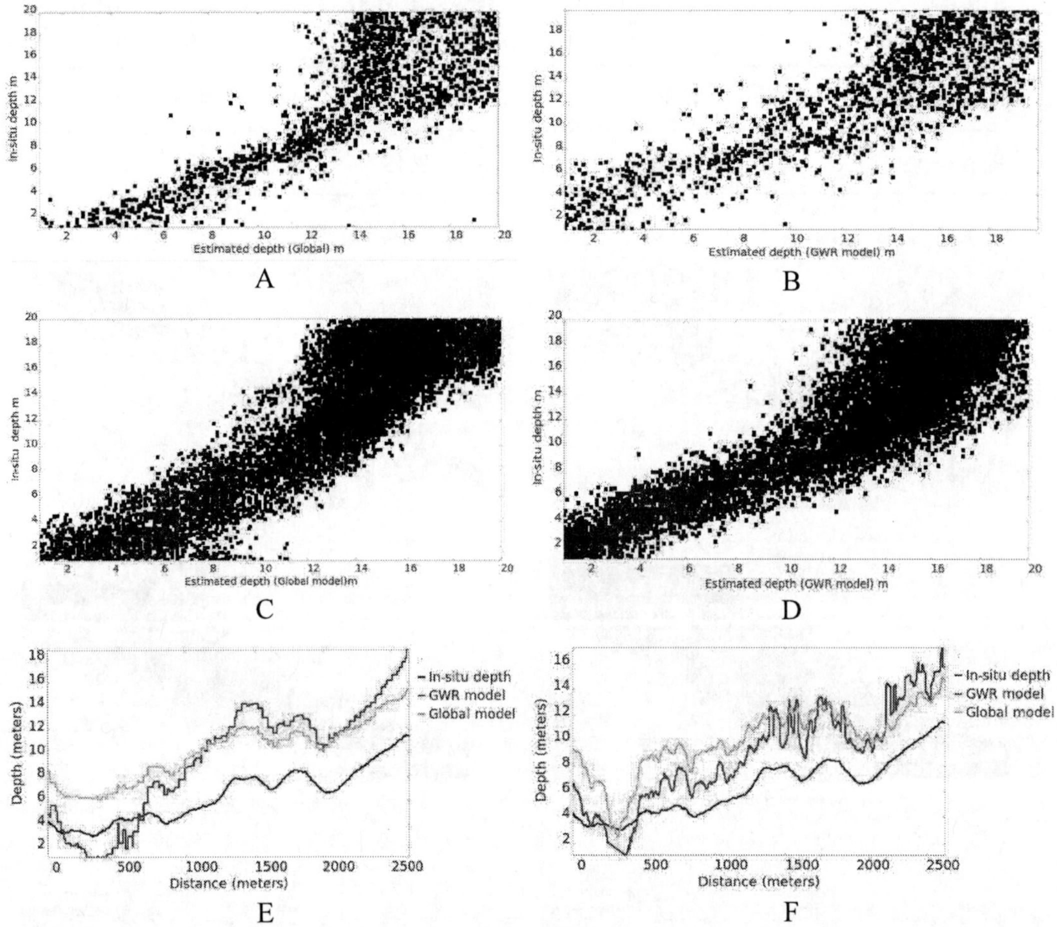

Figure 4. Evaluation of estimated depth derived from Landsat 8 and RapidEye data with reference depth for Scenario 4. Scatter plot of Landsat 8 (Global (A), GWR(B)), scatter plot of RapidEye (Global (C), GWR(D)) and profile of drawn on Landsat 8 (E) and RapidEye (F).

Scenario 4: The randomly distributed 20000 reference depth points for RapidEye and 3300 for Landsat 8 were collected from small area to extrapolate/estimate depth for entire study area even where no reference depth depth available. The method proposed by this study is efficient to extrapolate depth from bottom class based coefficients. The comparative quantitative analysis shown in the Table 4 indicate that results are better than global model in terms of R, R^2 and RMSE. Figure 4 (B, D) is showing strong linear relationship with reference depth than global model (A, C). Even though the quantitative analysis and bivariate scatter plot shows improvement in accuracy of GWR model, profile of the GWR model is not showing such improvements than global model.

Many previous studies have demonstrated that optical remote sensing imagery can be employed to derive water depth estimates. The widely used approach is to calibrate one global inversion model for the entire image scene and then apply the same inversion model to every image pixel for bottom depth retrieval. The intention of the study was to explore the efficacy of GWR model in addressing spatial non-stationarity in the near-shore water region in order to estimate water depth. The evaluation of the results derived by four different scenarios shows significant improvements in both quantitative and qualitative terms. The results provided by GWR model in Scenario 1, Scenario 2 and Scenario 3 shows the significant improvement over global model. In case of Scenario 4, a method originally proposed in this study is also providing better accuracy results in terms of quantitative analysis than global model. But the comparative study carried out by profile (Figure 4 (E, F) is showing that the results has to be improved further. Therefore the further research will be focused to estimate optimal number of bottom classes for improved depth estimation.

8. REFERENCES

Brunsdon, A. C., Fotheringham, S., and Charlton, E. M., 1996. *Geographically Weighted Regression: A Method for Exploring Spatial Nonstationarity*, Geographical Analysis, 28 (2), pp. 283-297.

Lyzenga, D. R., Malinas, N. R., and Tanis,F. J., 2006. *Multispectral bathymetry using a simple physically based algorithm*, IEEE Transaction on Geoscience and Remote Sensing., 44(8), 2251–2259.

Stoffle R. W., and Halmo,D. B., 1991.*Satellite monitoring of coastal marine ecosystems: A case from the Dominican Republic*, Consortium for Integrated Earth Science Information Network (CIESIN). Sagin aw, Michigan.

Stumpf, R. P., Holderied, K., and Sinclair, M., 2004. *Determination of water depth with high-resolution satellite imagery over variable bottom types*. Limnology and Oceanography., 48 (2), 547–556

Vinayaraj, P., Raghavan,V., Masumoto, S., Johnson, G., and Manessa, M. D. M., 2014. I*nvestigation of algorithm to estimate shallow water bathymetry from Landsat-8 images*, International symposium on Geinformatics for Spatial-Infrastructure in Earth and Allied Sciences GIS-IDEAS, Vietnam, 465-470,

Free and Open Source Software for Geospatial (FOSS4G) Conference Proceedings

Volume 15 *Seoul, South Korea* Article 51

2015

The Integration of Heterogeneous Open Source Software to Develop an Urban Simulation Model

Donghan Kim
Division of National Territorial Planning and Research Korea Research Institute for Human Settlements

AT2-02

The Integration of Heterogeneous Open Source Software to Develop an Urban Simulation Model

Donghan Kim, Ph.D.
Associate Research Fellow
Division of National Territorial Planning and Research
Korea Research Institute for Human Settlements
dhkim@krihs.re.kr

Recent development of open source geospatial software offers newopportunities for the spatial analysis and urban modeling fields. The use of open source software enables analysts and modelers to build dedicated and advanced modelsthrough computer programing.However, many open source geospatialsoftware usually provides building blocks forthe static data management, analysis, and visualisation. Hence, development of dynamic simulation model with open source geospatial software is not yet fully fledged.

The goals of this study are twofold. Firstly, it aims to develop a dynamic urban growth simulation model by using and integrating heterogeneous open source software. Secondly, by doing so, it aims to suggest a new way to use logistic regression model as a method for dynamic urban growth simulation.

The researchusesR and Processing to develop an urban growth simulation model. The former is a well-known open source statistical software, and the latter is an open source software for data visualisation. The integration of two open source software and the model development are carried outin a Java programming environment.

The reason of such integration is to build a dynamic urban growth simulation from a conventional binominal logistic regression model. Binominal logistic regression is well-known method to calculate a certain choice probability, and it has often been used to analyse the possibilityof future urban development. However, the result from such logistic regression by nature isstochastic and static. To make it as a method for urban growth simulation, what this research has done is the integration of following tasks: execution of logistic regression, extraction of coefficients from the result, calculation of development probability, iterative allocation of new development, and visualisation of such urban development.

The model was applied to a case study area, Busan Metropolitan Area, Korea in order to examine its usability.It has produced statistically meaningful outcome, and the model shows a new way of developingdynamic urban simulation model. However, all data processing and manipulation is done in a separateGIS environment, and it is not integrated into the model. A tight-coupling withopen source geospatial softwarecould be a possible future resarch.

Free and Open Source Software for Geospatial (FOSS4G) Conference Proceedings

Volume 15 *Seoul, South Korea* Article 52

2015

The Development of Web3D-based Open-pit Mine Monitoring System

Se-Yul Kim
Dept. of Civil Engineering, Sangji University

Dong-Gook Lee
Dept. of Civil Engineering, Sangji University

Jung-Bin Lee
Arts in Landscape Studies, The University of Hong Kong

Byung-Jin Jang

Ji-Ho You
Dept. of Civil Engineering, Sangji University

See next page for additional authors

Recommended Citation

Kim, Se-Yul; Lee, Dong-Gook; Lee, Jung-Bin; Jang, Byung-Jin; You, Ji-Ho; and Lee, Hyun-Jik, (2015) "The Development of Web3D-based Open-pit Mine Monitoring System," *Free and Open Source Software for Geospatial (FOSS4G) Conference Proceedings*: Vol. 15 , Article 52.
DOI: https://doi.org/10.7275/R5W09431
Available at: https://scholarworks.umass.edu/foss4g/vol15/iss1/52

The Development of Web3D-based Open-pit Mine Monitoring System

Authors
Se-Yul Kim; Dong-Gook Lee; Jung-Bin Lee; Byung-Jin Jang; Ji-Ho You; and Hyun-Jik, Lee

This paper is available in Free and Open Source Software for Geospatial (FOSS4G) Conference Proceedings:
https://scholarworks.umass.edu/foss4g/vol15/iss1/52

The Development of Web3D-based Open-pit Mine Monitoring System

Se-Yul Kim[1], Dong-Gook Lee[2], Jung-Bin Lee[3] Byung-Jin Jang[4], Ji-Ho You[5] , Hyun-Jik, Lee[6]

[1]Master`s Course, Dept. of Civil Engineering, Sangji University
83 Sangjidae-gil Wonju-si 220-702, Rep. of Korea
Email: winwopon@naver.com
[2]Master`s Course, Dept. of Civil Engineering, Sangji University
Email: leedg1210@naver.com
[3]Bachelor of Arts in Landscape Studies, The University of Hong Kong
Email: tomtem130@naver.com
[4]Team Leader, Gaia3D, Inc.,
702, SJ Techno Ville, Gasan-dong, Geumcheon-gu, Seoul-si 153-769, Rep. of Korea
Email: bjjang@gaia3d.com
[5]Part time Lecturer(Ph.D.), Dept. of Civil Engineering, Sangji University
Email: sjce96@hanmail.net
[6]Professor(Ph.D.), Corresponding Author, Dept. of Civil Engineering, Sangji University
Email: hjiklee@sangji.ac.kr

ABSTRACT

Large-scale open-pit mines are critical infrastructure for acquiring natural resources. However, this type of mine can experience environmental and safety problems during operations and thus requires continuous monitoring.

In this study, a web three-dimensional(3D)-based monitoring system is constructed using geospatial information open platform and open-source geospatial information software which targets open-pit mines in Gangwon-do, Korea. The purpose is to develop a monitoring system of open-pit mines that enables any person to monitor the topographic and environmental changes caused by mine operations and to develop and restore the area's ecology.

Open-pit mines were classified into active or inactive mines and monitoring items and methodologies were established for each type of mine. Cesium which is a WebGL-based open-source platform was chosen as it supports dynamic data visualization and hardware-accelerated graphics related to elapsed time which is the essential factor in the monitoring.

The open-pit mine monitoring system was developed based on the geospatial database which contains information required for mine monitoring as time elapses, and by developing the open-source-based system software. The geospatial information database for monitoring consists of digital imagery and terrain data and also includes vector data and the restoration plan datas. The basic geospatial information used in the monitoring includes high resolution ortho image(GSD 0.5 m or above) for all areas of the mines. This is acquired by periodically using an airborne laser scanning system and a LiDAR DEM (grid size 1m × 1 m). In addition, geospatial information data were acquired by using an UAV and terrestrial LiDAR for small-scale areas; these tools are frequently used for rapid and irregular data acquisition.

The geospatial information acquired for the monitoring of the open-pit mines represents various spatial resolutions and different terrain data. The database was constructed by converging this geospatial information with the Cesium-based geospatial information open platform of the ESRI World Imagery map and with SDK World Terrain meshes. The problems that resulted from the process of fusing aerial imagery and terrain data were solved in the Cesium-based open source environment.

The prototype menu for the monitoring system was designed according to the monitoring item which was determined by the type of mine. The scene of the mine and changes in terrain were controlled and analyzed using the raster function of PostGIS according to the elapsed time. Using the GeoServer, the aerial imagery, terrain and restoration information for each period were serviced using the web standard interface, and the monitoring system was completed by visualizing these elements in Cesium in 3D format according to the elapsed time.

This study has established a monitoring methodology for open-pit mines according to the type of mine and proposes a method for upgrading the imagery and terrain data required for monitoring. The study also showed the possibility of developing a Web3D-based open-pit mine monitoring system that is applicable to a wide range of mashup service developments.

1. INTRODUCTION

1.1 The purpose and necessity of research

Compared to the underground mining, open-pit mining tends to ensure considerably safer operation, while returning higher yield than the former type of mining, leading to lower prime cost, mass production and supply. Open-pit mining, however, has drawbacks as well, accompanying: forest damage, mine-waste, dust, wastewater and tailing discharge, slag sedimentation, ground subsidence and runoff, and land slide. These defects, which may cause environmental disaster, have been the primary factor for the regulation of open-pit mining.

Likewise, the massive open-pit mines of Taebaek Mountains in the Baekdoo Mountain range continues to incur civil petitions as the region's need for eco-friendly development and restoration is high due to its ecological value of nesting rarely found botanical colonies and animal communities original to the region.

The government, to address the problem, has established *Mining Damage Prevention and Restoration Act (2005)* to minimize mine damage of the environment, ecology and the forest caused by the development of open-pit mines, while ecologically restoring already damaged open-pit mines. Operating open-pit mines, with the above measure enacted, have to manage the topographical and environmental alteration. Within the management process, periodic and precise monitoring technique for the eco-friendly restoration of the mines is a necessity; in order to achieve this efficiently, development of a monitoring system is required.

Research on the method of restoration of open-pit mines, however, tends to be insufficient worldwide, while domestic level on management and accumulation of present state data regarding the topographical alteration incurred by the development and restoration of open-pit mines relies on digital topographic maps and site surveys.

This research will aim to establish a web3D-based monitoring system on open-pit mines in Gangwon province(South Korea), using geospatial information open platform and open-source GIS software; the monitoring system that would be easily approachable even to the public will support eco-friendly development and restoration of open-pit mines in the region, as well as indicate alterations in topography and environment occurred by operation of mines. Also, by suggesting methods on optimal sensing and upgrading of the system, the research will propose a scheme on servicing the system based on geospatial information and open platform, contributing to the preservation of environmental ecology of the Baekdoo Mountain range.

1.2 Trend of research

The research on the monitoring of open-pit mines was conducted by Froese et al. (2008), using airborne LiDAR (Light Detection and Ranging) and satellite InSAR (Interferometric SAR) technique, mapping and monitoring mines via calculation of quantitative information regarding geospatial patterns.

Brown et al.(2007), to monitor a nearly 2km wide open-pit mine, combined a GNSS receiver and robotic total station equipment during the process.

Kim et al.(2003) have designed fully automatic modified monitoring system software using RTK GPS, enhancing the conventional monitoring system.

Shong(2013) has conducted a research on the new V-World based environmental geographic information system for environment analysis. The author has utilized geospatial information open platform and open source software in this study.

Jang et al.(2013) have proposed a frame regarding an establishment of 3D geospatial information system that would provide service linkage with geospatial information open platform, thereby suggesting methods of servicing 3D geographic information by its category.

Singh et al.(2012) have utilized open source software and standard to develop a web-based GIS application for the information system on natural resources.

Yin et al.(2009) mentioned the adequacy of application layer, service layer, functioning layer and saving layer; they planned and established the visualization system and vector geospatial data management according to the framework.

Ji et al.(2012), using open source GIS software GeoServer and Open-Layers, have depicted the realization of web service of Seoul's MRDB (Multiple Representation DataBase).

The existing studies and researches have high interest on the utilization of geospatial information open platform and open source GIS and have applied to a variety of field, but the main aspect of these researches remain on the analysis of alterations by DEM (Digital Elevation Model) and on the suggestions concerning the management system of mines.

1.3 Scope and direction of research

This research has its base on assisting eco-friendly development and restoration of open-pit mines, and on providing easy access to the public on topographical and environmental changes regarding the operation of mines through the web3D-based monitoring system using open source GIS software.

Prior to the commence of the development of the monitoring system on open-pit mines, subject mines were divided into two categories of active mines and inactive mines to establish the methodology on monitoring and to classify monitoring areas to plan geospatial information database.

The geospatial information database consists of digital imagery and terrain data, and of which future restoration plans were formulated with vector data.

Basic geospatial information used in the open-pit mine monitoring system was accumulated with periodic acquirement of high-resolution ortho image (GSD 0.5m or above) of the mining region taken with airborne laser scanning system and LiDAR DEM (Digital Elevation Model: Grid size 1m x 1m) data.

Development of the open-pit mine monitoring system was conducted on different factors, of selecting adequate geospatial information open platform, of upgrading geospatial information database, and of developing a monitoring system based on open source.

The selection of geospatial information open platform was based on the prior research analyzing WWJ (World Wind Java) and Cesium in comparison, of which focused on dynamic-data visualization, support of hardware-accelerated graphics; in result, Cesium, a WebGL-based open source platform which met both criteria, was chosen as the platform for the monitoring system.

The development of the prototype of monitoring system was based on different types of mines' monitoring areas set up. Scenic and terrain change alterations throughout time were managed and analyzed using Raster mechanism of PostGIS, Periodic aerial image, geographical information and restoration information were serviced as a web-standard interface using GeoServer; all of the above combined was then integrated and visualized three-dimensionally into a service on Cesium, establishing the monitoring system capable of sensing periodic changes in the region.

Also, the geospatial information provided by Cesium based geospatial information open platform that this research has utilized lacks adequacy and precision for monitoring in terms of image resolution or size of DEM's grid. Therefore, this research has performed a periodic data acquisition using airborne laser surveying and non periodic acquisition in a small-scale region of interest using UAV system or terrestrial LiDAR, thereby enhancing the resolution of the image required for monitoring and reducing the size of the grid of DEM, upgrading geospatial information as a result.

Geospatial information acquired for monitoring of open-pit mines indicate information in a variety of geospatial resolution and geographical data; this information was integrated with original image (ESRI world imagery map) that consists Cesium-based geospatial information open platform and terrain data (STK World Terrain Meshes) to establish a geospatial information database for the monitoring. Various problems during the process of integrating image and terrain data were solved under Cesium-based open source environment.

The main process of the study described above is as follows in Figure 1:

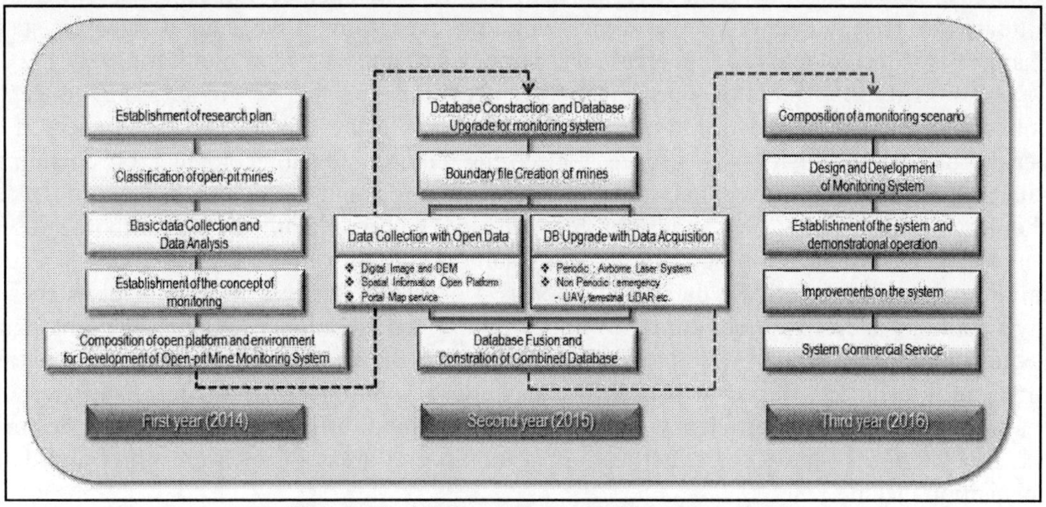

Figure 1. Main process of this research

2. TYPES OF OPEN-PIT MINES AND CONCEPT OF MONITORING

2.1 Types of open-pit mines

Gangwon province has approximately 116 open-pit mines, which of 64 are mining limestone. Open-pit mines, according to their state of operation, are categorized into either active or inactive mines. Active mines, in accordance with the mining act, are mines that are mining minerals or planning to mine minerals and are excavating ground, with the approval of the authority. Inactive mines are mines that have been put on the inactive status according to the approval and are not mining minerals.

Active mines, due to continuous mining and restoring processes, have a variety of factors concerning topographical changes. Inactive mines have been postponed from mining or have completed mining in the region; these mines are exposed to collapse or safety accidents.

This research aims to develop a monitoring system which targets active and inactive mines with considerable scale and need for monitoring. To establish the concept and definition of active mine, Lafarge Halla Cement mine located in Okgye-myeon of Gangneung; Dongyang Cement mine in Samcheok was selected as a mine representing the inactive state. Figure 2 is the change of image over time of the two representative mines.

2.2 Establishment of monitoring concepts for open-pit mines

To efficiently and systematically monitor open-pit mines, the establishment of monitoring concepts according to their types is crucial. In this research, the monitoring concepts of open-pit mines were classified as operating aspect and environmental ecological aspect. Operating aspects, factors which are created by operating active mines, should contain causes occurred by restoration and operation plans of mines, not to mention restoring and/or operating areas themselves. Environmental ecological aspects concern change to the environment caused by mining operation, restoration and other specific management plans.

Monitoring concept on operating aspects include mining area, disaster restoring area and ecological restoring area, while environmental ecological aspect consist of waste stowage area and ecological restored area. Mining area, waste stowage area and ecological restoring area are areas with continuous change in terrain, in need of periodic monitoring. Disaster restoring area, on the other hand, is vulnerable to sudden terrain changes triggered by natural or artificial inputs, making the speed and urgency of monitoring most critical. Ecological restored area is an area that has been reached to its former original ecological state; it thus has little or no terrain alteration over time, with the exception of vegetation. Due to such characteristic, ecological restored areas lack urgency and importance of monitoring.

To efficiently monitor open-pit mines according to the set monitoring concept, collection, acquisition and fusion of basic data for monitoring should be accompanied. If a periodical data acquisition is needed concerning environmental ecological or operating aspects of the mines, an annual, year-by-year restoration and cyclic mine operation plan should be established using annual or biennial aerial photogrammetry and airborne laser surveying that would aid processing a high-resolution digital ortho image and DEM of the mining region. Non periodic data acquisition is a necessary process in case of disasters or civil petitions addressing monitoring of mines. In a given situation where non periodic data acquisition is needed, UAV or terrestrial LiDAR, which are inexpensive and require less procedures, will be utilized to gain DEM and high-resolution ortho image of the area of interest to maximize the efficiency and potential of the monitoring system's ability to restore the area. Figure 2 shows the classification of monitoring areas of open-pit mines and data acquisition method.

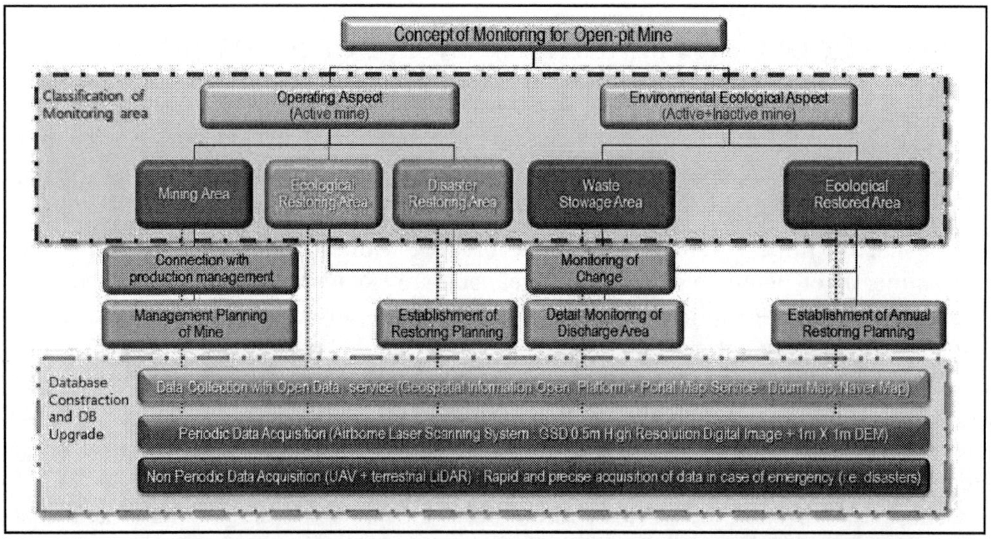

Figure 2. Classification of monitoring area and data acquisition method

2.3 Classification of monitoring area using basic geospatial information

Basic geospatial information is a basic data on the monitoring system of open-pit mines; with the information, the possibility of monitoring alterations in open-pit mines was tested using geospatial information analysis method. This research, to monitor the alteration of open-pit, active mines, has conducted airborne laser surveying twice in 2007 and 2014 of Lafarge Halla Cement mine. Monitoring analysis of the alteration of open-pit mines was divided into two basic frames, image-based analysis and DEM-based analysis. Differential image method, utilizing digital ortho image, was used in the image-based monitoring analysis, while earth-volume calculation and residual DEM was used in the DEM-based monitoring analysis. With application of the two methods of monitoring analysis, this research has verified the validity of the criteria of monitoring set up previously and has utilized such measures to the development of monitoring method.

2.3.1 *Image-based monitoring area classification analysis*

Differential image method which makes use of high-resolution ortho image, calculates difference between pixel changes in an image over different time and prints image. The differential between digital ortho image of 2007 and of 2014 of the active mine is as follows in Figure 3. As shown in Figure 4, differential image analysis indicated alterations in topography and enabled the classification of the region into ecological restoration area, disaster restoring area and waste stowage area. Figure 5 indicates results of specified areas such as mining area, disaster restoring area, ecological restored area, waste stowage area in different times as a comparison. Mining area, as shown in Figure 5(a), was described with light brown tone, as well as the disaster restoring area of Figure 5(b), since the area was originally restored but collapsed over time. Figure 5(c) is a ecological restoring area, which is visualized as light grey due to the stowage of waste rocks, resulting in changes of vegetation.

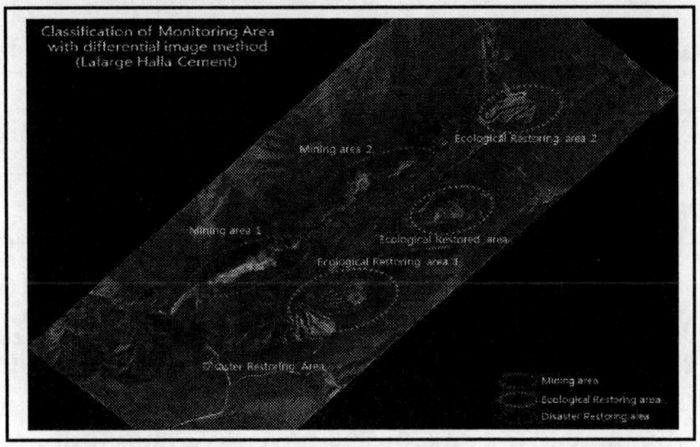

Figure 3. Classification of monitoring area using digital image

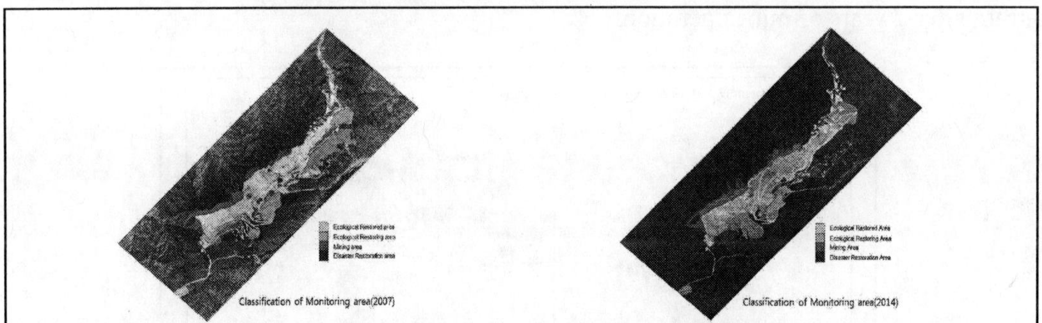

(a) Subdivision of Monitoring area with digital image(2007) (b) Subdivision of Monitoring area with digital image(2014)

Figure 4. Detail classification of monitoring area with digital image

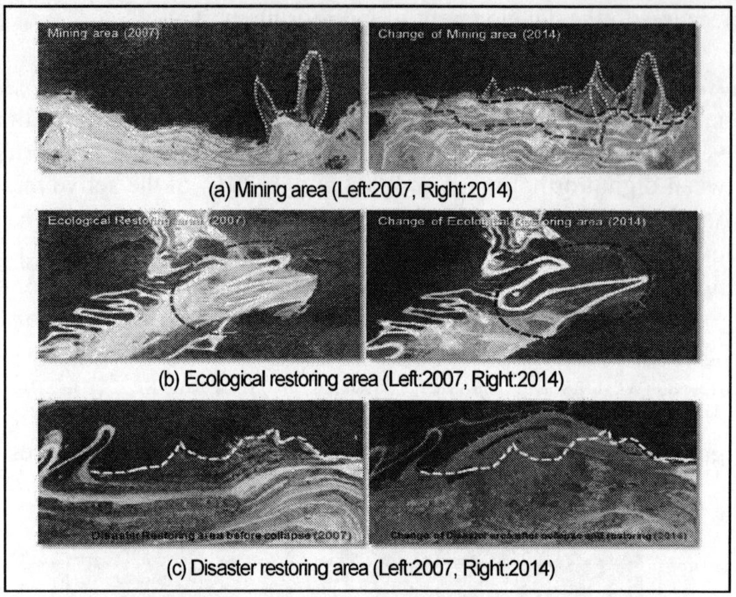

Figure 5. Change detection by each monitoring area with digital image

2.3.2 *DEM-base monitoring area analysis*

To comprehend topographical changes over time, 1m×1m DEM of the mining area was created via LiDAR data obtained from airborne laser surveying that had been conducted twice. Residual DEM was obtained between DEM of 2007 and 2014, which then changes were analyzed. In residual DEM analysis, unlike that of differential image analysis, three-dimensional visualization of terrain change was possible. Also, with quantitative analysis, operational areas of mining and waste stowage area were able to be specifically classified, with regards to the noticeable difference in earth-volume of the area. Figure 6 shows the residual DEM(1m×1m Grid Size) of an active mine, indicating considerable changes in terrain after 7 years of mine operation.

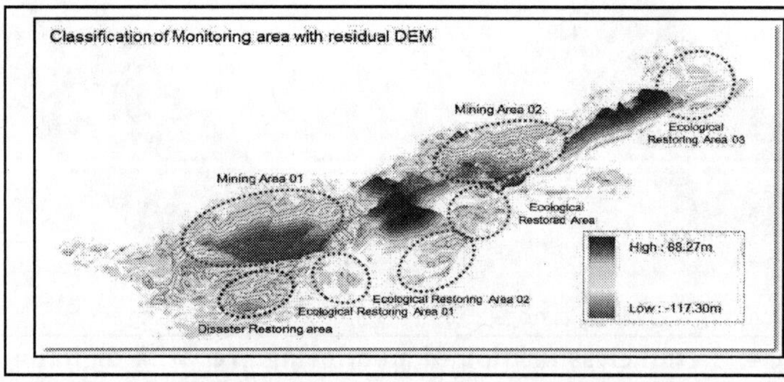

Figure 6. Residual DEM of active mine in elapse of time

Figure 7 is the residual DEM and the perimeters of mines of 2014 and 2007 overlapped. Figure 7(a) is the mining area of the active mine which has created additional volume of earth due to the extension of mining area in 2014. Figure 7(b) is an area where ecological restoration after mining has been performed. According to the restoration plan, waste rock and vegetation has been stowed and planted, showing little difference in DEM in elapse of time. Disaster restoring area of Figure 7(c) has been restored from collapse of transmission line; the restoration also has caused the alteration in topography and creation of earth volume.

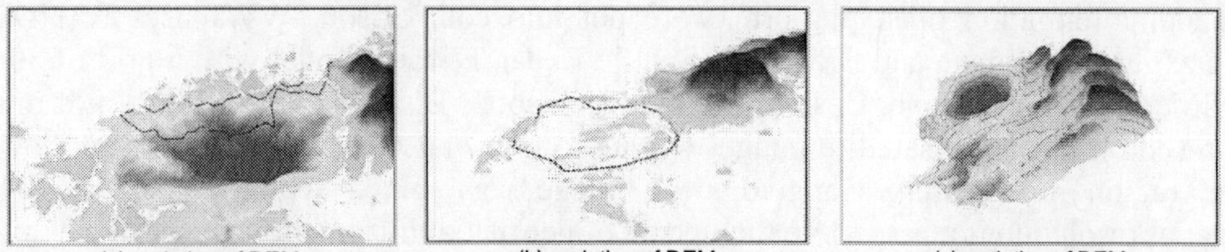

(a) variation of DEM (Mining area) | (b) variation of DEM (Ecological restoration area) | (c) variation of DEM (Disaster restoring area)

Figure 7. Change detection of monitoring area with residual DEM analysis

Earth-volume analysis in elapse of time of the active mine's monitoring area classified by image differential analysis as shown in Table 1. The mining area of active mine was calculated to be –31,110,170.7m³. 89.5% of the volume occurred from ground cutting, while 5.87% of ground stack occurred due to formation of operational roads. Continuous operation of mine resulted in the expansion of this area. Additional ground cutting of -1,453,513.61m³ incurred due to transmission line collapse in 2012, compared to the original ground-cut area in 2007 prior to the disaster. 78.41% of the volume occurred from ground cutting, while 8.14% of ground stack occurred due to formation of operational roads. Finally, in the ecological restoring area where mining has ended, ground stack was +404,167.78m³; of this, 27.12% of ground stack and 22.34% of ground cutting volume was utilized for restoration by stowing waste rocks or forming operational roads. 50.03% of the entire area showed no change, having difference of 0m.

Table 1. Changes in earth-volume of monitoring area with residual DEM analysis

Monitoring area	interval	Area(m^2)	ratio(%)	Sum of Volume(m^3)
Mining area	-50 m below	14,901	11.75	-3,110,170.7
	-50m ~ 0m	98,600	77.75	
	non change	5,868	4.63	
	0m ~ 30m	7,441	5.87	
Disaster Restoring area	-50m below	4,602	5.65	-1,453,513.6
	-50m ~ 0m	59,307	72.76	
	non change	10,961	13.45	
	0m ~ 10m	6,368	7.81	
	10m above	269	0.33	
Ecological Restoring area	-10m below	11,332	4.96	404,167.8
	-10m ~ 0m	40,900	17.89	
	non change	114,398	50.03	
	0m ~ 10m	36,504	15.97	
	10 m above	25,512	11.16	

3. DEVELOPMENT OF OPEN-PIT MINE MONITORING SYSTEM

3.1 Selection of geospatial information open platform

To effectively perform the development of open-pit mine monitoring system, a geospatial information open platform with high accessibility and usage should be selected as the base platform. 3D geospatial information open platform WWJ (WorldWindJAVA) and Cesium were compared to choose the most adequate platform suitable for open-pit mine monitoring system.

Both of WWJ and Cesium have 3D Viewing and provide the function with source, enabling developers to develop programs according to their needs

3D modeling is a crucial function on which key terrain features or restoration plans are expressed through geospatial information open platform. WWJ provides xml, kml format, while Cesium expresses animated 3D models using kml, GeoJSON and TopoJSON format.

Also, the primary image and quality of terrain features of Gangwon region provided by geospatial information open platform were put into comparison. WWJ uses LANDSAT7 image as its basic image and SRTMv2 (Shuttle Radar Terrain Model with a grid of 30m) as its DEM. On the other hand, Cesium utilizes Microsoft's Bingmap and ESRI World imagery map based on IKONOS satellite image with up to grid of 1.0m as its basic image; however, in the region of research, Gangwon, the image was mixed with comparatively lower geospatial resolution images, while the update of geospatial information was also lacking.

The DEM consists of STK World Terrain Meshes, which can express from 3m to 90m in range of the area. Table 2 is analysis of the characteristics of World Wind Java and Cesium in comparison.

Cesium comprises of basic data that has higher quality than World Win JAVA, with the advantage of offering a variety of 3D modeling format; such advantage has made Cesium the platform to be utilized in the research. The quality and renewal of geospatial information provided by Cesium, however, was not sufficient for utilization in the monitoring system that this research aims to create and develop; therefore, additional upgrading of image and topographical information was found to be urgent.

Table 2. Characteristics of geospatial information open platform WWJ and Cesium

item	WWJ(World Wind JAVA)	Cesium
Digital image	Bing - TerraColor : GSD 15m - LANDSAT 7 : GSD 30m	Bing - TerraColor : GSD 15m ESRI World Imagery map - IKONOS : GSD 1.0m
DEM	SRTM - global : 1km, USA : 30m SRTMv2 - don't provided in KOREA	STK World Terrain Meshes - 1km(global), 90m(Korea), 3m(USA) - National Elevation Dataset(NED) - EU-DEM - Australia SRTM-derived 1 Second DEM
Language	JAVA SDK	JAVAScript+HTML
3D model	XML format	KML, GeoJSON, TopoJSON format

3.2 Design and development of the system using open source programming

3.2.1 *Composition of development environment for open-pit mine monitoring system*

The open-pit mine monitoring system was developed with geospatial information open platform, along with open source-based GIS software. Open-pit mine monitoring system consists of client, middleware, server, DBMS and data and to be serviced through web browser.

Client is able to visualize changes and restoration blueprints in 3D, to overlap DEM terrain with aerial and satellite ortho images in different, free points of view. Middleware links the client and the server to interexchange data. Server provides the data client requests through middleware. DBMS manages and organizes data for effective data usage.

PostGIS and GeoServer were selected as DB server and middleware; Apache was chosen as the webserver. PostGIS receives general information, DEM and mine perimeters and provides them to Cesium through Geoserver and Apache in HTML and WMS format. DEM and ortho image is set to be provided on Cesium through Geoserver, the middleware, as WMS. The components of the open-pit mining monitoring system are as follows in Figure 8.

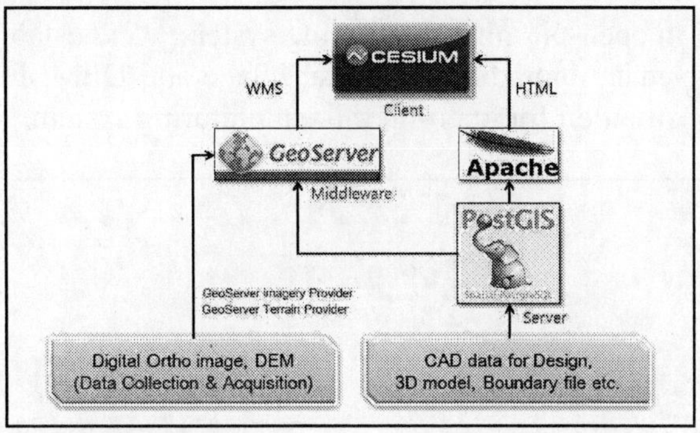

Figure 8. Development environment for monitoring system

3.2.2 *Design and realization of open-pit mine monitoring system*

The open-pit mine monitoring system should be composited in a way that anyone can easily use. Therefore, the intention was to design an intuitive and practical UI (User Interface).

Crucial functions considered to be required for the monitoring system were divided into basic function and analysis function. Basic function is a function that provides basic information concerning the monitoring mine and command to express changes of the mine in 3D. Analysis function is a function which allows users to analyze points of interest in the case that users are aware of a mine's operating and restoring plan and alterations caused by changes in elapse of time. The open-pit monitoring system is developed with an information window that provides a search engine that would find a mine of interest, operation status of the searched mine, type of mineral mined and the address of the mine. Also, functions showing mining, and restoration layouts, present and past forms, residual DEM, and true elevation and topographical angle of any set point in mines are being developed to enhance the effectiveness and efficiency of the system. The start-up screen of the monitoring system is planned to be visualized as shown in Figure 9. Current progress of the research focuses on 3D Viewing and comparison between DEM and mining region's elapsed image. More effort is put in to continuously improve the monitoring system.

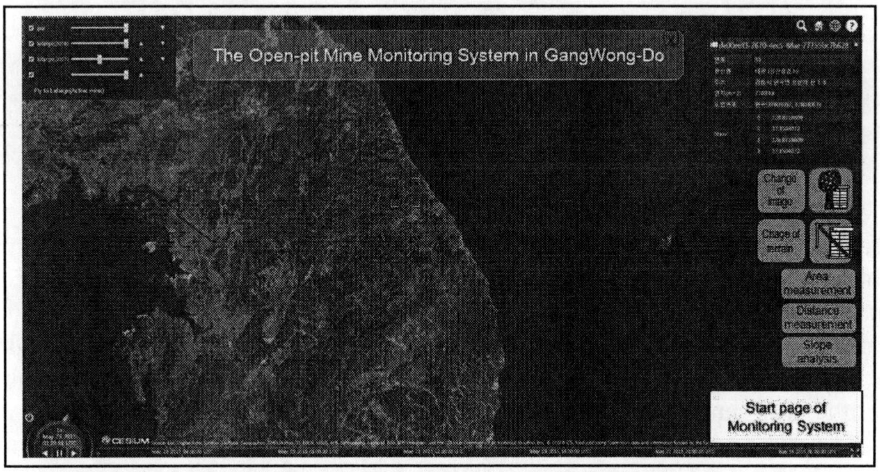

Figure 9. Composition of start-up screen of the monitoring system

3.3 Database upgrade for efficient monitoring

The open-pit mine monitoring system has to be updated at all times, in terms of digital ortho image and DEM, to detect and analyze change of open-pit mines, and should be capable of constant data upload and comparison with elapsed data. Therefore, methods concerning the upgrade of ortho image and DEM data on Cesium, the geospatial information open platform utilized in open-pit mine monitoring system, were established. Figure 10 is the process and method for data upgrading of digital image and DEM. Figure 11 is diagram of combined database construction for open-pit mine monitoring system.

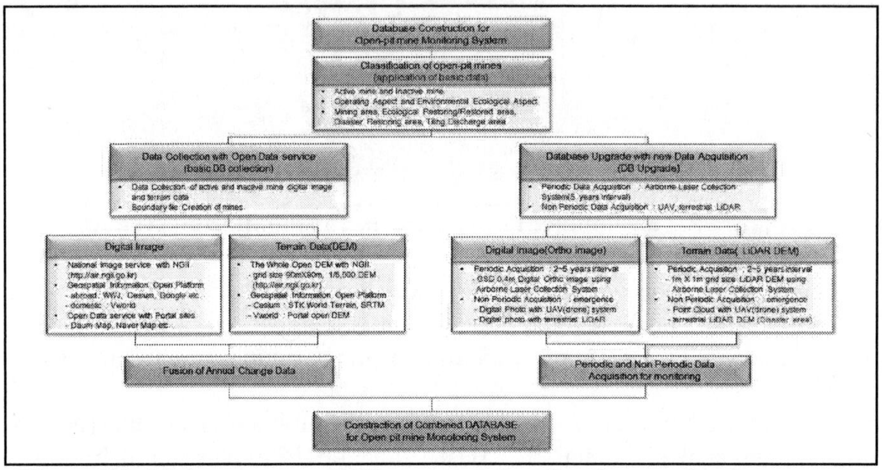

Figure 10. Main process of DB construction for the monitoring system

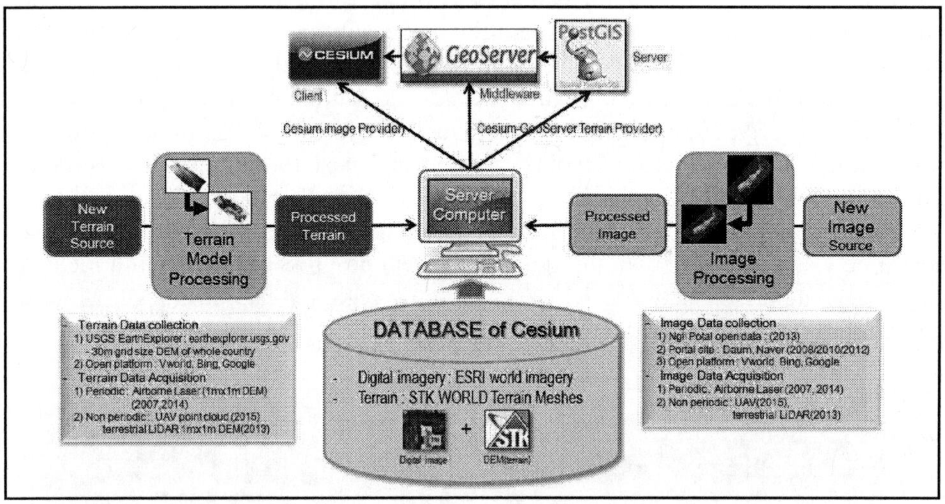

Figure 11. Construction of combined DB for monitoring system

3.3.1 *Methods of upgrading digital image*

The coordinate system of Cesium is WGS84, converted from coordinates of ortho-image utilized for upgrading the monitoring system. To register and issue the ortho-image with converted coordinates on GeoServer, work space and repository, along with a layer should be created, and the image should be issued in Geotiff format. Finally, using Image Provider, the ortho-image and GeoServer are linked to visualize newly registered image on Cesium. Figure 12 is a workflow of the process concerning the upload of ortho-images on Cesium, while Figure 13 shows an example of an uploaded ortho image expressed in Cesium. Due to the difference in color impression between uploaded and background images, the uploaded image needs refinement; such problem will be minimized as histogram and color matching are rebalanced.

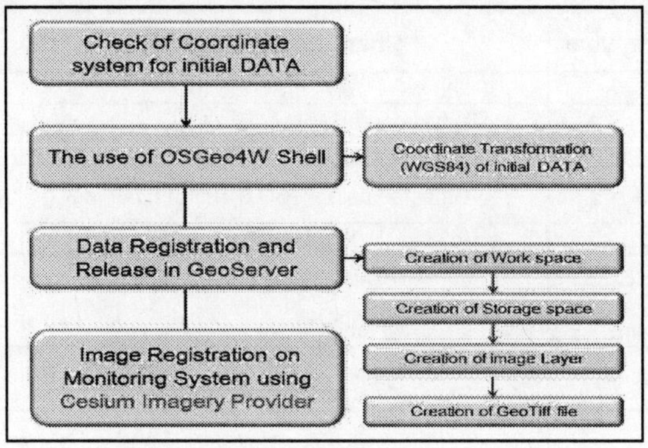

Figure 12. Process of ortho image uploading on monitoring DB

Figure 13. Digital ortho image newly uploaded on Cesium

3.3.2 *Methods of upgrading terrain DEM*

Cesium's DEM, unlike image upload, cannot be uploaded by registering topographical data on GeoServer ; because of such limitation, Cesium-GeoserverTerrainProvider written by kaktus40 on GitHub was taken into reference to upload DEM on Cesium. Cesium-GesoserverTerrainProvider expresses DEM through assigning height value on Raster DEM at GeoServer. The DEM uploaded on Cesium is based on 1second SRTM DEM and LiDAR data of 2014. To merge the SRTM DEM and LiDAR data of 2014, Raster Calculator provided by SAGA of QGIS was utilized. Figure 14 is the diagram regarding the upload of DEM on Cesium. CrossDomain error occurs when DEM is put on Cesium through GeoserverTerrainProvider; Jetty's CrossDomain solution was used to cope with the problem. DEM values on Cesium, however, was regarded as whole numbers on the platform; because of this, elevation difference such as that of Figure 15 occurred, fragmenting the terrain. This problem needs to be solved.

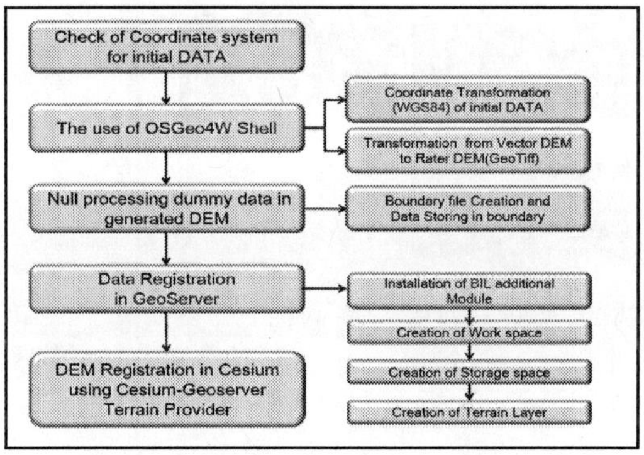

Figure 14. Process of DEM uploading on monitoring DB

Figure 15. Problem of uploading DEM on Cesium

3.3.3 *Change detection of monitoring areas*

The monitoring of open-pit mines is a monitoring which detects the topographical change over time; this research utilized digital image and DEM for the monitoring. Table 3 is the data inventory accumulated through this research, concerning the database of monitoring.

This research analyzed the change detection in the mining area, ecological restoring area and disaster restoring area in series of time, using digital image and terrain data of the open-pit mine monitoring database.

Table 3. Data inventory of monitoring Database

Data Acquisition Methods	Acquisition year	Characteristics of data
UAV surveying(non periodic)	2015	Fixed-wing UAV : GSD 0.3m
Airborne Laser surveying(periodic)	2014	Digital image : GSD 0.12m
		DEM : LiDAR 1m X 1m DEM
NGI Portal open data service (data collection)	2013	NGII Open data : GSD 0.5m (http://air.ngii.go.kr)
terrestrial LiDAR surveying (non periodic)	2012	gird size 1m X 1m DEM
Portal site open data service (data collection)	2012	Daum map Open data
	2010	- GSD 0.5m
	2008	- 2 years interval
Airborne Laser surveying(periodic)	2007	Digital image : GSD 0.4m
		DEM : LiDAR 1m X 1m DEM

Figure 16 and Figure 17 shows the analysis results of the mining area according to series of time. Figure 16(a) is the result of the 2D analysis with ortho image of the mining area of 2007 and 2014, ortho images collected from portal websites in 2008, 2010 and 2012 and ortho image acquired using UAV system in 2015. Figure 16 (b) describes 3D analysis using DEM of the mining area of 2007 and 2014, while Figure 17 is a result of a 3D modeling comparison between DEM of 2007 and 2014 acquired with airborne LiDAR surveying.

As shown in Figure 16 (a), the mining area tends to increase over time; this enlargement of mining area, Figure 16 (b) and 17 proves, occur massive change in topography. Such detection of changes can be utilized in the future operation and management of open-pit mines.

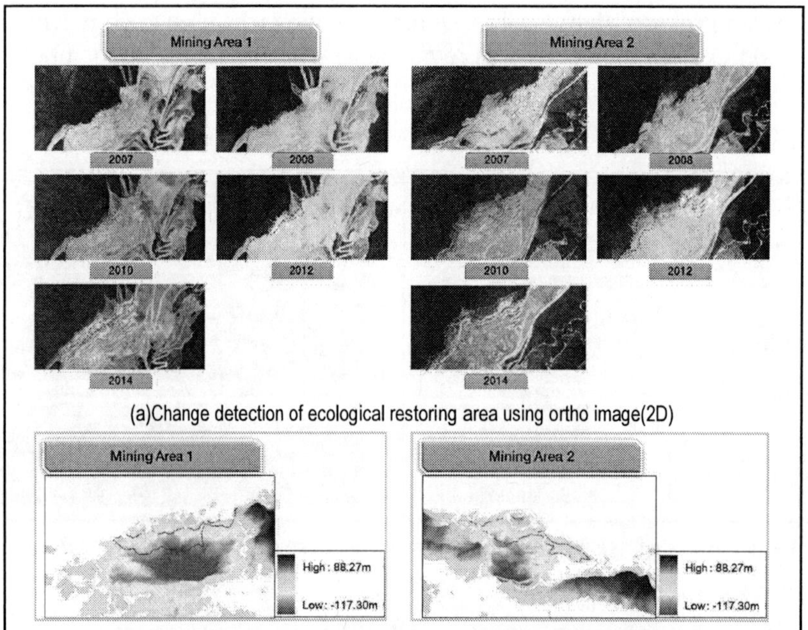

(a)Change detection of ecological restoring area using ortho image(2D)

(b)Change detection of ecological restoring area using residual DEM(3D)

Figure 16. Change detection of the mining Area

Figure 17. Change detection of mining area using 3D modeling

Figure 18 and Figure 19 shows the 2D and 3D analysis on change detection of ecological restoring area. Two-dimensional and three-dimensional changes of the area were analyzed using the same image and terrain information as did in the mining area.

Ecological restoring area is an area that is under ecological restoration according to restoring plan after the end of mining. Figure 18 (a) describes the change of ground coverage into vegetation under ecological restoration over time. In the areas of Figure 18 (b) and Figure 19, ecological and topographical restoration had taken place completely, showing no signs of three-dimensional topographical changes such as those of the mining area.

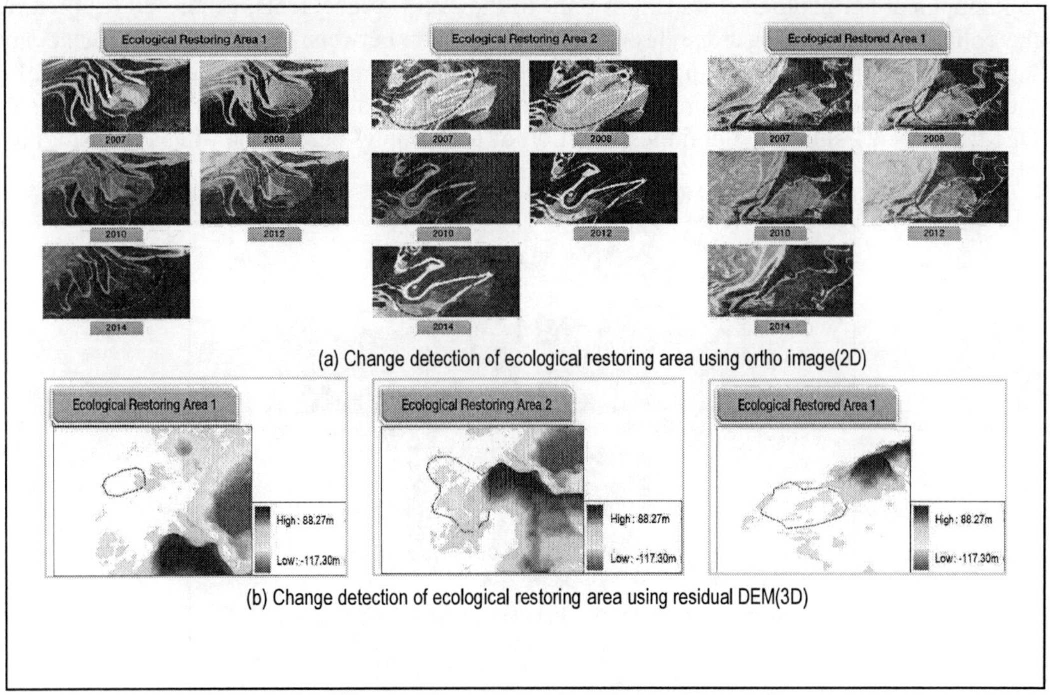

Figure 18. Change detection of ecological restoring area

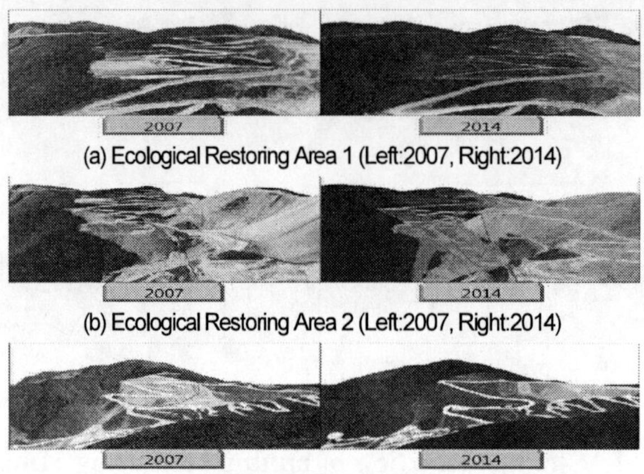

(a) Ecological Restoring Area 1 (Left:2007, Right:2014)

(b) Ecological Restoring Area 2 (Left:2007, Right:2014)

(c) Ecological Restoring Area 3 (Left:2007, Right:2014)

Figure 19. Change detection of ecological restoring area using 3D modeling

In 2013, terrestrial LiDAR surveying took place for disaster restoring area to understand the status quo and to establish a restoration plan of the disaster that had taken place in the area of interest. The topographical information of before and after the disaster were utilized for the change detection analysis. Figure 20 and Figure 21 are analyses of 2D and 3D changes of disaster restoring area.

Disaster restoring area, mining operation having finished, does not have any topographical nor ecological transition until in the event of accidents. Figure 20 (a) pictures the collapse of the mining slope through the comparison between 2012 and 2013 image, and the restoration in progress of the collapsed area with 2013 and 2014 image. Figure 20 (b) is the analysis of three-dimensional transition of mining area between 2007 and 2014 using DEM; Figure 21 shows the definite transition of topography before and after the restoration of collapse.

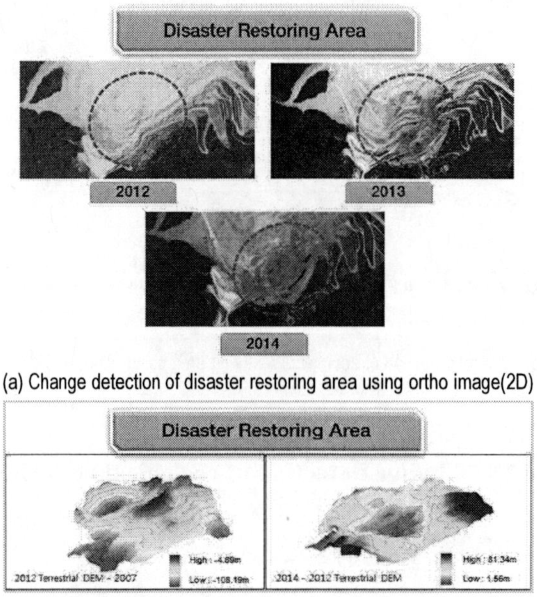

(a) Change detection of disaster restoring area using ortho image(2D)

(b) Change detection of disaster restoring area using residual DEM(3D)

Figure 20. Change detection of disaster restoring area

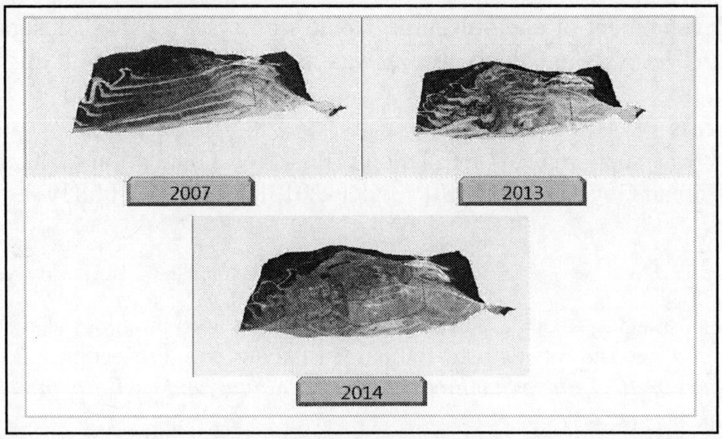

Figure 21. Change detection of disaster restoring area using 3D modeling

As a result of a time elapsed transition analysis of the mining, ecological restoring and disaster restoring area using the open-pit mine monitoring system the research has developed and digital image and DEM data from database, two-dimensional and three-dimensional change detection analysis methods were clearly noticed, enabling the efficient utilization of the monitoring system on operation and management of mines, ecological and environmental management and disaster control. The upgrade and enhancement of the monitoring system will continue as the effective and efficient usage of the open-pit mine monitoring system is the crucial concern of the research.

4. CONCLUSION

The primary objective of this research is to develop an effective monitoring system which monitors topographical change of open-pit mines using geospatial information open platform and open source programming. This research has reached to the following conclusion:

Firstly, concepts of monitoring according to the type of mines were established. In the process, acquisition and upgrade of required data and effective monitoring were sought.

1) Basic data of open-pit mines was collected through geospatial information open platforms and portal open data services from worldwide,
2) As a periodic acquisition of updated data, airborne laser surveying that provides high-resolution ortho image and LiDAR DEM was chosen for all of the monitoring area.
3) For non-periodic acquisition of data which requires speed as is the case in disaster areas, terrestrial LiDAR and UAV systems were applied to meet the urgency and the characteristic of the monitoring area.
4) Acquired data of the monitoring region was bundled in a series of time into a database; this database was merged with the database of Cesium, the platform which this research has chosen, and utilized in the development of the monitoring system.

Secondly, this research has chosen Cesium as the geospatial information open platform of the monitoring system that is effective on 2D and 3D visualization. As the development environment, PostGIS(server)-GeoServer(middleware)-Cesium(client) were established and the system was developed in an open source environment.

Thirdly, this research has established the method to utilize ortho image and DEM for the monitoring of open-pit mines. Applying the method onto specific monitoring areas, such method was efficient in terms of change monitoring.

This research is the result of the second year of the three-year research; therefore, the degree of completion and specification of the developed system lacks perfection. The possibility of development of open-pit mine monitoring system using geospatial information open platform and open source programming was realized, however, with this research.

Acknowledgments

This work was supported by the National Research Foundation of Korea (NRF) grant funded by the Korean Government (MSIP) (NRF-2013R1A2A2A01068391).

REFERENCES

Brown, N., Kaloustian, S., Roeckle, M., 2007, Monitoring of open pit mines using combined GNSS satellite receiver and robotic total stations. In Potvin (ed.), Proceedings, *2007 International Symposium on Rock Slope Stability in Open Pit Mining and Civil Engineering*, Perth. ACG, Perth, pp. 417-429

Froese, R., Mei, S., 2008, Mapping and monitoring coal mine subsidence using LiDAR and InSAR, *GeoEdmonton2008*, pp. 1127-1133

Jang, Woo Jin, Yu, Chang Ho, Jung, Dong Hun, 2013, Local government's administrative strategies for industrialization of local resources, *Korea Policy Research*, Vol. 13 No. 1 pp.125-147

Ji, Byoungsuk, Rho, Gonil, Park Woojin, Yu, Kiyun, 2012, Implementation of MRDB web map service using open source GIS, Journal of the Korean Society for Geospatial Information System, *Conference of Journal of the Korean Society for Geospatial Information System*, PP.66-67

KIGAM, 2013, Mining statistics, *Minerals Commodity Information* No. 257, pp.54-71

Kim, Langley R.B., Bone J., Chrzanowskiet A., 2003, Local deformation monitoring using GPS in an open pit mine: Initial study, *GPS Solutions, 7 (2003)*, pp. 176–185

Lee, Hyun Jik,, Yang, Seung Ryoung, Lee, Kyu Man, 2008, Ecological restoration monitoring of open-pit mines using airborne laser scanning, *Journal of the Korean Society for Geospatial Information System* Vol. 16, No. 4, pp.101-107

Lee, Ju Young, Han, Moo Young, Yang, Jung-Seok, Choi, Jaeyoung, 2009, A study on environmental.mine geographic information system approach for the sustainable mine management and prevention of mine hazards, *KISTI Connected Environmental Policy Research* Vol. 8, No. 1, pp. 129-143

Song, Dae-Ho, 2013, A study on measures to establish information supporting system of EIA through the open platform of spatial information, Master`s Degree, Department of Environmental Science Graduate School Keimyung Univerisy

Singh S., Chutia, D., Sudhakar, S.(2012), Development of a web based GIS application for spatial natural resources information system using effective open source software and standards, *Journal of Geographic Information System*, 2012, 4, pp. 261-266

YIN, F., Feng, M., 2009, A webGIS framework for vector geospatial data sharing based on open source projects, *Proceedings of the 2009 International Symposium on Web Information Systems and Applications* (WISA'90), pp. 124-127

Free and Open Source Software for Geospatial (FOSS4G) Conference Proceedings

Volume 15 *Seoul, South Korea* Article 53

2015

Development of Data Archiving And Distribution System For The Philippines' LIDAR Program Using Object Storage Systems

Ken Abryl Eleazar Salanio
Data Archiving and Distribution Component, DREAM/Phil-LiDAR 1 Program Rm 312-316, Juinio Hall, National Engineering Center, University of the Philippines,

Carlo Santos
Data Archiving and Distribution Component, DREAM/Phil-LiDAR 1 Program Rm 312-316, Juinio Hall, National Engineering Center, University of the Philippines

Ruby Magturo
Data Archiving and Distribution Component, DREAM/Phil-LiDAR 1 Program Rm 312-316, Juinio Hall, National Engineering Center, University of the Philippines

Gene Paul Quevedo
Data Archiving and Distribution Component, DREAM/Phil-LiDAR 1 Program Rm 312-316, Juinio Hall, National Engineering Center, University of the Philippines

Kenan Virtucio
Data Archiving and Distribution Component, DREAM/Phil-LiDAR 1 Program Rm 312-316, Juinio Hall, National Engineering Center, University of the Philippines

Development of Data Archiving And Distribution System For The Philippines' LIDAR Program Using Object Storage Systems

Authors
Ken Abryl Eleazar Salanio, Carlo Santos, Ruby Magturo, Gene Paul Quevedo, Kenan Virtucio, Kenneth Langga, Mark Edwin Tupas, and Enrico Paringit

This paper is available in Free and Open Source Software for Geospatial (FOSS4G) Conference Proceedings: https://scholarworks.umass.edu/foss4g/vol15/iss1/53

DEVELOPMENT OF DATA ARCHIVING AND DISTRIBUTION SYSTEM FOR THE PHILIPPINES' LIDAR PROGRAM USING OBJECT STORAGE SYSTEMS

Ken Abryl Eleazar Salanio[1], Carlo Santos[1], Ruby Magturo[1], Gene Paul Quevedo[1], Kenan Virtucio[1], Kenneth Langga[1], Mark Edwin Tupas[1], and Enrico Paringit[1]

[1]Data Archiving and Distribution Component, DREAM/Phil-LiDAR 1 Program
Rm 312-316, Juinio Hall, National Engineering Center, University of the Philippines, Diliman, Quezon City, Philippines
Email: kasalanio@dream.upd.edu.ph

ABSTRACT

The Philippines' Department of Science and Technology in collaboration with Higher Education Institutions, lead by the University of the Philippines, has embarked on a program for producing hazard maps on most major river systems in the Philippines. Realising the utility of LiDAR and its derived data sets, a concurrent program on resource assessment was also initiated, aimed to produce essential products that can be used in urban planning, resource planning, and other purposes these geospatial data might be able to provide.

As a result, both programs produce immensely sizable datasets due for storage and distribution. Compiling them as one large file, or a small number of large files is impractical, and storing these data sets in current spatial content management systems and geoportal solutions also pose a challenge.

In this study we present a system for storing and delivering LiDAR and LiDAR-derived data using Ceph object storage system and a spatial content management system derived from GeoNode. This approach hinges on our requirements of being scalable yet robust without much deviations from the current file system based storage structure.

Key Words: LiDAR, Data Archiving, Object Storage, Ceph, GeoNode

1. INTRODUCTION

The Philippines is situated in the Pacific Ring of Fire and the Pacific Typhoon Belt. Thus, it is prone to numerous natural hazards, particularly earthquakes, landslides, floods, and storm surges. This demands mapping of the Philippines' physical environment for preparedness and handling of natural disasters.

Mapping using Light Detection and Ranging (LiDAR) technology is suitable for these purposes due to the high resolution data it produces. With this, the University of the Philippines, through the support of the Department of Science and Technology (DOST) and in cooperation with Higher Education Institutions (HEIs) across the country, organized two concurrent programs that use LiDAR technology, drawing on the success of the Disaster Risk and Exposure Assessment for Mitigation (DREAM) LiDAR program. The Phil-LiDAR 1 program extends the original DREAM program, producing flood hazard and assessment maps, while the Phil-LiDAR 2 program produces resource maps from acquired and processed LiDAR data of the first program. These endeavors aim to produce essential products such as DEMs, Orthophotós and LAS that can be used for hazard mapping, urban planning, and resource planning among others.

The products of these programs are a large amount of data that need to be distributed and archived at a swift pace. As with other LiDAR operations, handling large swaths of spatial data is not an option, hence data sets are organized in contiguous blocks, subdivided by files and grouped by river systems and local government units, stored centrally in a file-based system.

Storing these data sets prove to be a challenge even for existing spatial content management systems and geoportal solutions. These systems have capabilities mostly for vector and raster data but do not natively support storage of point cloud data. There is also the problem with scaling due to the sheer volume of data produced. The common approach, like the program's legacy storage system, is to use a centralized Network Attached Storage (NAS) server to manage several attached storage disks such that it appears as one big file system. This file system is then shared over the network for access from each personnel's workstation.

To facilitate the storage and retrieval of data sets for the program's archiving and data distribution process, we developed a system for storing LiDAR and LiDAR-derived data using Ceph object storage system. Ceph's support for a distributed environment reduces the risk of a single point of failure. Consequently, it has methods for maintaining data integrity among its storage nodes which is a vital requirement for archiving. A spatial content management system derived from GeoNode provides a user interface for accessing available data.

2. RELATED WORK

Recent developments in digital acquisition technologies have enabled the collection of high-resolution spatial data sets. With the use of LiDAR, it is possible to collect billions of individual point measurements of the earth's surface (Crosby et al., 2011). Aside from mapping, these high-accuracy datasets have other several applications, including floodplain mapping, hydrology, geomorphology, forest inventory, urban planning, and landscape ecology (Chen, 2007).

To store these data, file-based storage systems are often used. In a file-based storage system, data sets are stored as files managed by an operating system. Each file has a unique path described in the file system. Navigation is done by browsing the directory structure or using a search tool or a file system navigator (like Nautilus or Windows Explorer) installed in the operating system. This system is usually implemented using a Network Attached Storage (NAS) platform to enable sharing of files over a network of computers. NAS is a specialized dedicated server designed to provide file access and storage to several client computers (Levine, 1998).

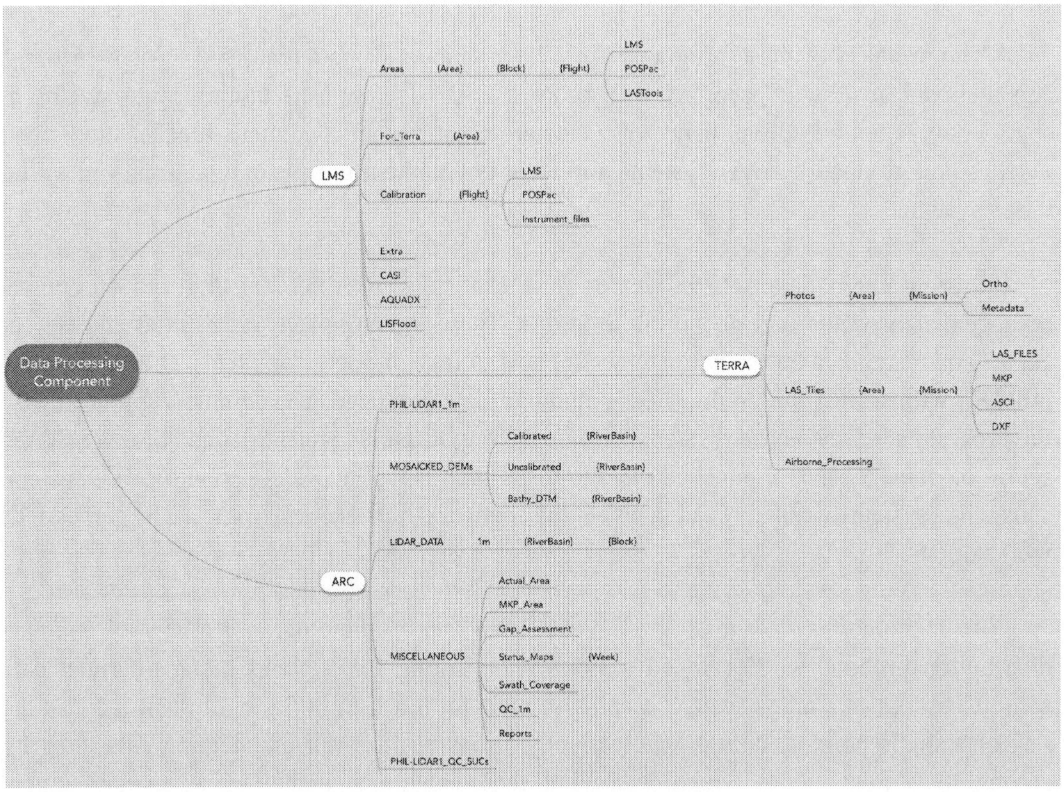

Fig 1. Sample File Directory Tree

The complexity of directory structure increases with the amount of data sets and the processes involved (e.g. raw, derived semi-processed data, processed data, versioning, etc.). Searching for the needed data set requires manually overlaying coverages and retrieving associated data based on folder and filename convention. Phil-LiDAR 1 uses a tiled approach for organizing and managing spatial data mentioned in a number of studies (Chen, 2007; David et al., 2008, Lewis et al., 2012).

GIS-enabled relational database management systems (RDBMS) can also be used for storing geospatial data sets (Fox et al., 2013). These are spatially enabled databases designed to index geographically-referenced datasets. The entries are stored in a table, with the spatial index created on a separate layer. This index is referenced when the database handles geospatial queries. Another approach is to design the RDBMs with specialized spatial columns. Some attempts to store LiDAR data includes Lewis et al (2012) who used Well Known Binary point geometry and Ramsey (2013) who devised a spatial datatype for storing point cloud clusters.

However, the acquired spatial data, as well as the derived ones, sum up to sizes reaching terabytes or petabytes worth of storage. Since the indexing is external to the data for traditional RDBMSs, updates incur additional indexing overhead. Such updates are still computationally intensive and affect performance adversely as data size increases (Fox et al., 2013). This produces scalability and query time issues (Al-Naami et al., 2014).

Another way of archiving spatial data sets is to use a combination of LiDAR flat tiles, RDBMS, and a cloud computing or distributed infrastructure. The RDBMS manages the metadata, while the LiDAR tiles are stored in a dedicated storage server. Processing of these datasets are done by high-performance compute servers. Both the storage and compute servers are provisioned by a cloud computing infrastructure. One example is the OpenTopography System (OpenTopography, 2015). It is built upon a multi-tier service oriented architecture (SOA) comprised of various software services and hardware resources. This consists of a front-end portal for serving data, as well as handling authentication. Actual data sets are stored as ASPRS LAS format on a dedicated storage server. Metadata is stored on an IBM DB2 database, and other datasets are stored on the SDSC Cloud platform. On-demand processing processing and visualization requests are handled by a dedicated large memory, multiprocessor system.

While OpenTopography's approach is highly appealing, implementing it in the Philippine setting raises several concerns. For one, the internet speeds in the Philippines are slow and erratic. While a high-performance data center may be set up, if access to it is sluggish, then its resources will not be fully utilized. Next, it requires the program to heavily upgrade its IT infrastructure, which can take time and valuable resources to set up.

Another alternative is to use Object-based storage systems. Object-based storage manages data as objects as opposed to a file hierarchy from a file system. Objects are storage containers with a file-like interface, effectively representing a convergence between file and block storage technologies, thus offering storage that is secure and easy to share across platforms, but also high-performance and scalable (Mesnier et al., 2003). Several object storage technologies exist, examples of which are Amazon S3 (Amazon Web Services, 2015), Openstack Swift (SwiftStack, Inc., 2015), and Ceph (Inktank Storage, Inc., 2015).

3. Significance and Objectives

With LiDAR datasets having such high value, securing the data in a more robust archiving solution is needed. One possible solution is migrating from a file-based storage system to a more apt distributed infrastructure. This requires the data sets must first be transferred into a storage system capable of scaling to terabytes and petabytes. Object storage systems are capable of filling this role. This paper discusses Phil-LiDAR's attempt at implementing a LiDAR archiving and distribution system using an object storage system that fills the need of it's unique organizational structure.

4. Methodology

Given the varied choices for object storage technologies, Ceph was chosen due to being open source, its compatibility with widely used cloud services such as OpenStack and Amazon's AWS, and the broad spectrum of programming languages it supports. It runs on commodity hardware, and is designed to be self-healing and self-managing (Inktank Storage, Inc., 2015).

To provide management for the data uploaded into the object storage, a geospatial content management solution is needed. GeoNode (GeoNode Development Team, 2013) is an open-source web-based application and platform for geospatial information systems (GIS) content management system written in Django. Built on top of GeoServer (Open Source Geospatial Foundation, 2014), it is able to store and share rasters and vectors and uses GeoExplorer (San Diego Supercomputer Center, 2015) to display the managed data sets. GeoNode not only provides a web interface for upload, display, and download of vectors and rasters but also allows sharing of these data sets using Open Geospatial Consortium (OGC) standard web services. Additional features can be developed on any open-source platforms it uses.

In section 4.1, this paper discusses the web portal and GeoNode-based spatial content management system we designed, the Lidar Portal for Archiving and Distribution. Section 4.2 describes how it interfaces with the Ceph object storage, as well as Ceph's architecture. Section 4.3 outlines the process flow for archiving spatial data sets.

4.1 Lidar Portal for Archiving and Distribution (LiPAD)

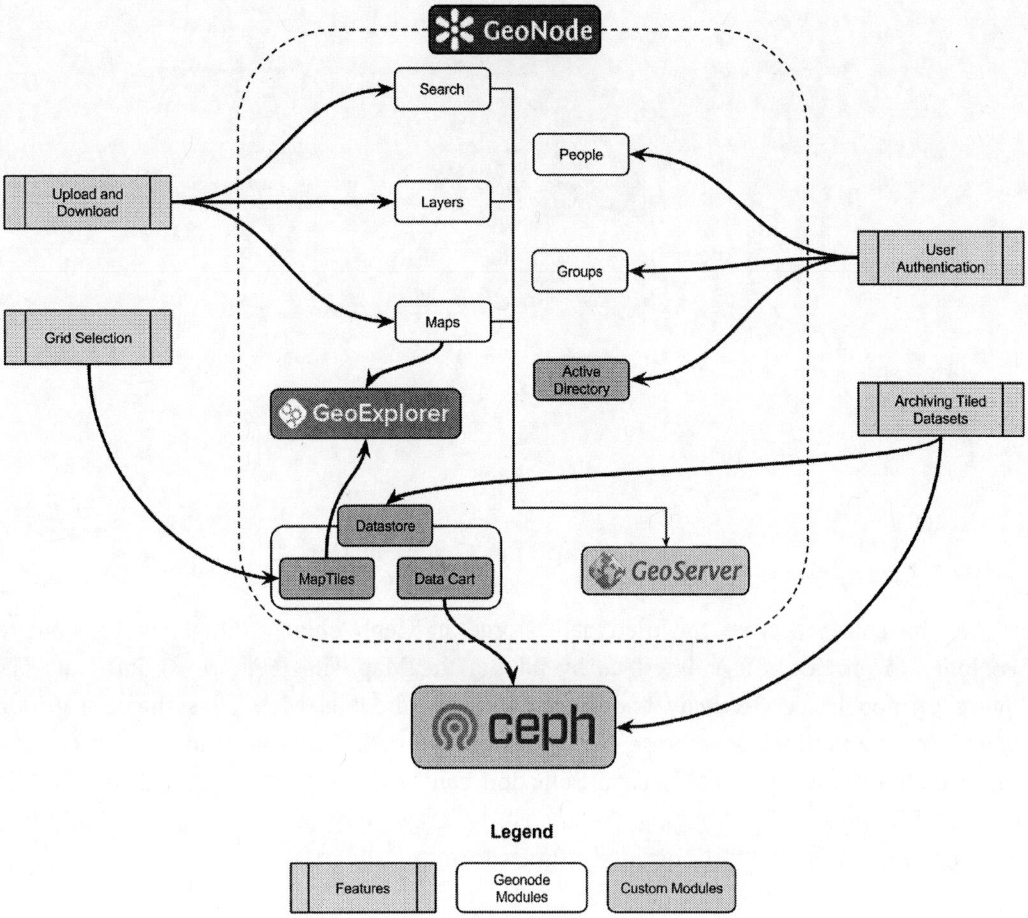

Figure 2. LiPAD Feature and Component Diagram

The Lidar Portal for Archiving and Distribution (LiPAD) is the customised GeoNode used for managing the archived data. GeoNode has native upload and download modules, but these are limited to rasters and vectors. Uploads are stored in the GeoServer running in the backend. These uploaded data can then be downloaded by users, via the Layers and Maps modules, with the permission of the owner. Users can also combine these layers, into a montage using the Map module. For user authentication and management, GeoNode has the People and Groups modules. These modules are then linked to the in-office Active Directory, a method of storing user credentials in a network of computers.

Figure 3. Map Tiles Grid Selection

To be able to access the tiled data stored in Ceph object storage, we customized GeoNode to provide a user interface by adding the Map Tiles (Figure 3) and Data Cart (Figure 4) modules, collectively under the Datastore module. Map Tiles shows a gridded shapefile of the Philippines, representative of the tiled data. The shapefile is color-coded to show which areas have available data. Selection can be done by clicking each tile, holding the SHIFT button while dragging the mouse to draw a bounding box, or submitting a bounding box in EPSG:900913 decimal degrees format.

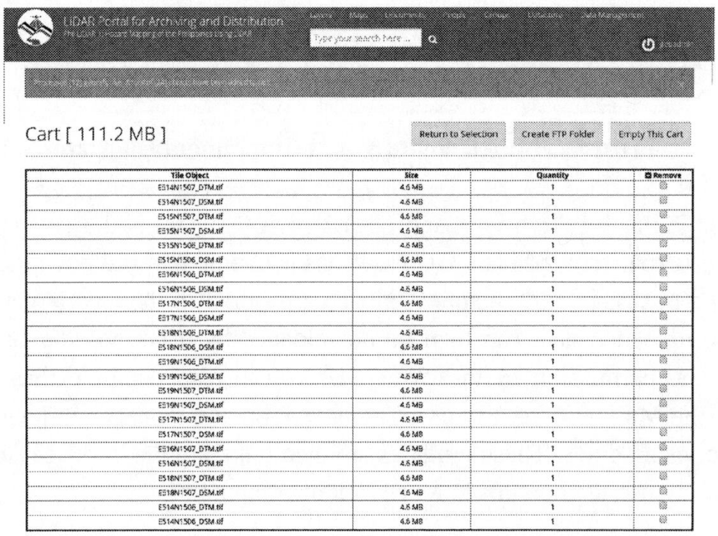

Figure 4. Data Cart Item List

Submitting the selected tiles adds all available data associated to each selected tile to the data cart. Once the user is satisfied with the cart's contents, a download request can be made with a click of a button. The request is then processed in the background by contacting Ceph via a RESTful API to retrieve the listed data.

4.2 Ceph Object Storage

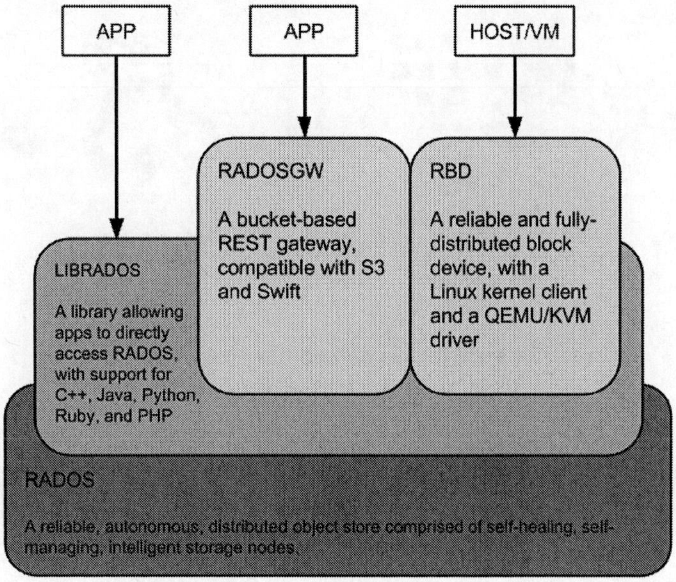

Figure 4. Ceph Object Storage Features

Ceph's REST API provides an interface for programmatically accessing the stored objects (e.g. download, upload, view details) . This interface is also known as the Object Gateway (OGW), which can be utilized using HTTP queries and Python scripts.

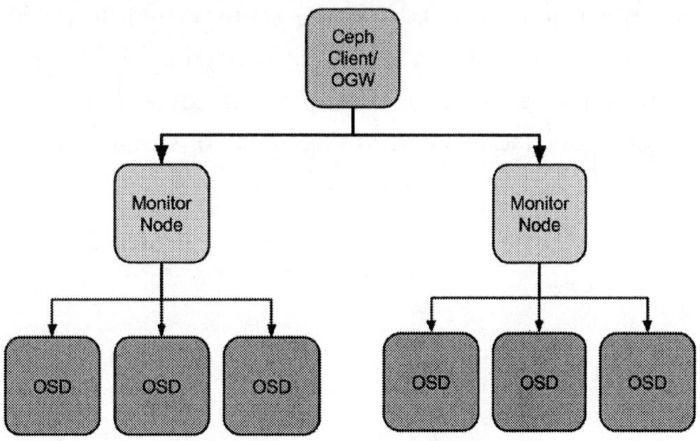

Figure 2. Ceph Architecture

The Ceph architecture comprises of multiple Object Storage Devices (OSDs) managed by at least one monitor node. The monitor node is responsible for ensuring high availability of the OSDs. A client node, which often is the object gateway, interfaces with the monitor nodes to handle requests for authentication, upload and download of objects, and status checks.

4.3 Archiving Process Flow

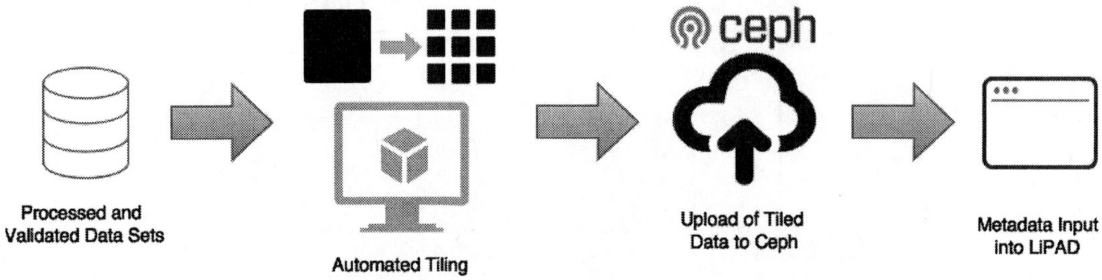

Figure 5. Archiving Process Flow

While acquired data goes through processing depending on the output desired, this section will focus on the archiving of LiDAR data and its derivatives. Once the LiDAR data has been processed and validated for archiving, the data is transferred to a virtualized host running a script that tiles the data into 1 km by 1km tiles. As a naming convention and for the purpose of spatial indexing, we use the EPSG:32651 Northing and Easting value of the upper left corner of the tile as well as the data type, as the object's name (e.g. E449N1271_LAZ.laz for LAZ, E449N1271_DTM.tif for DTM, E449N1271_DSM.tif for DSM, or E449N1271_ORTHO.tif for Orthophoto). These tiles can later be directly queried with their object name once uploaded to Ceph. These tiled data sets are then uploaded into Ceph using another script. The uploading script, once done, produces a CSV file containing the relevant metadata of each tile, which is then fed into the platform for metadata storage.

Automating the tiling and the upload in the same virtual host which houses the pre-existing file-based storage reduces the overhead introduced by working on the data sets in a workstation separate from the server. This overhead may come in the form of transferring or copying the files into the workstation's own storage disk, or simply opening the files over the network in a NAS.

5. FUTURE OUTLOOK

We envision the possible future scenario that the archiving and distribution capabilities of the program will be on a decentralized and distributed computing infrastructure to reduce the risk of a single point of failure. This will also enable the automation of some of the process flows, like the update of the gridded map data to reflect the objects stored in Ceph.

The use of Ceph allows the system to easily adapt to a distributed environment because of its compatibility with Amazon's AWS and OpenStack. Meanwhile, computing processes that can be parallelized, such as indexing and querying of data, can make use of technologies such as Hadoop (The Apache Software Foundation, 2014), a framework that allows for the distributed processing of large data sets across clusters of computers using simple programming models.

6. SUMMARY

Advances in geospatial technologies have enabled the acquisition of high-density, high-accuracy data sets with significantly detailed surface information of terrain and surface objects within a short period of time (Al-Naami et al., 2014). This results into large data sets that need to be stored for archiving, retrieval, and distribution. However, conventional storage systems, namely native file systems and spatial databases, have problems in scaling to the sizes of the datasets produced by the current GIS acquisition technologies (Fox et al., 2013).

In this paper, we presented the use of Ceph Object Storage coupled with customized GeoNode to store and manage LiDAR and LiDAR-derived data sets. The customized GeoNode serves as both the front-end interface and the content management platform, while Ceph stores the large data sets subdivided into 1 KM by 1 KM tiles. This setup paves the road ahead for migrating from a file-system based infrastructure into a distributed computing platform.

7. REFERENCES

Ackermann, F., 1999. Airborne laser scanning—present status and future expectations. *ISPRS Journal of Photogrammetry and Remote Sensing 54 (2-3)*, 64-67.

Al-Naami, K.M., and S. Seker, L. Khan, 2014. GISQF: An Efficient Spatial Query Processing System. *2014 IEEE 7th International Conference on Cloud Computing (CLOUD)*, pp. 681 - 688.

Amazon Web Services, 2015. *AWS | Amazon Simple Storage Service (S3) - Online Cloud Storage for Data & Files.* Retrieved July 2015, from https://aws.amazon.com/s3/

Boundless, 2015. *GeoExplorer — GeoExplorer.* Retrieved July 2015, from http://suite.opengeo.org/opengeo-docs/geoexplorer/

Chen, Q., 2007. Airborne lidar data processing and information extraction. *Photogrammetric Engineering & Remote Sensing, 73(2),* 91-95.

Crosby, C.J., Arrowsmith, J R., Nandigam, and V., Baru, C., 2011. *A Geoinformatics Approach To Online Access And Processing Of LIDAR Topography Data.* In R. Keller and C. Baru, Eds., *Geoinformatics: Cyberinfrastructure for the Solid Earth Sciences,* pp. 251-265. London: Cambridge University Press.

David, N., Mallet, C., and Bretar, F., 2008. Library concept and design for lidar data processing. *GEOgraphic Object Based Image Analysis (GEOBIA) Conference,* Calgary, Canada.

Fox, A., Eichelberger, C., Hughes, J., and Lyon, S., 2013. Spatio-temporal indexing in non-relational distributed databases. *2013 IEEE International Conference on Big Data,* pp. 291–299.

GeoNode Development Team, 2013. *About GeoNode — GeoNode 2.0 documentation.* Retrieved July 2015, from http://docs.geonode.org/en/master/organizational/about.html#about

Inktank Storage, Inc., 2015. *Welcome to Ceph — Ceph Documentation.* Retrieved July 2015, from http://ceph.com/docs/master/

Isenburg, M., 2012. LASzip:Lossless compression of LiDAR Data. *European LiDAR Mapping Forum.*

Jones, T., 2010. *Ceph: A Linux petabyte-scale distributed file system.* Retrieved July 2015, from http://www.ibm.com/developerworks/library/l-ceph/

Levine, R., 1998. *NAS Advantages: A VARs View.* Retrieved July 2015, from http://www.infostor.com/index/articles/display/55961/articles/infostor/volume-2/issue-4/news-analysis-trends/nas-advantages-a-vars-view.html

Lewis P., McElhinney C., and McCarthy T., 2012. LiDAR data management pipeline; from spatial database population to web-application visualization. *Conference Proceedings at Com.Geo 2012,* Washington DC, USA.

Mesnier, M., Ganger, G. R., and Riedel, E., August 2003. Object-Based Storage. *IEEE Communications Magazine*, pp. 84–90.

Open Source Geospatial Foundation, 2014. *About - GeoServer.* Retrieved July 2015, from http://geoserver.org/about/

OpenTopography, 2015. *NSF OpenTopography Facility | About.* Retrieved July 2015, from http://www.opentopography.org/index.php/about/

Ramsey, P., 2013. LIDAR in PostgresSQL with PointCloud. Available online: http://boundlessgeo.com/wp-content/uploads/2013/10/pgpointcloud-foss4-2013.pdf

San Diego Supercomputer Center, 2015. *SDSC Cloud.* Retrieved July 2015, from https://cloud.sdsc.edu/hp/index.php

SwiftStack, Inc., 2015. *OpenStack Swift | Enterprise Storage from SwiftStack.* Retrieved July 2015, from https://swiftstack.com/openstack-swift/

The Apache Software Foundation, 2014. *Welcome to Apache™ Hadoop®!.* Retrieved July 2015, from https://hadoop.apache.org/

Free and Open Source Software for Geospatial (FOSS4G) Conference Proceedings

Volume 15 *Seoul, South Korea* Article 54

2015

Usability Engineering For Successful Open Citizen Science

Jan Alexander Wirwahn
Institute for Geoinformatics, University of Münster

Thomas Bartoschek
University of Münster, Germany

AT3-01

USABILITY ENGINEERING FOR SUCCESSFUL OPEN CITIZEN SCIENCE

Jan Alexander Wirwahn and Thomas Bartoschek

Institute for Geoinformatics, University of Münster
Heisenbergstr. 2, 48149 Münster, Germany
Email: j.wirwahn@uni-muenster, bartoschek@uni-muenster.de

ABSTRACT

Do-it-yourself construction kits for sensor stations is a novel approach for citizen driven sensor networks. In this paper we present the development of SenseBox, a toolkit for open source sensor applications. We provide manuals and open source toolkits for hardware makers to establish a large scale sensor network in Germany, and additionally for schools to teach secondary school students programming in a playful and simple way based on open source microcontrollers. All collected data is open and being published on OpenSenseMap, our platform for open sensor data. After a test phase where we equipped citizens with construction kits for continuous measurements we discovered that most of the stations were disconnected after a few weeks. Following from that we performed a user study to reveal possible error sources during the wiring process, software installation and online registration on OpenSenseMap. Missing general computer skills led to larger problems than wiring of hardware parts.

1. INTRODUCTION

Citizen science can be explained as the engagement of non-professional scientists in collecting data, analyzing data, developing technologies and the publication of these on a voluntary basis. In a majority of citizen science projects, the data to be collected by citizens is geospatial and is being presented on maps. If the data is about environmental observations, this approach is often referred to as participatory sensing (Burke *et al.* 2006).

Our novel approach in this field is to equip citizens with do-it-yourself (DIY) environmental sensor stations to establish citizen driven sensor networks. This approach leads to a better data coverage but also contains motivational aspects, as the citizens build up their own devices and are therefor exposed to sensor technology as well. Most existing DIY sensor stations are not entirely open in terms of source code, data collection, hardware, educational documents or extensibility for other platforms (Uckelmann *et al.* 2011). In conclusion, user groups who are interested in the data or citizens who want to understand the algorithms are excepted from certain levels of those projects, even if they are willing to contribute.

In that context, the SenseBox project started at the Institute for Geoinformatics, University of Münster and is an ongoing open citizen science project. Based on open source hardware and open source software components citizens build their own Web of Things enabled sensor stations to collect environmental data about values of interest such as temperature, humidity, air pressure, loudness, illuminance and intensity of UV-light. The data is then being published as open data and visualized on a web based platform, the OpenSenseMap (OSeM). An educational edition of the SenseBox including hardware parts and didactical material are being introduced into secondary schools, where school students learn to code, measure environmental phenomena and work scientifically. The whole source code and all hardware components are open source, instructions are being published as open educational resources (OER) and models for a 3D-printed waterproof cases are available under open source licenses as well.

During a test deployment of around 50 SenseBox stations, we encountered a leak of reliability in terms of continuous data collection. In this paper we want to investigate, if usability problems were the cause of the low success rate and how the motivation of citizen scientist can be preserved to ensure a long-term data collection.

In the following, we give an overview on related work (Section 2) and introduce our development of SenseBox and OSeM (Section 3). After that, we present potential risks during different states of the development process and present a user study we performed (Section 4). Finally, we bring our approach in context to related work presenting a conclusion and an outlook to future work (Section 5).

2. RELATED WORK

The SenseBox development is closely related to the research field of Internet of Things (IoT), where physical objects and embedded devices are connected to the Internet and made uniquely addressable (Gershenfeld *et al.* 2004). In the Web of Things (WoT), the real world objects become accessible by an URI using REST principles and HTTP protocols (Guinard, Trifa, 2009). A connected world enables application scenarios of smart home and smart city. A major part for these fields are sensor networks, enabling the intelligent control of the environment. WoT enabled sensor stations are intelligent objects that are able to sense the environment, act on incidents or communicate with other sensor nodes or things (Bröring 2011).

Successful, large scale citizen science projects like Galaxy Zoo (Raddik *et al.* 2010) or eBird (Sullivan *et al.* 2014) have shown the power of the creative community as toolkit for science in different domains. In particular large scale projects were pioneers in this field, which enhanced prestige in the scientific community and awoke more interest in mainstream media. Sullivan *et al.* (2014) made aware that these projects vary in their quality of contribution for science, conservation and private policy. In their paper they present a novel approach of building cooperative partnerships in citizen science to increase participation by meeting community needs. Shirk *et al.* (2012) conceptualized a model for public participation in scientific research based on degree and quality of participation. They define the degree of participation as the influence of participants on the research process they are involved in. Formally, that can be expressed by the number of participants, duration of involvement or research effort. They consider key components for a high quality of participation as credibility, trust or responsiveness, which are important subjective factors. Their model about "public participation in scientific research" categorizes conceptual, contributory, collaborative, co-created and collegial projects. It is meant to identify the range of potential outcomes over the interactions between scientific and public interests.

3. BACKGROUND

In 2011 the SenseBox project was initiated during a study project at the Institute for Geoinformatics, at the University of Münster. The developed prototype was an WoT enabled sensor station. It focused on a generic and flexible approach for modular sensor setups in outdoor environments (Bröring *et al.* 2011). The core of the station was based on open source hardware Arduino microcontrollers. It was flexible in terms of sensor setup and application areas. The advantage of using Arduino as a basis, is the large and constantly growing community of DIY hardware prototype makers, organized in blogs, magazines and community meetings. Potential non-professional users can take advantage in the DIY community as they are not limited to specific tutorials and documentations. Currently, the SenseBox project offers two versions. Each of them is designed for a specific user group. Both of them are aiming to establish a large scale sensor network of open environmental data, collected by citizen scientists. In the following, we describe the current state of our project.

3.1 SenseBox Home: A DIY sensor station for citizen scientists

The SenseBox Home is a low cost starter kit for a DIY environmental measurement station, allowing non-professionals to collect and publish their own sensor data. Detailed construction manuals are available including descriptions for setting up the hardware, installing required software and registering the station on OSeM. As a core, the setup consists of an Arduino Uno microcontroller, a network adapter and the Arduino-compatible SenseBox Shield. The latter is designed for an user friendly setup of the four basic sensors, consisting of barometer, luxmeter, UV-light sensor and a combined temperature and humidity sensor. Because of its open source character, the basic setup can be enhanced with additional sensors and modules to cover as many application scenarios as possible. Mobile data logger as well as air- and water-quality sensor upgrades are currently in development. The aim is to set up and deploy a reliable sensing station which is pushing measurements to the a server continuously. Because of the motivational aspect, the DIY character of the construction kit is highlighted when supplying SenseBoxes to the citizens. Additionally we created a 3D model for a waterproof case, which is freely available on the Thingiverse platform[1] and can be 3D printed. On the left in Figure 1 below, an illustration of the 3D model is shown. The lamellae are used as sensor protection against raidiance. On the right, the assembled case without cover to show the wiring.

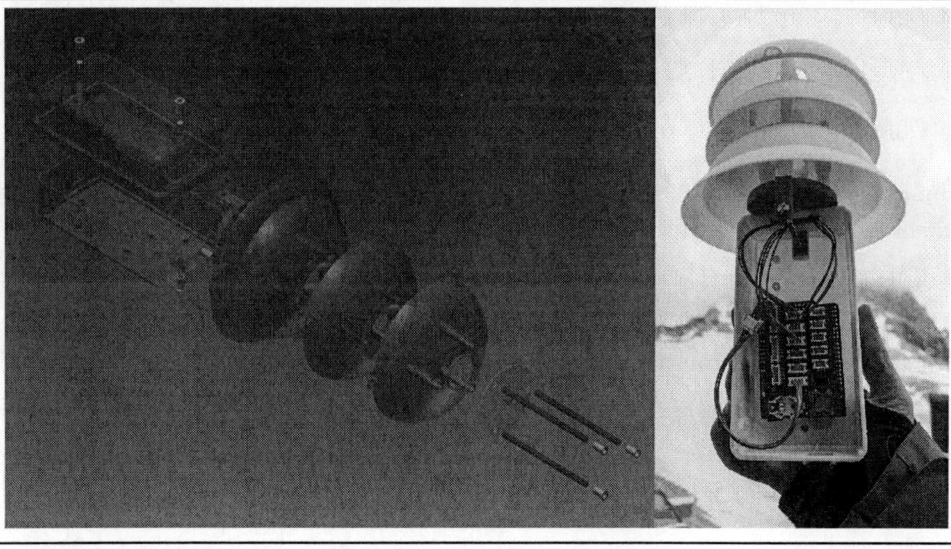

Figure 1. Our open 3D model can be used as casing for environmental sensors .

3.2 The SenseBox Edu: an educational toolkit

After the implementation of the hardware infrastructure described above, the SenseBox was used in practical workshops of ifgi's GI@School lab, providing secondary school students with a toolkit for learning to code with open source hardware. Evaluations of our

1 http://www.thingiverse.com/thing:728592

workshops have revealed that microcontroller programming implies a simple and playful way to teach children the basics of coding. The haptic feedback of intelligent I/O circuits controlled by microcontroller software is highly motivating for them when dealing with the complex topic of structured code writing. Construction kits for single projects, e.g. a weather station based on the SenseBox concept for schools, were developed afterwards. Beginning from scratch, in approximately four hours working with the SenseBox, small groups of two pupils are already able to program a car traffic counter with integrated data logger, controlled by ultrasonic sensors. In Figure 2 below, two school students are preparing a traffic counter in one of our workshops.

Figure 2. Pupils wiring and programming a traffic counter during a SenseBox workshop.

During the procedure of setting up the wiring and programming the microcontroller, they are instructed to use our construction manuals in combination with the online community as a support and therefor not only learn about how to code, but to solve technical problems using the knowledge of a creative community in the web.

After numerous hack events in our school laboratory and in the frame of project days in local secondary schools we selected experiments and projects developed in our workshops and created educational construction kits enabling a stand alone application in the schools. In that context, teachers can use the SenseBox in mathematics, computer science, natural science and technical related subjects to create interdisciplinary teaching units.

3.3 OpenSenseMap: A standard for collecting and sharing open sensor data

One result of the deployment of sensor station construction kits was the upcoming need of storing and sharing measurement data. Consequently, we developed an online web platform for that purpose: the OpenSenseMap. By analogy with OpenStreetMap where volunteers are mapping the environment by hand to fill blank areas with information, participants in our project are doing so by deploying and registering a new sensor station at a fixed location. In Figure 3 below, the map interface of OSeM is displayed with an example sensor station selected. Different colors of location markers are indicating different types of sensor setups.

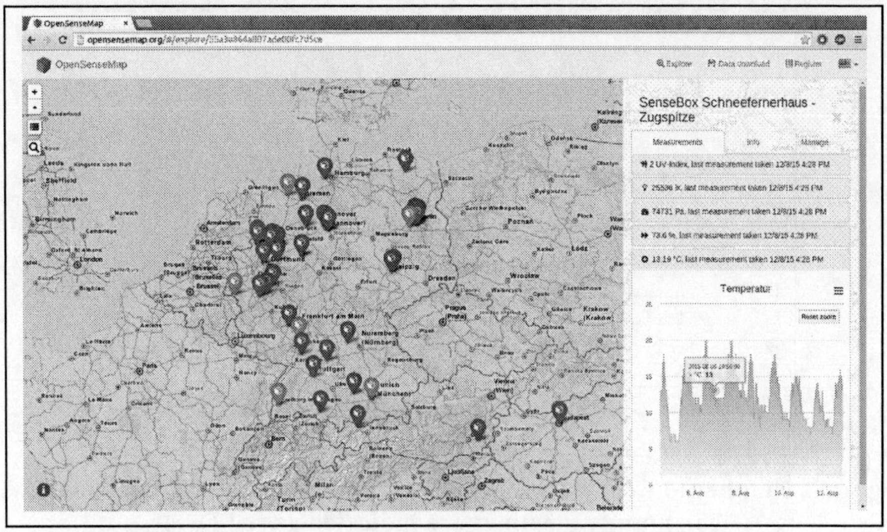

Figure 3. Screenshot of the OpenSenseMap interface.

The OSeM server application is based on a RESTful service architecture and has a document based database integrated. That allows data storage of any structure and makes the database highly scalable (Pfeil *et al.* 2015). Its open API is designed for simple access, enabling the storage of any possible measurement data. In addition, it is possible to subscribe to one ore more stations over the API or to download the datasets as CSV or GeoJSON. In a lightweight registration, SenseBox stations as well as custom stations can be instantiated in the database. As unique feature, we provide automatic software sketch generation for sensor stations based on open source Arduino microcontrollers. Primarily we want to establish a large scale, high resolution sensor network in Germany, driven by school projects and citizen scientists. The vision of OSeM is to become a standard for citizen science open measurement data.

3.4 Promoting the project

It is a crucial part for a project with public involvement to meet the corresponding user groups. Community meetings are focused on a specific user group and offering the possibility to draw attention on the project for potential users. Depending on size and publicity, exhibitions have the further advantage to reach people nation- or even worldwide. In our case the most important fairs we attended were the following three: Didacta, which is the most prominent educational fair worldwide with over 72000 visitors. Hannover Messe, which is the world's leading trade fair for industrial technology with over 220000 visitors (70000 of them from outside Germany). And at least the Maker Fair, which is a hot spot for hardware prototyping enthusiasts with more than 10000 visitors. On site, we presented the SenseBox on a stand, where we offered information material and simple interactive experiments to interested visitors. An indicator for the publicity are recorded sessions on our project website, shown in Figure 4 below.

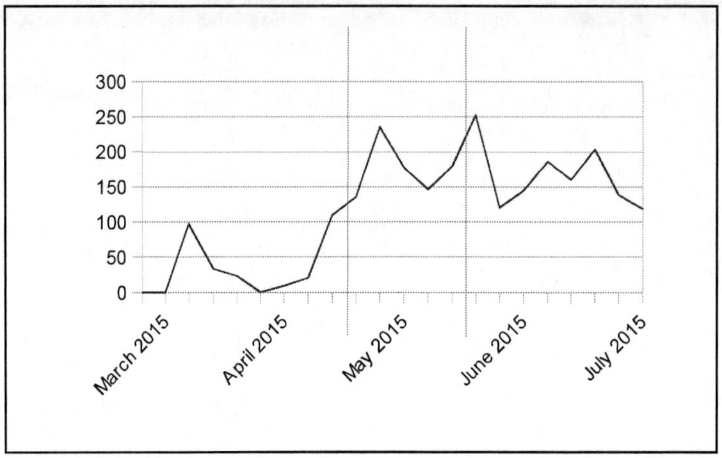

Figure 4. Number of weekly sessions on our project websites. Participations in community meetings are indicated as red lines.

3.5 Networking and funding

To reach more potential users for citizen science projects, an information network should be established to inform citizens about development state and product updates (Sullivan *et al.* 2014). The name of the project should be promoted and emphasized in online and offline media. As an example for the importance of mainstream media, an article about the deployment of a SenseBox station at a scientific station on the Zugspitze, which is the highest mountain in Germany, reaches 35000 readers of the Make Magazine as hard copies and many more online.

In Germany, a new citizen science era started with the establishment of the official Citizen Science platform *Bürger schaffen Wissen* (*Citizens Create Knowledge*)[2], established by the scientific community *Wissenschaft im Dialog*[3], in April 2014. Beginning with ten starter projects on their platform, there are now over 30 projects promoted. The platform has become the place to go for both, Citizen Science projects that want to become popular and interested parties which want to participate. SenseBox was one of the ten promoted starter projects and has become the most clicked project on the platform.

An important supporter in our network is the Federal Ministry of Education and Research (BMBF), which is particularly interested in the establishment of a nationwide sensor network of light measurmements for Germany within the scope of BMBF's "Make Light" initiative. Their funding enabled us to finalize our SenseBox prototypes and enhance our server infrastructure during the last months. As a state institution, they have connections to a large number of related institutions in Germany and access to events, like the three fairs mentioned in the last subsection. As a partner of the BMBF, we also draw additional attention in public media and gained a professional outward appearance in the public.

4. ENHANCING USABILITY OF DIY SENSOR STATIONS

In a first project phase, around 50 fixed SenseBox Home stations were deployed to citizens and schools distributed over Germany. Some participants had problems in the building and registration process of the SenseBox, others disconnected their SenseBox after some time. Even though, the number of registered stations increased constantly, the uploaded

2 http://www.buergerschaffenwissen.de/en
3 http://www.wissenschaft-im-dialog.de/en/about-us/

measurements did not (Figure 5).

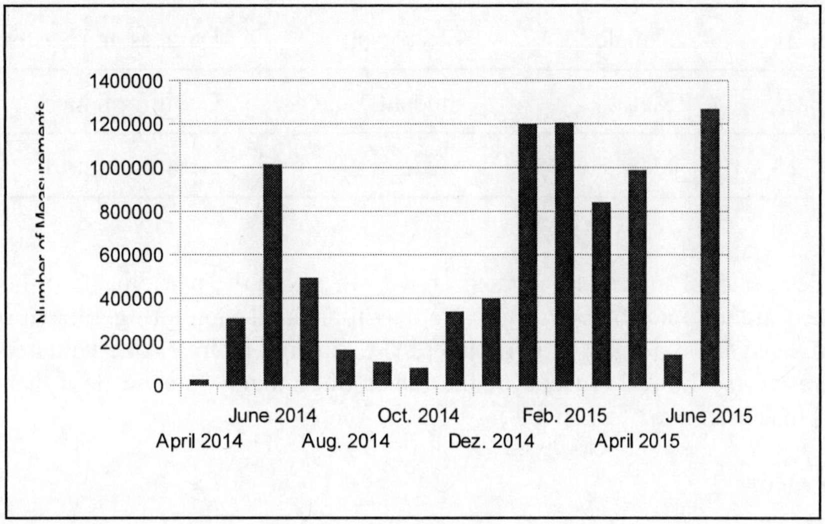

Figure 5. Uploaded measurements from SenseBox stations to OSeM per month.

Following from that, we looked for volunteers to perform a user study to find gaps and error sources of our approach. In this section of the paper we describe our methodology for selecting an appropriate user group and the results of the user study we performed at our institute.

4.1 Focus on user group(s)

User groups are varying depending on the domain of the project. In Section 3 we explained the two types of the SenseBox and their application areas in detail. Following from that, we focus on a user group of young teachers that are interested in technical project work for their classes (1) and on a user group of citizens that are in touch with DIY hardware prototyping or WoT development (2). In our case both of the user groups are interested or involved in geosciences, as we place special emphasis on the aspect of environmental sensing.

Table 1. Overview about personal information of the study participants.

ID	Age	Gender	Job	Area
A	33	Male	Teacher	Sports
B	29	Male	Student	Geography
C	27	Male	Student	Human Sciences
D	21	Female	Student	Geoinformatics
E	22	Female	Student	Maths / Geography
F	26	Female	Student Teacher	Human Sciences
G	33	Male	Teacher	Maths / Sports

H	27	Male	Student	Maths / Geography
I	21	Female	Electrician	Devices and Systems
J	28	Male	Student	Geoinformatics
K	25	Male	Student	Geoinformatics

Besides personal information stated above, we asked the participants in how far they are interested in the topics of DIY hardware, environmental monitoring, citizen science and the Web of Things. On a scale of 1 (no interest) to 5 (high interest) one volunteer was only partly interested (scale of 2) in one of the topics, the others were not less than interested (scale of 3) in at least two of the fields.

4.2 User study

We conducted the user study in a controlled environment at the Institute for Geoinformatics, University of Münster. The goal of the study was to reveal possible error sources in our documentation material, that was used in the initial testing mentioned above. We focused on non-professionals, that were at least interested in DIY hardware or in the domain of geosciences. In total, 30 volunteers were informed about our project. They were asked for participation if they showed interest for our topic. Finally, we chose 11 of them to further conduct in our study, which is enough for determining significant test results (Nielsen 1993). Each study took 45 minutes and as an inducement for participation, we randomly selected one out of the pool of volunteers and provided the person with a SenseBox Home construction kit worth ~100€. Personal information of the test subjects were anonymized, user names and e-mail addresses used for online registration, were deleted after each test run in the presence of the test subject.

The task of the study was to construct a SenseBox station (1), install all the required software (2), and integrate it into our sensor network by online registration on OSeM (3). All instructions were given in form of a printed version of our construction manual. Step 1 was video and sound recorded by a table top camera, in step 2 and 3 we captured the screen of the test computer used by the volunteers. A passive observer was present in our laboratory during the study and made notes about the subject's behavior during the test. In case that a volunteer was not able to continue because of wrong hardware handling, errors in software installation, or inappropriate handling of the construction kit, the observer intervened, solved the problem and noticed it. After the study the volunteers filled out a questionnaire, containing questions about personal information, background knowledge and interest. In addition three standardized NASA Task Load Index (TLX) forms were filled out for each of the construction steps, for subjective workload assessment (Hart, 1988).

4.3 Study results

Only four out of eleven subjects completed the study successfully without help of the passive observer. All problems that could not be solved by the participants themselves occurred in step 2, where subjects were asked to install required software and driver components. Even though we provided detailed step-by-step instructions in the manual, 90% of the mistakes happed during copying and unzipping required software libraries into a program folder on the hard disk. One participant was, as he mentioned, "confused" because of the high task load during the study. He overlooked one of the pages in the construction manual and after downloading the pre configured microcontroller code he selected a wrong file in the download folder after the registration in step 3. In Table 2 below, error sources of each participant are and time measurements illustrated.

Table 2. Error sources and construction time of the participants.

ID	Step 1	Step 2	Step 3	Time
A		Copy error	Forgot to upload	40:40
B				36:54
C		Copy error		38:53
D		Copy error		31:51
E	Loose cable			34:25
F		Unzip error		39:47
G		Driver installation error		31:23
H		Driver installation error	Selected wrong software file	34:11
I				38:43
J		Wrong installation sequence		25:17
K		Copy error		20:30

Results of the standart NASA TLX questionnaire underlined the problems occurred during the usabitlity test. On a scale from 0 (low) to 20 (high) the participants rated mental, physical and temporal effort, their own performance, effort in general and frustration. The results are illustrated in Figure 5 below.

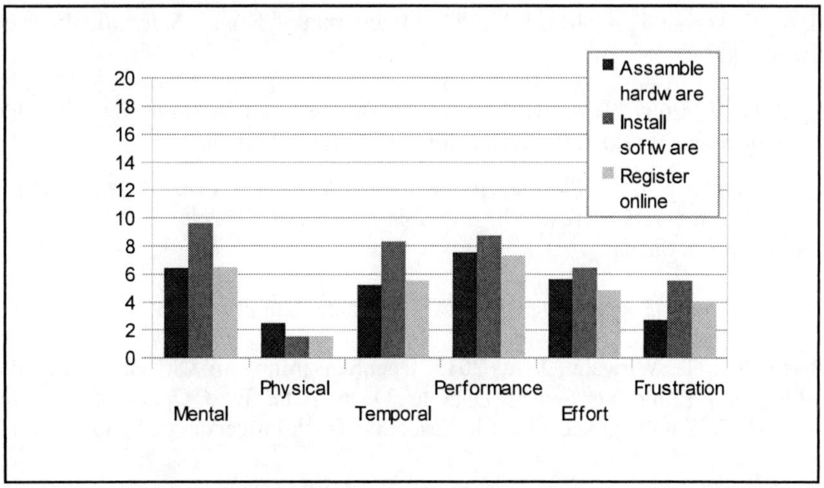

Figure 5. Average results of NASA TLX test of the three steps in the user study.

5. CONCLUSIONS AND FUTURE WORK

We have shown that open source hardware prototyping is a promising approach for environmental monitoring in the field of citizen science and education in secondary schools. In contrast to existing projects, we are aiming to provide introductions for DIY hardware, and microcontroller programming to bring methods of scientific data gathering to the public. We focus on the user groups of school teachers, hardware makers and citizens who want to build up professional stations by themselves to become "professionals". To get an inside look into sensor technology is a important motivation factor for citizens and school students when dealing with environmental monitoring. After less than one year of supplying SenseBox construction manuals, our sensor network grew up to 70 stations in Germany, with over eight million recorded measurements. To identify a user group is crucial to uncover potential risks in an early state of citizen science projects as well as for a high success rate in participation. More than 70 preorders of the educational version of the SenseBox are showing that there is a need for technical project work in German schools besides the citizen science movement. However, our user study showed that the provided documentation material must be complete and understandable for non-experts. We discovered that software and driver installation process are leading to high error rates, while hardware wiring was not a problem, even for beginners. Following from that we are implementing the study results into our construction manuals and develop a one-click software to fill this gap. In a next step we supply around 100 SenseBox stations nationwide in Germany, to proof our enhancements. After that, we are planning to expand our partnership to European level. By enlarging our network throughout Europe we are making a next step for OpenSenseMap to become the standard for sharing open sensor data.

6. REFERENCES

Bröring A., Remke A. and Lasnia D., 2011. SenseBox - A Generic Sensor Platform for the Web of Things. 8th Annual International Conference on Mobile and Ubiquitous Systems (MobiQuitous'11). *Computing, Networking and Services*. 186-196

Burke J., Estrin D., Hansen, M., Parker, A., Ramanathan, N., Reddy, S., 2006. Participatory Sensing. Workshop on World- Sensor-Web (WSW'06): *Mobile Device Sensor Networks and Applications*. 117-134

Gershenfeld, N., Krikorian, R., Cohen, D., 2004. The Internet of Things. *Scientific American* 291(4) (2004). 76– 81

Guinard, D., Trifa, V., 2009: Towards the Web of Things: Web Mashups for Embedded Devices. ACM International World Wide Web Conference, Madrid, Spain.

Hart, S. G., Staveland, L. E., 1988. Development of NASA-TLX (Task Load Index): Results of empirical and theoretical research. In *Human Mental Workload*. North Holland Press, Amsterdam, Netherlands.

Nielsen J., 1993. *Usability Engineering*. Morgan Kaufmann , San Francisco, USA.

Pfeil, M., Bartoschek, T., Wirwahn, J. A., 2015. OpenSenseMap – A Citizen Science Plattform for Publishing and Exploring Sensor Data as Open Data. Free Open Source Software for Geospatial (FOSS4G'15). Accepted for Academic Track Proceedings, Seoul, South Korea.

Raddik M. J., Bracey, G., Gay, P. L., Linott, C. J., Murray, P., Schawinski, K., Szalay, A., S., and Vandenberg, J., 2010. Galaxy Zoo: Exploring the Motivations of Citizen Science Volunteers. Astronomy Education Review 9(1): 18.

Sullivan, B. L., Aycrigg, J. L., Barry, J. H., Bonney, R. E., Bruns, N., Cooper, C. B., Damoulas, T., Dhondt, A. A., Dietrich, T., Farnsworth, A., Fink, D., Fitzpatrick, J. W., Fredricks, T., Gerbracht, J., Gomes, C., Hochachka, W. M., Iliff, M. J., Lagoze, C., La Sorte, F. A., Merrifield, M., Morris, W., Phillips, T. B., Reynolds, M., Rodewald, A. D., Rosenberg, K. V., Trautmann, N. M., Wiggins, A., Winkler, D. W., Wong, W., Wood, C. L. Yu, J., and Kelling, S., 2014. The eBird enterprise: an integrated approach to the development and application of citizen science. *Biological Conservation. Editor's Choice.* January 2014. 31-40.

Uckelmann D., Harrison M. and Michahelles F., 2011. *Architecting the Internet of Things*. Springer. Berlin, Germany.

Free and Open Source Software for Geospatial (FOSS4G) Conference Proceedings

Volume 15 *Seoul, South Korea* Article 55

2015

Object-Based Building Boundary Extraction from Lidar Data

You Shao
School of Civil and Environmental Engineering, UNSW

Samsung Lim
School of Civil and Environmental Engineering, UNSW,

Recommended Citation

Shao, You and Lim, Samsung (2015) "Object-Based Building Boundary Extraction from Lidar Data," *Free and Open Source Software for Geospatial (FOSS4G) Conference Proceedings*: Vol. 15 , Article 55.
DOI: https://doi.org/10.7275/R5891422
Available at: https://scholarworks.umass.edu/foss4g/vol15/iss1/55

AT3-03

Object-Based Building Boundary Extraction from Lidar Data

You Shao[1] and Samsung Lim[2]

[1]School of Civil and Environmental Engineering, UNSW, Sydney, NSW 2052, Australia
E-mail: you.shao@student.unsw.edu.au
[2]School of Civil and Environmental Engineering, UNSW, Sydney, NSW 2052, Australia
E-mail: s.lim@unsw.edu.au

ABSTRACT

Lidar is a remote sensing technology that uses laser beams to generate high-accuracy, three-dimensional (3D) information of the Earth. As urban areas are developing and expanding rapidly, lidar applications such as 3D building modelling and city mapping are of increasing importance. Hence building boundary extraction is one of the main applications of lidar in civil engineering and urban planning projects. In this paper, three boundary extraction algorithms including an alpha-shape algorithm, a modified concave hull algorithm and a grid-based algorithm are tested to assess their object-by-object accuracy. The alpha-shape algorithm generates reliable boundaries for most of sample buildings, while the grid-based algorithm shows less consistency in some cases. The concave hull algorithm performs moderately with a few limitations. Advantages and disadvantages of each algorithm are identified and addressed in this paper.

1. INTRODUCTION

Lidar is an active remote sensing technology that emits laser beams to detect distant objects and measure ranges. Airborne lidar has been widely used since the early stage of the technology development (Wehr and Lohr, 1999). Lidar is of advantage for rapid and accurate vertical measurements at a relatively low cost; thus it shows strength in large-scale mapping and data acquisition of ground surface such as city mapping and Digital Elevation Model (DEM) generation. In addition, lidar is also widely applied in feature extraction due to the high density and vertical accuracy of the data.

Urban areas are of increasing importance in most of the countries since they have been changing rapidly over time. Buildings are the main objects of these areas, and building boundaries are one of the key factors for urban mapping and city modelling. Accurate building extraction using lidar data has been a prevalent topic that many research efforts have been contributed to. However, the complexity of building shapes and irregularity of lidar point distribution make the task difficult to achieve. Although there are plenty of algorithms trying to solve the difficulties, it is not feasible for a single method to fit for all. Each can perform well under a certain situation and requirement only.

In this paper, several building boundary extraction algorithms including an alpha-shape algorithm, a grid-based algorithm, and a concave hull algorithm are assessed. The strengths and limitations of each algorithm are identified and addressed. The point cloud used in this research is derived from the airborne lidar data acquired over the main campus of the University of New South Wales (UNSW) Australia in 2005.

Typically, the boundary extraction algorithms are applied to the clusters of building points when lidar data is segmented and classified. Many approaches have been attempted to improve the extraction algorithms. The simplest way to extract a rough boundary is using the convex hull method which has been implemented by several researchers (Qihong et al., 2008). However, this algorithm only fits for buildings with regular convex shapes. In order to overcome the limitation of this method many researchers have modified and improved the algorithm and obtained more reliable boundaries (Sampath and Shan, 2007). Another prevalent and recent method is to use an alpha-shape algorithm based on two-dimensional Delaunay Triangulation (Verma et al., 2006; Wei et al., 2011). This method works for both concave and convex shapes, and even for some complicated shapes. Another approximation-based algorithm was introduced by Zhou and Neumann (2008) using watertight grids.

Although it is observed that the aforementioned algorithms work well in different scenarios, a quantitative comparison analysis on each algorithm's performance on an identical dataset is rarely reported. Aiming at evaluating and improving these algorithms, we implemented a mathematical framework to compare the algorithms on an object-by-object basis. This study compares the boundary points selected by different algorithms and the impact of the selection on the accuracy.

2. BUILDING BOUNDARY EXTRACTION ALGORITHMS

2.1. Alpha-Shape Algorithm

The alpha-shape algorithm applied to the boundary extraction is based on the 2D Delaunay Triangulation. The boundary derived by this algorithm is always a subset of the Delaunay Triangulation of the point set (Edelsbrunner and Mucke, 1994). For a set of points, the alpha-shape algorithm draws and moves circles of a radius α towards the point set, and assign two points as boundary points if the two points touch the edge of a circle at the same time and no other point lies within the circle as illustrated in Figure 1. When the radius α is close to 0, every point would be a boundary of itself. On the other hand, when α is approaching to infinity, the boundary will be the convex hull of S. Thus, if a suitable α value is chosen, detailed inner and outer boundaries will be extracted. Generally, an α value larger than the average point spacing L and less than $2L$ is preferred (Wei et al., 2011). In this research, an optimal α value that matches this criterion is chosen.

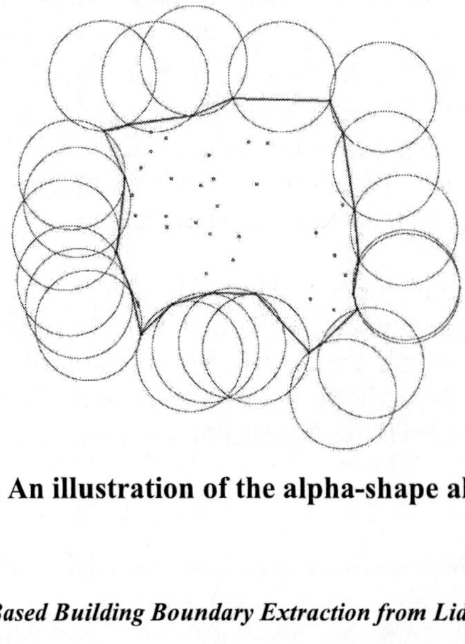

Figure 1. An illustration of the alpha-shape algorithm

2.2. A Grid-Based Algorithm

A grid-based algorithm is proposed by (Zhou and Neumann, 2008) based on the work of (Oliver et al., 2006) and (Verma et al., 2006). In principle, a uniform and watertight fishnet is applied to a 2D plane with all points within the cells. For every cell containing points, each side that does not overlap with others is regarded as an initial reference boundary. Then the points within the cell that are closest to the side are extracted as boundary points as illustrated in Figure 2. The cell size of the fishnet is the only control value for this algorithm and we believe that a value that makes the average point number in a single cell between 13 and 15 performs well for most cases. However, it can be seen from Figure 2 that some points may locate outside the extracted boundary. Therefore, the effect of these missing points will be demonstrated in this paper.

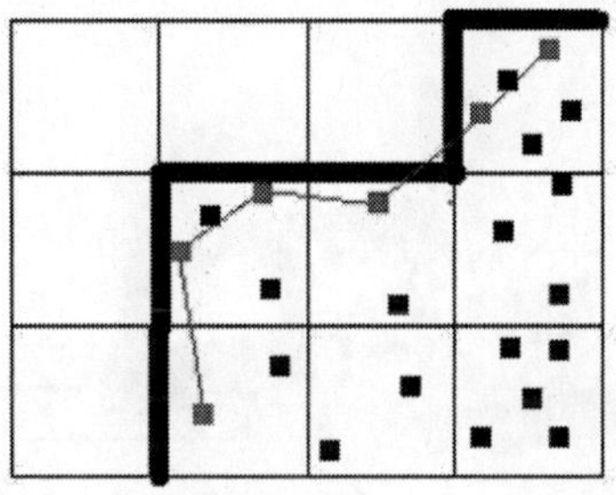

Figure 2. An illustration of the grid-based algorithm

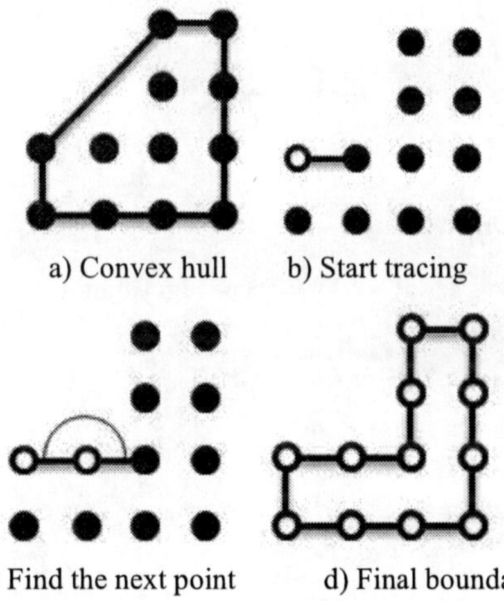

Figure 3. An illustration of the modified convex hull algorithm

2.3. A Modified Convex Hull Algorithm

This algorithm was proposed by (Sampath and Shan, 2007) based on the general convex hull algorithm. A search radius is added to the original method as a control value. For a set of irregularly distributed points the convex hull cannot represent the actual shape formed by these points properly (Figure 3a). In the modified algorithm, after the first boundary point (usually the left-most point) is selected, only the points within a certain distance are considered in the boundary tracing process (Figure 3b). In this process, the point with the smallest clockwise angle to the previous boundary point is picked to be the next one (Figure

3c). Comparing to the original convex hull algorithm, the modified convex hull algorithm provides a more reasonable boundary as seen in Figure 3d.

3. RESULTS

In this research, three algorithms mentioned in the previous section are tested with the last returns of lidar data over the UNSW campus. The average point spacing of the lidar data is 1.15 m, while the vertical accuracy is 8 cm. In total, 8 building-point clusters are extracted from the lidar data (Figure 4). The extraction method is based on the principle that points belong to the same building must cluster and have similar height. Most of the trees are removed using the last return points.

Figure 4. Distribution of building-point clusters

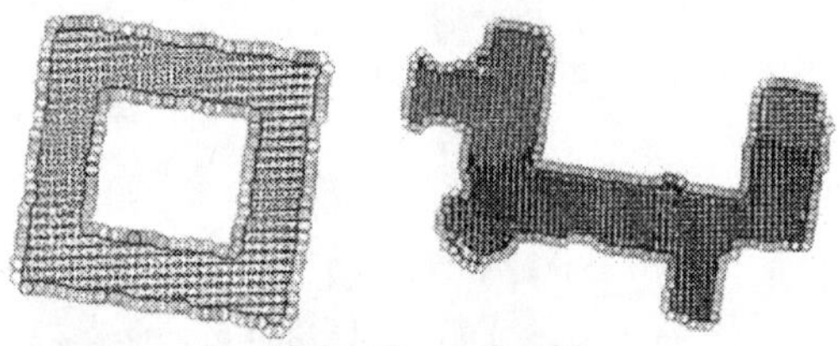

a) Alpha-shape algorithm

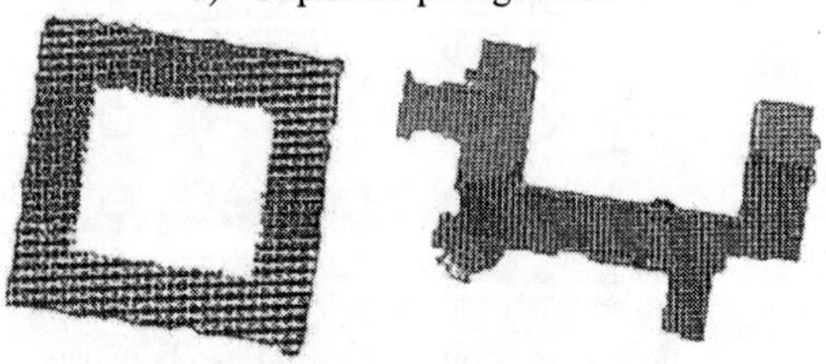

b) Modified convex hull algorithm

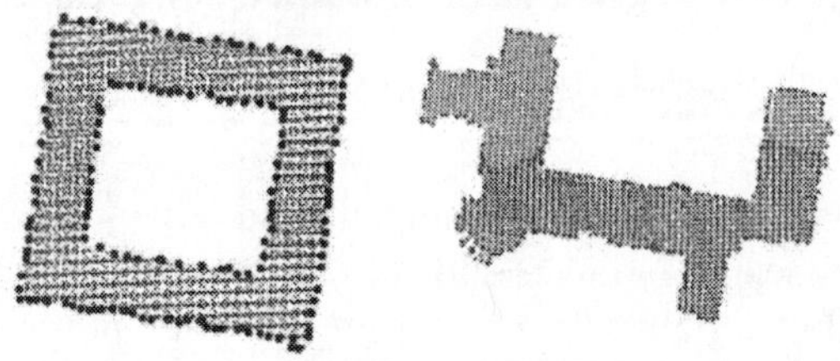

c) Grid-based algorithm

Figure 5. Extraction results

3.1 Boundary Extraction

The extracted boundaries of various types of building shapes are shown in Figure 5 where red points and lines represent the building boundaries. The radius α used in Figure 5a is 1.5 m which is between the average point spacing L=1.15 and 2L. The cell size used in Figure 5b is 3 m, which makes the average point number in a single cell between 13 and 15. The search radius used in Figure 5c is set to be 2α. It can be seen that all three algorithms work well for both convex and concave shapes. However, the modified convex hull algorithm fails to extract the inner boundary of buildings.

3.2 Assessment

Although there are many researches on developing new algorithms and improving existing ones, few have compared the performance of these algorithms. Our accuracy assessment depends on many factors such as the accuracy of lidar data, the accuracy of reference data, the classification quality, and the performance of extraction algorithms. Moreover, due to the limitation of a lidar system the errors caused by the system may become larger than the errors generated in the processing steps. Therefore, this research focuses on the difference between the extracted boundaries instead of the absolute accuracy of the algorithms. A modified object-based evaluation method based on (Awrangjeb et al., 2010) is proposed for this comparison. Instead of the evaluation between extracted boundary polygon and reference polygon used in (Awrangjeb et al., 2010), evaluation between extracted boundary points and reference polygon is applied. The reference polygons are extracted from a high-resolution aerial image acquired over the UNSW campus in 2002.

Table 1. RMSD of the distances from the extracted boundary points to reference polygon (unit in m)

	B1	B2	B3	B4	B5	B6	B7	B8
Alpha-shape	1.24	1.10	0.50	0.78	1.12	1.13	1.16	1.09
Modified convex hull	1.30	1.19	0.58	0.78	1.29	0.99	1.25	1.14
Grid-based	1.36	1.12	0.60	0.85	1.27	1.14	1.17	1.06

Table 2. Standard deviation of the distances from the extracted boundary points to reference polygon (unit in m)

	B1	B2	B3	B4	B5	B6	B7	B8
Alpha-shape	0.84	0.56	0.39	0.49	0.99	0.88	0.95	0.82
Modified convex hull	0.84	0.48	0.42	0.49	1.11	0.78	1.01	0.84
Grid-based	0.83	0.49	0.46	0.47	1.10	0.90	0.97	0.79

Root Mean Square Difference (RMSD) is calculated and shown in Table 1 since the main focus is to compare the difference rather than the absolute accuracy. Table 2 shows the standard deviation of the distance from extracted points to the reference polygons. It can be seen from Table 1 that all three algorithms can generate similar results for simple convex building shapes (Buildings 1-4). For concave building shapes (Buildings 6-8), the modified convex hull algorithm produces boundary with a significant difference to others. Building 6 results in a smallest RMSD. However, this is due to the missing inner boundary points.

Comparing the results from the alpha-shape algorithm and the grid-based algorithm, it can be seen that the missing points from the grid-based algorithm (see Subsection 2.2) do not affect the accuracy much. However, for some complicated building shapes such as Building 5, the grid-based algorithm shows an inconsistency due to the impact of the missing points. A more suitable cell size is suggested to resolve this problem. Since the alpha-shape algorithm performs consistently for all sample buildings, it is considered a suitable algorithm for most of the general extraction work. However, the efficiency of this algorithm is questionable when dealing with a huge dataset of high density since it will extract every possible boundary points. In that case, a grid-based algorithm with a large cell size can be used for initial extraction, followed by the alpha-shape algorithm.

4. CONCLUDING REMARKS

In this paper, three algorithms for building boundary extraction are assessed in an object-by-object basis. The alpha-shape algorithm generates reliable boundaries for most of sample buildings, while the grid-based algorithm shows a little inconsistency in some cases. The concave hull algorithm performs moderately with a few limitations. The alpha-shape algorithm is suggested for general building boundary extraction for its consistency and reliability. Future work is required to increase the efficiency of the alpha-shape algorithm because the basic principle of alpha-shape formation is to extract every possible boundary points which can be redundant for a high-density dataset.

REFERENCES

Awrangjeb, M., M. Ravanbakhsh and C. S. Fraser (2010). "Automatic detection of residential buildings using LIDAR data and multispectral imagery." ISPRS Journal of Photogrammetry and Remote Sensing 65(5): 457-467.

Edelsbrunner, H. and E. P. Mucke (1994). "Three-dimensional alpha shapes." ACM Transactions on Graphics 13(1): 43-72.

Oliver, W., S. K. Lodha and D. P. Helmbold (2006). A Bayesian approach to building footprint extraction from aerial LIDAR data. Third International Symposium on 3D Data Processing, Visualization, and Transmission (3DPVT'06), 14-16 June 2006, Piscataway, NJ, USA, IEEE.

Qihong, Z., L. Jiazhen, L. Xianhua, M. Jianhua and L. Xuefeng (2008). Simple building reconstruction from LIDAR point cloud. 2008 International Conference on Audio, Language and Image Processing, 7-9 July 2008, Piscataway, NJ, USA, IEEE.

Sampath, A. and J. Shan (2007). "Building boundary tracing and regularization from airborne lidar point clouds." Photogrammetric Engineering and Remote Sensing 73(7): 805-812.

Verma, V., R. Kumar and S. Hsu (2006). 3D building detection and modeling from aerial LIDAR data. 2006 IEEE Computer Society Conference on Computer Vision and Pattern Recognition, CVPR 2006, June 17, 2006 - June 22, 2006, New York, NY, United states, Institute of Electrical and Electronics Engineers Computer Society.

Wehr, A. and U. Lohr (1999). "Airborne laser scanning - An introduction and overview." ISPRS Journal of Photogrammetry and Remote Sensing 54(2-3): 68-82.

Wei, S., Z. Jin and Y. Feng (2011). A new algorithm of building boundary extraction based on LIDAR data. 2011 19th International Conference on Geoinformatics, 24-26 June 2011, Piscataway, NJ, USA, IEEE.

Zhou, Q.-Y. and U. Neumann (2008). Fast and extensible building modeling from airborne LiDAR data. 16th ACM SIGSPATIAL International Conference on Advances in Geographic Information Systems, ACM GIS 2008, November 5, 2008 - November 7, 2008, Irvine, CA, United states, Association for Computing Machinery.

Free and Open Source Software for Geospatial (FOSS4G) Conference Proceedings

Volume 15 *Seoul, South Korea* Article 56

2015

Automatic Improvement of Point-of-Interest Tags For OpenStreetMap Data

Sabine Storandt
Insitut für Informatik, Albert-Ludwigs-Universität Freiburg 79110 Freiburg, Germany

Stefan Funke
FMI, Universität Stuttgart 70569 Stuttgart, Germany

AT1-01

AUTOMATIC IMPROVEMENT OF POINT-OF-INTEREST TAGS FOR OPENSTREETMAP DATA

Sabine Storandt[1] and Stefan Funke[2]

[1]Insitut für Informatik, Albert-Ludwigs-Universität Freiburg
79110 Freiburg, Germany
Email: storandt@informatik.uni-freiburg.de
[2]FMI, Universität Stuttgart
70569 Stuttgart, Germany
Email: funke@fmi.uni-stuttgart.de

ABSTRACT

The user experience of any OpenStreetMap (OSM) based service heavily depends on the quality of the underlying data. If the service deals with points-of-interest (POIs), consistent and comprehensive tagging of the respective map elements is a necessary condition for a satisfying service.

In this paper, we develop methods that can automatically infer tags characterizing POIs solely based on the POI names. The idea being that many POI names already contain sufficient information for tagging. For example, 'Pizzeria Bella Italia' most certainly indicates an Italian restaurant. As the OSM data contains hundred of thousands POIs for Germany alone, we aim for a tool that can accomplish tag extrapolation in an automated way. In a first step, we automatically extract typical words and phrases that occur in names associated with a certain tag. For example, learning indicators for 'shop=hairdresser' on German OSM tags led to high scores for 'fris', 'cut', hair' and 'haar'. Having available such indicator phrases, we use standard machine learning techniques to derive the probability for a POI to exhibit a certain tag. If this probability exceeds a certain threshold, we assign the tag to the POI in an automated fashion. We used our extrapolation framework to create new amenity, shop, tourism, and leisure tags. The accuracy of our approach was over 85% for all considered tags. Moreover, for POIs tagged with amenity=restaurant, we aimed for extrapolating the respective cuisine tag. For more than 19 thousand out of 28 thousand restaurants in Germany lacking the cuisine-tag, our approach assigned a cuisine. In a random sample of those assignments 98% of these appeared to be true.

1. INTRODUCTION

In particular in the context of community-based or crowd-sourced data gathering efforts like OpenStreetMap (OSM), the issue of data quality is of utmost importance. Irrespectively, how sophisticated and advanced the employed algorithms are, the user experience when using services based on the gathered data is first and foremost affected by the quality of the underlying data. Competing commercial offerings will always claim an alleged superior data quality to justify their business. Sometimes, their typically much more centralized approach of data maintenance and collection indeed allows for an easier monitoring of quality issues, for crowed-based approaches like OSM this is much harder to achieve.

In the concrete case of OSM, data quality is very much dependent on the contributors' tagging discipline. While there are well thought out guidelines how to tag mapped elements, in principle every contributor can tag at his/her discretion. This freedom undisputedly has its advantages in terms of flexibility, it also creates some consistency problems, though.

For example, when querying a geo-search engine or a location-based service for points-of-interest (POIs) in a certain region or next to the current user location, one often asks for classes ('hotels New York', 'supermarkets Berlin', 'Italian restaurants London') rather than single points ('Hotel Belvedere New York'). In OpenStreetMap (OSM), one can specify the basic class along with every POI e.g. via the tourism tag (*tourism=hotel*), the shop tag (*shop=supermarket*), the amenity tag (*amenity=restaurant*), or several other specialized tags as the cuisine tag (*cuisine=italian*) for restaurants. Not providing the appropriate tags when mapping the respective element typically leads to omission of these elements in the result list for a class-based query. Moreover, class tags are useful to categorize search results. For example, when searching for 'Venice beach' the user should be informed that there are beaches, hotels, fitness studios and clothing stores with that name.

Unfortunately, in OSM, there are still plenty of POIs where the class is not provided via the appropriate tags. Many of those POIs exhibit a name tag (as e.g. 'Sunset Hotel', 'Wal Mart', 'Pizzaria Bella Italia'), though, which already contains some information about the respective class. In this paper, we investigate methods for automatic extrapolation of classes based on POI names. Using machine learning tools we extract typical words and phrases that occur in name tags associated with a certain class and learn respective POI classifiers. As a result we can augment the existing OSM data by inferred tags and improve the data quality. This can be done either fully autonomously or with humans in the loop who verify the augmentations suggested by our algorithms – again, a community-based approval mechanism for such changes might be an interesting option.

1.1 Related Work

Numerous papers use machine learning (ML) techniques to work on/for OSM data. In the following we pick a few examples without claiming completeness. Basically, ML can be employed either on the application level – leaving the underlying data pool untouched – or to verify and even augment the underlying data pool. For the former, e.g., (Hagenaur et al., 2012) propose the use of artificial neural networks and genetic algorithms to infer land-use patterns without directly feeding the results back into the OSM data pool. For the latter, (Jiliani et al., 2014) make use of machine learning techniques to assess correctness of highway tags. (Fathi et al., 2010) aim at inferring the structure of the road network by analyzing GPS traces using ML techniques.

1.2 Contribution

We describe a framework for automatic tag extrapolation based on POI names. We explain in detail how to process the OSM data and how to determine extrapolatable tags. Then we introduce a machine learning approach primarily based on k-grams of POI names. We apply our framework to extrapolate selected tourism, leisure, amenity, shop and cuisine tags for the dataset of Germany. Our experimental evaluation shows the ability of our framework to enrich the OSM data. For example, for cuisine, we can extrapolate more than 70% of missing tags with a precision of 98%.

2. POINTS-OF-INTEREST IN OSM

We are interested in nodes in the OSM data that potentially are POIs. Nodes in OSM come with a specific ID and geo-coordinates (lat/lon). In addition, tags in form of key-value-pairs (k,v) can be specified, as shown in this example:

<node id="2360485476" visible="true" version="5" changeset="21956784" timestamp="2014-04-26T15:28:19Z" user="q_un_go" uid="11374" lat="47.9955298" lon="7.8447728">
<tag k="name" v="Backshop"/>
<tag k="shop" v="bakery"/>
<tag k="wheelchair" v="yes"/>
</node>

In the following, we will list tags that indicate POIs of various kinds (like *shop*) and explain their taxonomy.

2.1 Restaurants, Cafes, Pubs and Fast-food Facilities

Let us first consider tags associated with going out to eat or drink. We differentiate restaurants, cafes, pubs and fast-food facilities in accordance with the OSM Wiki. The *amenity=restaurant* tag is by far the most frequent amenity tag. As specified in the Wiki, *amenity=restaurant* should be used 'for a generally formal place with sit-down facilities selling full meals served by waiters and often licensed (where allowed) to sell alcoholic drinks'. The *cuisine* tag can be used in addition to further refine what kind of restaurant it is. A cuisine can refer to the ethnicity of the food (*cuisine=chinese*), to the way of food preparation (*cuisine=wok* or *cuisine=grill*), to the food itself (*cuisine=pasta*) or to other classifications (*cuisine=fine_dining*).

Instead of *amenity=restaurant*, one should use *amenity=fast_food* 'for a place concentrating on very fast counter-only service and take-away food'. The cuisine tag is used in this context as well (e.g. *cuisine=burger*). Nevertheless, also *amenity=restaurant* and *cuisine=fast_food* or *cuisine=burger* are commonly used. *Amenity=cafe* should be used 'for a generally informal place with sit-down facilities selling beverages and light meals and/or snacks', including coffee-shops, tea shops and bistros. Again, combinations like *amenity=restaurant* and *cuisine=coffee_shop* are often used instead.

For drinking, *amenity=pub, amenity=bar* and *amenity=biergarten* are intended. All are used for establishments that sell 'alcoholic drinks to be consumed on the premises'. Hereby, a *pub* should indicate a facility where you can sit down, food is available and the atmosphere is rather relaxed. In contrast, a *bar* is assumed to be more noisy, with music and no meal-like food. A *biergarten* is like a pub, but outdoors. Also combinations like *amenity=pub* and *biergarten=yes* are possible.

Other amenity tags associated with eating and drinking are *bbq, drinking_water, food_court* and *ice_cream*.

Note, that there is overlap between all mentioned amenities, and tags are combined in various ways to classify places. Therefore we consider all of them together in our learning approach.

2.2 Shops, Services and Entertainment

Besides restaurants, cafes, pubs and similar facilities, there exists a large variety of other amenities that mark POIs. For example, places for entertainment as *cinema, theatre, casino* or *nightclub* fall into that category. But also *parking, post_office, post_box, fuel* (for gas stations), public *toilets, library, dentist* and other facilities that are public or provide some kind of (health) service are valid amenity tags. Usually, they are more easily to classify than facilities associated with eating and drinking. But there is some overlap with another important tag, namely the *shop* tag. It should be used for all kind of facilities where products are sold, as e.g. supermarkets, kiosks, bakeries, clothing stores, furniture stores, and many more. There are also nodes tagged with *amenity=shop, amenity=shopping* or variants thereof. For those the OSM Wiki encourages to check whether a *shop* tag can be used instead.

2.3 Hotels, Tourism Spots and Leisure Facilities

The *tourism* tag is used to describe possibilities for paid lodging as *hotel, hostel, motel, camp_site* and so on. But also in the context of sights and attractions the *tourism* tag should be used, including *zoo, museum* or *theme_park.* There is also an attraction tag for specification, e.g. *attraction=big_wheel.* Alternatively, one can tag a sight primarily according to its type, e.g. *waterway=waterfall* and then add *tourism=yes.*

The *leisure* tag applies to all kind of facilities where people can spend their spare time. Most prominent representatives are *playground* and *sports_centre.* When it comes to e.g. parks and gardens there is some overlap with the tourism tag, though.

3. EXTRAPOLATION FRAMEWORK

To be able to extrapolate missing tag information our overall plan is to identify characteristic properties (also called *features*) of the name tag value that are 'typical' for a certain a *amenity, cuisine, shop* ... tag. For those OSM nodes which should bear a respective tag, e.g. shop=*hairdresser*) but do not because either the information was not provided or added in a non-conformal way (e.g. as part of the informal *description* tag), we hope to infer the missing tag by examining its name tag for characteristic features typical for nodes actually bearing this tag. This section gives an overview of the necessary steps for this task.

3.1 Data Extraction and Processing

We only consider named nodes in the OSM data, i.e. the tag k=*"name"* is required. Most named OSM nodes refer to streets or parts of the public transport system (as e.g. *amenity=bus_station, name=Norris_Street*). Such nodes are not POIs according to our definition. They are excluded by checking for presence of *highway* and *public_transport* tags. Moreover, we pruned nodes tagged with *cemetery, power, fire_hydrant, historic, natural* and *man_made.*

In order to learn the correlation between names and certain tags, we need to have POIs with complete information, that is, a name and the tags we are interested in. These POIs will serve as training data in our machine learning approach. For extrapolation, we consider the POIs that potentially miss tags of a certain kind.

3.2 Selection of Extrapolatable Tags

Not all tags are suitable for extrapolation. First, there need to be sufficiently many POIs which exhibit a certain tag to allow the machine learning approach to work. There are plenty of tags in the OSM data which occur only once or very few times, either because they are over-specified (e.g. *cuisine=asian;curry;noodle*), too specific (e.g. *cuisine=self made cake*), home-brewed (e.g. *cuisine=german-bohemian*), exhibit spelling errors (e.g. *cuisine=chineese*), are not in English (e.g. *cuisine=bürgerliche_küche*), simply used wrong (e.g. *cuisine=music*) or indeed rare (e.g. *cuisine=israelian*). Therefore, we count how often a certain tag or a combination of tags occurs and only further consider tags whose respective count exceeds 200. Second, there are tags which subsume each other or overlap in terms of their semantics. For example, *cuisine=asian* is used but also *cuisine=japanese, chinese, vietnamese, thai* amongst others. To accommodate for such dependencies, we first group tags and specify their relations manually. If we consider two tags to be interchangeable like *cuisine=steakhouse* and *cuisine=steak*, we merge them into one. For a class subsuming several others, we check whether the subclasses are large on their own. If that is the case, we try to learn the more specific group. Otherwise, we cumulate the names of all subgroups and try to learn the more general group.

3.3 Feature Extraction

Once we fixed the set of classes/tags, we need to specify suitable *features* (characteristic properties of the name tag) that allow to learn the correlation between names and tags. We want to identify words and phrases that are typical for certain classes. Consider for example this list of names of hairdressers of a city in Germany:

Claudia's Frisierstube, Cut & Color, Der Goldene Schnitt, emporio, Freiseur Ryf, Frerich, Friseur Ganter, Friseur Roth, Friseur Ryf, Friseur Salon H.Jonas, Frisör Charisma, Frisörsalon Annette, Frisuren-Atelier, Gutjahr Hairlounge, Haar-WG, HaarBalance, HaarBar, Haarstudio Burger, Haarstudio Marina Lindle, Haarstudio Marita, Hair Body Soul, Hair Saloon, hairkiller, HairSpeed, Helbling, Horst Fischer Friseursalon, Nölle, Power Hair Styling, Salon Carmen, Salon Haargenau, Toni & Guy, Via Style

We observe that e.g. 'fris', 'seur', 'haar' (the German word for hair), 'hair', 'styl', 'salo' and 'studio' appear multiple times and therefore might be good indicators for *shop=hairdresser*. Determining indicator phrases manually for thousands of POIs in hundreds of classes is impractical, though. To automatize the process, we proceed as follows. Let N be the list of names associated with a certain tag (e.g., *shop=hairdresser*). For each name in N we construct all k-grams for k between 3 and 10. A k-gram of a string/word is a consecutive substring of length k. For example, all 4-grams for 'Hair Styling' would be 'Hair', 'air ', ir S', 'r St', ' Sty', 'Styl', 'tyli', 'ylin', 'ling'. We count for each k-gram how often it occurs in N. We consider a k-gram to be significant when at least two percent of names in N share this k-gram. If a significant k-gram is a substring of another significant k-gram with a similar count (e.g. considering *cuisine=burger*, 'onald' and 'McDonald's' both appear 753 times), we prune the smaller k-gram as we assume it has no significance on its own. As a counter-example, 'burger' appears more often in the list than 'Burger King', therefore both k-grams are kept. After this pruning step, we have for each class a final list of indicator phrases (k-grams) at hand, each with a percentage specifying the fraction of nodes of this class exhibiting the respective k-gram.

Then we construct for each name a so-called *feature vector*. A feature vector of a name is a vector with as many real-valued entries as there are class/significant k-gram combinations. The entry corresponding to a certain indicator phrase and class is set to the length of the phrase multiplied by the percentage of nodes in the class containing this k-gram. Here the intuition is that long shared sequences between the name and the names in N, as well as a shared sequence with many names in N indicate a high correlation with the class. Standard machine learning machinery is then applied to the derived feature vectors of the names.

3.4 Machine Learning

We use the *Random Forest* (Breiman, 2001) approach for learning the classifier, as it allows to take care of dependencies between the feature vector entries. We expect to learn a classifier for POI names that can decide which tag (from a given set) should be assigned. As it might very well be the case that no tag is suitable, we have to accommodate for that.

Therefore, we not only aim for the classification itself but rather for a probability distribution over the classes. So for each name to classify, we derive a probability for every class denoting how likely it is that the name belongs to this class. The sum over all class probabilities for a name always equals 1. If no class has a significantly higher probability than the others, it can be assumed that none of the classes fit.

3.5 Evaluation

In order to check whether the selected features allow for an accurate classification, we first perform a 5-fold cross validation. Here the training data (the set of POIs with known tags) is split into five equal sized parts P1,P2,P3,P4,P5. Then for each part Pi, we train on the other four parts and classify the feature vectors in Pi on that basis. So we train on P2,P3,P4,P5 and check whether the resulting classifier works as intended for set P1 (for which we know the correct classification), and repeat for P2 vs P1,P3,P4,P5, as well as P3 vs P1,P2,P4,P5, etc.

As a quality measure we compute the following statistical standard machine learning quantities (averaged over the 5 experiments):

- Recall: for a specific category - let's say amenity=hair_dresser -, we consider the ratio (#items correctly classified by our algorithm)/(#items that really have amenity=hair_dresser items)
- Precision: for a specific category we consider the ratio (#items correctly classified by our algorithm)/(#total number of items that are classified as amenity=hair_dresser by our algorithm)

A perfect precision score of 1.0 (or 100%) means that every item classified as having *amenity=hair_dresser* by our algorithm is indeed a hairdresser (but does not imply that every hairdresser was found). On the other hand, a perfect recall score of 1.0 means that all hairdressers were actually classified as having *amenity=hair_dresser* by our algorithm.

Furthermore, we consider the accuracy of the classification, that is, the percentage of correctly classified POIs among all considered POIs.

4. EXPERIMENTS

We implemented the described framework using C++ and Python. For the machine learning part, we relied on the scikit-learn package (Pedregosa, 2011). Our experiments were conducted on a single core of an Intel i5-4300U CPU with 1.90GHz and 12GB RAM. The Germany data set extracted from OSM as basis for all our experiments contains 771,325 named nodes. Among those, we identified 84,618 with insufficient tagging (about 12,000 contained only the name tag, the others only non-classifying additional tags as e.g. *wheelchair=yes/no*, *opening_hours*, website or Wikipedia references and address information).

4.1 Amenity and Cuisine Tags for Eating and Drinking

4.1.1 Restaurants, Fast Food Facilities, Cafes, Pubs, Bars and Biergartens

Filtering our data set for eating and drinking related amenities, the following distribution was observed: 60,819 POIs with *amenity=restaurant,* 18,823 with *amenity=cafe*, 18701 with *amenity=fast_food*, 14,484 with *amenity=pub*, 3,862 with *amenity=bar*, 2,078 with *amenity=biergarten*, 786 with *amenity=ice_cream*, 746 with *amenity=drinking_water* and 391 with *amenity=bbq*.

Conducting a cross-validation on this data, we observed that *pub*, *bar* and *biergarten* are not sufficiently separable with our basic approach as many bars and biergartens are indeed tagged with *amenity=pub*. Also *pub* and *restaurant* were confused frequently. Therefore, we inserted an additional step: We first learned a classifier for *bar* and *biergarten* and applied it to all POIs with *amenity=pub*. Then we excluded those classified as *biergarten* or *bar* from the training data for *pub* and re-run the experiment. The precision increases from 68% to 82%. Moreover, we used the classifiers for *pub*, *bar* and *biergarten* to prune the training data for *restaurant* and the *ice_cream* classifier to prune *cafe* names. Based on the remaining training data, we learned the final classifier. In the cross-validation, the overall accuracy was 76%.

Next, we applied the learned classifier to data with missing tags. We only assigned a tag automatically when the classification probability was 100%. In that way, we created 461 new tags. For 100 of these, we checked the correctness by using the OSM search engine and Google on the name (and possibly further associated tags). In 85% of the instances, the assigned tag was valid. Examples for misclassification are e.g. 'kaffeemaschinenservice kafas' classified as *cafe* (but should be a *shop*), 'uh80 ga weingarten' classified as *biergarten* but really is a *fire_hydrant*, and 'lind haustechnik' classified as *restaurant* (because 'haus' as part of 'gasthaus' occurs quite frequent in German restaurant names) but is a building service.

4.1.2 Cuisine Tags

In total, about 1500 different *cuisine* tags among POIs with *amenity=restaurant* were contained in the data set. Many of those occurred only once or very few times. The most frequent ones are listed in Table 1. They all either indicate ethnicity or type of food. If a *cuisine* tag contained multiple entries (as *cuisine=pizza;kebab*), we counted the POI in both categories.

Table 1: Overview of cuisines.

ethnicty	frequency	type of food	frequency
italian	7,365	pizza	3,275
asian	1,808	sandwich	568
bavarian	532	burger	1,491
german	6,753	kebab	2,926
chinese	1,847	chicken	125
greek	3,002	ice_cream	1,921
japanese	281	coffee_shop	723
mexican	394	fish	244
regional	6,673	seafood	116
indian	661	vegetarian	108
french	240	steak_house	310
thai	582	sushi	305
international	787		
turkish	1,299		
spanish	369		
croatian	108		
american	163		
vietnamese	281		

We performed the following modifications manually to increase the performance and meaningfulness of our approach. We merged *regional* and *german* into one group as they are both too diverse to easily tell them apart. Also *bavarian* was integrated into this group. In contrast, *japanese, chinese, thai* and *vietnamese* were considered each on their own and not accumulated into the *asian* group. We excluded *international* as we do not expect to identify consistent phrases and words that indicate this cuisine. Regarding type of food, we merged *fish* and *sea_food* into *sea_food* as they were used synonymously and the OSM Wiki recommends to use *sea_food* for both. Furthermore, we excluded *vegetarian*, as it should not be a *cuisine* tag but a *diet* tag instead. The *croatian, american* and *chicken* groups do not contain enough POIs for consideration. So in total, we distinguish 12 ethnicity cuisines and 9 cuisines related to food type.

We first performed a cross-validation, subdivided by ethnicity and food type. For food type, the results are presented in Figure 1.

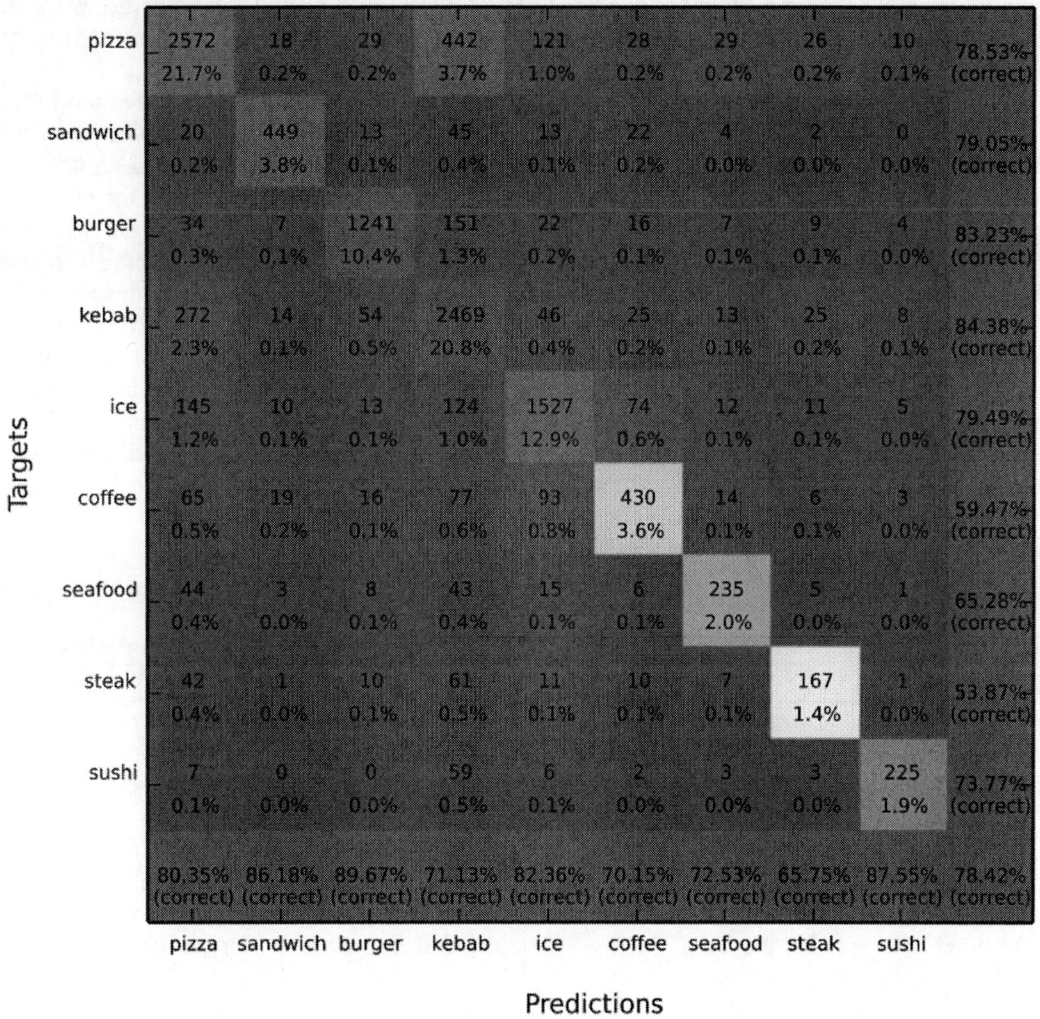

Figure 1. Accuracy of the learned food type classifier in a cross-validation.

The overall accuracy is about 78%. We observe, that the accuracy is worse for groups with a small number of representatives as *coffee, seafood, steak* and *sushi.* In addition, there are some natural mix-ups as *pizza* and *kebab*, or *ice* and *coffee* which often occurred together in cuisine tags of our input data. For ethnicity, we achieved an overall accuracy of about

81%. For *cuisine=german,* the precision was even above 91% and the recall about 94%. Again, for smaller groups the results were worse. Main number of mix-ups occurred for *german/italian, mexican/spanish, thai/chinese* and *greek/turkish*.

For our evaluation on unclassified data, we extracted 28,218 POIs tagged with *amenity=restaurant* but without a *cuisine* tag. We tried to classify those POIs by food type and ethnicity. We only assigned an ethnicity tag when the probability for a certain class exceeded 75%, and a food type when the classifier was 100% sure. The reason for the different percentages being that we expect most POIs to belong to none of the food types in question. But the classifier creates a probability distribution over the classes with the probabilities summing up to 1. With only nine classes to consider, the chance of a false classification would be too high otherwise. In contrast, for ethnicity, we expect most POIs indeed to belong to one of the classes we consider. For 19,671 out of the 28,128 restaurants, our approach assigned an ethnicity cuisine with a sufficient probability, and for 1,460 a food type was matched. Some POIs received both an ethnicity and a food type cuisine, with the most popular combinations being *pizza;italian, kebab;turkish, ice_cream;italian* and *sushi;japanese.*

We manually checked 250 extrapolated cuisines for ethnicity and 250 for food type (by having a look at the restaurant's website). We first selected 10 examples for each considered class randomly (if possible). The remaining samples were selected completely randomly among all classified POIs. Table 2 shows an excerpt of 30 samples for ethnicity and food type cuisines assigned by our framework. For food type, two examples for misclassification can be seen: 'rosenburger hof' and 'speisekammer' both serve German food. But as those names contain 'burger' and 'eis' (the German word for ice) respectively, they get assigned *cuisine=burger* and *cuisine=ice_cream* with high confidence. Nevertheless, for the 500 samples in total, the classification accuracy was 98%. As observable in Table 2, even spelling errors as 'kebap' could be taken care of with our k-gram based approach, as well as names borrowed from places or persons as 'delphi', 'dschingis khan' and 'café mallorca'. The reason for the better precision on real data than in the cross-validation is due to only assigning a class to a POI with unknown cuisine when the probability for that class is high enough. In the cross-validation, every POI gets assigned the class with the highest probability automatically.

Table 2. Result excerpts for cuisine classification. Red entries indicate misclassification.

food type	ethnicity
pizzahaus c = pizza	schwaben-bräu c = german
fischerklause c = seafood	ginnheimer wirtshaus c = german
la stella c = pizza	china imbiss drache c = chinese
pizzeria capriccio c = pizza	pizzeria capriccio c = italian
eiscafé rialto c = ice_cream	bauernstübchen c = german
pizzeria italia c = pizza	gameiro pizza-express c = italian
pizzeria venezia c = pizza	taverna ilios griechisches restaurant c = greek
50's diner c = burger	zur feurigen bratwurst c = german
block house c = steak_house	pizzeria italia c = italian
fischhaus c = seafood	kartoffelhaus c = german
calimero c = ice_cream	pizzeria venezia c = italian
ristorante pizzeria isola d'ischia c = pizza	einkehr c = german
nordsee c = seafood	gasthof pension drexler c = german
pizzeria marino c = pizza	brauhaus am schlössle c = german
rosenburger hof c = burger	winzerhof weinstuben c = german
nazar kebap stube c = kebab	schusterstübchen c = german

chilli peppers rock cafe c = coffee_shop	sushi for friends c = japanese
eis-cafe da vinci c = ice_cream	il capriccio c = italian
steakhouse cheyenne c = steak_house	deutscher hof c = german
eiscafé dolce vita c = ice_cream	delphi c = greek
fischkombüse c = seafood	zum bembelsche c = german
baguetterie filou c = sandwich	mykonos c = greek
classic western steakhouse c = steak_house	rhodos c = greek
shaki sushi c = sushi	zum neuen schwanen c = german
cafe kamps c = coffee_shop	sausalitos c = mexican
trattoria la grappa c = pizza	my thai c = thai
sakura sushi & grill c = sushi	mr. kebab c = turkish
speisekammer c = ice_cream	dschingis khan c = chinese
piccola italia c = pizza	el paso c = mexican
döner haus c = kebab	café mallorca c = spanish

4.2 Other Amenity and Shop Tags

In total, the data set contained 938 different *amenity* and 1,853 *shop* tags. The five most frequent *amenity* tags not related to eating and drinking are *bank* (18,765 times), *pharmacy* (16,256), *place_of_worship* (14,309), *parking* (10,853) and *kindergarten* (10,174). The most prominent shop tags are *bakery* (22,634 times), *supermarket* (17,655), *clothes* (14,440), *hairdresser* (13,310) and *butcher* (6,862). Overall, we identified 67 reasonable *amenity* and 73 reasonable *shop* classes. The cross-validation revealed a classification accuracy of 84%. Applied to real data, we got 4,212 new tags for previously unclassified POIs. We manually checked for each of the 140 considered classes two extrapolated POIs with that class for correctness. The accuracy was about 76%. Considering only the ten most frequent classes listed above, and 10 examples each, the accuracy was 88%, though.

Table 3 lists the most frequent k-grams for the main shop tags. Reconsidering our example *shop=haidresser*, the main k-grams extracted by our program are close to what one would select manually. Interestingly, for supermarket, the k-grams almost exclusively equal supermarket chains. We observed a similar result for gas station chains. Nevertheless, for almost all classes we identified k-grams that occurred in over ten percent of the respective class names. This fact, and the overall good classification accuracy, shows that indeed many names contain classification information.

Table 3. K-grams and their for selected shops.

bakery	supermarket	clothes	hairdresser	butcher
38.75 bäcker	12.01edeka	10.87 mode	25.45 fris	54.84 erei
38.71 rei	11.72 netto	7.94 haus	19.91 friseur	51.62 erei
33.27 bäckerei	11.42 markt	7.13 kik	15.91 haar	42.67 ger
11.36 back	10.67 rewe	5.61 textil	15.10 salon	35.08 metzger
11.20 sch	10.03 aldi	4.41 family	13.84 hair	34.99 metzgerei
5.62 ste	6.69 lidl	2.91 s.oliver	8.96 studio	24.50 fleisch
4.60 mann	6.51 penny	2.85 jeans	8.20 friseur	16.13 fleischere
2.80 konditorei	5.72 kauf	2.31 peek	5.00 haarstudio	16.13 leischerei
2.13 backstube	3.81 norma	2.20 kleid	4.62 cut	4.83 land

4.3 Tourism and Leisure Tags

We identified 168 different *tourism* tags of which 16 occurred more than 200 times. *Information* (45,879) and *hotel* (12,228) and *attraction* (9,404) had the highest counts. The others are *viewpoint, artwork, hostel, museum, alpine_hut, picnic_site, camp_site, guest_house, caravan_site, chalet, theme_park, apartment,* and *zoo*. For *leisure*, 153 different tags were contained in the data. Only 9 of them exhibit a high frequency: *sports_centre* (4,622), *playground* (3,108), *marina* (1,734), as well as *park, water_park, pitch, stadium, slipway* and *nature_reserve*. We excluded *artwork*, as due to its nature where is little hope for consistent indicator phrases. Furthermore, we excluded attraction as this class is too diverse and the extracted k-grams were too general. The remaining 23 classes were fed in our classifier. The first cross-validation indicated too much mix-up between information and hotel. Therefore, we first created a hotel classifier in order to prune the information data. After this step, the overall accuracy improved from 62% to 73%.

For the real data, we newly augmented 3,452 POIs with a *tourism* or *leisure* tag. Computing the precision by manually looking up samples was not so easy in this case, as *information* tagged entities are often simply signs next to hiking trails. Moreover, most classes were not assigned at all. Therefore, we restricted ourselves in the precision calculation to *hotel, playground, marina,* and *sports_centre*. We checked 50 examples for each class. The overall accuracy was 92%. For sports_centre, we even achieved 98% (e.g. 'tennishalle görner', 'willy-lemkens-sportpark', 'eissporthalle', 'the strike bowlingcenter', 'tanzsportzentrum', 'turnhalle herringhausen' are correct examples).

5. CONCLUSIONS AND FUTURE WORK

We showed the potential of OSM name tags to serve as basis for extrapolating tags that indicate the class of a POI. Our machine learning approach for automatic tag extrapolation was proven to work well on real data. The accuracy was significantly over 80% for most considered tags. And especially for *cuisine*, a significant fraction of missing tags was identified with our approach.

In future work, other tags beside name tags could be considered to improve the results further. For example, the *opening_hour* tag could help to distinguish between restaurants and pubs. The *brand* tag could be helpful when it comes to supermarkets, gas stations, dealerships, clothing stores and so on. Also the free text tags *note* and *description* could be parsed for that purpose. Maybe it is possible to learn also the set of indicator tags for each class in an automated way. Furthermore, other countries besides Germany should be investigated. Some tags only occur in certain parts of the world, and the indicator phrases as well as their frequencies for certain tags are expected to change significantly for other countries.

6. REFERENCES

Pedregosa, F., Varoquaux, G., Gramfort, A., Michel, V., Thirion, B., Grisel, O., Blondel, M., Prettenhofer, P., Weiss, R., Dubourg, V., Vanderplas, J., Passos, A., Cournapeau, D., Brucher, M., Perrot, M., and Duchesnay, E., 2011, Scikit-learn: Machine learning in Python. *Journal of Machine Learning Research*, 12:2825–2830

Jilani, M., Corcoran, P., and Bertolotto, M., 2014, Automated Highway Tag Assessment of OpenStreetMap Road Networks. *Proc. 22nd ACM SIGSPATIAL Int. Conference on Advances in Geographic Information Systems*

Hagenaur, J., Helbich, M.. 2012. Mining urban land-use patterns from volunteered geographic information by means of genetic algorithms and artificial neural networks. *Int. J. Geogr. Inf. Sci.* 26, 6 (June 2012), 963-982.

Fathi, A., Krumm, J., 2010. Detecting Road Intersections from GPS Traces. *Proc. 6th Int. Conf. On Geographic Information Science, LNCS, Vol. 6292*

Breiman, Leo. "Random forests." *Machine learning* 45.1 (2001): 5-32.

AUTHOR INDEX